AF839162

Introduction to Optical Metrology

This book describes both the theory and practice of optical techniques to measure various parameters encountered routinely in science and engineering.

Introduction to Optical Metrology, Second Edition, examines the theory and practice of various measurement methodologies utilizing both the corpuscular and the wave nature of light. The book begins by introducing the subject of optics and then addresses the propagation of laser beams through free space and optical systems. It discusses interferometry, holography, speckle metrology, the moiré phenomenon, photoelasticity, and microscopy. The remaining chapters describe techniques and methods of measurements of refractive index, thickness, radii of curvature, angle, velocity, pressure, length, optical testing, and fiber-optic-based methods. Apart from these, this edition includes a chapter on temperature measurement, sections on fringe unwrapping methods, testing of free-form optics, shearography, etc. Featuring new and updated exercise problems at the end of each chapter, this edition provides an applied understanding of essential optical measurement concepts, techniques, and procedures.

The primary audience for this book is undergraduate and graduate students who specialize in optics. It will also be useful to researchers and professionals working on optical testing and fiber-optic-based and MEMS-based measurements. A solutions manual and figure slides are available for adopting professors.

Rajpal S. Sirohi served as Professor of Physics at IIT Madras for more than two decades and as Director of IIT Delhi and Vice Chancellor to several Universities in India. He was a Distinguished Scholar at the Rose-Hulman Institute of Technology, Terre Haute, Indiana; Chair Professor at Tezpur University, Assam, India; and Faculty at Alabama A&M University, Huntsville, Alabama. Professor Rajpal S. Sirohi is now retired and spends his time reading books on the history of science and spends mornings and evenings with his grandchildren. His research areas are optical metrology, optical instrumentation, laser instrumentation, holography, and speckle phenomenon.

OPTICAL SCIENCES AND APPLICATIONS OF LIGHT

Founding Editor: Brian J. Thompson
University of Rochester, Rochester, New York

Tunable Laser Applications, Second Edition
Edited by F. J. Duarte

Laser Beam Shaping Applications, Second edition
Fred M. Dickey, Todd E. Lizotte

Lightwave Engineering
Yasuo Kokubun

Laser Safety: Tools and Training
Edited by Ken Barat

Computational Methods for Electromagnetic and Optical Systems
John M. Jarem, Partha P. Banerjee

Biochemical Applications of Nonlinear Optical Spectroscopy
Edited by Vladislav Yakovlev

Optical Methods of Measurement: Wholefield Techniques, Second Edition
Rajpal Sirohi

Fundamentals and Basic Optical Instruments
Daniel Malacara Hernández

Advanced Optical Instruments and Techniques
Daniel Malacara Hernández

Entropy and Information Optics: Connecting Information and Time, Second Edition
Francis T.S. Yu

Handbook of Optical Engineering, Second Edition, Two Volume Set
Daniel Malacara Hernández

Optical Materials and Applications
Moriaki Wakaki

Photonic Signal Processing: Techniques and Applications
Le Nguyen Binh

Practical Applications of Microresonators in Optics and Photonics
Edited by Andrey B. Matsko

Near-Earth Laser Communications, Second Edition
Edited by Hamid Hemmati

Nonimaging Optics: Solar and Illumination System Methods, Design, and Performance
Roland Winston, Lun Jiang, Vladimir Oliker

Introduction to Optical Metrology
Rajpal Sirohi

For more information about this series, please visit: https://www.crcpress.com/Optical-Science-and-Engineering/book-series/CRCOPTSCIENG

Introduction to Optical Metrology

Second Edition

Rajpal S. Sirohi

CRC Press
Taylor & Francis Group
Boca Raton London New York

CRC Press is an imprint of the
Taylor & Francis Group, an **informa** business

Designed cover image: Rajpal Sirohi

MATLAB® and Simulink® are trademarks of The MathWorks, Inc. and are used with permission. The MathWorks does not warrant the accuracy of the text or exercises in this book. This book's use or discussion of MATLAB® or Simulink® software or related products does not constitute endorsement or sponsorship by The MathWorks of a particular pedagogical approach or particular use of the MATLAB® and Simulink® software.

Second edition published 2026
by CRC Press
2385 NW Executive Center Drive, Suite 320, Boca Raton FL 33431

and by CRC Press
4 Park Square, Milton Park, Abingdon, Oxon, OX14 4RN

CRC Press is an imprint of Taylor & Francis Group, LLC

© 2026 Rajpal S. Sirohi

First edition published by CRC Press 2015

Reasonable efforts have been made to publish reliable data and information, but the author and publisher cannot assume responsibility for the validity of all materials or the consequences of their use. The authors and publishers have attempted to trace the copyright holders of all material reproduced in this publication and apologize to copyright holders if permission to publish in this form has not been obtained. If any copyright material has not been acknowledged please write and let us know so we may rectify in any future reprint.

Except as permitted under U.S. Copyright Law, no part of this book may be reprinted, reproduced, transmitted, or utilized in any form by any electronic, mechanical, or other means, now known or hereafter invented, including photocopying, microfilming, and recording, or in any information storage or retrieval system, without written permission from the publishers.

For permission to photocopy or use material electronically from this work, access www.copyright.com or contact the Copyright Clearance Center, Inc. (CCC), 222 Rosewood Drive, Danvers, MA 01923, 978-750-8400. For works that are not available on CCC please contact mpkbookspermissions@tandf.co.uk

Trademark notice: Product or corporate names may be trademarks or registered trademarks and are used only for identification and explanation without intent to infringe.

ISBN: 978-1-032-87279-7 (hbk)
ISBN: 978-1-032-87282-7 (pbk)
ISBN: 978-1-003-53184-5 (ebk)

DOI: 10.1201/9781003531845

Typeset in Times
by KnowledgeWorks Global Ltd.

Dedication to the Almighty form that transcends all of us.

Contents

Series Preface

Optical and photonics are enabling technologies in many science and engineering fields. The purpose of the Optical Sciences and Applications of Light series is to present the state of the art of basic science of optics, applied optics, and optical engineering, as well as the applications of optics and photonics in a variety of fields, including healthcare and life sciences, lighting, energy, manufacturing, information technology, telecommunication, sensors, metrology, defense, and education. This new and exciting material will be presented at a level that makes it useful to the practicing scientists and engineers working in a variety of fields. The books in this series cover topics that are a part of the rapid expansion of optics and photonics in various fields all over the world. The technologies discussed impact numerous real-world applications, including new displays in smartphones, computers, and televisions; new imaging systems in cameras; biomedical imaging for disease diagnosis and treatment; and adaptive optics for space systems for defense and scientific exploration. Other applications include optical technology for providing clean and renewable energy, optical sensors for more accurate weather prediction, solutions for more cost-effective manufacturing, and ultrahigh-capacity optical fiber communications technologies that will enable further growth of the Internet. The complete list of areas in which optics and photonics are involved is very long and constantly growing as well.

James C. Wyant
University of Arizona

Preface

If your experiment needs statistics, you ought to have done a better experiment.

Ernest Rutherford

Optical metrology is the science and technology of using light for the purpose of measurement of a variety of variables and quantities. Light exhibits a dual character—both particle and wave nature. Both these characters, however, are not exhibited simultaneously. All the measurement techniques discussed in this book utilize the wave nature of light. The measurand, that is, the quantity to be measured, can alter any of the wave properties of light, namely, amplitude, phase, wavelength, frequency, and polarization. A large number of measurement techniques have been developed over the period utilizing light both as a sensor and as a carrier of information. The purpose of this book is to describe both the theory and the practice of various measurement techniques to students and researchers. In that sense, this book is different from many books available on optical metrology. Also, several chapters have been included to make the book self-contained.

For example, Chapter 1 introduces the subject of optics, which summarizes the entire book. Chapter 2 introduces the propagation of laser beams through free space and through optical systems. Because most of the techniques discussed in this book use a laser as a source, it is essential to discuss how a Gaussian beam propagates and how to set up a collimator to get a collimated beam for experimentation.

The optical signal from the measurand is to be detected or recorded. Earlier, photographic emulsions were used to record the optical signals but they have now been replaced by array detectors. For many applications, a photodetector or a photomultiplier tube is used. Chapter 3 has been devoted to the description of light sources and the subject of detection and recording of optical signals. As mentioned earlier, any one of the wave characteristics could be used for measurement, but phase measurement is often used as it can be used for both point-wise and whole-field measurements and gives results with high accuracy. Because detectors in the optical regime are essentially energy detectors, the phase information is converted into intensity information, which is then recorded or measured over a certain time interval. The technique that converts phase information into intensity information is referred to as interferometry. There are several variants of the technique hence Chapter 4, is entirely devoted to interferometry. This chapter also contains phase-shifting methods that have been incorporated into interferometry. Over the past several decades, numerous whole-field measuring techniques have been researched and applied for routine and industrial measurements. The theory and experimental details of these techniques are discussed in Chapter 5, the longest chapter by far. The techniques under this heading are hologram interferometry, speckle metrology, moiré phenomenon, photoelasticity, and microscopy. A basic understanding of these techniques is required to understand the principles and functions of various measurement techniques described in the subsequent chapters.

Chapter 6 describes the various principles used for the measurement of refractive indices of solids, liquids, and gases. The refractive index is measured using the angle of deviation, critical angle, and Brewster angle. Interference from thin film is also used to measure the refractive index. Apart from the refractive index of the material, an optics designer needs to know the methods for measuring the curvature of surfaces and also the focal length associated with the image-forming optical elements. Although a routine measurement of these parameters is quite simple, the complexities increase when the range of measurement along with the accuracy is quite large.

Chapter 7 deals with the various methods of measurement of these two parameters. Various methods of optical testing are described in Chapter 8. Optical testing usually refers to measuring the shape of the wave-front emerging from an optical system. An optical system may be a simple plate, a prism, a complex lens, or a combination thereof. This chapter presents descriptions of methods to measure surface errors of concave and convex surfaces, plane surfaces, inhomogeneity of the material, and so on. The angle between the two surfaces using either an autocollimator or an interferometer is measured with varying degrees of accuracy. These methods are described in Chapter 9.

Thin films on optical surfaces modify the optical properties of the elements. These are also used to protect the surface. Methods to measure the thickness of the film, over a large range, are described in Chapter 10. These include interferometric and polarimetric methods. Chapter 11 gives a glimpse of velocity-measuring methods, which include Doppler-based methods and particle image velocimetry. Pressure measurement using optical techniques is described in Chapter 12. All the conventional methods of measurement can also be carried out using fiber optics, the use of which not only miniaturizes the device but also gives it all the other advantages of fiber optics. These methods are covered in Chapter 13. Chapter 14, the final chapter, discusses length measurement. Length is the main parameter of measurement in optical metrology, which is measured to great precision using optical methods.

Furthermore, to have a consistent treatment throughout the book, a wave propagating in the positive z-direction is described by $e^{i(\omega t - kz)}$. This is in contrast to the treatment in several books, in which a wave propagating in the positive z-direction is described by $e^{i(kz - \omega t)}$. Often, both representations are used in several books, resulting in confusion.

In summary, this book includes a range of topics and measurement techniques and procedures and is expected to be of considerable interest to students, teachers, and researchers.

Acknowledgments

> The nature of light is a subject of no material importance to the concern of life or to the practice of the arts, but it is in many other respects extremely interesting.
>
> *Thomas Young*

While the topics covered in this book on optical metrology have been well known and discussed in several other textbooks and research publications in one form or another, I present them here in a consolidated fashion from my unique perspective. I have been active in the research and teaching of optics over a period of several decades and have benefited from the interactions and collaborations with a large number of optical scientists and groups. An acknowledgments section is indeed the right spot to place on record the learning experiences I have had with my mentors, notably Dr. K.-J. Rosenbruch, Professor K. D. Hinsch, Professor Hans Tiziani, Professor Brian Thompson, Professor J. C. Wyant, Professor P. Hariharan, Professor I. Yamaguchi, Professor T. Tschudi, Professor P. K. Katti, and Professor M. S. Sodha. I am grateful to them for teaching me optics.

There is another group, consisting of my colleagues, who are numerous in number and scattered all over the globe. I express my thanks for the many fruitful interactions and conversations with my colleagues, namely, Professor W. Osten, Professor M. Takeda, Professor T. Yatagai, Professor M. Kuzawinska, Professor Kehar Singh, Professor L. N. Hazra, Professor Chandra Vikram, and Professor M. P. Kothiyal. The exchanges and conversations I had with them were most exciting. There are also many of my students who did some of the most exciting work in optical science and technology and currently hold important positions. They include Professor C. Joenathan, Professor Chandrashakher, Professor A. R. Ganesan, Professor P. Senthilkumaran, Professor C. Narayanamurthy, and Professor M. V. Murukeshan. I also give my special thanks to Professor J. C. Wyant who permitted me to use the problem sets that he uses at the University of Arizona. I have, however, certainly missed many names; I extend my apologies for the same and express my gratitude for the support of those not mentioned. It has always been a pleasure to work with Ashley Gasque at CRC Press, whose support I acknowledge. As the book has been written while I was working at the Rose–Hulman Institute of Technology, Terre Haute, Indiana, and Tezpur University, Assam, India, it is my pleasure to thank the institutions for allowing me to engage in this useful activity. Writing a book always encroaches on the time spent with loved ones, and I mention the support given to me by my wife, Vijayalaxmi Sirohi, and the proofreading done by my son, Jayant Sirohi.

Finally, if there are any shortcomings in the content of the textbook, the responsibility is mine.

Rajpal S. Sirohi
Tezpur University

Preface to the Second Edition

Due to the suggestion of a reviewer, Chapter 5 in the first edition has been recast into several smaller chapters, and a chapter on "Temperature Measurement" has been added. The figures of all the chapters have been reworked. Also, one or two sections have been added in some chapters. I would like to thank Professor K. Ramesh for providing a set of questions on photoelasticity. Professors Wolfgang Osten and Chandra Shakher extended help in preparing a short note on the testing of aspheric and free-form optics. Professor Partha Banerjee provided me with a note on the phase unwrapping procedure. Further, I have used irradiance in place of intensity throughout the book.

1 Introduction to Optics

1.1 INTRODUCTION

Metrology is the science and technology of making measurements and drawing significant conclusions from the set of data. Optical metrology uses light-based techniques for measurement. The majority of measurements involve length in one form or another; hence most of the techniques reported in this book will relate to length measurement. Also, there are some physical parameters of direct relevance in optics; so the techniques of measurement of these parameters are included in the text as well.

Length is expressed in meters. Prior to 1889, the meter was defined as 1/10,000,000 the distance between the North Pole and the equator along the line of longitude passing through Paris, France. Platinum bar of length equal to the meridional definition of meter was to be the prototype meter (end standard). The end standard meter was found impracticable due to wear of the ends to ensure international reproducibility; hence, it was replaced by the line standard. The standard meter is defined as a length between two fine lines engraved on a platinum-iridium bar of X cross section maintained and measured under very accurately controlled conditions. It remained in force from 1889 to 1960. However, because of the ever-present threat of natural disaster damaging the physical artifact, the standard based on the platinum-iridium bar was replaced by a standard based on the wavelength of light. The length of 1,650,763.73 times the wavelength of the orange-red line in the emission spectrum of ^{86}Kr was accepted as a standard meter; this definition remained in force from 1960 to 1983.

During this period several other lasers became the contenders for the length standard. In 1983, the 17th General Conference on Weight and Measures (CGPM) decided to adopt the speed of light in a vacuum as the definition of meter. The standard meter is defined as "*the length of the path travelled by light in vacuum during a time interval of 1/299,792,458 of a second.*" In practice, secondary standards of length are used for calibration. However, a specific process has been established to ensure traceability to the definition of a standard meter.

During and prior to Newton's period, light was considered as a stream of particles (corpuscles) emanating from the luminous bodies and striking the eye. During the Newtonian period, through vigorous scientific experimentation and forceful justifications, light came to be considered as wave motion. This description was further supported by Maxwell's theory of electrodynamics, as well as experiments by Hertz, bringing the light into the ambit of the electromagnetic spectrum. Later the explanation of the photoelectric effect in terms of exchange of energy between photons and electrons revived the corpuscular description of the nature of light. Light thus has a dual nature; predominance of one over the other depends on the nature of experiment and the measurable

DOI: 10.1201/9781003531845-1

quantity. Fortunately, the wavelike characteristics of light can also be explained on the basis of photons. Therefore, the subject of light is studied as (1) geometrical optics, wherein the light propagates in straight lines and the passage is described in terms of rays; (2) physical optics, where the phenomena studied strongly depend on the wavelike nature of light, and the wave propagation is the propagation of wavefronts; (3) electromagnetic optics dealing with guided waves and wave propagation in structures that are fine; and (4) quantum optics dealing with the interaction of light with matter, involving energy exchanges. Optical metrology mostly utilizes the wave nature of light. At the same time, while dealing with many topics involving measurements, some concepts borrowed from other descriptions of light are frequently used.

1.2 LAWS OF REFLECTION

It is known that the direction of light beam propagation can be changed by reflection as well as by refraction. Both of these phenomena have been studied in great detail, and laws of reflection and refraction have been formulated based on the observations. The law of reflection can be stated in two parts:

1. The angle of reflection is equal to the angle of incidence.
2. The incident ray, the normal to the reflecting surface at the point of incidence, and the reflected ray lie in a plane.

Both these parts of the law can be put in vector form as

$$\hat{n}_2 = \hat{n}_1 - 2(\hat{n}_1 \cdot \hat{s})\hat{s} \tag{1.1}$$

where $\hat{n}_1$, $\hat{n}_2$, and $\hat{s}$ are the unit vectors of the incident ray, the reflected ray, and the normal to the surface. Under paraxial approximation, $\hat{n}_1 \cdot \hat{s} \approx -1$; hence, the law of reflection takes the form $\hat{n}_2 = \hat{n}_1 + 2\hat{s}$.

1.3 LAWS OF REFRACTION

Refraction takes place when the light travels from one medium to another. An isotropic and homogeneous medium is characterized by an index of refraction or refractive index μ, which is the ratio of the velocity of light in a vacuum to that in the medium. The law of refraction is stated into two parts as follows

1. $\mu_1 \sin\theta_i = \mu_2 \sin\theta_r$.
2. The incident ray, the normal to the surface separating the media at the point of incidence, and the refracted ray lie in one plane.

Here θ_i and θ_r are the angles of incidence and refraction, respectively. The refractive indices of the two media are μ_1 and μ_2, and the incident ray is in the medium of refractive index μ_1.

Both parts of the law of refraction can also be cast in a single vectorial equation as

$$\mu_2\hat{n}_2 = \mu_1\hat{n}_1 - \mu_1(\hat{n}_1\cdot\hat{s})\hat{s} - \left(\sqrt{(\mu_2)^2 - (\mu_1)^2 + (\mu_1)^2(\hat{n}_1\cdot\hat{s})^2}\right)\hat{s} \tag{1.2}$$

Under paraxial approximation, the angle θ_i is small; hence, $\hat{n}_1\cdot\hat{s} \approx -1$. Therefore, the laws of refraction can be expressed as

$$\mu_2\hat{n}_2 = \mu_1\hat{n}_1 - (\mu_2 - \mu_1)\hat{s} \tag{1.3}$$

It is seen that the angle of refraction is always smaller than the angle of incidence when the light rays enter from the rarer (smaller refractive index) to the denser (larger refractive index) medium. On the other hand, if the ray enters from the denser medium to the rarer medium, the angle of refraction is always greater than the angle of incidence. Therefore, there is an angle of incidence for which the refracted ray makes an angle of 90° with the normal to the surface: the refracted ray travels along the interface. This angle of incidence is called the critical angle. If the angle of incidence is increased beyond the critical angle, there is no refracted ray, just a reflected ray: the incident ray is totally reflected. This phenomenon is known as *total internal reflection* and plays a major role in the design of several optical components.

Image formation by mirrors and lenses is described by the application of laws of reflection and refraction. Under paraxial approximation, formulae for image formation by mirrors and lenses are routinely used.

1.4 INTERFERENCE

Sometimes it is said that light plus light equals darkness. The explanation for this statement lies in the wonderful phenomenon of interference. When two waves are superposed, redistribution of light *irradiance* takes place in the region of superposition. We will first define the term *irradiance.* It is the average value of the Poynting vector and is expressed in W/m^2. Optical detectors, like the eye, photographic emulsions, and photodetectors cannot follow the amplitude variations in the light beam and only respond to energy or irradiance. For this reason, the interference phenomenon becomes more important as it renders phase information into irradiance information to which the detectors respond. Let us first consider a wave represented as

$$u(r;t) = u_0 \cos\left(\omega t - \bar{k}\cdot\bar{r}\right) \tag{1.4}$$

This is a scalar wave of frequency ω propagating in an arbitrary direction. Here u_0 is the amplitude of the wave, $\bar{k}$ is the propagation vector, $\bar{r}$ is the radius vector, and t is the time. A detector will measure an average value of $u^2(r;\,t)$. When two scalar waves of the same frequency are superposed, the detector will measure an average value of $\left[u_1(r;t)+u_2(r;t)\right]^2$ resulting in redistribution of irradiance; this is an interference pattern having bright and dark regions called fringes. Light is a small part of the electromagnetic spectrum and hence the light waves are transverse waves in which electric field vector $\bar{E}$, magnetic field vector $\bar{B}$, and propagation vector $\bar{k}$ form an orthogonal triplet. Further, the photo effects are attributed to electric field vector so we express the light wave as an oscillating electric field. In order to explicitly express the dependence of fringe contrast on the polarization of waves, we consider the interference between two polarized waves. Let the two waves be represented by

$$\bar{E}_1(r;t) = \bar{E}_{01}\cos\left(\omega t - \bar{k}\cdot\bar{r}_1\right) \tag{1.5a}$$

and

$$\bar{E}_2(r;t) = \bar{E}_{02}\cos\left(\omega t - \bar{k}\cdot\bar{r}_2\right) \tag{1.5b}$$

When these two waves are superposed, the detector will measure the resultant irradiance, which is proportional to the average value of $\left|\bar{E}_1 + \bar{E}_2\right|^2$. Therefore, the resultant irradiance $I(\delta)$ is expressed as

$$I(\delta) = I_1 + I_2 + 2\bar{E}_{01}\cdot\bar{E}_{02}\cos\delta \tag{1.6}$$

where $\delta = \bar{k}\cdot(\bar{r}_2 - \bar{r}_1)$ is the phase difference between the two waves and I_1, I_2 are the irradiances of the individual waves. The irradiances I_1 and I_2 are: $I_1 = \left|E_{01}\right|^2$ and $I_2 = \left|E_{02}\right|^2$. It is thus seen that the resultant irradiance consists of two terms, $(I_1 + I_2)$ and $2\bar{E}_{01}\cdot\bar{E}_{02}\cos\delta$. The first term is the DC bias, which may be constant or varying slowly over the *xy*-plane/observation plane, whereas the second term is the interference term, which varies rapidly with δ. Therefore, the resultant irradiance in the superposed region varies between the minimum and maximum values. This variation arises due to the phenomenon of interference, which redistributes the light irradiance. The light distribution on the *xy*-plane is called a fringe pattern or an interference pattern. The regions of maximum brightness constitute the bright fringes and those of minimum brightness appear as the dark fringes. It may be noted that the fringes are loci of constant phase difference δ. The consecutive bright (or dark) fringes are separated by 2π in phase. Following Michelson, we define the visibility of fringes as

$$V = \frac{I_{max} - I_{min}}{I_{max} + I_{min}} \tag{1.7}$$

where I_{max} and I_{min} are the maximum and minimum values of the resultant irradiance. Substituting these values in the expression for the visibility, we obtain

$$V = \frac{2\bar{E}_{01}\cdot\bar{E}_{02}}{I_1 + I_2} \tag{1.8}$$

The visibility depends on two parameters, the polarization states of the two beams and the irradiance ratio, which is not obvious in this form. The visibility will be a maximum if the interfering beams have the same state of polarization, i.e., the ***E*** vectors in both the beams are directed in the same direction. If the beams are orthogonally polarized, no fringe pattern forms. In other words, the interference term is zero. Therefore, the optical setups are accordingly configured. For example, the **E** vector in the beam is arranged to be vertical when the experiment is set up on a horizontal table so that the successive reflections at the mirrors do not rotate the plane of polarization. Now the visibility V can be expressed as

$$V = \frac{2\sqrt{I_1 I_2}}{I_1 + I_2} = \frac{2\sqrt{I_1 / I_2}}{1 + (I_1 / I_2)} \tag{1.9}$$

The visibility is obviously unity when the two beams are of equal irradiance. Further, the fringe visibility is quite good even if one of the beams is very weak. This explains why the contrast of spurious fringes due to dust particles, scratches, etc., in an interference pattern is very good. In this analysis it has been tacitly assumed that both the beams are coherent. The beams are therefore derived from the same parent beam. The real light sources emit partially coherent light; hence, the degree of coherence also appears in the definition of the visibility.

The interfering beams can be derived from the parent beam either by (1) the wave front division or (2) the amplitude division. One of the very well-known examples of wave front division is the famous double-slit experiment by Young. A pair of slits samples a wave front at two spatial locations. These sampled wave fronts are diffracted by the slits thereby expanding their spatial extent so that they superpose at a plane some distance away from the slit plane. Alternately the pair of slits could be considered a pair of secondary sources excited by the parent wave. The phase difference between the two waves at any point on the xy-plane distant z from the slit plane, under first-order approximation, is

$$\delta = k\frac{x\ z}{d} \tag{1.10}$$

where $k = 2\pi / \lambda$, d is the separation between the slits and the x-axis is taken normal to the pair of slits. The irradiance distribution in the interference pattern, however, is given by

$$I(\delta) = 2I_0 \mathrm{sinc}^2(2bx / \lambda z)\left[1 + \cos(\delta)\right] \tag{1.11}$$

where I_0 is the irradiance on axis due to each slit, and $\mathrm{sinc}(2bx / \lambda z) = \sin(\pi 2bx / \lambda z) / (\pi 2bx / \lambda z)$ is the amplitude distribution due to diffraction at the slit of width $2b$. Therefore, the irradiance distribution in the interference pattern is modulated by the diffraction term: if the slit is very narrow the diffracted irradiance is distributed over a very wide angular range. When more than two slits are illuminated by a collimated wave, the irradiance distribution in the interference pattern gets further modified. An element that contains a large number of equi-spaced slits is a grating, which creates multiple beams by wave front division.

One of the earliest instruments based on amplitude division is the Michelson interferometer. The path matching between the two arms of the interferometer is accomplished by introducing a compensator so that low coherence length sources could also be used. A plane parallel plate (PPP) interferometer is another example employing amplitude division, but it requires moderate coherence length of the source. All polarization interferometers utilize amplitude division. The Fabry-Perot interferometer and thin film structures utilize multiple-beam interference by amplitude division.

Let us first consider interference between two plane waves. One plane wave is incident normally at the xy-plane, whereas the other plane wave lies in the xz-plane and makes an angle θ with the z-axis. The phase difference δ between the waves at a plane parallel to xy-plane is given by

$$\delta = k\ x\ \sin\theta \tag{1.12}$$

The fringes run parallel to the y-axis and the fringe spacing is $\bar{x} = \lambda / \sin\theta$. The fringe pattern is a system of straight lines running parallel to the y-axis with constant spacing. If the plane waves were incident symmetrically about the z-axis and enclosed an angle θ between them, the fringe spacing will be $\bar{x} = \lambda / 2\ \sin(\theta / 2)$. For small θ, the fringe width is almost equal in both the cases.

Let us now consider the interference between a plane wave along the z-axis and a spherical wave emanating from a point source on the origin. The phase difference δ is given by

$$\delta = \delta_0 + k\frac{x^2 + y^2}{2z} \tag{1.13}$$

where δ_0 is a constant phase difference and the distance between the observation plane and the point source is z. This expression is valid under paraxial approximation.

The bright fringes are formed when

$$x^2 + y^2 = 2m\lambda z: \quad m = 0,\ 1,\ 2,\ 3 \tag{1.14}$$

The fringe pattern consists of circular fringes with radii proportional to the square root of the natural number and the brightness of the zeroth order fringe depends on δ_0. If the plane wave is added at an angle, the fringes will appear as arcs of the circles.

1.5 DIFFRACTION

While discussing the topic of imaging, one encounters the propagation of light beam from object plane to image plane via imaging lens, stops, etc. The beam therefore suffers diffraction during this process. The term diffraction has been very conveniently described by Sommerfeld as "*any deviation of light rays from rectilinear paths which cannot be interpreted as reflection or refraction.*" It is often stated in a different form as "*bending of rays near corners.*" From Fourier optics principles, a wave suffers diffraction when its transverse extent is limited. Due to the finite size of optical elements or other restrictions, the beam always suffers diffraction during propagation.

Grimaldi first observed the presence of bands in the geometrical region of an obstacle. Huygens provided a satisfactory explanation of this observation by introducing the concept of secondary waves. Fresnel improved upon the intuitive approach of Huygens and the theory is known as the Huygens–Fresnel theory of diffraction.

Kirchhoff placed the Huygens–Fresnel theory on a firmer mathematical foundation. He based his mathematical formulation on two boundary conditions that were found to be inconsistent with each other. Kottler attempted to resolve these contradictions by reinterpreting Kirchhoff's boundary value problem as a "Saltus" problem: Saltus signifies a discontinuity or a jump of a function. Kirchhoff's theory was later modified by Sommerfeld who eliminated one of the boundary conditions by making use of the theory of Green's functions. This is known as the Rayleigh–Sommerfeld diffraction theory.

1.5.1 Propagation of a Monochromatic Wave

We consider a scalar monochromatic wave. The light disturbance at any position P and time t is represented by $u(P, t)$, where

$$u(P,t) = U(P)\cos(\omega t + \phi(P)) = Re\left[\boldsymbol{U}(P)\, e^{i\omega t}\right] \tag{1.15}$$

with $\boldsymbol{U}(P) = U(P)\, e^{-i\phi(P)}$. $\boldsymbol{U}(P)$ is a complex function of position and is often referred to as a phasor. *Re* refers to real part. Because light is a wave motion, it should obey the wave equation at each source point. Thus, we have

$$\nabla^2 u - \frac{1}{c^2}\frac{\partial^2 u}{\partial t^2} = 0 \tag{1.16}$$

where ∇^2 is the Laplacian operator. Substituting for $u(P, t)$, we obtain

$$(\nabla^2 + k^2)\boldsymbol{U}(P) = 0 \tag{1.17}$$

where k is the wave number defined as $k = \omega / c = 2\pi / \lambda$.

Equation (1.17) is known as the Helmholtz equation. The complex amplitude of any monochromatic wave propagating through free space must obey the Helmholtz equation.

1.5.2 Kirchhoff Theory of Diffraction

Consider a spherical wave from a point source at P_2, incident on an aperture, which has an opening Σ that diffracts the light as shown in Figure 1.1. It is desired to obtain the amplitude of the wave at the observation point P_0. The amplitude of the spherical wave at a point P_1 at the opening is expressed as

$$\boldsymbol{U}(P_1) = A\frac{e^{-ikr_{21}}}{r_{21}} \tag{1.18}$$

where r_{21} is the distance between points P_2 and P_1, and A is the amplitude at unit distance from point P_2.

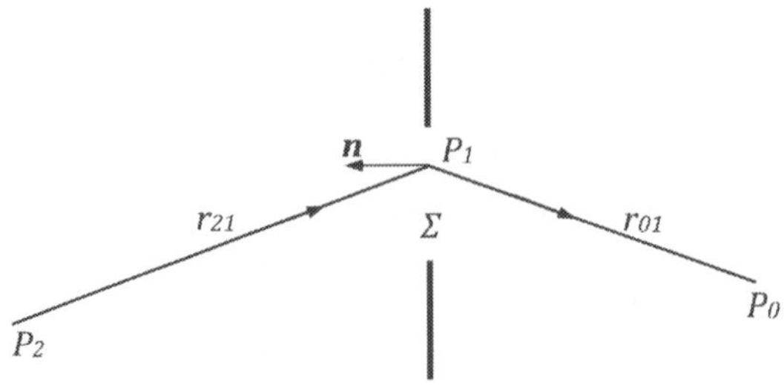

FIGURE 1.1 Geometry to calculate the diffracted field at point P_0.

Kirchhoff made use of Green's theorem to derive an expression of the amplitude of the wave at the observation point P_0. Green's function $\boldsymbol{G}(P_1)$ at point P_1 is taken as the unit amplitude wave expanding from point P_0. It is expressed as

$$\boldsymbol{G}(P_1) = \frac{e^{-ikr_{01}}}{r_{01}} \tag{1.19}$$

Making use of Green's theorem, it can be shown that the amplitude of the wave at point P_0 can be expressed as

$$\boldsymbol{U}(P_0) = \frac{1}{4\pi} \iint_S \left(\frac{\partial U}{\partial n} \boldsymbol{G} - \boldsymbol{U} \frac{\partial G}{\partial n} \right) ds \tag{1.20a}$$

where the integration is carried out over the surface that surrounds point P_0, $\boldsymbol{U}$, and $\frac{\partial U}{\partial n}$ are the amplitude of the wave from point P_2 and its first derivative at the surface of the opening, and n is taken along the outward normal. $\boldsymbol{G}$ and $\frac{\partial G}{\partial n}$ are Green's function and its derivative at the surface of the opening. Therefore, this is rewritten as

$$\boldsymbol{U}(P_0) = \frac{1}{4\pi} \iint_S \left\{ \frac{\partial U}{\partial n} \left[\frac{e^{-ikr_{01}}}{r_{01}} \right] - U \frac{\partial}{\partial n} \left[\frac{e^{-ikr_{01}}}{r_{01}} \right] \right\} ds \tag{1.20b}$$

Further, using the Sommerfeld radiation condition and Kirchhoff's boundary conditions, it can be shown that integration needs to be carried out on the surface area Σ of the diffracting aperture. Therefore, the field at point P_0 is expressed as

$$\boldsymbol{U}(P_0) = \frac{1}{4\pi} \iint_\Sigma \left\{ \frac{\partial U}{\partial n} \left[\frac{e^{-ikr_{01}}}{r_{01}} \right] - U \frac{\partial}{\partial n} \left[\frac{e^{-ikr_{01}}}{r_{01}} \right] \right\} ds \tag{1.21}$$

Taking the derivatives and using the conditions $k \gg 1/r_{01}$, and $k \gg 1/r_{21}$, the amplitude at point P_0 can be expressed as

$$\boldsymbol{U}(P_0) = \frac{iA}{\lambda} \iint_\Sigma \frac{e^{-ik(r_{01}+r_{21})}}{r_{01}r_{21}} \left[\frac{\cos(n, r_{01}) - \cos(n, r_{21})}{2} \right] ds \tag{1.22a}$$

This is known as the Fresnel–Kirchhoff diffraction formula. The formula holds valid for a scalar field at a distance that is very large compared with the wavelength of light.

The boundary conditions imposed by Kirchhoff are inconsistent with each other. These inconsistencies were removed by an appropriate choice of Green's function.

The inconsistency is removed either by choosing Green's function such that it is zero everywhere on the aperture, but its derivative exists or Green's function exists but its derivative is identically zero on the diffracting aperture.

If Green's function is chosen such that it represents two diverging spherical waves oscillating in anti-phase, and originating from mirror-image points, the field distribution at point P_0, under the approximation $k \gg 1/r_{01}$, is given by

$$U(P_0) = \frac{iA}{\lambda} \iint_{\Sigma} \frac{e^{-ik(r_{01}+r_{21})}}{r_{01} r_{21}} \cos(n, r_{01})\, ds \tag{1.22b}$$

This expression is known as the Rayleigh–Sommerfeld diffraction formula. When compared with the result of Kirchhoff's theory, the difference may be seen in the form of an obliquity factor introduced by Fresnel. The field distributions obtained by these two formulae are equally valid.

1.5.3 Small Angle Approximation

We assume that the spatial frequency content in the object is low so that the light is diffracted into a small angle, i.e., $\cos\theta \sim 1$, and $\sin\theta \sim \theta$. We also assume that the illumination source is nearer to the optical axis. Under these conditions, we can take the obliquity factor as unity. Both the formulae, Equations (1.22a) and (1.22b), now reduce to

$$U(P_0) = \frac{iA}{\lambda} \iint_{\Sigma} \frac{e^{-ik(r_{01}+r_{21})}}{r_{01} r_{21}}\, ds \tag{1.23}$$

Here $Ae^{-ikr_{21}} / r_{21}$ is the amplitude of the spherical wave at any point on the opening. The secondary wave, as postulated by Huygens, will have its amplitude proportional to that of the exciting wave, and the field at the observation point is thus proportional to $Ae^{-ikr_{21}} / r_{21}$. Let us now consider an opening, which has a certain characteristic transmittance. A plane wave traveling along the optical axis is incident on the aperture as shown in the Figure 1.2.

The amplitude $U(P_0)$ is then given by

$$U(P_0) = \frac{i}{\lambda} \iint_{\Sigma} U(x_1, y_1) \frac{e^{-ikr_{01}}}{r_{01}}\, ds = \frac{i}{\lambda} \iint_{\Sigma} U(x_1, y_1) \frac{e^{-ikr_{01}}}{r_{01}}\, dx_1\, dy_1 \tag{1.24}$$

where $U(x_1, y_1)$ is the field exiting just after the aperture.

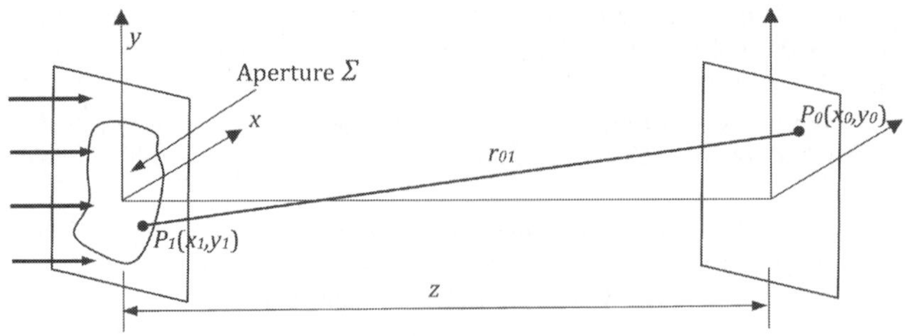

FIGURE 1.2 Diffraction at the aperture Σ.

1.5.4 Fresnel Approximation

To simplify this formula, we express r_{01} as

$$r_{01} = \sqrt{z^2 + (x_0 - x_1)^2 + (y_0 - y_1)^2} \tag{1.25}$$

After binomial expansion and retaining the first three terms, we have

$$r_{01} = z + \frac{1}{2z}\left[(x_0 - x_1)^2 + (y_0 - y_1)^2\right] - \frac{1}{8z^3}\left[(x_0 - x_1)^2 + (y_0 - y_1)^2\right]^2 \tag{1.26}$$

In the binomial expansion of r_{01}, if we drop the second-order term, we are in the Fresnel approximation. The field $U(x_0, y_0)$ is now

$$U(x_0, y_0) = \frac{i}{\lambda z} e^{-ikz} \iint_{\Sigma} U(x_1, y_1)\, e^{-i(k/2z)\left[(x_0 - x_1)^2 + (y_0 - y_1)^2\right]}\, dx_1\, dy_1 \tag{1.27}$$

It may be noted that the Fresnel region begins at some distance z from the opening. This distance z is obtained from the inequality

$$\frac{1}{8z^3}\left[(x_0 - x_1)^2 + (y_0 - y_1)^2\right]^2 \ll \lambda \tag{1.28}$$

This condition, for the field on axis ($x_0 = y_0 = 0$), gives

$$z^3 \gg \frac{(x_1^2 + y_1^2)^2}{8\,\lambda} \tag{1.29a}$$

If the diameter of the opening is $2D$, then

$$z \gg \sqrt[3]{(D^4 / 8\lambda)} \tag{1.29b}$$

The low spatial content in the object implies that $U(x_1, y_1)$ is a slowly varying function of spatial coordinates. In fact $U(x_1, y_1)$ may characterize the transmittance function of a diffracting object when illuminated with a unit amplitude plane wave.

Equation (1.27) is expressed as

$$U(x_0, y_0) = \frac{i\, e^{-ikz}}{\lambda z} e^{-i(k/2z)(x_0^2+y_0^2)} \iint_{\Sigma} U(x_1, y_1) e^{-i(k/2z)(x_1^2+y_1^2)}\, e^{i(k/z)(x_0 x_1 + y_0 y_1)}\, dx_1\, dy_1 \tag{1.30}$$

This, therefore, represents the Fourier transform of $U(x_1, y_1)\, e^{-i(k/2z)(x_1^2+y_1^2)}$ multiplied by a quadratic phase factor and a complex constant.

1.5.5 Fraunhofer Approximation

We now impose another condition, i.e.

$$(k/2z)\left(x_1^2 + y_1^2\right)_{\max} \ll 2\pi \tag{1.31a}$$

or

$$z \gg \frac{1}{2}\frac{D^2}{\lambda} \tag{1.31b}$$

where D is the radius of the aperture. The field $U(P_0)$ is now expressed as

$$U(x_0, y_0) = \frac{i\, e^{-ikz}}{\lambda z} e^{-i(k/2z)(x_0^2+y_0^2)} \iint_{\Sigma} U(x_1, y_1)\, e^{i(k/z)(x_0 x_1 + y_0 y_1)}\, dx_1\, dy_1 \tag{1.32}$$

Because $U(x_1, y_1)$ is finite only inside the aperture, we can extend the limits to $\pm\infty$. Thus

$$\begin{aligned} U(x_0, y_0) e^{i(k/2z)(x_0^2+y_0^2)} &= \frac{i\, e^{-ikz}}{\lambda z} \iint_{-\infty}^{\infty} U(x_1, y_1) e^{i(k/z)(x_0 x_1 + y_0 y_1)}\, dx_1\, dy_1 \\ &= \frac{i\, e^{-ikz}}{\lambda z}\, \mathfrak{F}\left[U(x_1, y_1)\right]_{x_0/\lambda z,\ y_0/\lambda z} \end{aligned} \tag{1.33}$$

where the symbol $\mathfrak{F}$ stands for Fourier transform operation. Thus the Fourier transform of $U(x_1, y_1)$ evaluated at spatial frequencies $x_0/\lambda z$, $y_0/\lambda z$ exists on the spherical surface. Thus, when the condition as expressed in Equation (1.31a) is satisfied, the field distribution is obtained as a Fourier transform of the field existing on the aperture plane. In practice a lens is used to bring the far-field diffraction pattern to its back focal plane. When the transparency is placed anywhere in front of the lens, its Fourier transform multiplied by a phase factor is obtained at the back focal plane. However, when the transparency is placed at the front focal plane, the phase factor cancels out and a pure Fourier transform is obtained at the back focal plane. There are also other geometries that can be used to obtain pure Fourier transforms.

1.6 POLARIZATION

Light constitutes a tiny portion of the electromagnetic spectrum. It contains both electric field and magnetic field oscillating in phase when it propagates in free space or a vacuum. The electric field vector $\boldsymbol{E}$, magnetic field vector $\boldsymbol{B}$ and propagation vector $\boldsymbol{k}$ form an orthogonal triplet. The average value of the Poynting vector, $(\boldsymbol{E}\times\boldsymbol{B})/\mu_0$, gives the irradiance of the wave and the direction of the Poynting vector is along the direction of propagation. Polarization generally means "orientation." The direction of polarization is by convention taken along the direction of electric field in the wave.

Consider a monochromatic wave of frequency ω traveling in the + z-direction as

$$\boldsymbol{E}(z;t)=\boldsymbol{E}_0\ cos(\omega t-k\ z) \tag{1.34}$$

The $\boldsymbol{E}$ vector lies in the xy-plane. We can express $\boldsymbol{E}$ as

$$\boldsymbol{E}=E_x\ \hat{i}+E_y\ \hat{j} \tag{1.35}$$

1.6.1 POLARIZATION ELLIPSE

We can write the monochromatic wave in its components along the x- and y-direction as

$$E_x=E_{0x}\ \cos(\omega t-k\ z+\phi_x) \tag{1.36a}$$

$$E_y=E_{0y}\ cos(\omega t-k\ z+\phi_y) \tag{1.36b}$$

where ϕ_x and ϕ_y are the phases associated with the x- and y-components of the electric field, respectively. Both E_x and E_y are functions of time and are oscillating with frequency ν (= $\omega/2\pi$). On eliminating the time dependence, we obtain

$$\frac{E_x^2}{E_{0x}^2}+\frac{E_y^2}{E_{0y}^2}-2\frac{E_x}{E_{0x}}\frac{E_y}{E_{0y}}\cos\phi=\sin^2\phi \tag{1.37}$$

where the phase difference ϕ is $\phi=\phi_x-\phi_y$. This is the equation of an ellipse in a nonstandard form. The tip of the electric field vector traces an ellipse at any fixed z-position; this is called an elliptically polarized light. However, the state of polarization is determined by the shape of the ellipse, that is, by the direction of the major axis and the ellipticity. The size of the ellipse is determined by the irradiance of the wave. The tip can trace the ellipse by moving clockwise or anticlockwise. While looking at the beam, if the tip rotates clockwise, it is termed as "right-handed elliptically polarized" or vice versa.

Depending on the magnitudes of the components and their phase differences, we also get linearly and circularly polarized light. When either of the components E_{0x} or E_{0y} is zero, or the phase difference ϕ is 0 or π, we have linearly polarized light.

In the former case, it is polarized either along the *y*- or *x*-direction, and in the latter case, it is linearly polarized in an arbitrary direction depending on the magnitudes of the components. The direction of the electric field and the direction of propagation determine the plane of polarization. When both the components, E_{0x} and E_{0y}, are equal and their phase difference is either $\pi/2$ or $3\pi/2$, we obtain circularly polarized light.

1.6.2 Representation of Polarization

1.6.2.1 Jones Vector

A two-element column vector, known as the Jones vector, can describe any state of polarization. The Jones vector is represented as

$$\bar{J} = \begin{bmatrix} E_x \\ E_y \end{bmatrix} \tag{1.38}$$

Given the Jones vector, the irradiance of the wave is obtained as $I = \left(E_x^2 + E_y^2\right)/2\eta$, where η is the impedance of the medium in which the wave propagates. While writing the Jones vectors of various states of polarization, the wave irradiance is normalized to unity. Table 1.1 gives the Jones vectors for several states of polarization.

When an optical wave is reflected or transmitted through an optical system, its state of polarization, in general, gets modified. Assuming the system to be linear, so that the principle of superposition can be applied, the evolution of the polarization state can be described by a 2 × 2 matrix, known as the Jones matrix. The Jones matrices for several elements are given in Table 1.2.

TABLE 1.1
Jones Vectors

	State of Polarization	Jones Vector
1	Linearly *x*-polarized	$\bar{J} = \begin{bmatrix} 1 \\ 0 \end{bmatrix}$
2	Linearly *y*-polarized	$\bar{J} = \begin{bmatrix} 0 \\ 1 \end{bmatrix}$
3	Linearly polarized making an angle θ with the *x*-axis	$\bar{J} = \begin{bmatrix} \cos\theta \\ \sin\theta \end{bmatrix}$
4	Right circularly polarized	$\bar{J} = \frac{1}{\sqrt{2}}\begin{bmatrix} 1 \\ i \end{bmatrix}$
5	Left circularly polarized	$\bar{J} = \frac{1}{\sqrt{2}}\begin{bmatrix} 1 \\ -i \end{bmatrix}$

TABLE 1.2
Jones Matrices for Several Optical Elements

	Optical Elements	Jones Matrix T
1	Linear polarizer along x-direction	$T=\begin{bmatrix}1 & 0\\ 0 & 0\end{bmatrix}$
2	Linear polarizer along y-direction	$T=\begin{bmatrix}0 & 0\\ 0 & 1\end{bmatrix}$
3	Linear polarizer making an angle 45° with x-axis	$T=\frac{1}{2}\begin{bmatrix}1 & 1\\ 1 & 1\end{bmatrix}$
4	Linear polarizer making an angle –45° with x-axis	$T=\frac{1}{2}\begin{bmatrix}1 & -1\\ -1 & 1\end{bmatrix}$
5	Linear polarizer making an angle θ with x-axis	$T=\begin{bmatrix}\cos^2\theta & \cos\theta\sin\theta\\ \cos\theta\sin\theta & \sin^2\theta\end{bmatrix}$
6	Polarization rotator	$T=\begin{bmatrix}\cos\theta & -\sin\theta\\ \sin\theta & \cos\theta\end{bmatrix}$
7	Wave retarder, fast axis along x-direction, delays y-component by Γ	$T=\begin{bmatrix}1 & 0\\ 0 & e^{-i\Gamma}\end{bmatrix}$
8	Quarter-wave plate – fast axis along y-direction	$T=e^{i\pi/4}\begin{bmatrix}1 & 0\\ 0 & -i\end{bmatrix}$
9	Quarter-wave plate – fast axis along x-direction	$T=e^{i\pi/4}\begin{bmatrix}1 & 0\\ 0 & i\end{bmatrix}$
10	Half-wave plate	$T=\begin{bmatrix}1 & 0\\ 0 & -1\end{bmatrix}$
11	Right-handed circular polarizer	$T=\frac{1}{2}\begin{bmatrix}1 & i\\ -i & 1\end{bmatrix}$
12	Left-handed circular polarizer	$T=\frac{1}{2}\begin{bmatrix}1 & -i\\ i & 1\end{bmatrix}$

Let a wave of Jones vector $\bar{J}_1$ be incident on an optical system described by the Jones matrix T, then the Jones vector $\bar{J}_2$ of the exiting wave is obtained as

$$\bar{J}_2 = T\bar{J}_1 \tag{1.39}$$

Let us consider an example in which a linearly polarized plane wave is represented by the Jones vector $\bar{J}_1=\begin{bmatrix}E_x\\ E_y\end{bmatrix}$. This wave is incident on the

x-polarizer followed by the y-polarizer. The Jones vector of the transmitted wave is obtained as

$$\bar{J}_2 = T_y\, T_x\, \bar{J}_1 = \begin{bmatrix} 0 & 0 \\ 0 & 1 \end{bmatrix}\begin{bmatrix} 1 & 0 \\ 0 & 0 \end{bmatrix}\begin{bmatrix} E_x \\ E_y \end{bmatrix} = \begin{bmatrix} 0 & 0 \\ 0 & 0 \end{bmatrix}\begin{bmatrix} E_x \\ E_y \end{bmatrix} \tag{1.40}$$

There is no light transmitted. Let us now consider another example. The linearly polarized plane wave passes first through the x-polarizer and then through a polarizer whose polarization direction makes an angle θ with the x-direction. The Jones vector of the transmitted light is given by

$$\begin{aligned} \bar{J}_2 &= \begin{bmatrix} \cos^2\theta & \cos\theta\,\sin\theta \\ \cos\theta\,\sin\theta & \sin^2\theta \end{bmatrix}\begin{bmatrix} 1 & 0 \\ 0 & 0 \end{bmatrix}\begin{bmatrix} E_x \\ E_y \end{bmatrix} \\ &= \begin{bmatrix} \cos^2\theta & 0 \\ \cos\theta\,\sin\theta & 0 \end{bmatrix}\begin{bmatrix} E_x \\ E_y \end{bmatrix} = \begin{bmatrix} E_x\,\cos^2\theta \\ E_x\,\cos\theta\,\sin\theta \end{bmatrix} \end{aligned} \tag{1.41}$$

The irradiance $I(\theta)$ transmitted by the arrangement is

$$\begin{bmatrix} E_x\,\cos^2\theta & E_x\,\cos\theta\,\sin\theta \end{bmatrix} \cdot \begin{bmatrix} E_x\,\cos^2\theta \\ E_x\,\cos\theta\,\sin\theta \end{bmatrix} = E_x^2\left(\cos^4\theta + \cos^2\theta\,\sin^2\theta\right) = I_0\cos^2\theta \tag{1.42}$$

where I_0 is the irradiance of the wave incident on the polarizer oriented at an angle θ with the x-direction. This is the statement of the Malus law. This has application in the control of irradiance of light.

When a randomly polarized wave of irradiance I_0 is incident on an x-polarizer, it transmits half its irradiance and polarizes the wave along the x-direction. The Jones vector of the light leaving the polarizer is given by

$$\bar{J} = \sqrt{\frac{I_0}{2}}\begin{bmatrix} 1 \\ 0 \end{bmatrix} \tag{1.43}$$

1.6.2.2 Stokes Vector

To treat fully, partially, or randomly polarized light we use the Stokes vector. This is a four-element column vector and is defined as

$$S = \begin{bmatrix} S_0 \\ S_1 \\ S_2 \\ S_3 \end{bmatrix} \tag{1.44}$$

All four elements of the Stokes vector can be measured. Suppose we have four detectors, three of them having polarizers in front of them. The detector without a polarizer measures the total irradiance I_0 of the wave, the detector with the x-polarizer measures the irradiance I_1, the detector with the polarizer oriented at +45° measures the irradiance I_2, and the detector with a right circular polarizer in front measures the irradiance I_3. The Stokes elements are related with the measured irradiances as

$$S_0 = I_0, \quad S_1 = 2I_1 - I_0, \quad S_2 = 2I_2 - I_0, \text{ and } S_3 = 2I_3 - I_0.$$

In other words, we can express the Stokes vector as

$$\begin{bmatrix} S_0 \\ S_1 \\ S_2 \\ S_3 \end{bmatrix} = \begin{bmatrix} E_x^2 + E_y^2 \\ E_x^2 - E_y^2 \\ 2E_xE_y \cos\phi \\ 2E_xE_y \sin\phi \end{bmatrix} = \begin{bmatrix} Irradiance\ I_0 \\ I(0^0) - I(90^0) \\ I(45^0) - I(135^0) \\ I(RCP) - I(LCP) \end{bmatrix} \tag{1.45}$$

where $I(\theta°)$ is the irradiance measured when the linear polarizer is oriented at an angle θ with respect to the x-direction, and $I(RCP)$ and $I(LCP)$ are the irradiances measured with the right-circular polarizer and the left-circular polarizer in front of the detector.

It may be noted that various polarization states can be described in terms of the Stokes parameters as

Linear polarization $(S_1 \neq 0,\ S_2 \neq 0,\ S_3 = 0)$,
Circular polarization $(S_1 = 0,\ S_2 = 0,\ S_3 \neq 0)$
Fully polarized light $(S_0^2 = S_1^2 + S_2^2 + S_3^2)$
Partially polarized light $(S_0^2 > S_1^2 + S_2^2 + S_3^2)$
Unpolarized light $(S_1 = 0,\ S_2 = 0,\ S_3 = 0)$

We can define several degrees of polarization now. The degree of polarization $\mathcal{P}$ is defined as

$$\mathcal{P} = \frac{\sqrt{S_1^2 + S_2^2 + S_3^2}}{S_0} \tag{1.46a}$$

Degree of linear polarization is

$$\mathcal{P}_{lin} = \frac{\sqrt{S_1^2 + S_2^2}}{S_0} \tag{1.46b}$$

Degree of circular polarization is

$$\mathcal{P}_{cir} = \frac{S_3}{S_0} \tag{1.46c}$$

The Poincaré sphere is also used to represent Stokes parameters and hence the state of polarization. When a wave with the Stokes vector $\overline{S}_{inc}$ is incident on an optical element, then the Stokes vector $\overline{S}_{out}$ of the exiting wave can be described as

$$\overline{S}_{out} = M\overline{S}_{inc}$$

where M is a 4×4 matrix associated with the optical element and is called the Mueller matrix. The Mueller matrices for various components are described in Table 1.3.

TABLE 1.3
Mueller Matrices of Various Optical Elements

	Optical Elements	Mueller Matrix M
1	Linear polarizer along x-direction	$M = \frac{1}{2}\begin{bmatrix} 1 & 1 & 0 & 0 \\ 1 & 1 & 0 & 0 \\ 0 & 0 & 0 & 0 \\ 0 & 0 & 0 & 0 \end{bmatrix}$
2	Linear polarizer along y-direction	$M = \frac{1}{2}\begin{bmatrix} 1 & -1 & 0 & 0 \\ -1 & 1 & 0 & 0 \\ 0 & 0 & 0 & 0 \\ 0 & 0 & 0 & 0 \end{bmatrix}$
3	Linear polarizer making an angle 45° with x-axis	$M = \frac{1}{2}\begin{bmatrix} 1 & 0 & 1 & 0 \\ 0 & 0 & 0 & 0 \\ 1 & 0 & 1 & 0 \\ 0 & 0 & 0 & 0 \end{bmatrix}$
4	Linear polarizer making an angle –45° with x-axis	$M = \frac{1}{2}\begin{bmatrix} 1 & 0 & -1 & 0 \\ 0 & 0 & 0 & 0 \\ -1 & 0 & 1 & 0 \\ 0 & 0 & 0 & 0 \end{bmatrix}$
5	Linear polarizer making an angle θ with x-axis	$M = \frac{1}{2}\begin{bmatrix} 1 & \cos 2\theta & \sin 2\theta & 0 \\ \cos 2\theta & \cos^2 2\theta & \sin 2\theta \cos 2\theta & 0 \\ \sin 2\theta & \sin 2\theta \cos 2\theta 0 & \sin^2 2\theta & 0 \\ 0 & 0 & 0 & 0 \end{bmatrix}$
6	Quarter-wave plate: – fast axis along y-direction	$M = \begin{bmatrix} 1 & 0 & 0 & 0 \\ 0 & 1 & 0 & 0 \\ 0 & 0 & 0 & -1 \\ 0 & 0 & 1 & 0 \end{bmatrix}$
7	Quarter-wave plate: – fast axis along x-direction	$M = \begin{bmatrix} 1 & 0 & 0 & 0 \\ 0 & 1 & 0 & 0 \\ 0 & 0 & 0 & 1 \\ 0 & 0 & -1 & 0 \end{bmatrix}$

(*Continued*)

TABLE 1.3 (*Continued*)
Mueller Matrices of Various Optical Elements

	Optical Elements	Mueller Matrix M
8	Right-handed circular polarizer	$M=\frac{1}{2}\begin{bmatrix}1&0&0&1\\0&0&0&0\\0&0&0&0\\1&0&0&1\end{bmatrix}$
9	Left-handed circular polarizer	$M=\frac{1}{2}\begin{bmatrix}1&0&0&-1\\0&0&0&0\\0&0&0&0\\-1&0&0&1\end{bmatrix}$

Let us consider an example in which a randomly polarized wave is incident on an x-polarizer. The Stokes parameters of the transmitted wave are given by

$$\begin{bmatrix}S_0\\S_1\\S_2\\S_3\end{bmatrix}=\frac{1}{2}\begin{bmatrix}1&1&0&0\\1&1&0&0\\0&0&0&0\\0&0&0&0\end{bmatrix}\begin{bmatrix}1\\0\\0\\0\end{bmatrix}=\frac{1}{2}\begin{bmatrix}1\\1\\0\\0\end{bmatrix} \tag{1.47}$$

This represents a x-polarized wave.

Let us now consider another example in which a randomly polarized wave is incident on an x-polarizer followed by a linear polarizer oriented at an angle θ with the x-direction. The Stokes parameters are now obtained as

$$\begin{bmatrix}S_0\\S_1\\S_2\\S_3\end{bmatrix}=\frac{1}{2}\begin{bmatrix}1&\cos 2\theta&\sin 2\theta&0\\\cos 2\theta&\cos^2 2\theta&\sin 2\theta\cos 2\theta&0\\\sin 2\theta&\sin 2\theta\cos 2\theta 0&\sin^2 2\theta&0\\0&0&0&0\end{bmatrix}\frac{1}{2}\begin{bmatrix}1&1&0&0\\1&1&0&0\\0&0&0&0\\0&0&0&0\end{bmatrix}\begin{bmatrix}1\\0\\0\\0\end{bmatrix}$$

$$=\frac{1}{4}\begin{bmatrix}1+\cos 2\theta\\\cos 2\theta(1+\cos 2\theta)\\\sin 2\theta(1+\cos 2\theta)\\0\end{bmatrix} \tag{1.48}$$

The irradiance of the transmitted wave is given by $I(\theta)\propto ½(1+\cos 2\theta)=\cos^2\theta$. As the incident irradiance is taken unity, this is the statement of the Malus law.

1.7 FRESNEL EQUATIONS

Let us consider a linearly polarized plane wave incident on an interface between two dielectric media of refractive indices μ_1 and μ_2, respectively. The interface is taken in the xy-plane. Let the xz-plane be the plane of incidence. The wave is incident at angle θ_i in the medium of refractive index μ_1 and the angle of the transmitted wave is θ_t. The incident linearly polarized wave can be decomposed into two waves: one wave is polarized in the plane of incidence (referred to as ||), and other is polarized in a plane perpendicular to the plane of incidence (referred to as $\perp$). Applying the boundary conditions and using Snell's law of refraction, the reflection and the transmission coefficients can be obtained. Writing $\mu = \mu_2 / \mu_1$, the reflection and the transmission coefficients for parallel and perpendicular polarizations are given as

$$r_{\parallel} = \frac{-\mu \cos\theta_i + \cos\theta_t}{\mu \cos\theta_i + \cos\theta_t} = \frac{-\mu^2\cos\theta_i + \sqrt{\mu^2 - \sin^2\theta_i}}{\mu^2\cos\theta_i + \sqrt{\mu^2 - \sin^2\theta_i}} \tag{1.49a}$$

$$r_{\perp} = \frac{\cos\theta_i - \mu \cos\theta_t}{\cos\theta_i + \mu \cos\theta_t} = \frac{\cos\theta_i - \sqrt{\mu^2 - \sin^2\theta_i}}{cos\theta_i + \sqrt{\mu^2 - \sin^2\theta_i}} \tag{1.49b}$$

and

$$t_{\parallel} = \frac{2\ \mu \cos\theta_i}{\mu^2\cos\theta_i + \sqrt{\mu^2 - \sin^2\theta_i}} \tag{1.49c}$$

$$t_{\perp} = \frac{2 \cos\theta_i}{\cos\theta_i + \sqrt{\mu^2 - \sin^2\theta_i}} \tag{1.49d}$$

Further, it can be shown that $\mu\ t_{\parallel} = 1 - r_{\parallel}$ and $t_{\perp} = 1 + r_{\perp}$. The fraction of power of the incident wave reflected and transmitted is called the reflectance and the transmittance, respectively. The reflectance and the transmittance of the interface for parallel and perpendicular polarizations are given by

$$\mathcal{R}_{\parallel} = \left|r_{\parallel}\right|^2 \tag{1.50a}$$

$$\mathcal{R}_{\perp} = \left|r_{\perp}\right|^2 \tag{1.50b}$$

and

$$\mathcal{T}_{\parallel} = \mu\left(\cos\theta_t / \cos\theta_i\right)\left|t_{\parallel}\right|^2 \tag{1.50c}$$

$$\mathcal{T}_{\perp} = \mu\left(\cos\theta_t / \cos\theta_i\right)\left|t_{\perp}\right|^2 \tag{1.50d}$$

An examination of Equation (1.49a) of the reflection coefficient for parallel polarization reveals that at some angle θ_B the reflection coefficient $r_{\parallel}$ is zero, that is

$$\left(-\mu^2\cos\theta_i + \sqrt{\mu^2 - \sin^2\theta_i}\right)_{\theta_i=\theta_B} = 0 \tag{1.51}$$

This gives $\mu = \tan\theta_B$. The angle θ_B is called the Brewster angle. This equation suggests a simple method to measure the refractive index of a transparent substance with respect to air. Further, the reflected light is linearly polarized. Its plane of polarization is perpendicular to the plane of incidence. This is thus a clever method to find the direction of polarization of elements like Polaroids, etc.

1.8 THIN FILM OPTICS

Reflection and transmission properties of the surfaces can be drastically modified by depositing thin films of appropriate thicknesses and materials. The surface could be made anti-reflecting, high reflecting, or of a tailored reflectivity. Almost all surfaces in image-forming optics are coated with broadband anti-reflection coatings. The coated surface can alter the spectral transmittance of an element. The purpose of this section is to study the physics of thin films.

Let us consider a thin film of thickness t bounded by plane interfaces 1 and 2, respectively. A plane wave is incident obliquely. It is assumed that the electric field vector of the wave is perpendicular to the plane of incidence (transverse electric [TE] wave). The directions of electric and magnetic field vectors in the three regions are shown in Figure 1.3.

The boundary conditions dictate that the tangential components of the electric (E) and magnetic (B) fields must be continuous at the interface, that is, their magnitudes on either side of the interface must be equal. The magnitudes of the incident, reflected, and transmitted fields have been indicated in Figure 1.3.

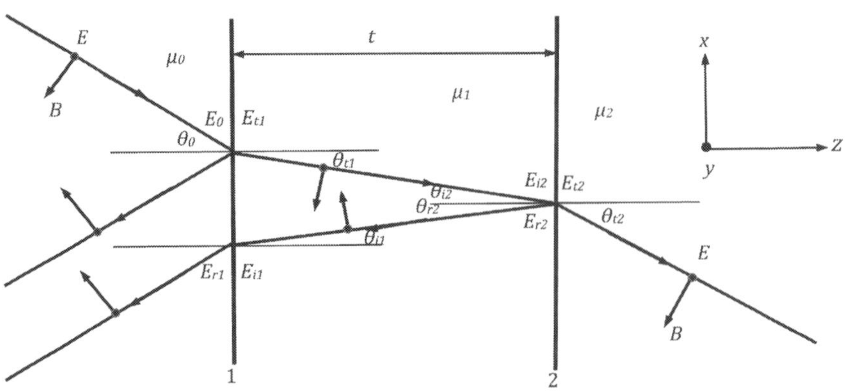

FIGURE 1.3 Optics of thin film.

Let E_1 and B_1 be the magnitudes of the electric and magnetic fields just on the left side of interface 1. Then

$$E_1 = E_0 + E_{r1} = E_{t1} + E_{i1} \quad (1.52a)$$

$$B_1 = B_0\cos\theta_0 - B_{r1}\cos\theta_0 = B_{t1}\cos\theta_{t1} - B_{i1}\ \cos\theta_{t1} \quad (1.52b)$$

It may be noted that the angles $\theta_{t1} = \theta_{i2} = \theta_{r2} = \theta_{i1}$. Let E_2 and B_2 be the magnitudes of electric and magnetic fields on the right side of interface 2. Then

$$E_2 = E_{i2} + E_{r2} = E_{t2} \quad (1.53a)$$

$$B_2 = B_{i2}\cos\theta_{i2} - B_{r2}\ \cos\theta_{r2} = B_{t2}\cos\theta_{t2} \quad (1.53b)$$

It is known that the electric and magnetic fields are related through the velocity of light in the medium. Thus $B_0 = (E_0 / v_0) = \mu_0\, E_0 / c$ and similar expressions for other magnetic field components. Here v_0 is the velocity of the wave in the medium of refractive index μ_0 and c is the velocity in a vacuum. We now rewrite Equations (1.52a) and (1.52b) as

$$E_1 = E_0 + E_{r1} = E_{t1} + E_{i1} \quad (1.54a)$$

$$B_1 = (E_0 - E_{r1})\frac{\mu_0\cos\theta_0}{c} = (E_{t1} - E_{i1})\ \frac{\mu_1\cos\theta_{t1}}{c} \quad (1.54b)$$

Defining $\beta_0 = \frac{\mu_0\cos\theta_0}{c}$, $\beta_1 = \frac{\mu_1\cos\theta_{t1}}{c}$ and $\beta_2 = \frac{\mu_2\cos\theta_{t2}}{c}$, Equation (1.54b) is written as

$$B_1 = (E_0 - E_{r1})\beta_0 = (E_{t1} - E_{i1})\ \beta_1 \quad (1.54c)$$

Similarly at interface 2, we have

$$E_2 = E_{i2} + E_{r2} = E_{t2} \quad (1.55a)$$

$$B_2 = (E_{i2} - E_{r2}\ )\beta_1 = E_{t2}\beta_2 \quad (1.55b)$$

As the wave propagates from interface 1 to interface 2, it acquires a phase δ, where $\delta = \kappa\mu_1 t\cos\theta_{t1}$. Thus the field E_{i2} and E_{i1} can be expressed as

$$E_{i2} = E_{t1}\ e^{-i\delta} \quad (1.56a)$$

and

$$E_{i1} = E_{r2}\ e^{-i\delta} \quad (1.56b)$$

Substituting these in Equations (1.55a) and (1.55b), we obtain

$$E_2 = E_{t1}e^{-i\delta} + E_{i1}e^{i\delta} \quad (1.57a)$$

$$B_2 = (E_{t1}e^{-i\delta} - E_{i1}e^{i\delta})\beta_1 \quad (1.57b)$$

Solving for E_{t1} and E_{i1}, we obtain

$$E_{t1} = \frac{\beta_1 \; E_2 + B_2}{2\beta_1} \; e^{i\delta} \tag{1.58a}$$

and

$$E_{i1} = \frac{\beta_1 \; E_2 - B_2}{2\beta_1} \; e^{-i\delta} \tag{1.58b}$$

We will insert these values in Equations (1.54a) and (1.54b) at interface 1, that is

$$E_1 = E_{t1} + E_{i1} = E_2 \cos\delta + B_2 \frac{i \; \sin\delta}{\beta_1} \tag{1.59a}$$

and

$$B_1 = \left(E_{t1} - E_{i1}\right) \beta_1 = E_2 \; \beta_1 i \; \sin\delta + B_2 \; \cos\delta \tag{1.59b}$$

These two equations can be put in matrix form as

$$\begin{bmatrix} E_1 \\ B_1 \end{bmatrix} = \begin{bmatrix} \cos\delta & \dfrac{i \; \sin\delta}{\beta_1} \\ i \; \beta_1 \; \sin\delta & \cos\delta \end{bmatrix} \begin{bmatrix} E_2 \\ B_2 \end{bmatrix} \tag{1.60}$$

Equation (1.60) relates the electric and the magnetic fields just before the film to that just after the film. The 2 × 2 matrix is called the transfer matrix, M, of the film. In the general form it can be written as

$$M = \begin{bmatrix} m_{11} & m_{12} \\ m_{21} & m_{22} \end{bmatrix} \tag{1.61}$$

Instead of a single film, if there are n films, each with a transfer matrix M_n, then the transfer matrix M of the film structure (multilayer) is given by

$$M = M_1 M_2 M_3 M_4 \ldots\ldots M_n \tag{1.61}$$

The fields are related as

$$\begin{bmatrix} E_1 \\ B_1 \end{bmatrix} = M \begin{bmatrix} E_n \\ B_n \end{bmatrix} \tag{1.62}$$

We can now return to Equations (1.54a) and (1.54b) for a single film and insert the values of the fields into Equation (1.60), which then becomes

$$\begin{bmatrix} E_0 + E_{r1} \\ (E_0 - E_{r1})\beta_0 \end{bmatrix} = \begin{bmatrix} \cos\delta & \dfrac{i\ \sin\delta}{\beta_1} \\ i\ \beta_1\ \sin\delta & \cos\delta \end{bmatrix} \begin{bmatrix} E_{t2} \\ \beta_2\ E_{t2} \end{bmatrix} = \begin{bmatrix} m_{11} & m_{12} \\ m_{21} & m_{22} \end{bmatrix} \begin{bmatrix} E_{t2} \\ \beta_2\ E_{t2} \end{bmatrix} \quad (1.63)$$

This gives

$$E_0 + E_{r1} = m_{11}E_{t2} + m_{12}\beta_2\ E_{t2} \quad (1.64a)$$

$$(E_0 - E_{r1})\beta_0 = m_{21}E_{t2} + m_{22}\beta_2\ E_{t2} \quad (1.64b)$$

Defining the reflection coefficient $r_\perp$ and transmission coefficient $t_\perp$ as

$$r_\perp = \frac{E_0 + E_{ri}}{E_0} \quad (1.65a)$$

and

$$t_\perp = \frac{E_{t2}}{E_0} \quad (1.65b)$$

we obtain,

$$1 + r_\perp = m_{11}t_\perp + m_{12}\beta_2 t_\perp \quad (1.66a)$$

and

$$(1 - r_\perp)\beta_0 = m_{21}t_\perp + m_{22}\beta_2 t_\perp \quad (1.66b)$$

These can be solved for $r_\perp$ and $t_\perp$ to yield

$$r_\perp = \frac{\beta_0 m_{11} + \beta_0\ \beta_2\ m_{12} - m_{21} - \beta_2\ m_{22}}{\beta_0 m_{11} + \beta_0\ \beta_2\ m_{12} + m_{21} + \beta_2\ m_{22}} \quad (1.67a)$$

$$t_\perp = \frac{2\beta_0}{\beta_0 m_{11} + \beta_0\ \beta_2\ m_{12} + m_{21} + \beta_2\ m_{22}} \quad (1.67b)$$

When we consider parallel polarization (transverse magnetic [TM] polarization), we obtain the same expressions but the βs are defined differently. For the parallel (||) component, the β is represented as

$$\beta_1 = \frac{\mu_1}{c\ \cos\theta_{t1}} \quad (1.68)$$

Note that the polarization states (|| or ⊥) are indistinguishable for normal incidence as $\cos\theta_{t1} = 1$. For oblique incidence, the reflection coefficients for both the polarizations are to be computed. The reflectance of the film, when a randomly polarized wave is incident on the film, is then obtained as an average of the reflectance for the parallel ($\mathcal{R}_{\|}$) and the perpendicular ($\mathcal{R}_{\perp}$) polarizations, that is

$$\mathcal{R} = \frac{\mathcal{R}_{\|} + \mathcal{R}_{\perp}}{2} \tag{1.69}$$

As an example, we consider a quarter-wave ($\delta = \pi/2$) layer (thickness $\lambda/4$) on a substrate and assume normal incidence. The transfer matrix then takes the form

$$\begin{bmatrix} 0 & \dfrac{i\,c}{\mu_1} \\ \dfrac{i\,\mu_1}{c} & 0 \end{bmatrix} \tag{1.70}$$

Thus $m_{11} = m_{22} = 0$, $m_{12} = \frac{i\,c}{\mu_1}$ and $m_{21} = \frac{i\,\mu_1}{c}$. Further, $\beta_0 = \frac{\mu_0}{c}$ and $\beta_2 = \frac{\mu_2}{c}$. On substituting these values in the expression for reflection coefficient, we obtain

$$r_{\perp} = r = \frac{\beta_0\,\beta_2\,m_{12} - m_{21}}{\beta_0\,\beta_2\,m_{12} + m_{21}} = \frac{\dfrac{\mu_0}{c}\,\dfrac{\mu_2}{c}\,\dfrac{i\,c}{\mu_1} - \dfrac{i\,\mu_1}{c}}{\dfrac{\mu_0}{c}\,\dfrac{\mu_2}{c}\,\dfrac{i\,c}{\mu_1} - \dfrac{i\,\mu_1}{c}} = \frac{\mu_0\mu_2 - \mu_1^2}{\mu_0\mu_2 + \mu_1^2} \tag{1.71}$$

The reflectance, $\mathcal{R}$, is given by

$$\mathcal{R} = \left(\frac{\mu_0\mu_2 - \mu_1^2}{\mu_0\mu_2 + \mu_1^2}\right)^2 \tag{1.72}$$

The reflectance will be zero when $\mu_1 = \sqrt{\mu_0\mu_2}$. In this case, the waves reflected from the two interfaces meet out of phase leading to destructive interference. We can study multilayer structures using the same approach, and the insight gained into the process is better when the individual layers are taken either as a quarter-wave or half-wave coupled with the fact that there is phase change of π on reflection when the reflection takes place from rare to denser media and no phase change when the reflection takes place from denser to rarer media.

1.9 OPTICAL COMPONENTS

Optical components are usually used for beam manipulation, which may include change of size, curvature, and direction. Some components either singly or in multi-element configurations are used for image formation. These can also be grouped as reflective, refractive, and diffractive components. We give a brief description in the following sections.

1.9.1 Reflective Components

Reflective components are plane and curved (both spherical and aspherical), usually front-coated for optical applications, and are termed as plane mirrors or spherical mirrors, in general. Plane mirrors are used to bend the beams, change the direction, fold the beams, etc. The spherical mirrors can perform the dual function of change of direction and change of curvature and in particular image formation.

1.9.1.1 Mirrors

Mirrors are used both for changing the direction of light propagation and imaging. A plane mirror makes a perfect image of an object. The image is reversed, of unit magnification, and always virtual. A pair of plane mirrors enclosing an angle α between them makes multiple images with different orientations: the number of images is given by $\left(\frac{360^0}{\alpha}-1\right)$, where α is in degrees. A pair of mirrors parallel to each other makes an infinite number of images with decreasing irradiance due to the reflectance of the mirrors being less than unity. Plane mirrors are often used to bend the beam in an optical experiment.

Concave mirrors are used mostly for imaging but in some cases perform the dual function of imaging and bending the direction of a beam. A concave mirror produces both real and virtual images of magnification varying over a large range. On the other hand, a convex mirror always produces a virtual, erect image with magnification less than unity. It covers a wide field of view and often is used at places requiring seeing on a large field of view.

The equation that governs the image formation by mirrors, with proper sign convention, is given by

$$\frac{1}{p}+\frac{1}{q}=\frac{2}{R} \tag{1.73}$$

where p is the object distance measured from the vertex and is taken positive when the object lies in front of the mirror, q is the image distance measured from the vertex and is taken positive when the image is in front of the mirror, and R is the radius of curvature of the mirror and is taken positive for concave and negative for the convex mirror. This equation is valid under paraxial approximation ($\sin\theta \approx \theta$) and the focal length f of the mirror is then given by $f = R/2$.

Along with spherical mirrors, there are paraboloidal, ellipsoidal, and hyperboloidal mirrors with their unique imaging properties.

1.9.2 Refractive Components

Functioning of these elements is based on Snell's law of refraction.

To start with we take a plane interface. The propagation of a ray of light is governed by Snell's law of refraction. There are several optical components that are bound by plane surfaces including PPP, wedge plate, prisms, etc.

1.9.2.1 Plane Parallel Plate (PPP)

A plate bounded by two plane/flat surfaces, which are parallel to each other, constitutes a PPP. It has two main functions, namely, to provide two beams of nearly the same amplitude but with a path difference and to shift the beam laterally or longitudinally. PPPs are used as beam splitters, compensators, and shear devices in interferometers. The path difference between the two consecutive reflected or transmitted beams introduced by the PPP is given by

$$\Delta = 2\mu t\cos\theta_t \tag{1.74}$$

where t is the thickness, μ is the refractive index of the material of the plate, and θ_t is the angle of refraction. The PPP is a very convenient element to produce sheared beams for interferometric applications.

The lateral shift, d, of the beam due to PPP is given by

$$d = t\sin\theta_i\left(1 - \frac{\mu_i}{\mu_p}\frac{\cos\theta_i}{\cos\theta_t}\right) \tag{1.75}$$

where θ_i and θ_t are the incidence and the refracted angles at the first interface; μ_i and μ_p are the refractive indices of the surrounding and the material of the plate, respectively; and t is the plate thickness. For small angles, this reduces to

$$d = t\ \theta_i\left(1 - \frac{\mu_i}{\mu_p}\right) \tag{1.76}$$

The shift of the beam, therefore, increases linearly with the angle of incidence. Further, if the plate is placed in a convergent beam, then the focus gets shifted away from the plate. The shift of focus, under paraxial approximation, is given by

$$d_f = t\left(1 - \frac{\mu_i}{\mu_p}\right) \tag{1.77}$$

The focal shift is entirely governed by the thickness and refractive index of the plate.

1.9.2.2 Wedge Plate

A wedge plate functions either as a beam splitter or a shear plate or a beam deviation prism. When two such plates are used, and rotated relative to each other, the beam can be effectively steered. Such prisms are known as Risley prisms. A wedge plate of refractive index μ_p in a medium of refractive index μ_i deviates the beam by an angle γ given as

$$\gamma = \left(\frac{\mu_p - \mu_i}{\mu_i}\right)\alpha \tag{1.78}$$

where α is the apex angle of the prism, which is assumed small. Essentially the line of sight deviates by an angle γ. When two such prisms with angles α_1 and α_2 are used, the resultant angular deviation is given by

$$\bar{\gamma} = \bar{\gamma}_1 + \bar{\gamma}_2 \qquad \gamma = \sqrt{\gamma_1^2 + \gamma_1^2 + 2\gamma_1\gamma_2\cos\beta} \tag{1.79}$$

where γ_1 and γ_2 are the deviations produced by the wedges of apex angles α_1 and α_2, respectively, and β is the angle between the two wedges. Obviously, when the wedges are parallel, their deviation angles add and when they are anti-parallel, the deviation angles subtract. When one of the prisms/wedges is rotated, the line of sight goes in a circle.

A collimated beam incident at an angle θ at the front surface will be angularly sheared on reflection from the front and back surfaces of a wedge. The angular shear is given by

$$\Delta\theta = \sin^{-1}\left[\sin\theta_i + \frac{1}{\mu_i}\left(\mu_p^2 - \mu_i^2\sin^2\theta\right)^{1/2} 2\alpha\right] - \theta_i \tag{1.80}$$

where θ_i is the angle of incidence. For a small angle of incidence, the angular shear is given by

$$\Delta\theta = \frac{\mu_p}{\mu_i} 2\alpha \tag{1.81}$$

A wedge plate is also used for collimation testing. If the wedge angle is small so that the beams reflected from the front and the back surfaces enclose a very small angle, the straight-line fringes running parallel to the wedge edge are formed. Curvature in the incident beam rotates these fringes.

1.9.2.3 Prisms

Prisms are used both for deviation and dispersion. With polychromatic light, both the functions occur simultaneously. However, prism systems can be designed that deviate the beam without dispersion. With monochromatic light, prisms provide a robust way to deviate the beam through a desired angle, rotate the image, retro-reflect the beam, rotate the plane of polarization, etc. A functional prism has to have four or more plane surfaces. Some surfaces in a prism are to be coated. In some prisms, a surface carries a roof. Unfolded prism can be shown in the form of a tunnel diagram. Prisms like Rochon, Wollaston, etc., made from doubly refracting materials are used for special functions.

1.9.3 Diffractive Components

The function of these components is governed by the diffraction of light. These are lightweight and can be formed either on the surface of a refractive component, or a reflective component, or be stand-alone elements. Gratings of a constant or variable period fall under this category. These components exhibit large dispersion and hence are to be used with monochromatic or nearly monochromatic radiation. Ronchi grating, zone-plate, and computer-generated elements are realized by mechanical means, whereas sinusoidal grating and Gabor zone plate and other specialized elements are realized by holography.

1.10 REFRACTION AT CURVED INTERFACE

Consider a block of glass whose one face has been ground and polished to a spherical shape of radius of curvature R as shown in Figure 1.4. An object is placed a distance p from the vertex of this spherical surface in a medium of refractive index μ_1. Its image is formed at a distance q. The image formation is governed by the equation

$$\frac{\mu_1}{p} + \frac{\mu_2}{q} = \frac{\mu_2 - \mu_1}{R} \tag{1.82}$$

where μ_2 is the refractive index of the glass and the spherical interface is between the media of refractive indices μ_1 and μ_2, respectively.

Equation (1.82) is valid under paraxial approximation. The magnification is given by $m = -\frac{\mu_1}{\mu_2}\frac{q}{p}$. Depending on the location of the object, both the real and the virtual images can be formed. For a plane refracting surface, $q = -\frac{\mu_2}{\mu_1}p$ and the magnification is $m = +1$. The image is always a virtual one. For example, if the object is in water and is seen by an observer in air, it appears nearer to the water interface.

1.10.1 LENSES

Usually, a lens is a piece of transparent material with at least one surface curved. However, there are lenses with both flat surfaces: the power to bend the rays is either built in through diffractive structures or by varying refractive index.

There are six main configurations of refractive lenses: three each with positive power and the remaining three with negative power. These are identified as double-convex (bi-convex), plano-convex, meniscus (convex-concave), double concave

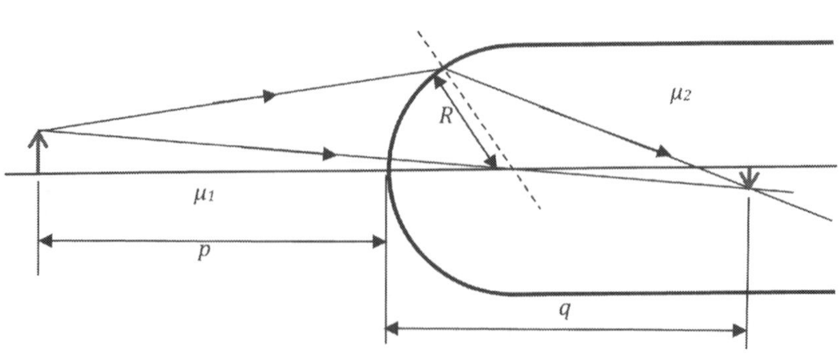

FIGURE 1.4 Refraction at the spherical surface.

(bi-concave), plano-concave, and meniscus (concave-convex). The image formation is governed by the equation

$$\frac{1}{p}+\frac{1}{q}=(\mu-1)\left(\frac{1}{R_1}-\frac{1}{R_2}\right) \tag{1.83}$$

where p and q are the object and image distances, and R_1 and R_2 are the radii of curvatures of the two surfaces bounding the transparent material of refractive index μ in air. The ray paths for image formation by a positive lens are shown in Figure 1.5.

Distances measured from the vertex of the lens in object space for object position and in image space for image position are positive, and the radius of curvature to the right of the surface is positive.

The magnification m is defined as $m=-\frac{q}{p}$. This is also called lateral magnification. In addition to lateral magnification, axial magnification m_a is also used. It is related to lateral magnification by the relation $m_a=m^2$. F_1 and F_2 are the primary and secondary focal points. The focal length is defined as

$$\frac{1}{f}=(\mu-1)\left(\frac{1}{R_1}-\frac{1}{R_2}\right) \tag{1.84}$$

The focal length is negative for concave lenses and positive for convex lenses. Further, a concave lens always makes a virtual image, which is upright and has a magnification less than unity. A convex lens makes both real and virtual images with varying magnifications.

A single lens suffers both from chromatic and monochromatic aberrations. Chromatic aberrations arise due to the dependence of the refractive index of the material of the lens on the wavelength resulting in wavelength-dependent focal length and magnification. The monochromatic aberration exists even though the lens is being used with a monochromatic light. It has been classified as spherical aberration, astigmatism, coma, curvature, and distortion. A lens can be optimized to possess minimum spherical aberration by changing its shape. Non-spherical surfaces may result in the elimination of spherical aberration. However, multi-element lenses are designed for improved performance by minimizing the overall aberration function. A large number of lens design software is commercially available.

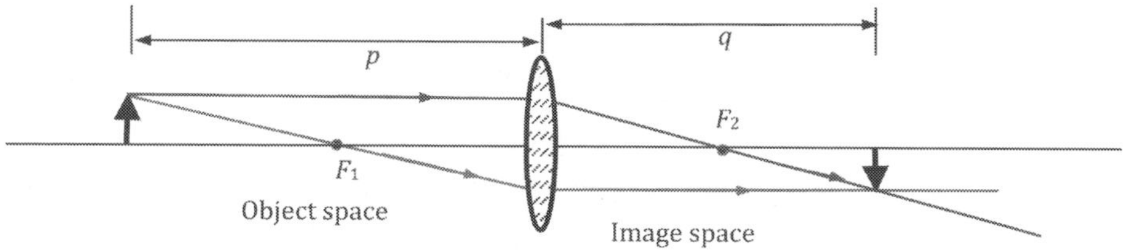

FIGURE 1.5 Image formation by a positive lens.

1.11 PARAXIAL OPTICS

Paraxial optics is often required to trace the rays through an optical system. In general, the law of refraction, the law of reflection, and translation from one surface to another are required to trace the rays. However, when the rays propagate very close to the optical axis and make a very small angle such that the laws of reflection and refraction are linearized, we are in the paraxial domain. It is often convenient to use matrix methods for ray tracing. A ray is described by a two-component vector $\mathbf{r} = \begin{bmatrix} r \\ \theta \end{bmatrix}$, where r is the height from the optical axis at a plane and θ is the angle it makes with the perpendicular at the plane. The ray propagation from one plane to another is described by a 2 × 2 matrix, often called an ABCD matrix and is defined as $M = \begin{bmatrix} A & B \\ C & D \end{bmatrix}$. Let $\mathbf{r}_0 = \begin{bmatrix} r_0 \\ \theta_0 \end{bmatrix}$ be the vector at the input plane and $\boldsymbol{r}_1 = \begin{bmatrix} r_1 \\ \theta_1 \end{bmatrix}$ be the vector at the output plane. These are connected through

$$\begin{bmatrix} r_1 \\ \theta_1 \end{bmatrix} = \begin{bmatrix} A & B \\ C & D \end{bmatrix} \begin{bmatrix} r_0 \\ \theta_0 \end{bmatrix} \tag{1.85}$$

From this

$$r_1 = Ar_0 + B\theta_0$$

$$\theta_1 = Cr_0 + D\theta_0$$

When the matrix approach is applied to an imaging system, the element $C = -1/f$. We can, however, examine some special cases:

When $A = 0$, we have $r_1 = B\theta_0$. The component r_1 does not depend on r_0. The output plane is the second focal plane.
When $B = 0$, we have $r_1 = Ar_0$. The component r_1 does not depend on θ_0. The input and output planes are conjugate planes and $-A$ is the magnification of the system.
When $C = 0$, we have $\theta_1 = D\theta_0$. The component θ_1 does not depend on r_0 implying that the system is telescopic wherein parallel rays enter the system and parallel rays leave.
When $D = 0$, we have $\theta_1 = Cr_0$. The component θ_1 does not depend on θ_0. The input plane must be the first focal plane of the system.

If the matrices for different components are available, the ray trace is obtained simply by the matrix multiplication. Let us construct the matrices for some simple cases. We shall first construct a translation matrix.

1.11.1 Translation Matrix

Let us consider that the two planes P_0 and P_1 are separated by a distance t. The ray hits the plane P_0 at height r_0 and makes an angle of θ_0 with the normal and hits the

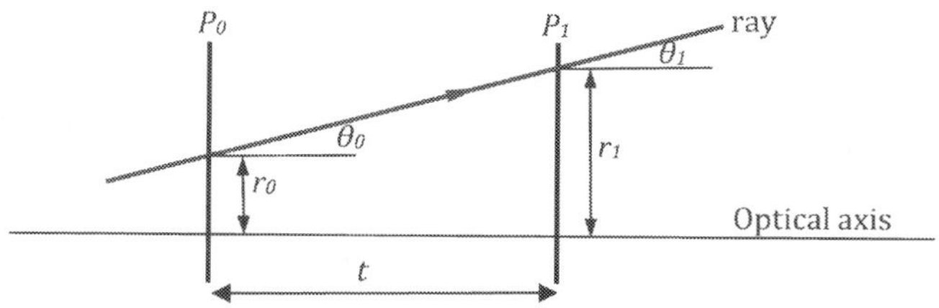

FIGURE 1.6 Translation matrix for the ray.

plane P_1 at height r_1 and makes an angle θ_1 with the normal to the plane as shown in the Figure 1.6.

From the Figure 1.6, we obtain

$$r_1 = r_0 + t\theta_0 \tag{1.86a}$$

and

$$\theta_1 = \theta_0 \tag{1.86b}$$

Further, using the matrix approach, the vectors $\boldsymbol{r}_1$ and $\boldsymbol{r}_0$ are related as

$$\begin{bmatrix} r_1 \\ \theta_1 \end{bmatrix} = \begin{bmatrix} A & B \\ C & D \end{bmatrix} \begin{bmatrix} r_0 \\ \theta_0 \end{bmatrix}$$

This gives

$$r_1 = Ar_0 + B\theta_0 \tag{1.87a}$$

and

$$\theta_1 = Cr_1 + D\theta_0 \tag{1.87b}$$

Comparing the Equations (1.86) and (1.87), we obtain that $A = 1$, $B = t$, $C = 0$, and $D = 1$. Therefore, the translation matrix is $\begin{bmatrix} 1 & t \\ 0 & 1 \end{bmatrix}$.

1.11.2 Refraction Matrix

1.11.2.1 Plane Dielectric Interface

Let us consider a plane interface between two dielectric media of refractive indices μ_0 and μ_1, respectively. A ray is incident at an angle θ_0 as shown in the Figure 1.7

For this ray, $r_1 = r_0$ and $\theta_1 = \frac{\mu_0}{\mu_1}\theta_0$. This leads to $A = 1$, $B = 0$, $C = 0$, and $D = \frac{\mu_0}{\mu_1}$. The ABCD matrices for the dielectric interface is $\begin{bmatrix} 1 & 0 \\ 0 & \frac{\mu_0}{\mu_1} \end{bmatrix}$.

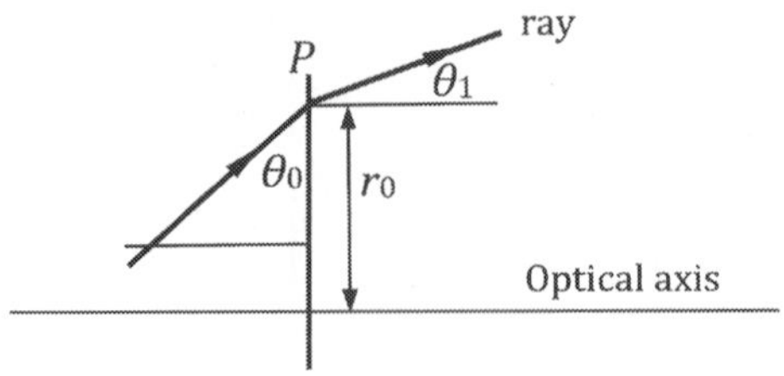

FIGURE 1.7 Calculation for the refraction matrix.

1.11.2.2 Spherical Dielectric Interface

In Figure 1.8, *AB* is the incident ray, *BG* is the refracted ray, and *EC* is the normal to the spherical surface. *C* is the center of curvature of the spherical surface and hence $BC = R$, where *R* is the radius of curvature. Further, $<ABD = \theta_0$, $<GBF = \theta_1$, $<ABE = \theta_i$ (angle of incidence), and $<CBG = \theta_r$ (angle of refraction), and the angle $<OCB = \psi$. Using Snell's law of refraction, we have $\mu_0\theta_i = \mu_1\theta_r$. Further, $\theta_0 = \theta_i - \psi$ and $\theta_1 = \theta_r - \psi$.

Substituting for θ_i and θ_r in the linearized Snell's law, we have

$$\mu_0(\theta_0 + \psi) = \mu_1(\theta_1 + \psi)$$

This can be rearranged as

$\theta_1 = \frac{\mu_0}{\mu_1}\theta_0 + \left(\frac{\mu_0 - \mu_1}{\mu_1}\right)\psi$. Writing for ψ as $\psi = \frac{r_0}{R}$, we obtain

$\theta_1 = \frac{\mu_0}{\mu_1}\theta_0 + \left(\frac{\mu_0 - \mu_1}{\mu_1}\right)\frac{r_0}{R}$. Further, $r_1 = r_0$. The matrix elements are thus obtained as $A = 1, B = 0, C = \left(\frac{\mu_0 - \mu_1}{\mu_1 R}\right)$, and $D = \frac{\mu_0}{\mu_1}$. The matrix for the spherical dielectric interface is now expressed as $\begin{bmatrix} 1 & 0 \\ \left(\frac{\mu_0 - \mu_1}{\mu_1 R}\right) & \frac{\mu_0}{\mu_1} \end{bmatrix}$.

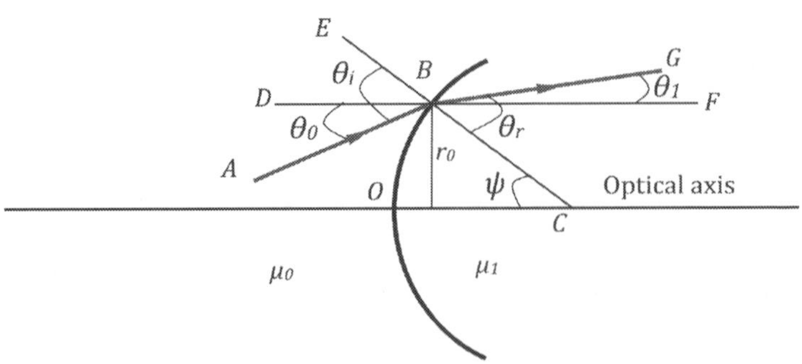

FIGURE 1.8 Refraction matrix at spherical surface.

1.11.3 Matrix for a Thin Lens

Figure 1.9 shows a thin lens of focal length f on which a ray originating from a point object at A is incident at a height r_0. The lens images the point object at B: the lens changes the direction of the incident ray.

The image formation is governed by the equation

$$\frac{1}{p}+\frac{1}{q}=\frac{1}{f}$$

On multiplying by r_0, this equation becomes

$$\frac{r_0}{p}+\frac{r_0}{q}=\frac{r_0}{f} \tag{1.88}$$

Because the ray leaving the lens makes an angle θ_1, which is taken as negative by the sign convention, Equation (1.88) can be rewritten as $\theta_1=\theta_0-\frac{r_0}{f}$. Further, $r_1=r_0$. Therefore, the matrix elements are $A=1$, $B=0$, $C=-\frac{1}{f}$, and $D=1$. The matrix for the lens is thus $\begin{bmatrix} 1 & 0 \\ -\frac{1}{f} & 1 \end{bmatrix}$. This is valid for both the positive and negative lenses.

The ABCD matrices for some of the commonly used interfaces and optical components are given in Table 1.4.

1.11.4 Matrix for a Thick Lens

Let us apply the matrix approach to obtain the focal length of a thick lens of refractive index μ_1, which is shown in Figure 1.10. The lens bounded by spherical surfaces of radii of curvature R_1 and R_2, and central thickness t is in media of refractive indices μ_0 and μ_2.

This involves three operations, namely (1) refraction matrix at R_1, (2) translation matrix, and (3) refraction matrix at R_2. The system matrix can therefore be written as

$$M=M_3M_2M_1=\begin{bmatrix} 1 & 0 \\ \left(\dfrac{\mu_1-\mu_2}{\mu_2R_2}\right) & \dfrac{\mu_1}{\mu_2} \end{bmatrix}\begin{bmatrix} 1 & t \\ 0 & 1 \end{bmatrix}\begin{bmatrix} 1 & 0 \\ \left(\dfrac{\mu_o-\mu_1}{\mu_1R_1}\right) & \dfrac{\mu_0}{\mu_1} \end{bmatrix}$$

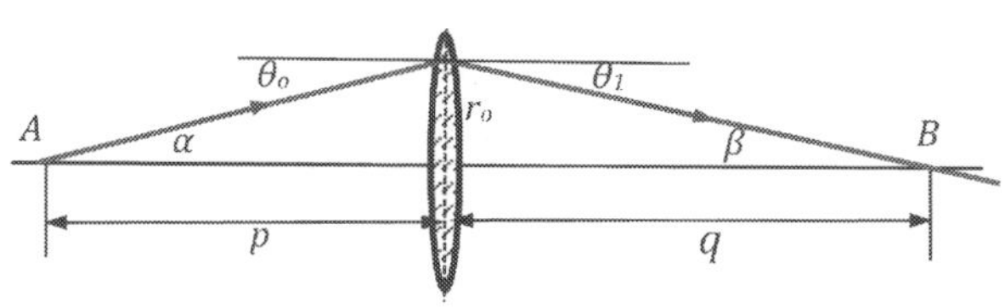

FIGURE 1.9 Matrix for a lens.

TABLE 1.4
ABCD Matrices for Some Elements

S. No.	Matrix	Formula
1	Translation matrix	$\begin{bmatrix} 1 & t \\ 0 & 1 \end{bmatrix}$
2	Refraction matrix (Plane interface)	$\begin{bmatrix} 1 & 0 \\ 0 & \frac{\mu_0}{\mu_1} \end{bmatrix}$
3	Refraction matrix (Spherical interface)	$\begin{bmatrix} 1 & 0 \\ \left(\frac{\mu_0-\mu_1}{\mu_1 R}\right) & \frac{\mu_0}{\mu_1} \end{bmatrix}$
4	Thin-lens matrix [focal length f where $\frac{1}{f}=\frac{\mu_1-\mu_0}{\mu_0}\left(\frac{1}{R_1}-\frac{1}{R_2}\right)$]	$\begin{bmatrix} 1 & 0 \\ -\frac{1}{f} & 1 \end{bmatrix}$: (+$f$: positive, –$f$: negative lens)
5	Reflection matrix (surface of radius of curvature R)	$\begin{bmatrix} 1 & 0 \\ \frac{2}{R} & 1 \end{bmatrix}$: (+$R$: convex, –$R$: concave)

$$M = \begin{bmatrix} 1+\left(\dfrac{\mu_0-\mu_1}{\mu_1 R_1}\right)t & \dfrac{\mu_0}{\mu_1}t \\ \left(\dfrac{\mu_1-\mu_2}{\mu_2 R_2}\right)+\dfrac{\mu_1}{\mu_2}\left(\dfrac{\mu_0-\mu_1}{\mu_1 R_1}\right)+\left(\dfrac{\mu_0-\mu_1}{\mu_1 R_1}\right)\left(\dfrac{\mu_1-\mu_2}{\mu_2 R_2}\right)t & \left(\dfrac{\mu_1-\mu_2}{\mu_2 R_2}\right)\dfrac{\mu_0}{\mu_1}t+\dfrac{\mu_1}{\mu_2}\dfrac{\mu_0}{\mu_1} \end{bmatrix}$$

The focal length f of the lens is obtained as

$$-\frac{1}{f}=\left(\frac{\mu_1-\mu_2}{\mu_2 R_2}\right)+\frac{\mu_1}{\mu_2}\left(\frac{\mu_0-\mu_1}{\mu_1 R_1}\right)+\left(\frac{\mu_0-\mu_1}{\mu_1 R_1}\right)\left(\frac{\mu_1-\mu_2}{\mu_2 R_2}\right)t$$

When the lens of refractive index μ ($\mu = \mu_1$) is in the air $(\mu_0 = \mu_2 = 1)$, the focal length f is then given by

$$\frac{1}{f}=(\mu-1)\left(\frac{1}{R_1}-\frac{1}{R_2}\right)+\frac{(\mu-1)^2}{R_1 R_2}\left(\frac{t}{\mu}\right)$$

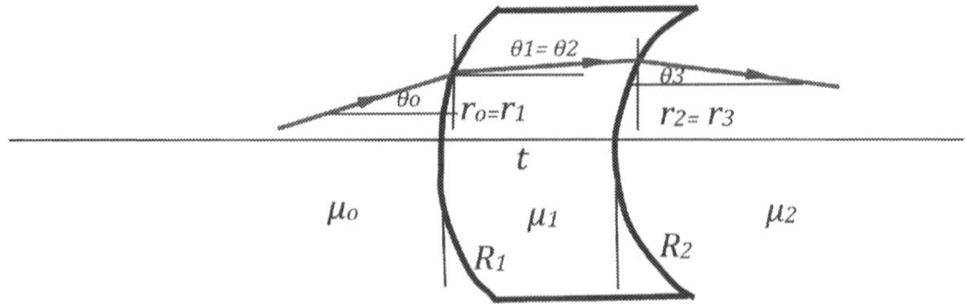

FIGURE 1.10 Geometry of a thick lens.

This reduces to the lens maker's formula when the lens is considered thin ($t = 0$).

QUESTIONS

1.1 What is the usual distance equation that specifies the requirement for far-field diffraction?

1.2 Explain the following statement: The far-field diffraction pattern of an aperture is the Fourier transform of its transmittance function.

1.3 Show that a grating made up of alternately transparent and opaque strips of equal width eliminates all the even orders (except $m = 0$).

1.4 Show that the dispersion $D(= d\theta / d\lambda)$ of a grating is $D = \tan\theta / \lambda$, where θ is the angle of diffraction for wavelength λ.

1.5 Show that the irradiance distribution in the interference pattern of a three-slit grating is given by

$$I(\delta) = \frac{1}{9} I(0)\left(1 + 4\cos\delta + 4\cos^2\delta\right),$$

where $\delta = 2\pi d \sin\theta / \lambda$ is the phase difference between consecutive slits of separation d and $I(0)$ is the irradiance when $\delta = 0$. The slits are assumed to be extremely narrow.

1.6 Manufacturers of wire (and other objects of small dimension) sometimes use a laser to continuously monitor the thickness/size of the product. Suppose a He-Ne laser, with a wavelength of 632.8 nm, illuminates a wire, and the diffraction pattern at a screen distance of 2.60 m from the wire is observed. If the desired wire diameter is 1.37 mm, what is the observed distance between the two tenth-order minima (one on each side of the central maximum).

1.7 Show that, in a Newton's rings experiment, the difference in radii between the adjacent bright rings (maxima) is given by

$$\Delta r = r_{m+1} - r_m \approx \frac{1}{2}\sqrt{\lambda R / m},$$

assuming $m \gg 1$ and R is the radius of curvature of the spherical surface. Further, show that the area between adjacent bright rings is given by $A = \pi\lambda R$, assuming $m \gg 1$. Note that the area is independent of m.

1.8 In a double-slit arrangement to produce interference fringes, the slits of unequal width are used: the width of one slit is double that of the other. Obtain an expression for the irradiance distribution.

1.9 Thin glass plates of refractive indices 1.50 and 1.65, respectively, are placed before each slit of a double-slit arrangement. The central maximum now shifts to the position, that was occupied by the fifth bright fringe ($m = 5$) before the plates were introduced. Assuming that λ = 480 nm and that the plates have the same thickness, find the thickness of the plates.

1.10 Two glass plates 5.0 cm long are held together at one end and a 0.1-mm diameter wire is inserted at the other end thereby forming a wedge of air. If the light of wavelength λ = 632.8 nm is used in reflection to observe the interference pattern, what is the fringe width? If the air wedge is filled with water (μ =1.33), what is the new fringe width?

1.11 A block of glass with a cavity of 50 μm is placed in the collimated beam. The length of the block is 2.0 cm and its refractive index is 1.52. Assuming that it is illuminated with a wavelength λ = 632.8 nm, what is the path difference introduced by the block of glass. Express this path difference in terms of wavelength.

1.12 A thin film of refractive index μ and thickness d is in air and is illuminated normally by light with wavelengths λ_1 and λ_2. When illuminated by λ_2, irradiance maximum is observed in reflected light, whereas irradiance minimum is observed when illuminated by λ_1. Show that the integer value m is given by $m = \lambda_1 / 2(\lambda_2 - \lambda_1)$. Determine the thickness of the film if μ = 1.4, λ_1 = 500 nm, and λ_2 = 370 nm.

1.13 A piece of transparent material having an index of refraction μ is cut into the shape of a wedge. The angle of the wedge is small, and a monochromatic light is incident normally on its surface. If the thickness difference between the thick and thin ends of the wedge is t and its length is h, show that the bright fringes, in reflection, occur at positions $x_m = \lambda h(m + ½) / 2\mu t$, where x_m is measured from the thin end.

1.14 A plano-convex lens having a radius of curvature of R_1 = 4.0 m is placed on a concave surface having a radius of curvature of R_2 = 12.0 m. Determine the radius of the 20th bright ring when illuminated with light of a wavelength 632.8 nm normally.

1.15 A plano-cylindrical glass (μ = 1.52) lens with a central thickness of 2.0 cm and the radius of the cylindrical surface of 1.0 m is placed with its cylindrical surface on a flat glass plate and illuminated normally with a light wave of wavelength 632.8 nm. What is the nature of the fringes and the fringe width at the 10th fringe? If this lens is placed on a convex glass surface with a radius of 4.0 m, what is the nature of the fringes and the fringe width at the 10th fringe?

1.16 Slit 1 of a double-slit arrangement is wider than slit 2 so that the light from slit 1 has an amplitude 3.0 times that of the light from slit 2. Show that the irradiance distribution is given by $I = \frac{I_{\max}}{4}\left(1 + 3\cos^2\frac{\delta}{2}\right)$, where δ is the phase difference.

1.17 Show that the irradiance distribution in the double-slit interference pattern when the amplitude due to one slit is twice that of the other is given by $I = I_0\left(\frac{5}{9} + \frac{4}{9}\cos\delta\right)$, where I_0 is the maximum irradiance.

1.18 A slit is cut in the metal sheet, and its diffraction pattern is viewed by illuminating it with light that has a wavelength 500 nm. The screen is 8.0 m away from the sheet. The central maximum is 10.0 cm wide. The temperature of the sheet is raised by 80° resulting in the change of width of central maximum by 0.002 cm. What is the linear coefficient of expansion of the sheet?

1.19 A thin slit is illuminated by light with a wavelength λ and the first diffraction minima in air are observed at ±38.2°. When the entire apparatus is immersed in an unknown liquid, the first-order minima occur at ±17.4°. What is the index of refraction of the liquid?

1.20 Show that the Brewster angle θ_{Bag} when the light is reflected from an air-glass interface, and the Brewster angle θ_{Bga} when the light is reflected from a glass-air interface are complimentary, i.e., $\theta_{\text{Bag}} + \theta_{\text{Bga}} = \pi/2$. What is the significance of the Brewster angle?

1.21 What is the thickness of a soap film ($\mu = 1.33$), which appears green ($\lambda = 550$ nm) when illuminated with white light normally?

1.22 An x-polarizer is followed by an analyzer whose polarization direction is oriented at an arbitrary direction α with respect to the polarizer. A wave plate with phase retardation δ whose fast axis makes an angle ϕ with the polarizer is placed between them. Show that the irradiance transmitted is given by

$$I = I_0\left[\cos^2\alpha\cos^2\frac{\delta}{2} + \cos^2(\alpha - 2\phi)\sin^2\frac{\delta}{2}\right]$$

where I_0 is the irradiance incident on the polarizer. How could you use this to find the orientation of the fast axis of the wave plate?

1.23 An optical system consisting of a convex lens, a concave lens, and a mirror is assembled as shown in the Figure 1.11. For the data given in the figure, obtain the location of the final image using both the conventional method and the matrix method. All distances are in centimeters.

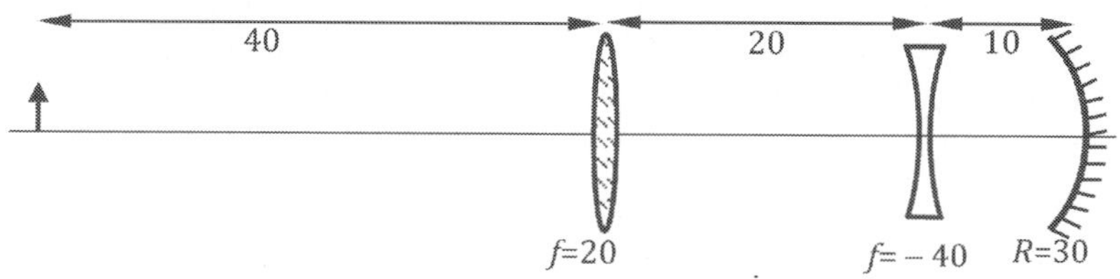

FIGURE 1.11

1.24 Show that the equation $\left(-\mu^2\cos\theta_i + \sqrt{\mu^2 - \sin^2\theta_i}\right)_{\theta_i=\theta_B} = 0$ leads to the Brewster law, i.e., $\tan\theta_B = \mu$.

2 Laser Beams

2.1 GAUSSIAN BEAMS

The laser beam is an obvious choice in many metrological applications. When the laser is oscillating in transverse electromagnetic mode (TEM_{00}), the beam has a Gaussian profile, and its spatial extent is very small. It is important to understand its propagation and manipulation by the lens or any other image-forming component. Because light is a wave motion, we begin with the three-dimensional wave equation and search for a solution, which would resemble a laser beam. The wave equation is given by

$$\nabla^2 u - \frac{1}{c^2}\frac{\partial^2 u}{\partial t^2} = 0 \tag{2.1}$$

where u is the wave amplitude and $\nabla^2 = \frac{\partial^2}{\partial x^2} + \frac{\partial^2}{\partial y^2} + \frac{\partial^2}{\partial z^2}$ is the Laplacian. We will consider a monochromatic wave traveling along positive z-direction and express the solution as $u(x, y, z : t) = U(x, y, z)e^{i\omega t}$, where ω is the angular frequency of the wave. Substituting the amplitude in the wave equation [Equation (2.1)] yields the Helmholtz equation:

$$\nabla^2 U + k^2 U = 0 \tag{2.2}$$

where the wave vector k is given by $k = \frac{\omega}{c}$. Because $U(x,y,z)$ represents the amplitude of the wave of wave vector k traveling in the positive z-direction, the amplitude $U(x,y,z)$ is expressed as $U(x, y, z) = U_0(x, y, z)e^{-ikz}$. On substituting this in the Helmholtz equation [Equation (2.2)], we obtain

$$\frac{\partial^2 U_0}{\partial x^2} + \frac{\partial^2 U_0}{\partial y^2} + \frac{\partial^2 U_0}{\partial z^2} - 2ik\frac{\partial U_0}{\partial z} = 0 \tag{2.3}$$

Because the wave is traveling along the positive z-direction, its amplitude may decrease slowly over the distance. We may therefore set $\frac{\partial^2 U_0}{\partial z^2} = 0$. Thus the Helmholtz equation [Equation (2.2)], on substitution and weak approximation, takes the form

$$\frac{\partial^2 U_0}{\partial x^2} + \frac{\partial^2 U_0}{\partial y^2} - 2ik\frac{\partial U_0}{\partial z} = 0 \tag{2.4}$$

This equation is called the paraxial wave equation. We now search for the solution of this equation, which represents a laser beam.

We assume a solution of the type

$$U_0(x, y, z) = Ae^{-i\frac{k(x^2+y^2)}{2q(z)}} e^{-ip(z)} \tag{2.5}$$

DOI: 10.1201/9781003531845-2

where A is a constant and $q(z)$ and $p(z)$ are general functions. This solution partly has its genesis in the propagation of a spherical wave. It may be noticed that x, y, and z dependences have now been separated. On substitution in the paraxial wave equation [Equation (2.4)], we obtain

$$U_0\left[\frac{k^2}{q^2(z)}\left(1-\frac{dq}{dz}\right)(x^2+y^2)+2k\left(\frac{i}{q(z)}+\frac{dp}{dz}\right)\right]=0 \tag{2.6}$$

Equation (2.6) should be satisfied for all values of x and y, hence

$$1-\frac{dq}{dz}=0 \tag{2.7a}$$

and

$$\frac{i}{q(z)}+\frac{dp}{dz}=0 \tag{2.7b}$$

Solutions of Equations (2.7a) and (2.7b), respectively, give

$$q(z)=z+iz_0 \tag{2.8a}$$

and

$$p(z)=-i\ \ln\left(1+\frac{z}{iz_0}\right) \tag{2.8b}$$

where z_0 is a constant.

Substituting the values of $q(z)$ and $p(z)$ from Equations (2.8a) and (2.8b) into Equation (2.5), we have

$$U_0(x,y,z)=Ae^{-i\frac{k(x^2+y^2)}{2(z+iz_0)}}e^{-i\left[-i\ \ln\left(1+\frac{z}{iz_0}\right)\right]}=Ae^{-i\frac{k(x^2+y^2)(z-iz_0)}{2(z^2+z_0^2)}}e^{-\ln\left(1-\frac{iz}{z_0}\right)} \tag{2.9a}$$

We can express $ln\left(1-\frac{iz}{z_0}\right)$ as

$$\ln\left(1-\frac{iz}{z_0}\right)=\ln\left(\sqrt{1+\frac{z^2}{z_0^2}}\ e^{-i\ \tan^{-1}\frac{z}{z_0}}\right)=\ln\sqrt{1+\frac{z^2}{z_0^2}}-i\ \tan^{-1}\frac{z}{z_0}$$

Therefore, Equation (2.9a) can be written as

$$U_0(x,y,z)=Ae^{-i\frac{k(x^2+y^2)(z-iz_0)}{2(z^2+z_0^2)}}e^{-\ln\sqrt{1+\frac{z^2}{z_0^2}}}e^{i\ \tan^{-1}\frac{z}{z_0}}$$

or

$$U_0(x,y,z)=\frac{A}{\sqrt{1+\left(\frac{z}{z_0}\right)^2}}\,e^{-i\frac{k(x^2+y^2)}{2z\left(1+\frac{z_0^2}{z^2}\right)}}\,e^{-\frac{\pi}{\lambda}\frac{(x^2+y^2)}{z_0\left(1+\frac{z^2}{z_0^2}\right)}}\,e^{i\tan^{-1}\left(\frac{z}{z_0}\right)} \tag{2.9b}$$

Defining the radius of curvature of the beam as $R(z)=z\left[1+\left(\frac{z_0}{z}\right)^2\right]$, half beam-waist as $\omega_0=\sqrt{\frac{z_0\lambda}{\pi}}$, half beam-spot size as $\omega(z)=\omega_0\sqrt{1+\left(\frac{z}{z_0}\right)^2}$ and the Gouy phase $\phi(z)=\tan^{-1}\frac{z}{z_0}$, we can express the amplitude of a laser beam as

$$U_0(x,y,z)=A\frac{\omega_0}{\omega(z)}e^{-i\frac{k(x^2+y^2)}{2R(z)}}\,e^{-\frac{(x^2+y^2)}{\omega^2(z)}}\,e^{i\phi(z)} \tag{2.10a}$$

Alternately it can be expressed as

$$U_0(x,y,z)=A\frac{\omega_0}{\omega(z)}e^{-i\frac{k(x^2+y^2)}{2q(z)}}\,e^{i\phi(z)} \tag{2.10b}$$

We may now identify $q(z)$ through the relation $\frac{1}{q(z)}=\frac{1}{R(z)}+\frac{i\lambda}{\pi\omega^2(z)}$ and may call it the complex radius of curvature of the laser beam. The irradiance of the laser beam can be expressed as

$$I(x,y,z)=A^2\frac{\omega_0^2}{\omega^2(z)}e^{-\frac{2(x^2+y^2)}{\omega^2(z)}}=I_0(z)\,e^{-\frac{2(x^2+y^2)}{\omega^2(z)}} \tag{2.11}$$

The irradiance distribution at any plane normal to the beam propagation is a Gaussian with the spot size $2\omega(z)$.

It is interesting to see how $q(z)$ varies as we translate the plane of observation. Let $q(z_1)$ ($= q_1$) be its value at plane $z = z_1$ and $q(z_2)$ ($= q_2$) at $z = z_2$. It is thus seen from Equation (2.8a) that $q_2 = q_1 + z_2 - z_1 = q_1 + \Delta z$. The complex radius of curvature therefore varies according to translation in optics. Let us further examine its behavior on refraction. We may note that a lens converts a diverging wave front from a point away from its focal plane to a converging wave front that converges to a point according to the lens formula

$$\frac{1}{q}+\frac{1}{p}=\frac{1}{f} \tag{2.12}$$

Therefore, we may assume that the complex radius of curvature would also vary as

$$\frac{1}{q_{\text{left}}}=\frac{1}{q_{\text{right}}}+\frac{1}{f} \tag{2.13a}$$

where q_{left} and q_{right} are the q values on the left and right sides of the lens. Alternately

$$q_{\text{right}} = \frac{q_{\text{left}}}{-\left(\dfrac{q_{\text{left}}}{f}\right)+1} \tag{2.13b}$$

2.2 ABCD LAW FOR GAUSSIAN BEAMS

We just examined the way $q(z)$ behaves under translation and refraction. We may now examine whether we could use the ABCD matrix method as applied to paraxial optics for the propagation of the laser beam, particularly the Gaussian beam. We write an expression for the evolution of q as $q' = \frac{Aq+B}{Cq+D}$. Let us apply this evolution of q to the ABCD matrices of translation and refraction, namely,

$$T = \begin{bmatrix} 1 & d \\ 0 & 1 \end{bmatrix} \tag{2.14a}$$

and

$$R = \begin{bmatrix} 1 & 0 \\ -\Phi & 1 \end{bmatrix} = \begin{bmatrix} 1 & 0 \\ -1/f & 1 \end{bmatrix} \tag{2.14b}$$

On substitution of matrix elements, A, B, C, and D of translation matrix T, the evolution of q in free space propagation gives,

$$q' = \frac{q+d}{0+1} = q+d \tag{2.15a}$$

Similarly, the propagation through a lens of focal length f results in

$$q' = \frac{q+0}{-q/f+1} = \frac{q}{-q/f+1} \tag{2.15b}$$

These equations [(2.15a) and (2.15b)] are the same results that we obtained earlier, thus justifying the expression for the evolution of q. We can therefore model the propagation of the laser beams using the ABCD matrix method, which is applicable in the paraxial region. Further, let us examine whether the ABCD matrix method could also be applied to multiple elements so that the propagation of a beam through free space and lenses could also be studied. Let there be two lens-like media in succession. We evolve q_2 after the first medium using the relation,

$$q_2 = \frac{A_1 q_1 + B_1}{C_1 q_1 + D_1} \tag{2.16a}$$

Similarly, we evolve q_3 after the second medium, which is given by

$$q_3 = \frac{A_2 q_2 + B_2}{C_2 q_2 + D_2} = \frac{A_2 \dfrac{A_1 q_1 + B_1}{C_1 q_1 + D_1} + B_2}{C_2 \dfrac{A_1 q_1 + B_1}{C_1 q_1 + D_1} + D_2} = \frac{(A_1 A_2 + C_1 B_2) q_1 + B_1 A_2 + D_1 B_2}{(A_1 C_2 + C_1 D_2) q_1 + B_1 C_2 + D_1 D_2} = \frac{A_3 q_1 + B_3}{C_3 q_1 + D_3} \tag{2.16b}$$

where the matrix $\begin{bmatrix} A_3 & B_3 \\ C_3 & D_3 \end{bmatrix} = \begin{bmatrix} A_1A_2 + C_1B_2 & B_1A_2 + D_1B_2 \\ A_1C_2 + C_1D_2 & B_1C_2 + D_1D_2 \end{bmatrix} = \begin{bmatrix} A_2 & B_2 \\ C_2 & D_2 \end{bmatrix}\begin{bmatrix} A_1 & B_1 \\ C_1 & D_1 \end{bmatrix}$.

Alternately $M_3 = M_2\, M_1$. In general, if there are n elements, the matrix M_s for the system can be written as

$$M_s = M_n\; M_{n-1}\; \ldots\ldots\ldots M_2\; M_1 \tag{2.17}$$

We now apply the ABCD matrix method to some simple cases.

2.2.1 Free Space Propagation

It is intended to find out how the beam-spot size and beam radius of curvature vary as the Gaussian beam propagates in free space. Let q_0 and q_z be the q values at $z = 0$ and z = z planes. We, therefore, have $q_z = q_0 + z$. Further, we assume that the beam-waist lies at the $z = 0$ plane. Thus,

$$\frac{1}{q_0} = \frac{i\lambda}{\pi\omega_0^2} \tag{2.18a}$$

and

$$\frac{1}{q_z} = \frac{1}{R(z)} + \frac{i\lambda}{\pi\omega^2(z)} \tag{2.18b}$$

where ω_0 is the beam-waist radius at the $z = 0$ plane, $\omega(z)$ is the beam-spot radius, and $R(z)$ is the radius of curvature of the Gaussian beam at the $z = z$ plane. Rewriting $q_z = q_0 + z$ as $\frac{1}{q_z} = \frac{1}{q_0}\frac{1}{1+\frac{z}{q_0}}$ and inserting the values of q_0 and q_z, we have

$$\frac{1}{R(z)} + \frac{i\lambda}{\pi\omega^2(z)} = \frac{i\lambda}{\pi\omega_0^2}\,\frac{1}{1+\dfrac{i\lambda z}{\pi\omega_0^2}} \tag{2.19}$$

Rearranging and equating the real and the imaginary parts, we have

$$R(z) = \frac{\left(\pi\omega_0^2\right)^2}{\lambda^2 z} + z = z\left[1 + \left(\frac{z_0}{z}\right)^2\right] \tag{2.20a}$$

and

$$\pi\omega^2(z) = \frac{(\pi\omega_0^2)^2 + z^2\lambda^2}{\pi\omega_0^2} \Rightarrow \omega(z) = \omega_0\left[1+\left(\frac{z}{z_0}\right)^2\right]^{\frac{1}{2}} \tag{2.20b}$$

Here we have defined $z_0 = \frac{\pi\omega_0^2}{\lambda}$. The distance z_0 is the distance from the beam-waist plane to the plane where the beam-spot diameter becomes $\sqrt{2}$ times the beam-waist diameter. The distance, $\pm z_0$ is called the Rayleigh range in which the beam-spot diameter is considered practically constant.

2.2.2 Propagation through a Lens

2.2.2.1 Beam-Waist Lies at Plane 1 at the Lens

Beam-waist of the input beam lies at plane 1 as shown in Figure 2.1. Hence at plane 1, we have $\omega = \omega_{01}$ and $R_1 = \infty$ so that

$$\frac{1}{q_1} = \frac{1}{R_1} + i\frac{\lambda}{\pi\omega_{01}^2} = i\frac{\lambda}{\pi\omega_{01}^2} = \frac{i}{z_{01}} \tag{2.21}$$

Transferring to plane 2 gives

$$q_2 = \frac{q_1}{-q_1/f+1} \Rightarrow \frac{1}{q_2} = \frac{1}{q_1} - \frac{1}{f} = i\frac{\lambda}{\pi\omega_{01}^2} - \frac{1}{f} \tag{2.22}$$

Beam parameter q_2 at plane 2 can be rewritten as

$$q_2 = \frac{\pi f\omega_{01}^2}{i\lambda f - \pi\omega_{01}^2} = -\frac{\pi f\omega_{01}^2(i\lambda f + \pi\omega_{01}^2)}{(\lambda f)^2 + (\pi\omega_{01}^2)^2} = a + ib \tag{2.23}$$

where

$$a = -\frac{f(\pi\omega_{01}^2)^2}{(\lambda f)^2 + (\pi\omega_{01}^2)^2} \tag{2.24a}$$

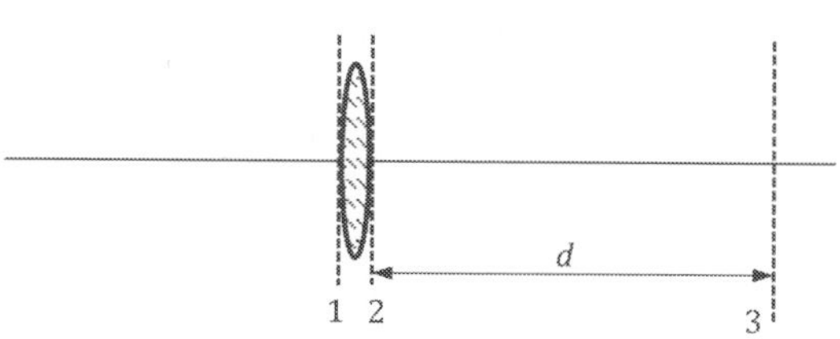

FIGURE 2.1 Propagation of beam-waist through a lens: beam-waist at plane 1.

and

$$b = -\frac{f^2 \lambda \pi \omega_{01}^2}{(\lambda f)^2 + (\pi \omega_{01}^2)^2} \tag{2.24b}$$

The beam parameter q_3 at plane 3 distant d from the lens is

$$q_3 = q_2 + d = a + d + ib \tag{2.25a}$$

We may express q_3 as

$$\frac{1}{q_3} = \frac{1}{R_3} + i\frac{\lambda}{\pi \omega_3^2} = \frac{a + d - ib}{(a+d)^2 + b^2} \tag{2.25b}$$

Comparing the real and imaginary part in Equation (2.25b), we have

$$\frac{1}{R_3} = \frac{a+d}{(a+d)^2 + b^2} \Rightarrow R_3 = \frac{(a+d)^2 + b^2}{a+d} \tag{2.26a}$$

and

$$\frac{\lambda}{\pi \omega_3^2} = \frac{-b}{(a+d)^2 + b^2} \tag{2.26b}$$

If plane 3 has to be the plane where the beam-waist lies, then $R_3 = \infty$ and $\omega_3 = \omega_{03}$. For R_3 to be infinite, $a + d = 0$. Substituting for a from Equation (2.24a) in this expression we get

$$-\frac{f(\pi \omega_{01}^2)^2}{(\lambda f)^2 + (\pi \omega_{01}^2)^2} + d = 0 \tag{2.27a}$$

This can be expressed as

$$d = \frac{f}{1 + \left(\dfrac{\lambda f}{\pi \omega_{01}^2}\right)^2} = \frac{f}{1 + \left(\dfrac{f}{z_0^2}\right)^2} \tag{2.27b}$$

The size of the beam-waist is obtained from Equation (2.26b) by setting $a + d = 0$ as

$$\frac{\lambda}{\pi \omega_{03}^2} = \frac{-1}{b} \tag{2.28a}$$

On substituting b from Equation (2.24b) in Equation (2.28a) and rearranging, we obtain,

$$\frac{\omega_{03}^2}{\omega_{01}^2} = \frac{1}{1 + \left(\dfrac{\pi \omega_{01}^2}{\lambda f}\right)^2} = \frac{1}{1 + \left(\dfrac{z_0}{f}\right)^2} \tag{2.28b}$$

Examination of Equation (2.27b) reveals that the beam comes to focus closer to the lens when the Gaussian beam is incident on the lens. Further, the beam-waist size is controlled by both initial beam-waist and focal length of the lens.

2.2.2.2 Beam-Waist Lies in Front of the Lens

We assume that the beam-waist lies at plane 1 at a distance p from the lens as shown in Figure 2.2. The goal is to find out the plane at which the beam-waist lies after the Gaussian beam propagates through the lens of focal length f. Let the beam parameter at plane 1 be q_1, that is, $\frac{1}{q_1} = \frac{1}{R_1} + i\frac{\lambda}{\pi\omega_{01}^2} = i\frac{\lambda}{\pi\omega_{01}^2}$, where ω_{01} is the beam-waist at this plane.

The beam parameter at plane 4 can be obtained using Equation (2.17) as

$$q_4 = \frac{A_T q_1 + B_T}{C_T q_1 + D_T} \tag{2.29}$$

where the matrix $\begin{bmatrix} A_T & B_T \\ C_T & D_T \end{bmatrix}$ is given by

$$\begin{bmatrix} A_T & B_T \\ C_T & D_T \end{bmatrix} = \begin{bmatrix} 1 & d \\ 0 & 1 \end{bmatrix}\begin{bmatrix} 1 & 0 \\ -1/f & 1 \end{bmatrix}\begin{bmatrix} 1 & p \\ 0 & 1 \end{bmatrix} = \begin{bmatrix} 1-d/f & p+d-pd/f \\ -1/f & 1-p/f \end{bmatrix} \tag{2.30}$$

Substituting the matrix elements in Equation (2.29), we obtain

$$q_4 = \frac{(1-d/f)q_1 + p + d - pd/f}{(-1/f)q_1 + 1 - p/f} = \frac{-i(f-d)z_{01} + (p+d)f - pd}{iz_{01} + f - p} \tag{2.31}$$

Rewriting Equation (2.31), we have

$$\frac{1}{q_4} = \frac{(f-p)\left[pf + (f-p)d\right] - (f-d)z_{01}^2 + iz_{01}\left[(f-p)(f-d) + pf + d(f-p)\right]}{\left[pf + (f-p)d\right]^2 + (f-d)^2 z_{01}^2} \tag{2.32}$$

On equating real and imaginary parts of Equation (2.32), we obtain

$$R_4(z) = \frac{\left[pf + (f-p)d\right]^2 + (f-d)^2 z_{01}^2}{(f-p)\left[pf + (f-p)d\right] - (f-d)z_{01}^2} \tag{2.33a}$$

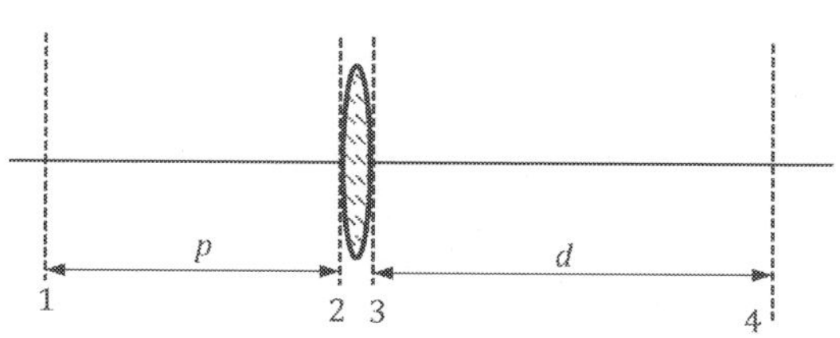

FIGURE 2.2 Propagation of beam-waist through a lens: beam-waist in front of lens at plane 1.

and

$$\frac{\lambda}{\pi\omega_4^2} = \frac{z_{01}\left[(f-p)(f-d)+pf+d(f-p)\right]}{\left[pf+(f-p)d\right]^2+(f-d)^2 z_{01}^2} \tag{2.33b}$$

For the beam-waist to lie at plane 4 distant d from the lens, $R_4 = \infty$, i.e.,

$$(f-p)\left[pf+(f-p)d\right]-(f-d)z_{01}^2 = 0 \tag{2.34a}$$

This can be reorganized as

$$\frac{1}{d} = \frac{(p-f)\left[(p-f)+z_{01}^2/(p-f)\right]}{f(p-f)\left[p+z_{01}^2/(p-f)\right]} = \frac{\left[p+z_{01}^2/(p-f)\right]-f}{f\left[p+z_{01}^2/(p-f)\right]} = \frac{1}{f} - \frac{1}{\left[p+z_{01}^2/(p-f)\right]} \tag{2.34b}$$

This equation is like $\frac{1}{p}+\frac{1}{q}=\frac{1}{f}$, provided d is identified with q, and p is identified with $\left[p+z_{01}^2/(p-f)\right]$. In this sense, it can be said that a lens does imaging of the beam-waist of a Gaussian beam.

The distance d can also be expressed as

$$d = f + \frac{f^2(p-f)}{(p-f)^2+z_{01}^2} \tag{2.34c}$$

Similarly, the expression for the beam-waist ($\omega_4 = \omega_{04}$ when $R_4 = \infty$) can be written as

$$\frac{\lambda}{\pi\omega_{04}^2} = z_{01}\frac{(f-p)(f-d)+\dfrac{f-d}{f-p}z_{01}^2}{\left(\dfrac{f-d}{f-p}z_{01}^2\right)^2+(f-d)^2 z_{01}^2} = \frac{1}{z_{01}}\frac{(f-p)}{(f-d)} = \frac{\lambda}{\pi\omega_{01}^2}\frac{(f-p)}{(f-d)} \tag{2.34d}$$

This gives

$$\frac{\omega_{04}^2}{\omega_{01}^2} = \frac{(f-d)}{(f-p)} \tag{2.34e}$$

Substituting for d from Equation (2.34c) in Equation (2.34e), we obtain

$$\frac{\omega_{04}^2}{\omega_{01}^2} = \frac{f - \dfrac{f\left[p(p-f)+z_{01}^2\right]}{(p-f)^2+z_{01}^2}}{(f-p)} \tag{2.35}$$

On simplifying the expression [Equation (2.35)], we obtain

$$\frac{\omega_{04}}{\omega_{01}} = \frac{f}{\sqrt{(p-f)^2 + z_{01}^2}} \tag{2.36}$$

This is the magnification factor, i.e., the beam-waist after passage through the lens is magnified by this factor.

2.2.2.3 Beam Focusing

Using a strong positive lens or a microscope objective, the Gaussian laser beam can be focused to a smaller size, i.e., then $\omega_{04} \ll \omega_{01}$. It can be shown that ω_{04} is now given by

$$\omega_{04} = \frac{\lambda f}{\pi\ \omega_{01}}$$

The beam-waist occurs close to the focal plane of the lens.

2.3 LASER COLLIMATOR

Before we discuss the design of a laser collimator, let us find out the total power in the Gaussian beam. This would be used to determine the size of the collimating lens. The total power P_0 in the beam is obtained by integrating the irradiance in the beam over an infinite surface area, i.e.,

$$P_0 = \iint I(x,y,z)\, dA = A^2 \frac{\omega_0^2}{\omega^2(z)} \iint e^{-2\frac{x^2+y^2}{\omega^2(z)}}\, dx\, dy = A^2 \frac{\omega_0^2}{\omega^2(z)} 2\pi \int_0^\infty e^{-2\frac{\rho^2}{\omega^2(z)}}\, \rho\, d\rho = A^2 \frac{\pi\omega_0^2}{2} \tag{2.37}$$

where $\rho^2 = x^2 + y^2$. The total power in the beam is $P_0 = A^2\left(\frac{\pi\ \omega_0^2}{2}\right)$. The total power in the beam is independent of $\omega(z)$. However, as the beam propagates in space, the beam keeps on expanding. If this beam is captured by a lens of an aperture of radius a, then the power P_a captured by the aperture is

$$P_a = A^2 \frac{\omega_0^2}{\omega^2(z)}\, 2\pi \int_0^a e^{-2\frac{\rho^2}{\omega^2(z)}}\, \rho\, d\rho = A^2\left(\frac{\pi\ \omega_0^2}{2}\right)\left[1 - e^{-2\frac{a^2}{\omega^2(z)}}\right] \tag{2.38}$$

The fraction Φ_p of total power captured is

$$\Phi_p = \frac{P_a}{P_0} = \left[1 - e^{-2\frac{a^2}{\omega^2(z)}}\right] \tag{2.39}$$

Table 2.1 gives the power captured as a function $a/\omega(z)$.

It is thus seen that a lens of an aperture, which is 1.5 times the spot size will capture 99% of the power in the beam. It may also be noted that the irradiance in

TABLE 2.1
Variation of Power with $a/\omega(z)$

$a/\omega(z)$	Φ_p
1.0	0.8647
1.25	0.9561
1.5	0.9889
2.0	0.9997

the beam is not uniform. Because the beam profile is Gaussian, it is impossible to obtain uniform irradiance over the aperture of the lens. It is the degree of uniformity that could be varied at the expense of the power captured. The beam is, therefore, overexpanded and only the central portion of the beam is captured by the lens when nearly uniform irradiance is desired. Thus, there is always a tradeoff between the power captured and the irradiance uniformity.

A collimator consists of a microscope objective, which focuses the incident beam on a pinhole of an appropriate diameter. The pinhole is placed on axis at the focus of a long focal length well-corrected lens, called the collimating lens. Figure 2.3 shows a schematic of a laser collimator.

Let us now calculate the beam-spot size at the collimating lens. The beam with semi-divergence angle θ and beam-waist radius ω_0 is incident on the microscope objective. The laser beam is focused to a spot by a microscope objective. The radius of the beam-waist ω_f at the focal plane of the microscope objective of focal length f is $\omega_f = \frac{\lambda f}{\pi\, \omega_0}$. The collimating lens of focal length f_c is placed such that its front focal plane lies at the beam-waist.

The spot radius at the lens aperture is

$$\omega(f_c) = \omega_f \left[1 + \left(\frac{f_c}{z_{0f}}\right)^2\right]^{\frac{1}{2}} \tag{2.40}$$

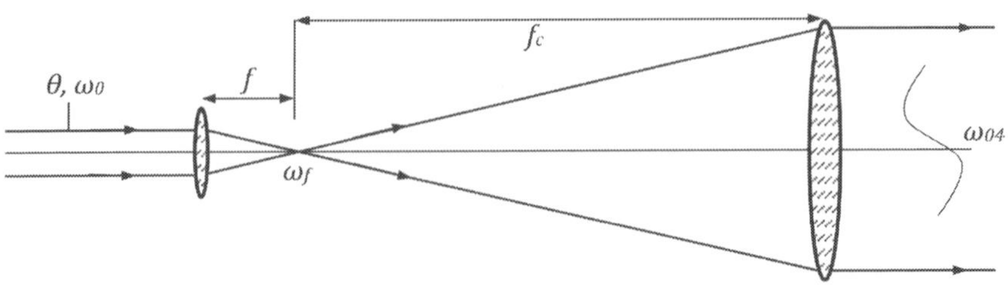

FIGURE 2.3 Schematic of a laser collimator.

Because the Rayleigh range z_{0f} is very small compared with the focal length f_c of the collimating lens, we can drop 1 in comparison with $\left(\frac{f_c}{z_{0f}}\right)^2$. Thus,

$$\omega(f_c) = \omega_f \frac{f_c}{z_{0f}} = \omega_f \frac{\lambda\ f_c}{\pi\omega_f^2} = \frac{\lambda\ f_c}{\pi\ \omega_f} = \frac{f_c}{f}\ \omega_0 \tag{2.41}$$

Beam-waist at the back focal plane is

$$\omega_{04} = \frac{\lambda f_c}{\pi\ \omega_f} = \frac{f_c}{f}\ \omega_0 \tag{2.42}$$

Thus, the beam radius after the collimator is practically the same. The Rayleigh range z_{04} is given by

$$z_{04} = \frac{\pi\omega_{04}^2}{\lambda} = \frac{\pi\omega_0^2}{\lambda}\left(\frac{f_c}{f}\right)^2 = z_{01}\left(\frac{f_c}{f}\right)^2 \tag{2.43}$$

Let us consider an example. A beam from a He-Ne laser (λ = 632.8 nm) is to be collimated to a 100-mm diameter with a collimating lens of aperture 100 mm and focal length 1000 mm. The laser beam has a diameter of 1 mm and semi-divergence angle of 0.9 mrad.

The waist radius ω_0 is obtained from the semi-divergence $\theta\left(\theta = \frac{\lambda}{\pi\omega_0}\right)$ as

$$\omega_0 = \frac{\lambda}{\pi\ \theta} = \frac{632.8\ \text{x}\ 10^{-9}\ \text{x}\ 10^3}{\pi\ x9\ \text{x}\ 10^{-4}}\text{mm} = 0.45\ \text{mm}.$$

Focal length of the microscope objective is taken as 10 mm (~20X). The radius of the focal spot is $\omega_f = \frac{\lambda f}{\pi\omega_0} = \frac{632.8\ \text{x}\ 10^{-9}\ \text{x}\ 10}{\pi\ \text{x}\ 0.45}\text{m} = 4.5\ \mu\text{m}$.

Radius of the beam-spot size at the surface of the lens is obtained

$$\omega(f_c) = \frac{\lambda\ f_c}{\pi\ \omega_f} = \frac{f_c}{f}\ \omega_0 = \frac{1000}{10}\ \text{x}\ 0.45 = 45\ \text{mm}$$

The lens allows 91.5% of the laser power. The beam-waist at the back focal plane of the lens is

$$\omega_{04} = \frac{\lambda f_c}{\pi\ \omega_f} = \frac{\lambda f_c}{\lambda f}\omega_0 = \frac{\text{f}_c}{f}\ \omega_0 = \frac{1000}{10}\ \text{x}\ 0.45 = 45\ \text{mm}$$

This collimator will capture nearly 92% of the laser power passing through the pinhole in front of the microscope objective.

2.4 VORTEX BEAMS

Beams with a helical phase profile are known as vortex beams or optically singular beams. The beams are mathematically described by

$$u(r,\phi,z:t) = U(r,z)e^{im\phi}\ e^{i(\omega t-kz)} \tag{2.44}$$

where $u(r,\phi,z:t)$ is the complex amplitude of a monochromatic light wave with frequency ω, wavelength λ, and propagating along $+z$-axis, $k = \frac{2\pi}{\lambda}$ is the propagation vector and m is the topological charge. As the wave propagates, the equi-phase surfaces (wave front) trace a helicoid given by

$$m\,\phi - k\,z = \text{constant} \tag{2.45}$$

After one round trip of the wave front around the z-axis, there is a continuous transition into the next wave front sheet separated by $m\phi$, which results in a continuous helicoidal wave surface. The sign of the topological charge is positive or negative, depending upon whether the wave surface has a right handed or left handed helicity.

Vortex beams have found a number of applications in metrology. They are used for collimation testing and spiral interferometry. The hills and valleys can be identified from a single interferogram. There are various ways to generate these beams. Phase plates (q-plates) are used to make these beams, but they are wavelength specific. Computer-generated holograms are an easy way to generate these beams with different helicities.

2.5 BESSEL BEAMS

A Gaussian laser beam has an inherent diffraction limited divergence $\theta\left(=\lambda/\pi\omega_0\right)$. The divergence can be decreased by increasing the beam-waist. Further, the beam size may be considered practically of the same size over a distance $\pm z_0$, which is the Rayleigh range. Actually, the beam-spot size increases to $\sqrt{2}$ times the beam-waist over the Rayleigh range or its cross-sectional area becomes double. It is evocative to search for beams that do not diverge due to diffraction. There are beams that do not diverge; these are called Bessel beams. They can be described by

$$U_0\left(r,\phi,z\right) = A_0 e^{-ik_z z}\, J_n\left(k_r r\right) e^{\pm in\phi}$$

where J_n is the nth-order Bessel function, k_z and k_r are the longitudinal and radial wave vectors, with $k = \left(k_z^2 + k_r^2\right)^{1/2} = 2\pi/\lambda$ and r, ϕ, and z are radial, azimuthal, and longitudinal coordinates.

The zeroth-order Bessel beam has a bright core surrounded by a number of rings, whereas the higher-order Bessel beams have a phase singularity on axis and hence a black spot. One of the simplest ways to generate a zeroth-order Bessel beam is to diffract a laser beam at a ring aperture (circular slit). Other methods include using a refractive axicon, and holographic or diffractive optical elements. The Bessel beams possess the remarkable property of healing, i.e., if an obstacle is placed in its path, it regenerates itself past the obstacle. The J_0 Bessel beam is well suited in applications requiring high pointing accuracy.

QUESTIONS

2.1 Following Figure 2.2, show that $\frac{1}{\omega_{04}^2} = \frac{1}{\omega_{01}^2}\left(1-\frac{p}{f}\right)^2 + \frac{1}{\omega_{01}^2}\frac{z_{01}^2}{f^2}$. Further, show that when $p = f$, the beam-waist radius is given by $\omega_{04} = \frac{\lambda f}{\pi \omega_{01}}$.

2.2 A laser diode emits 3 mW power at 670 nm. The beam-spot size at a distance from the laser is 2 mm and its radius of curvature is 2 m. The beam-spot size becomes 3 mm when the screen is moved away by 1 m. Find the beam-waist and its location. How much power is received at this plane on an area of a 2 mm radius?

2.3 A Gaussian beam from a He-Ne laser is incident on a convex lens of focal length of 10 cm. The beam-waist at the front surface of the lens is 2 mm. What is the size of the focal spot? If the beam has 5 mW power, what is the irradiance at the focal point? If a pinhole of a 0.5 mm diameter is placed at the location of the beam-waist, what is the irradiance at the focal point?

2.4 A Gaussian beam from a laser diode emitting a wavelength of 600 nm travels along the z-axis. Its beam-waist of 1 mm diameter is located at $z = 0$. Calculate the distance z where the radius of curvature R of the phase front is $R = 2z$.

2.5 A stable resonator is composed of a plane output mirror and a concave mirror of radius of curvature R separated by d. Show that the radius of beam-waist is given by

$$\omega_0 = \left(\frac{\lambda\, d}{\pi}\right)^{1/2}\left(\frac{R}{d}-1\right)^{1/4}$$

2.6 Consider a 5 mW beam from a He-Ne laser incident on a screen 2 m away from the exit face where the beam-waist of 1.0 mm lies. A lens with a 10 mm focal length is placed 10 cm from the laser.

a. What is the beam-spot size on the screen?
b. What is the irradiance at the focal plane?
c. A lens with a 20 cm focal length is placed such that the beam-waist after the first lens with a 10 mm focal length lies at the front focal plane of the lens. Where is the beam-waist behind the lens and what is its size?

2.7 Show that the optical thickness of a glass plate for producing a singular beam of charge m is given by

$$t(\theta) = t_0 + \frac{m\lambda\theta}{2\pi(\mu-1)}$$

where t_0 is the base thickness and μ is the refractive index at wavelength λ.

2.8 A Gaussian beam from a He-Ne laser (632.8 nm) is incident on a lens with a focal length of 10 cm. At the lens, before being refracted, the radius of curvature of the wave is infinite and the diameter of its beam-waist is 2 mm. The total power of the beam is 5 mW.

a. What is the maximum irradiance of the beam (in W/m^2) at the focal point of the lens?
b. How much power will be transmitted through a 0.2 mm diameter aperture placed a distance of 1 cm from the smallest beam-waist of the beam (focal point)?

2.9 A Gaussian beam of wavelength 600 nm travels along the z-axis. Its beam-waist with a diameter 2mm is located at the $z = 0$ plane.

a. At what value of z will the radius of curvature R of the phase front be equal to $2z$ ($R = 2z$)?
b. What is the irradiance at point $z = z_0$, $r = 2\omega(z)$ relative to the irradiance at $z = 0$, $r = 0$ (r is the distance from the axis of the beam).
c. Calculate the Gouy phase at $z = z_0$.
d. Calculate the phase difference between the point ($z = z_0$, $r = 0$) and ($z = z_0$, $r = 2\omega(z)$).
e. If the power of the beam is 5 mW, what is the maximum irradiance at $z = 5z_0$?

3 Sources, Detectors, and Recording Media

3.1 INTRODUCTION

This chapter deals with the sources of radiation and the detectors that are used in optical metrology, where the light incident on the object is to be measured. The subject matter of measurement of radiation falls under radiometry, which includes the wavelength range from 10 nm to 1000 μm (the regions commonly called ultraviolet [UV], visible, and infrared [IR]). On the other hand, photometry is the detection of light in the visible spectrum, or that part of the electromagnetic spectrum to which eye is sensitive. This corresponds to the wavelengths in the range from 360 to 830 nm. Radiometry includes the complete optical spectrum, whereas photometry is limited to the visible spectrum as defined by the response of the eye. Before discussing the light sources, a brief description of radiometric and photometric units is given in the following section.

3.2 RADIOMETRIC UNITS

The following are commonly used radiometric units:

Radiant energy, Q, is a derived SI quantity. It is the energy traveling in the form of electromagnetic waves and is measured in joules. Radiant power, Φ, also called radiant flux, is the rate of change of radiant energy, measured in watts, i.e., $\Phi = dQ / dt$. Radiant flux density at a surface, $M = E = d\Phi / dA$, is the radiant flux at a surface divided by the area of the surface. Radiant exitance or emittance, M, is the radiant flux per unit area emitted from the surface. Irradiance, E, is the radiant flux density incident on a surface. Both the radiant exitance and the irradiance are measured in $\mathrm{W / m^2}$. Note that the term *intensity* instead of *irradiance* is used in physical optics; however, intensity in radiometry is defined differently. The SI unit for intensity is Candela (Cd). Radiant intensity I is defined as $I = d\Phi / d\Omega = dQ / dt\ d\Omega$. It is the radiant flux proceeding from the source per unit solid angle in the direction considered and is measured in W / sr.

Radiance L is the derivative of radiant power with respect to the solid angle and the projected area and is expressed as $L = d^2\Phi / d\Omega\ dA \cos\theta$, where θ is the angle between the surface normal and the specified direction. Often called brightness, it is conserved on passage through an optical system. Radiance of an image cannot exceed that of the object.

Photometric units are the same as the radiometric units but are weighted by the spectral response of the eye. They carry a subscript v, which stands for visible.

DOI: 10.1201/9781003531845-3

TABLE 3.1
Various Radiometric and Photometric Quantities

Quantity	Radiometric Symbol, Unit	Photometric Symbol, Unit
Power	Φ, Watt (W)	Φ_v, lumen (lm)
Power per unit area	M, E, W/m^2	E_v, lm/m^2 = lux
Power per unit solid angle	I, W/sr	I_v, lm/sr = cd
Power per unit solid angle per unit area	L, W/m^2 sr	L_v, lm/m^2 sr = cd/m^2 = nit

Lumen, a derived SI unit of luminous flux, is abbreviated as lm and is represented as Φ_v. The lumen is derived from the candela and is the luminous flux emitted into unit solid angle by an isotropic source having a luminous intensity of 1 Cd. In other words, the lumen is the product of the luminous intensity and the solid angle and has a unit cd sr. An isotropic source emits uniformly in all direction, and a small spherical surface approximates to a point isotropic source. Candela is one of the seven SI units and is defined as "the luminous intensity, in a given direction, of a source that emits monochromatic radiation of frequency 5.40×10^{14} Hz (555 nm) and that has a radiant intensity in that direction of 1/683 watt per steradian." Thus, 1 W of radiant power is equal to 683 lm of luminous power.

Illuminance is another SI-derived quantity, which denotes luminous flux density. The unit has a special name, the lux, which is lm/m^2, and the symbol is E_v. Luminance is analogous to radiance and is a derivative of luminous power with respect to solid angle and the projected area, and its symbol is L_v. Its unit has a special name, the nit, which is cd/m^2 or lm/m^2 sr.

Table 3.1 summarizes these definitions and units.

3.3 BLACKBODY

Light is a carrier of information and illumination in optical metrology; therefore, measuring instruments are equipped with light sources, which could be incoherent or coherent light sources. The incandescent bulb is an example of an incoherent light source. The question before us is "How is the light produced?"

It is known that a body when heated high enough begins to glow and gives out light. It is observed that the amount of radiation emitted by a hot body varies with temperature. Josef Stefan in 1879 established that the total energy radiated by a hot body increased as the fourth power of its temperature. In fact, this law applies strictly to a blackbody (a body that absorbs the entire radiation incident on it). For a real body this is to be multiplied by a constant, called the emissivity of the body. Boltzmann provided the mathematical description of a blackbody radiator, and showed that the total power $E(T)$ radiated by a blackbody per unit area over its whole spectrum is given by

$$E(T) = \sigma T^4 \tag{3.1}$$

This is the Stefan–Boltzmann law and σ is called the Stefan-Boltzmann constant ($\sigma = 5.67 \times 10^{-8}$ W/m^2 K^4) and T is the absolute temperature of the blackbody.

Wien carried out experiments on blackbodies at different temperatures and showed that the radiation measured as a function of wavelength exhibited a maximum, which shifts toward the lower wavelength when the temperature is increased. Mathematically Wien's law is described as

$$\lambda_{max} T = 2.897 \text{ x } 10^{-3} \text{ m K} \tag{3.2}$$

where λ_{max} is the wavelength measured in meters at which the maximum in the energy density versus wavelength plot occurs when the blackbody is at temperature T.

Planck developed the quantum theory of blackbody radiation by considering the exchange of energy between radiation and matter in the form of quanta of energy. The energy radiated at the frequency ν is expressed as

$$\rho(\nu) = \frac{8\pi h \nu^3}{c^3} \frac{1}{e^{h\nu/kT} - 1} \tag{3.3}$$

where h is the Planck's constant, and k is the Boltzmann constant.

Planck's law describes the spectral radiant emittance of a perfect blackbody radiator as a function of its temperature and the wavelength of the emitted radiation,

$$M(\lambda, T) = \frac{C_1}{\lambda^5 \left(e^{C_2/\lambda T} - 1\right)} \left(\text{W / m}^2 \ \mu\text{m}\right) \tag{3.4}$$

where λ is in micrometers and the constants $C_1 = 2\pi hc^2 = 3.74177 \text{ x } 10^8 \left(\text{W } \mu\text{m}^4 / \text{m}^2\right)$ and $C_2 = hc / k = 1.43878 \text{ x } 10^4 \left(\mu\text{m K}\right)$

By integrating $M(\lambda, T)$ over all the wavelengths, we obtain the Stefan–Boltzmann law,

$$M(T) = \int_0^\infty M(\lambda, T) \, d\lambda = \sigma \, T^4 \tag{3.5}$$

where $\sigma = \left(2\pi^5 \ k^4\right) / \left(15 \ c^2 h^3\right) = 5.672 \text{ x } 10^{-8} \text{ W / m}^2 \text{ K}^4$. Blackbody is used for the calibration of the light sources.

3.4 LIGHT SOURCES

Optical radiation is emitted when accelerated electrons make inelastic collisions with atoms, ions, molecules, or the lattice structure of the solids. The radiation in UV, visible, and IR arises due to transitions between energy levels of the materials.

Incoherent light sources include incandescent lamps, discharge lamps, and light-emitting diodes (LEDs). In incandescent lamps, the radiation is emitted from the surface of a body. In discharge lamps, the gas is ionized, and the light emission is due to the collisions between accelerated electrons and ions resulting in electronic transitions between energy levels of the material of the ionized gas.

3.4.1 Incandescent Tungsten Lamps

In incandescent lamps, tungsten wire/filament in the form of a coil or coiled coil is the hot body whose temperature is raised by Joule heating. The filament is enclosed in a glass shell, which is evacuated or filled with an inert gas. The melting point of tungsten is 3680 K, and it can be drawn into fine wires to make the filament of an incandescent lamp. The lamp is operated around 2800 K in order to reduce the evaporation loss, thus increasing the life of the lamp. When used at a higher temperature to obtain increased optical radiation, tungsten evaporates and deposits on the glass shell, thus reducing the light output and shortening the life of the lamp. The rate of tungsten evaporation varies as $e^{-10,500/T}$, while the brightness of the lamp is proportional to T^4. This means that if a tungsten lamp is run at 10% higher than rated voltage, it would have three times the shorter life and about 30% higher output. Alternately, running it at 10% lower than rated voltage will give five times the longer life at about 30% less output. Evaporated tungsten collects on the glass shell, resulting in progressive dimming with time. Rate of evaporation can be lowered by running the lamp with its filament temperature reduced and/or by filling the glass shell with an inert gas at a higher pressure, which retards the diffusion of tungsten vapor away from the filament. The incandescent lamps are available in a variety of sizes, electrical input voltages, and power outputs.

3.4.2 Tungsten-Halogen Lamps

The shell of tungsten-halogen lamps contains a fraction of a μmol / cm^3 of bromine or iodine. This results in a halogen chemical cycle, which allows the filament to be operated at a higher temperature. The evaporated tungsten reacts with the halogen in the cooler part of the lamp forming tungsten halide, which through diffusion reaches the hot filament where it is dissociated, thus depositing the tungsten back on the filament. This results in cleaner shell; hence, its size can be reduced. In other words, the tungsten-halide lamps are compact and brighter. Further, by filling the bulb with inert gas at higher pressure, the rate of evaporation can be reduced. The tungsten-halogen lamps filled with Kr or Xe at high pressure can be operated at temperatures up to about 3500 K.

For optical applications, compact lamps with smaller size and often with flat and tightly wound filament are required. The tungsten-halide lamps are more compact and can be operated even at higher temperatures. The development of tungsten-halogen lamps has resulted in many products that are compact and suitable for optical applications. In some applications, tungsten-halide lamps with back reflectors are required. Because more than 50% of the radiation from these lamps is in IR, it could be utilized by concentrating it back on the filament where it is absorbed and thus raises the temperature of the filament. This has been possible by depositing IR-reflecting multilayer coating on the reflector. These lamps, therefore, can be operated with lower electrical power resulting in savings. One of the advantages of tungsten-halogen lamps is their stability, which is often required in optical applications.

3.4.3 Discharge Lamps

Discharge lamps can be grouped into two categories, namely low pressure (LP) and high pressure (HP). Incandescent lamps essentially produce a continuum of radiation, whereas the LP discharge lamps produce discrete narrow spectral lines and the HP discharge lamps produce broadened spectral lines followed by a continuum. A discharge lamp consists of a glass bulb filled with an inert gas like Ar, Kr, or Ne with a trace of an active element like Hg or Na and have a cathode and an anode. The cathode is normally hot and is a source of electrons, which are accelerated toward the anode. In the process, they strike the atoms of the inert gases and ionize them, thus forming plasma. This establishes conduction current. The motion of the electrons is random. The collisions are more frequent in HP discharge lamps than in LP discharge lamps. The collisions raise the atoms/molecules of the active elements to higher energy states; the radiation is emitted when they relax to lower lying states.

3.4.4 Coherent Sources

Lasers are the coherent sources, and the most common is the He-Ne laser. The laser action has been obtained in gases, liquids, solids, semiconductors, etc.; hence, we have a variety of lasers covering a large spectral range. The three main components of a laser are (1) the amplifying medium, (2) the pump, and (3) the resonator. The amplifying medium consists of a collection of atoms or ions or molecules that act as an amplifier for the radiation. In thermal equilibrium, the number of atoms at the lower energy state is always higher than that of the excited energy state resulting in absorption of the wave propagating through it. To obtain amplification, the number of atoms/molecules in the excited state must be higher than those in the lower energy state. A pump is therefore required to create this population inversion and is a source of energy to maintain population inversion. A resonator in the form of a pair of mirrors containing the amplifying medium and other accessories provides the necessary feedback to the medium.

3.4.4.1 He-Ne Laser

The He-Ne laser is a four-level laser with favorable dynamics, has low threshold, and operates in the continuous wave (CW) mode. It consists of a tube filled with a mixture of He and Ne (10:1) at a pressure of approximately 1 torr to 3 torr depending on the tube diameter. The tube has a small bore (~2 mm) and is 12 cm to 1 m long and may deliver from 1 mW to 120 mW output. The tube is either closed at the end with mirrors or with Brewster windows. Most commercial He-Ne lasers are a one-piece design featuring integral mirrors aligned during manufacture. The output is an unpolarized beam. Some lasers have the Brewster plates inside and hence their output is polarized. Laboratory or research He-Ne lasers have external mirrors; the tube is either closed by antireflection-coated windows attached perpendicular to the bore of the tube or by the Brewster windows. The tube is excited by high-voltage electrical discharge. The population inversion is established by the electron collisions of the second kind and the atom collisions of the second kind. The inversion is established between 3s and 3p levels, 3s and 2p levels, and 2s and 2p levels of Ne. He is used to excite the Ne atoms to 3s and 2s levels by a resonant transfer of energy. The laser emits in IR at 3.39 μm ($3s_2$ to $3p_4$) and 1.15 μm ($2s_2$ to $2p_4$), and in visible region at 632.8 nm ($3s_2$ to $2p_4$). The gain is very high for a 3.39 μm transition; hence, He-Ne lasers feature borosilicate glass windows instead of quartz because borosilicate has higher absorption at this wavelength. Even the borosilicate glass windows may not suppress the oscillations at this wavelength so magnets are placed along the length of the tube to suppress this transition through Zeeman splitting. Most used radiation from the laser is at 632.8 nm in visible region. It can also lase at several other lines in visible region but with reduced power levels. For example, the gain for green 543.5 nm transition is very small and hence special low-loss optics are required to obtain oscillations at this wavelength.

The discharge is maintained by supplying high DC voltage across the anode and the cathode. The anode is small, whereas the cathode has a large surface area and is placed in a large gas ballast volume. The cathode is made usually of high purity aluminum. The He-Ne laser has a modest current requirement of 5 mA to 6 mA for an average tube and voltage of 1.5 kV to 2 kV with a striking voltage up to 10 kV. The tube exhibits negative resistance so a ballast resistance is added in series.

Until 1990, the He-Ne laser was the only option for many applications in pointing, alignment, and scanning. After 1990, these applications began to be serviced by much more compact and reliable semiconductor lasers. However, in applications requiring good beam quality, beam collimation, and coherence, He-Ne laser is still the preferred source.

3.4.4.2 Argon-Ion Laser

The argon ion laser uses argon gas that has been ionized. The gas is ionized by passing a discharge current up to 40 A through a low-pressure (<1 torr) plasma tube. The laser uses a small-bore diameter tube, typically 0.5 mm to 2mm, to increase the current density because the output power increases as the square of current density. In the case of a 1.5-mm diameter bore tube with 40 A discharge current, the current density is over 2260 A/cm^2 compared with 0.3 A/cm^2 for the small tube of a He-Ne laser. At such large current densities, the temperature of plasma is very high, exceeding 5000 K. This requires complex tube design and power supply design. There are a limited number of materials for the construction of the plasma tube to withstand high temperature and erosion due to high-energy plasma. Shorter tubes are made from beryllium oxide, a ceramic, and a few other refractory materials including tungsten and graphite. The tube length ranges from 30 cm to more than 1 m and the tube ends are enclosed by the quartz Brewster windows. Longer tubes often use beryllium oxide with the graphite or the tungsten disks inserted into the tube. The holes drilled through the disks form the bore of the plasma tube. In larger tubes the energetic plasma can easily damage the cathode and surface of the tube on contact. Therefore, magnetic confinement is invariably used. The magnets are placed co-axial to the plasma tube and are cooled along with the plasma tube by flowing water. As the tube operates, the argon atoms are pushed toward the anode, increasing pressure at the anode. A return path is provided to equalize the gas pressure in the tube.

Argon ion (Ar^+) has several energy levels in two bands: nine levels in the upper laser level centered around 36eV and two in the lower laser level centered around 33 eV. There are 10 laser transitions spanning the violet to green region of the visible spectrum. Out of these 10 transitions, two at 514.5 nm and 488 nm have the high gain. The laser can deliver 5 W to 6 W of optical power at these wavelengths. It is obvious that the efficiency of Ar^+ laser is poor as almost 36 eV energy must be injected into an argon atom to produce a photon with energy of 2.5 eV.

A small, air-cooled Ar^+ laser is operated with a discharge current of about 10 A and 90 V across the plasma tube. This thus dissipates 900 W of power. Therefore, the design of the argon ion laser power supply is radically different than that of the He-Ne laser power supply.

The argon ion laser has an inherently larger spectral width (5 GHz) and hence oscillates in many longitudinal modes. An etalon made of quartz is usually used to obtain a single mode operation. For an etalon made from quartz and a thickness of 10 mm, the free spectral range is 10.3 GHz. When placed in the cavity, it is impossible for two longitudinal modes to oscillate simultaneously. Because it is a multiline laser, a wavelength selector comprising of a prism and high reflectivity mirror is used to tune it for a specific wavelength.

3.4.4.3 Nd:YAG/Nd:Glass Laser

The Nd:YAG is a four-level laser. The active ions are Nd^{+3} ions in yttrium-aluminum-garnet (YAG) or glass as a host. Usual concentrations of Nd^{+3} ions are 1% to 1.5% in YAG and several times these values in glass. The concentration of Nd^{+3} ions for CW operation is between 0.5% and 0.8%, whereas for Q-switched operation it lies between 1% and 1.4%. The laser medium could be in the form of a rod or a slab; the most common diameter range is 4 mm to 8 mm. The active medium in the form of a rod is pumped by a Kr lamp in an elliptical cavity, whereas the slab geometry is pumped by a semiconductor laser. The Nd:YAG laser has multiple pump levels, allowing the active medium to absorb energy at a variety of wavelengths in the ranges 525 nm to 585 nm, 730 nm to 750 nm, and 790 nm to 810 nm. All of these pump levels have short lifetimes of around 100 ns, and hence decay rapidly to the upper laser level. The upper laser level is long-lived with a lifetime of 1.2 ms compared with the lower laser level, which decays to the ground level in 30 ns. There are several transitions, but the population inversion occurs quickly for a 1064 nm transition.

The Nd:YAG laser has high gain, narrow line width, low threshold and favorable physical properties. On the other hand, the spectral width of the Nd:YAG laser is around 0.5 nm. The spectral width of the Nd:glass laser is very large and at least 30 to 50 times larger than that of the Nd:YAG laser; hence, it is used for producing mode-locked pulses.

The output of an Nd:YAG laser can be frequency doubled and tripled using potassium dihydrogen phosphate (KDP) crystal in succession.

3.4.4.4 Semiconductor Lasers

The semiconductor lasers, often called diode lasers, are compact, provide output over a large wavelength range as well as power-outputs over a large range, and require simple low-voltage power supplies. There are a large number of structures used to obtain laser action. Basically, a laser diode consists of a forward biased p–n junction of a suitably doped direct band gap semiconductor material. Usually, the ends are cleaved to form mirror-like end faces. Because the refractive index of the semiconductor is large, the large refractive index difference at the cleaved interface (semiconductor-air) provides a reflectance of about 30%, which is good enough to sustain laser oscillations due to large gain coefficients in most diode lasers. Homostructure lasers neither have the carrier confinement nor the optical confinement. The active layer is about 1 μm thick. Therefore, they require large current densities for their operation.

When the active layer is sandwiched between two layers of different semiconductors exhibiting a larger band gap, it is called a heterostructure diode laser. The inner layer, roughly 0.1 μm thick, forms two heterojunctions at the two interfaces between the semiconductors. The inner layer has a smaller band gap. This leads to the formation of a potential well that can effectively trap the electrons and the holes and hence the current carriers injected into the layer get confined. Fortunately, the higher band gap semiconductors have a slightly smaller refractive index than that of the inner layer; hence, optical radiation emitted in the layer also gets confined. The confinement of current carriers by potential step and optical radiation by refractive index leads to a drastic reduction in the current density for sustaining laser operation.

The radiation is confined in the direction perpendicular to the junctions, but there is no confinement in the plane of the junction. Due to this spreading of radiation, the current densities required may be large and the emission pattern may not be stable with variation in current. To overcome these defects, lasers that also confine the radiation laterally have been developed. These lasers have a lower refractive index of the surrounding regions compared with that of the active layer.

In another class of diode laser called a quantum well semiconductor laser, the active layer is about 10 nm thick compared with 100 nm for the heterostructure laser. The active layer is sandwiched between layers of a higher band gap semiconductor. The active region may be made of a single quantum well. The optical radiation in such a laser extends well beyond the quantum well, which cannot guide the optical wave efficiently. Therefore, a separate structure for optical confinement is provided in quantum well lasers.

The spectral width of the diode laser is very large and hence the laser oscillates in a large number of longitudinal modes. Many applications, however, require a single mode output. This is achieved either by using a cleaved couple cavity or distributed feedback. In the former case, the laser consists of two independent cavities that are optically coupled. The mode that is sustained in both the cavities provides the output radiation. In the second case, a periodic variation of thickness of the layer surrounding the active region is used to select a particular mode. Optical radiation is reflected back from each period and the Bragg condition thus is satisfied for one mode only, which is sustained. This periodic variation is provided at both ends of the active region and such lasers are called distributed Bragg reflector lasers.

The lasers described so far emit the radiation from the edge and are collectively called edge-emitting lasers. There is another class of lasers that emit light vertically from the surface and are called vertical cavity surface emitting lasers (VCSELs). The active layer is sandwiched between two high reflectivity mirrors. These mirrors may consist of over 120 layers. They act as Bragg reflectors and hence the output of these lasers is essentially single mode. Further, a large number of such lasers can be grown on the substrate. Because the reflectivity of the Bragg reflectors can approach 100%, the threshold current is significantly reduced.

3.4.4.5 Fiber Lasers

An optical fiber is a low-loss optical cavity. Usually, the core is doped by rare-earth elements (Yb^{+3}, Er^{+3}, Tm^{+3}, Nd^{+3}). A fiber laser can have either a linear resonator or a ring resonator. A linear fiber laser resonator consists of two reflectors set at the opposite ends of the doped fiber, which is the gain medium, and inter-cavity elements to control the temporal and spatial properties. In a linear resonator light bounces back and forth, whereas in a ring resonator there is a unidirectional flow of light. There is no spatial hole burning in a fiber ring laser. The absence of the spatial hole-burning effect helps in achieving the single longitudinal mode operation.

Fiber lasers possess unique characteristics such as high-power confinement, high beam quality, low insertion loss, compactness, and ruggedness. These can be operated both in CW and pulsed modes (Q-switching and mode locking). They are attractive for different application areas such as material processing, telecommunication, medicine, and optical sensing.

The fiber lasers have a long gain medium of a very small diameter making them susceptible to nonlinear effects. Pumping is usually axial mostly by semiconductor diode lasers. The typical diode wavelengths used for pumping applications are 793 nm, 915 nm, 980 nm, 1480 nm, and 1530 nm. Power output is limited when conventional-doped (single-clad) fibers are used. However, with the use of double-clad fibers (DCFs) along with the clad pumping technique, high power output is obtained.

Q-switching in fiber lasers is performed with cavity loss modulation. Q-switched pulses with a repetition rate of a few kilohertz to 100 kHz can be produced. Q-switched fiber lasers are attractive in several applications such as medicine, optical time-domain reflectometry, terahertz generation, industrial material processing, remote sensing, and optical instrumentation.

There are also dual wavelength fiber lasers (DWFLs) in which two wavelengths are simultaneously produced in a single cavity. These can also be used both in CW and pulsed modes. One of their applications is in microwave and terahertz generation.

3.5 DETECTORS

3.5.1 The Eye

Detectors detect the radiation and usually produce an electrical output that is processed using additional devices. The eye under this definition may be considered as a packaged detector, which contains a lens, light control arrangement (iris), and photodetector (retina). The eye produces an inverted image on the retina, which is processed by the brain. The human visual system responds to a very narrow range of the electromagnetic spectrum ranging from 3.8×10^{14} Hz (~780 nm in vacuum) to 7.9×10^{14} Hz (~380 nm in vacuum). The sensitivity of the retina to different frequencies varies, but it peaks nearly at the peak of the daylight spectrum. The retina is actually a compound detector having different types of sensing elements. In the central part, the cones abound and detect the light with high resolution and color sensitivity. There may be three kinds of cones that respond separately to red, green, and blue light. The peripheral region contains rods, which have no color sensitivity and have low resolution but higher sensitivity to brightness. The color response of the human eye to light is rather complex. For example, light that contains a 50–50 mixture of red light and green light appears yellow, even though no yellow frequency component is present. In addition, certain pairs of frequencies evoke the same response in the visual system as that produced by broadband white light. The perception of whether an object is white or not is also context sensitive. The human visual system is highly sophisticated and capable of broad dynamic range in input power (to 10^6), broad color discrimination (greater than 10^3), self-correction to some degree, and long operating life (10^2 years).

3.5.2 Photoelectric Detectors

Photoelectric detectors can be grouped under two categories, namely the photon detectors and the thermal detectors. Both types give electrical output when the radiation is incident on them, but there are distinctions in the functioning of the two. Photon detectors include photo-emissive, photoconductive, and photovoltaic (photodiode) detectors. These are fabricated from semiconducting materials. The thermal detectors produce an electrical signal in response to the change in its bulk temperature. Thermocouples, thermopiles, bolometers, thermistors, etc., fall under this category.

The detectors are characterized by several factors:

Spectral sensitivity
Responsivity
Quantum efficiency
Dark photocurrent
Rise time

Spectral sensitivity depends on the material composition of the detector. Essentially the absorption coefficient of the material varies with wavelength being smaller at the longer wavelengths. Ideally an incident photon should generate one electron, but due to losses arising from reflection, material imperfections, etc., one photon/sec does not create one electron/sec. Responsivity takes into account all the factors and hence is a measure of how 1 W of optical power is converted to an electrical current and is expressed in units of amperes per watt (A/W). Responsivity increases with wavelength unlike the absorption coefficient. Because the number of electrons created is proportional to the number of photons, the same watt of optical power has more photons at larger wavelengths than at shorter wavelengths. Quantum efficiency is the ratio of the number of electrons generated by the detector to the number of photons incident on the detector. Dark current is the small current that flows through the photodetector even when there is no radiation incident on it. The magnitude of dark current depends on temperature and contributes to the noise in the electrical output of the detector. Rise time of a detector is a measure of its response to an impulse and is measured by the time it takes to reach 63.2% of its steady-state value.

Photon detectors include photo-emissive, photoconductive, and photovoltaic (photo diodes) detectors, and these detectors rely on photoelectric effect.

3.5.2.1 Photo-Emissive Detectors

The simplest form of a photo-emissive detector is a photocathode inside a vacuum tube as shown in Figure 3.1. The incident photon releases an electron from the photocathode, which is accelerated toward the anode. The photocathode could be a metal or a semiconductor, and the process of release of an electron is different in each case.

Let us consider that n photons per second are incident on the photocathode, which releases n electrons per second, generating a current I given by

$$I = e\eta n \tag{3.6}$$

where η is the quantum efficiency and e is the electronic charge. If we represent the incident flux as $n\,h\nu = n\,hc/\lambda$, the responsivity R is given by

$$R = \frac{e\,\eta\,n}{n\,h\,c/\lambda} = \frac{e\,\eta\,\lambda}{h\,c} \tag{3.7}$$

These detectors are capable of time resolution in the sub-nanosecond range.

3.5.2.2 Photomultipliers

A PMT consists of a photocathode, a chain of dynodes comprised of 8–12 dynodes, and an anode in an evacuated tube as shown in Figure 3.2. The tube may contain additional electron optics to focus the electrons on the dynode. The cathode could be alkali halides or a semiconductor depending on the choice of application. The cathode is kept at a very low negative potential of ~1500 V. The potential difference between successive dynodes can be in the range of 100 V. A photon incident on a transparent cathode releases an electron. This photoelectron is very difficult to detect and hence is amplified by about a million times. This is achieved within the PMT without the use of external amplifiers. The photoelectron from the cathode is

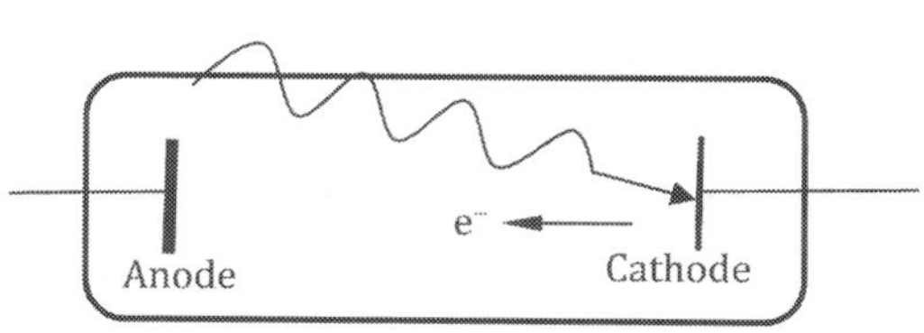

FIGURE 3.1 Photo-emissive detector.

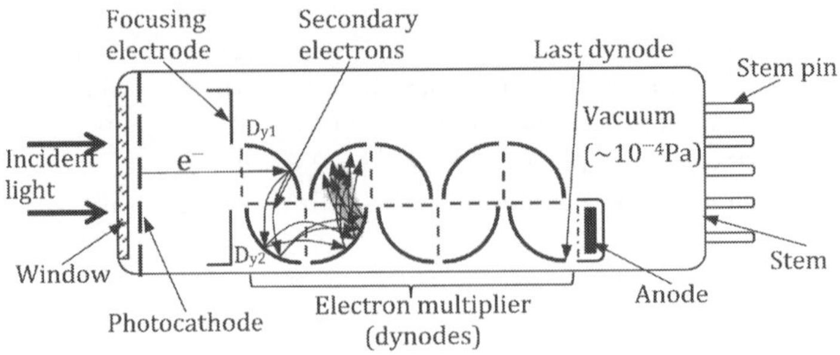

FIGURE 3.2 A schematic of a photomultiplier.

accelerated toward the first dynode D_{y1}, which is at a positive potential of about 100 V with respect to the cathode. The accelerated photoelectron strikes the first dynode and produces five or six secondary electrons, which are in turn accelerated to the second dynode D_{y2} and produce 25 to 36 secondary electrons. This process continues until the last dynode and the anode thereafter collects the electron beam. If the amplification at each dynode is m and there are n dynodes, then the number of photoelectrons at the anode due to the single photoelectron from the cathode is m^n. Such a large gain is obtained with practically no noise, as the noise figure for a typical PMT is 1.2, whereas that of an ideal amplifier is 1. Another feature is that the gain is obtained without sacrificing bandwidth. The PMTs can have bandwidths approaching gigahertz and pulse width could be as small as 1 ns. PMTs with a very small size are available and are the preferred detectors in the range of 140 nm to 1200 nm. In short, the PMT is a versatile and sensitive light detector that can be used in many applications, especially when the light levels are low and fast response is desired.

3.5.2.3 Photoconductive Detectors

When light strikes some semiconductors, electron-hole pairs are generated, and their internal electrical conductivity increases. This phenomenon is called the "photoconductive effect." Photoconductive detectors are divided into two groups, namely intrinsic detectors and extrinsic detectors doped with impurities. When light enters the photoconductor, its internal resistance changes, changing the current flowing as shown in Figure 3.3. This change over the bias is detected as signal. Various

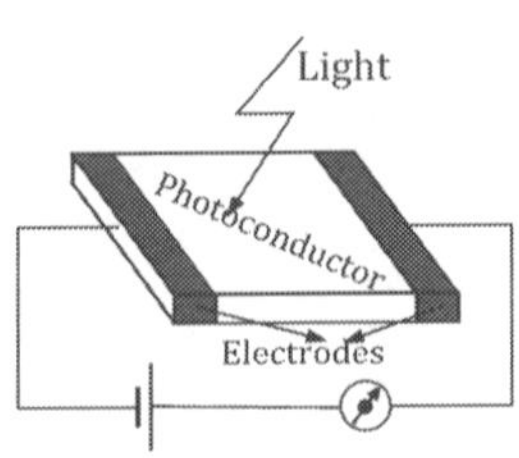

FIGURE 3.3 Schematic of a photoconductor.

materials are used to fabricate photoconductive detectors. For example, CdS is used for visible light detection and PbS, PbSe, InAs, HgCdTe, and PbSnTe are mainly used to detect IR radiation. PbS detectors are among the most sensitive uncooled detectors in the 1.3 μm to 3 μm range. Though these detectors are slow, they have been widely used due to their other useful features such as small size, lightweight, and response over a very wide spectral range from visible to far IR.

3.5.2.4 Photovoltaic Detectors

While the PMTs and phototubes utilize the external photoelectric effect, the photodiodes utilize the internal photoelectric effect. A silicon photodiode is essentially a *p*–*n* junction consisting of positively doped *p*-region and negatively doped *n*-region as shown in Figure 3.4(a). Between these two regions there is a region of neutral charge known as the depletion region. The *p*-layer is formed by selective diffusion of boron to a thickness of approximately 1 μm. The spectral response and the frequency response of the photodiode can be controlled by varying and controlling the thickness of the *p*-layer, substrate *n*-layer, and the bottom n^+-layer along with the doping concentration.

When light is incident on the diode and the photon energy is higher than the band gap, the electrons are excited to the conduction band leaving the holes behind in the valence band. The electron-hole pairs are generated throughout the device, that is, in the *p*-layer, in the depletion layer, and in the *n*-layer as shown in Figure 3.4(b). Those generated in the depletion layer diffuse toward their respective electrodes. This process results in positive charge buildup in the *p*-layer and negative charge buildup in the *n*-layer. The electron-hole pairs generated are proportional to the amount of incident light. When an external circuit is connected to the electrodes, the electrons will flow from the *n*-layer and the holes from the *p*-layer toward the opposite electrode, respectively. The current-voltage (*I*-*V*) characteristics of the *p*–*n* junction of the diode, when not illuminated, are given by the diode equation,

$$I = I_0\left[e^{eV/k_BT} - 1\right] \tag{3.8}$$

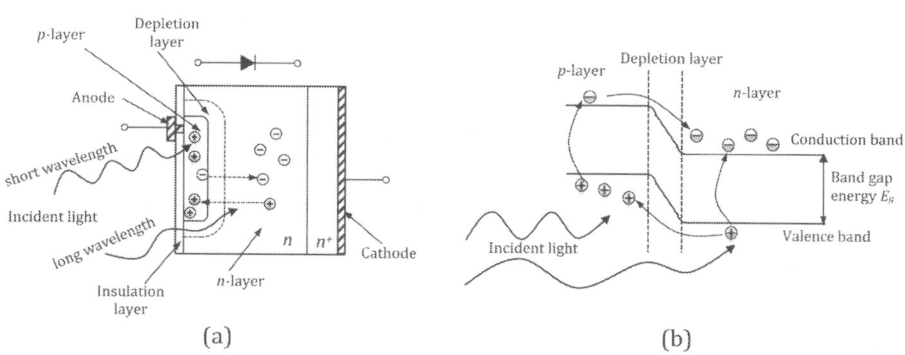

FIGURE 3.4 (a) Cross section of a photodiode. (b) Photodiode *p*–*n* junction state.

where I is the injection current with forward bias, V, and I_0 is the saturation current due to thermally generated carriers that flow through the junction. When the diode is illuminated, the diode equation becomes

$$I = I_0\left[e^{eV/k_BT} - 1\right] - I_p \tag{3.9}$$

where $-I_p$ is the photocurrent when the junction is illuminated by light; this is also called short-circuit ($V = 0$) current. I_p is proportional to the photon flux incident on the diode. The open circuit voltage when $I = 0$ is the photovoltage V_p, which is given by $V_p = \frac{k_BT}{e}\ln\left(\frac{I_p}{I_0} + 1\right)$. The open circuit voltage increases logarithmically with the photon flux.

The p–n photodiodes are fast, compact, rugged, cheap, and operate at low voltage.

3.5.2.5 Avalanche Photodiode

Avalanche diode is a p–i–n diode, in which an intrinsic i-region has been sandwiched between heavily doped p- and n-regions. The intrinsic region can be lightly doped but has to be highly resistive. To make the low resistance contact, a highly doped n^+-layer is added as shown in Figure 3.5(a).

The basic element of the structure is the p–n junction. Under reverse bias, an electric field exists in a certain volume close to the junction and this volume becomes depleted of free charge carriers. These charge carriers drift under the electric field toward the corresponding electrodes. The avalanche photodiode (APD) differs from the photodiode due to a buildup of a very high electric field (>10^5 V/cm) in the depleted region shown in Figure 3.5(b). The charge carriers traversing this region acquire enough energy to create electron-hole pairs by impact ionization. These secondary charge carriers further collide with atoms creating more charge carriers and so on as shown in Figure 3.5(c). Thus, there is an avalanche of charge carriers

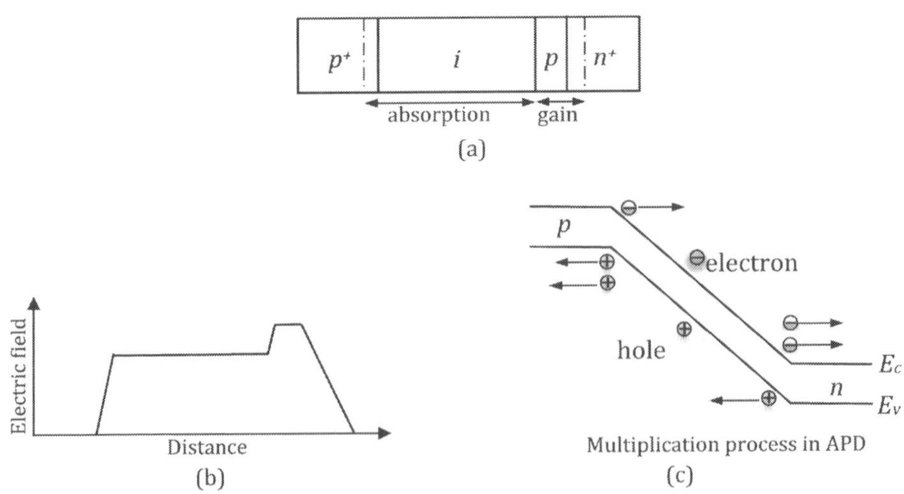

FIGURE 3.5 (a) Cross section of an avalanche photodiode (APD). (b) Electric field variation. (c) Multiplication process.

moving through the detector. An ideal avalanche diode is a *p–i–n* diode with an internal gain. The APDs require high voltage power supplies for their operation. The voltage may range from 30 V to 70 V for a GaAs APD and 300 V for a silicon APD.

3.5.3 Thermal Detectors

A detector, which produces an electrical signal or a change in output in response to a change in its bulk temperature, is classified a thermal detector. Thermal detectors absorb radiation, which produces a temperature change that in turn changes a physical or electrical property of the detector. Because a change in temperature takes place, thermal detectors are generally slow in response and have relatively low sensitivity compared with other detectors. The thermal detector works by heating the detector chip. Therefore, a heat equation needs to be examined to model the thermal detector mathematically.

Thermocouples, thermistors, bolometers, and thermopiles are thermal detectors. Pyroelectric detectors are also grouped with thermal detectors. The thermal detectors generally lack the sensitivity of semiconductor detectors, but they can provide flatter wavelength response and a variety of options in packaging, size, and cost. Most of them are passive devices, requiring no bias. Because of their stable properties, thermopiles are frequently used as calibration standards.

Primary noise sources for thermal detectors are white noise (the noise associated with a blackbody) and noise due to random thermal fluctuations (Johnson noise).

3.5.4 Pyroelectric Detectors

These detectors are based on the pyroelectric effect, in which a surface charge is generated when the temperature of the pyroelectric material changes due to the change in the spontaneous polarization. Because the pyroelectric detector responds to the change in temperature, the incoming radiation must be chopped. A chopper is a rotating or oscillating shutter in front of the detector to provide AC rather than DC output. Because relatively weak AC signals are conveniently handled by the conditioning circuitry, these detectors are used in pyrometers to measure temperatures of relatively less hot bodies

3.6 RECORDING MEDIA

Under this section, we will discuss materials and devices that record an image. These include photographic/holographic films and plates, dichromatic gelatin, photoresists, photopolymers, thermoplastics, photochromics, and ferroelectric crystals. Their properties have been summarized in Table 3.2. In recent past, the photographic plates/films were most-widely used media with high sensitivity but these have been replaced by charge-coupled devices (CCDs), spatial light modulators (SLMs), etc. For the sake of continuity, we cover all these recording media first and then take up electronic recording media.

TABLE 3.2
Recording Media for Photography and Holography

S. No.	Class of Material	Spectral Range (nm)	Recording Process	Spatial Frequencies (l/mm)	Types of Grating	Processing	Readout Process	Diffraction Efficiency (η_{max} %)
1	Photo-graphic materials	400–700 (<1300)	Reduction to Ag metal grain. Bleached to silver salts	>3000	Plane/vol. amplitude	Wet chemical	Density change	5
					Plane/vol. phase	Wet chemical	Refractive index change	20–50
2	Dichromated gelatin	250–520 and 633	Photo cross-linking	>3000	Plane phase	Wet chemical followed by heat	Refractive index change	30
					Volume phase			>90
3	Photo-resists	UV-500	Photo-cross linking or photopolymerization	<3000	Surface relief/ phase blazed reflection	Wet chemical	Surface relief	30–90
4	Photo-polymers	UV-500	Photopolymerization	~200–1500 band pass	Volume phase	None or post exposure and post heating	Refractive index change or surface relief	10–85
5	Photo-plastics/ photoconductor thermoplastics	Nearly panchromatic for PVK TNK photoconductor	Formation of an electrostatic latent image with electric-field-produced deformation of heated plastic	400–1000 band pass	Plane phase	Corona charge and heat	Surface relief	6–15
6	Photochromics	300–450	Generally, photo induced new absorption bands	>2000	Volume absorption	None	Density change	1–2
7	Ferroelectric crystals	400–650	Electro-optic effect/ Photorefractive effect	>3000	Volume phase	None	Refractive index change	90

3.6.1 Photographic/Holographic Plates and Films

A photographic plate/film consists of silver halide grains distributed uniformly in a gelatin matrix deposited in a thin layer on a transparent substrate of either a glass plate or an acetate film. When the photographic emulsion is exposed to light, the silver halide grains absorb optical energy and undergo a complex physical change, that is, a latent image is formed. The exposed film is then developed. The development process converts the halide grains that have absorbed sufficient optical energy into metallic silver. The film is then fixed, which removes the unexposed silver halide grains while leaving the metallic silver. The silver grains are largely opaque at the optical frequency. Therefore, the processed film would exhibit spatial variation of opacity depending on the density of silver grains in each region of the transparency.

Photographic emulsions (silver halides) are widely used for photography (incoherent recording) and holography (coherent recording). They are sensitive over a wide spectral range and offer a very wide resolution range and good dynamic response. An apparent drawback of the photographic materials is that they require wet development and fixing processing followed by drying. However, development process provides gain of the order of a million, which amplifies the latent image formed during the exposure.

It is mentioned earlier that the spatial variation of light irradiance incident on a photo-emulsion is converted into the variation of density of metallic silver grains; consequently, its transmission becomes spatially variable. The transmission function τ is related to the photographic density D, the density of the metallic silver grains per unit area, by the relation

$$D = -\log\tau = -\log\frac{I_t(x,y)}{I_0} = \log\frac{I_0}{I_t(x,y)} \tag{3.10}$$

where $I_t(x,y)$ is the local transmitted irradiance and I_0 is the incident irradiance. The reflection losses at the interfaces are ignored. The transmission is averaged over a very tiny area around the point (x,y).

One of the commonly used descriptions of the photosensitivity of a photographic film is the Hurter–Driffield curve or the H&D curve. It is the plot of the density D versus the logarithm of the exposure E. The exposure E is defined as the energy per unit area incident on the film and is expressed as $E = I_0T$ where I_0 is the incident irradiance and T is the exposure time. Figure 3.6 illustrates a typical H&D curve for a photographic negative. If the exposure is below a certain level, the density is independent of the exposure. This minimum density is usually referred to as gross fog. In the "toe," the density begins to increase with log E. There follows a region where the density increases linearly with logarithm of exposure; this is the linear region of the curve, and the slope of this linear region is referred to as the film gamma, γ. Finally, the curve saturates in a region called the "shoulder." There is no change in density with logarithm of exposure after the "shoulder." However, with very large exposures, solarization takes place.

The linear region of the H&D curve is generally used in conventional photography. A film with a large value of γ is called a high-contrast film, whereas a film with

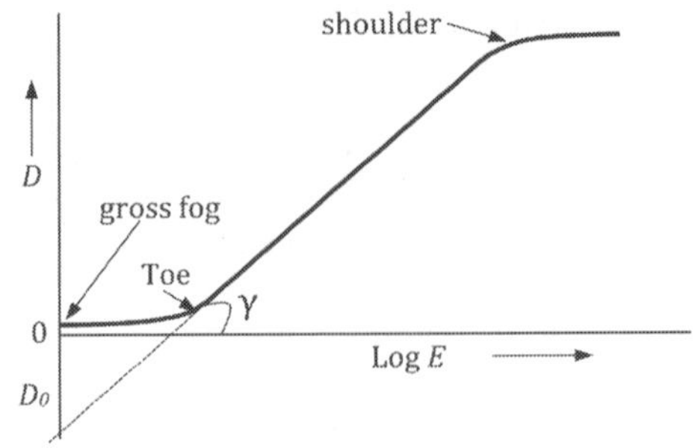

FIGURE 3.6 The Hurter–Driffield (H&D) curve.

low γ is a low-contrast film. The γ of the film also depends on the development time and the developer. It is thus possible to obtain a prescribed value of γ by the judicious choice of film, developer, and development time.

As the emulsion is usually used in the linear portion of the H&D curve, the density D when the film is given an exposure E can be expressed as follows:

$$D = \gamma_n \log E - D_0 = \gamma_n \log(I_0 T) - D_0 \tag{3.11}$$

where the subscript n means that a negative film is being used and D_0 is the intercept. Equation (3.11) can be written in terms of the transmission function τ_n of the negative film as follows:

$$\tau_n = K_n\, I_0^{-\gamma_n} \tag{3.12}$$

where $K_n = 10^{D_0}\, T^{-\gamma_n}$ is a positive constant. Equation (3.12) relates the incident irradiance to the transmission function of the film after development. It is seen that the transmission function is a highly nonlinear function of the incident irradiance for any positive value of γ_n. In many applications, it is required either to have a linear or a power law relationship. This, however, requires a two-step process. In the first step, a negative transparency is made in the usual fashion, which will have a gamma of γ_{n1}. The transmission τ_{n1} of this negative transparency is given by

$$\tau_{n1} = K_{n1}\, I_0^{-\gamma_{n1}} \tag{3.13}$$

In the second step, the negative transparency is illuminated by uniform irradiance I_{01} and the light transmitted is used to expose a second film, which will have a gamma of γ_{n2}. This results in a positive transparency with transmission τ_p, which is given by

$$\tau_p = K_{n2}\,(I_{01}\tau_{n1})^{-\gamma_{n2}} = K_{n2}\, I_{01}^{-\gamma_{n2}}\, K_{n1}^{-\gamma_{n2}}\, I_0^{\gamma_{n1}\gamma_{n2}} = K_p\, I_0^{\gamma_p} \tag{3.14}$$

where $K_p\ (= K_{n2}\, I_{01}^{-\gamma_{n2}}\, K_{n1}^{-\gamma_{n2}})$ is another positive constant and $\gamma_p = \gamma_{n1}\gamma_{n2}$ is the overall γ of the two-step process. Evidently, a positive transparency does provide a linear mapping of irradiance when the overall γ is unity ($\gamma_p = 1$).

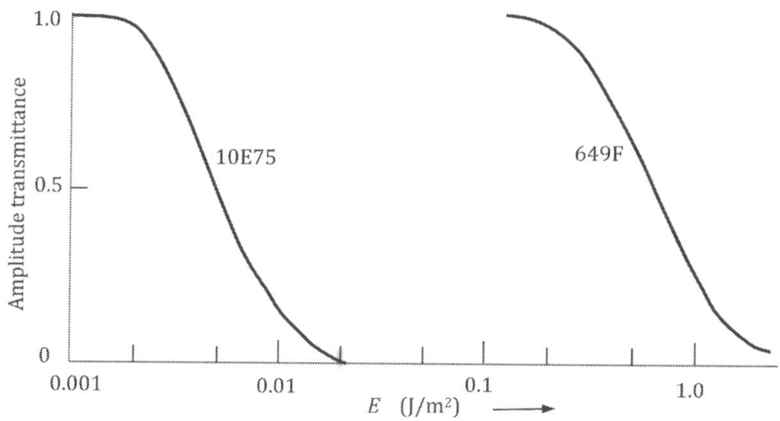

FIGURE 3.7 Plots of amplitude transmittance |*t*| versus exposure *E* for two holographic emulsions.

When photographic emulsions are used for holography or in general for coherent optical systems, the H&D curve is never used. Instead, the plot of amplitude transmittance $t(x,y)$ versus exposure E is used. The amplitude transmittance is usually complex as the film introduces both the amplitude and the phase variations to the incident plane wave. However, if the film is used in a liquid gate, the phase variations can be eliminated. The amplitude transmittance then is given by the square root of the transmission function, i.e., $|t(x, y)| = \sqrt[+]{\tau(x, y)}$. Typical plots of $|t(x, y)|$ versus E for two holographic emulsions are shown in Figure 3.7. Holographic recording is generally carried out in the linear region of the $|t(x, y)|$ versus E curve.

Because the formation of latent image does not cause any changes in the optical properties during exposure, it is possible to record several holograms in the same photographic emulsion without any interaction between them. The information can be recorded either in the form of transmittance variations or phase variations. For recording in the form of phase variations, the amplitude holograms are bleached. Bleaching converts the metallic silver grains back to the transparent silver halide crystals. Also, holographic information can be recorded in the volume provided the emulsion is sufficiently thick and has enough resolution.

3.6.2 Dichromated Gelatin

Dichromated gelatin is, in some respects, an ideal recording material for volume phase holograms as it has a large refractive index modulation capability, high resolution, and low absorption and scattering. The gelatin layer can be deposited on a glass plate and sensitized. Alternately, the photographic plates can be fixed, rinsed, and sensitized. Ammonium, sodium, and potassium chromates have been used as sensitizers. Most often, ammonium dichromate is used for sensitization. Sensitized gelatin thus obtained is called dichromated gelatin. Dichromated gelatin exhibits sensitivity in the wavelength range of 250 nm to 580 nm. The sensitivity at 514 nm is about a fifth of that at 488 nm. Gelatin can also be sensitized at the red wavelength of the He-Ne laser by the addition of methyl blue dye.

The exposure causes cross-linking between gelatin chains and alters swelling properties and solubility. Treatment with warm water dissolves the unexposed gelatin, thereby forming a surface relief pattern. However, much better holograms can be obtained if the gelatin film is processed to obtain a modulation of the refractive index. During processing, rapid dehydration of the gelatin film is carried out in an isopropanol bath at an elevated temperature. This creates a very large number of small vacuoles in the gelatin layer and hence modulates the refractive index. It is also suggested that the formation of complexes of a Cr^{3+} compound, gelatin, and isopropanol in the hardened areas is also partly responsible for the refractive index modulation. The phase holograms in gelatin are very efficient, directing more than 90% of the incident light into the useful image.

3.6.3 Photoresists

Photoresists are organic materials that are sensitive in the UV and the blue region of the visible spectrum. They are all relatively slow and thus require very long exposure. Usually, a thin layer ($\approx$1 μm) is obtained either by spin coating or spray coating on the substrate. This layer is then baked at around 75°C. On exposure, one of the three processes takes place: formation of an organic acid, photo cross-linking, or photopolymerization.

There are two types of photoresists, negative and positive. In negative photoresists, the unexposed regions are removed during development, whereas in positive photoresists the exposed regions are removed during development. Therefore, a surface relief recording is obtained. A grating recorded on photoresist can be blazed by ion bombardment. Blazed grating can also be recorded by optical means. The relief recording offers the advantage of replication using thermoplastics.

When exposure is made in a negative photoresist from the air-film side, the layer close to the substrate is the last to photolyze. This layer will be simply dissolved away during development, as it has not photolyzed fully. This non-adhesion of negative photoresists in holographic recording is a serious problem. To overcome this problem, the photoresist is exposed from the substrate film side to photolyze the resist better at the substrate film interface. On the other hand, positive photoresists do not have this problem and hence are preferred. One of the most widely used positive photoresists is Shipley AZ-1350. The recording can be done at the 488-nm wavelength of an Ar^+ laser or at the 442-nm wavelength of a He-Cd laser.

3.6.4 Photopolymers

Photopolymers are also organic materials. Those for use in holography can be in the form of either a liquid layer enclosed between glass plates or a dry layer. Exposure causes photopolymerization or cross-linking of the monomer, resulting in refractive index modulation, which may or may not be accompanied by surface relief.

Photopolymers are more sensitive than photoresists and hence require moderate exposure. They also possess the advantage of dry and rapid processing. Thick polymer materials like polymethyl methacrylate (PMMA) and cellulose acetate butyrate (CAB) are excellent candidates for volume holography as the refractive index change on exposure could be as large as 10^{-3}. Two photopolymers are commercially available: Polaroid DMP-128 and Du Pont OmniDex. Polaroid DMP-128 uses dye-sensitized photopolymerization of a vinyl monomer incorporated in a polymer matrix, which is coated on a glass or plastic substrate. Coated plates or films can be exposed with blue, green, and red light. Du Pont OmniDex film consists of a polyester base coated with a photopolymer and is used for contact copying of master holograms with UV radiation.

The response of photopolymers is band-limited because of the limitation imposed by the diffusion length of the monomer at the lower end, and the length of the polymer at the higher end of the response curve.

3.6.5 Thermoplastics

A thermoplastic is a multilayer structure having a substrate coated with a conducting layer, a photoconductor layer, and a thermoplastic layer. A photoconductor that works well is the polymer poly-N-vinyl carbazole (PVK) to which is added a small amount of electron donor 2,4,7-trinitro-9-fluorenone (TNF). The thermoplastic is a natural tree-resin, Staybelite. The thermoplastics for holographic recording combine the advantages of high sensitivity and resolution, dry and nearly instantaneous in situ development, erasability, and high readout efficiency.

The recording process involves several steps. First, a uniform electrostatic charge is established on the surface of the thermoplastic in the dark by means of a corona discharge assembly. The charge is capacitively divided between the photoconductor and the thermoplastic layers. In the second step, the thermoplastic is exposed; the exposure causes the photoconductor to discharge its voltage at the illuminated regions. This does not cause any variation in the charge distribution on the thermoplastic layer; the electric field in the thermoplastic layer remains unchanged. In the third step, the surface is charged again by the corona discharge assembly. In this process, the charge is added at the exposed regions. Therefore, an electric field distribution, which forms a latent image, is now established. In the fourth step, the thermoplastic is heated to its softening point, thereby developing the latent image. The thermoplastic layer undergoes local deformation because of the varying electric field across it and becomes thinner wherever the field is higher (the illuminated regions) and thicker in the unexposed areas. Cooling quickly to room temperature freezes the deformation; the recording is now in the form of surface relief. The recording is stable at room temperature but can be erased by heating the thermoplastic to a temperature higher than that used for development. At the elevated temperature, surface tension evens out the thickness variation and hence erases the recording. This is the fifth step called the erasure step. Figure 3.8 shows the whole recording process. The thermoplastic can be reused several hundred times. The response of these devices is band-limited and is dependent on the thickness of the thermoplastic layer and other factors.

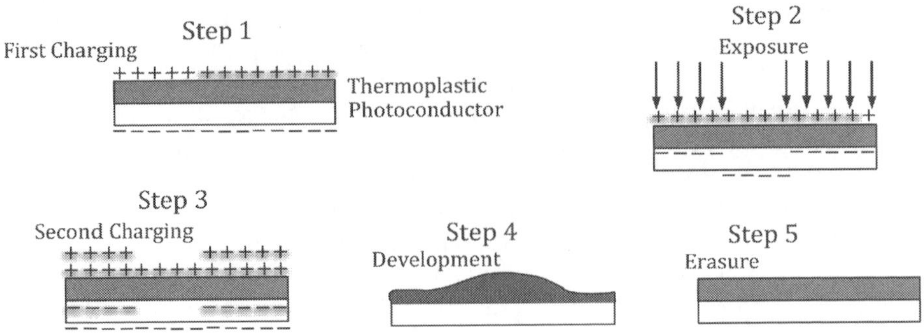

FIGURE 3.8 The record-erase cycle (Steps 1–5) for a thermoplastic material.

3.6.6 Photochromics

Materials that undergo a reversible color change on exposure are called photochromic materials. Photochromism occurs in a variety of materials, both organic and inorganic. Organic photochromics have a limited life and are prone to fatigue. However, organic films of spiropyrane derivative have been used for hologram recording in darkening mode at 633 nm. Inorganic photochromics are either crystals or glasses doped with selected impurities: photochromism is due to reversible charge transfer between two species of electron traps. Recording in silver halide photochromic glasses has been done in darkening mode at 488 nm and in bleaching mode at 633 nm. Doped crystals of CaF_2 and SrO_2 have been used in bleaching mode at 633 nm. The sensitivity of photochromics is very low because the chemical reaction occurs at the molecular level. For the same reason, they are essentially grain-free and have resolution in excess of 3000 lines/mm. Inorganic photochromics have large thicknesses and hence several holograms can be recorded in them. They do not require any processing and can be reused almost indefinitely. Despite all these advantages, these materials have limited applications due to their low diffraction efficiency (<0.02) and low sensitivity.

3.6.7 Ferroelectric Crystals

Certain ferroelectric crystals like lithium niobate ($LiNbO_3$), lithium tantalate ($LiTaO_3$), barium titanate ($BaTiO_3$), strontium barium niobate (SBN), etc., exhibit small changes in refractive index when exposed to intense light. The photo-induced refractive index change can be reversed by an application of heat and light. The mechanism of recording in these crystals is as follows: exposure to light frees trapped electrons, which then migrate through the crystal lattice and are again trapped in adjacent unexposed or low irradiance regions. The migration usually occurs through diffusion or an internal photovoltaic effect. This produces a spatially varying net space-charge distribution and corresponding electric field distribution. The electric field modulates the refractive index through the electro-optic effect and creates a volume phase grating. These are real-time recording materials; the records are stable as the charges are bound to the localized traps. The recording can, however, be erased by illuminating it with a light beam of wavelength that can release the trapped electrons.

$LiNbO_3$, particularly Fe-doped, has been used for holographic interferometry and data storage. Recording is fixed by temperature. The disadvantage of this crystal is that it is rather slow. Higher sensitivity is obtained with photoconductive electro-optic crystals like bismuth silicon oxide (BSO) and bismuth germanate (BGO) by an application of an external electric field. These crystals are available in the form of thin slices several centimeters in diameter. $BaTiO_3$ crystal is used in holographic interferometry and speckle photography. Recording can be made over a very wide spectral range.

3.7 IMAGE DETECTORS

In earlier sections, we have discussed several media that are used to record an image or a spatially varying irradiance distribution. An image can also be recorded electronically. Area array detectors are used for this purpose. It is also possible to record an image with a linear array detector by scanning. There are three mechanisms at play in imaging with detector arrays. First, the image is intercepted by the elements of the array and converted to electrical charge distribution, which is proportional to the irradiance in the image. Second, the charges are read out as elements of an image while retaining correlation with the position on the array where each charge was generated. Finally, the image information is displayed or stored.

3.7.1 CCD and CMOS

The area or focal plane arrays can be organized from a single detector, but this implementation would require many wires and processing electronics. The concept of CCDs makes the retrieval of detector signals easy and eliminates the need for a maze of wires. The CCD consists of several thousands to several millions of light-sensitive cells or pixels that produce an electrical charge proportional to light incident on them. Each pixel is essentially a metal oxide semiconductor (MOS) capacitor, usually of the buried channel type. A schematic cross section of a buried channel capacitor is shown in Figure 3.9(a). It uses a p-doped silicon substrate approximately 300 μm thick over which approximately a 1 μm thick n-type layer is formed. Next, a thin silicon dioxide layer of approximately 0.1 μm thickness is grown. A transparent electrode is then deposited over the silicon dioxide as a gate to form a tiny capacitor. When the device is reverse biased, a potential well just below the electrode is formed. Light incident on the device generates electron-hole pairs in the depletion region and, due to the applied voltage, the electrons migrate toward the n-type silicon layer and are trapped in the potential well. The number of electrons (charge) in the well is a measure of the incident light irradiance. In a CCD, the charge is generated by the incident photons, which is passed between spatial locations and is detected at the edge of the CCD. The charge position in the MOS array of capacitors is electrostatically controlled by voltage levels. Under proper application of these voltage levels and their relative phases, the capacitor can be used to store and transfer the charge packet across the semiconductor substrate in a controlled manner.

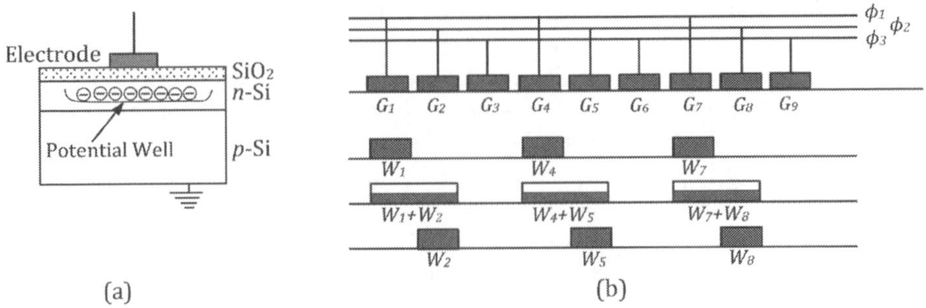

FIGURE 3.9 (a) Structure of a CCD pixel. (b) Mechanism of charge transfer in a three-phase clocking layout.

Figure 3.9(b) illustrates the principle of charge transfer through the propagation of potential wells under a three-phase clocking layout. In phase ϕ_1, gates G_1, G_4, and G_7 are turned on, while all other gates are turned off. Hence, electrons are collected in wells W_1, W_4, and W_7. In phase ϕ_2, gates G_1, G_2, G_4, G_5, G_7, and G_8 are turned on. Therefore, wells W_1 and W_2, and W_4 and W_5, and W_7 and W_8 merge into wider wells. In phase ϕ_3, gates G_1, G_4, and G_7 are turned off, while G_2, G_5, and G_8 are left on. The electrons earlier stored in W_1 are now shifted to W_2. Similarly, the electrons stored in W_4 are now shifted to W_5, and electrons stored in W_7 are shifted to W_8. By repeating this process, all charge packets will be transferred to the edge of the CCD where they are read by external electronics.

There are several ways to transfer the charge; of these two, the interline transfer (IT) and frame transfer (FT) are described. An IT device has columns of photosensitive elements separated by columns of light-shielded registers. At the end of exposure, all the photosensitive elements simultaneously transfer their accumulated charge to the readout register. Due to the column of shielded registers, about 40% of the area is not light sensitive. An FT device has an area array of light-sensitive pixels and an area array of light-shielded storage pixels. After the exposure, the captured image is quickly transferred to the adjacent storage section.

As a rule, the pixels in CCD sensors are built by MOS capacitors in which the electrons generated by photon absorption during the exposure are stored. The maximum number of electrons that can be stored in a pixel is the full well capacity. In the presence of bright objects in the scene and hence in the image, the potential wells overflow, i.e., the electrons flow to the surrounding potential wells and saturate the pixels. This gives rise to an effect called blooming, which needs to be controlled. If it is not controlled, the image suffers from overexposed regions.

Area detector arrays with pixels in numbers of 256×256 to 4096×4096 are commonly available. However, the largest CCD has 9000×7000 pixels for astronomy work. The center-to-center distances of the pixels range from 10 μm to 40 μm, but commonly used devices have a pixel separation of 12.7 μm. The commercially available smallest pixel size is 4.65 μm × 4.65 μm. A dynamic range of 1000 to 1 is quite common. The sensitivity of the video rate CCD cameras is of the order of 10^{-8} W/cm^2, which corresponds to about 0.05 lux. CCD cameras are available with exposure times as small as 1/10,000 second, achieved by electronic shuttering, and a signal-to-noise ratio (SNR) of 50 dB.

Complementary metal oxide semiconductor (CMOS) sensors like CCDs also have columns and rows of pixels. Unlike CCDs, a pixel in CMOS sensors contains a photodiode, a capacitor, and up to three transistors. Prior to the exposure or integration period, the capacitor is charged to a known voltage. When the exposure begins, the charge on the capacitor is allowed to drain away through the photodiode: the rate of drain is directly proportional to the incident light irradiance. At the end of exposure, the charge on the capacitor is read out and digitized. It is also possible to use another kind of circuit, in which the capacitor is charged during exposure time rather than drained.

CMOS sensors have two types of pixels, passive and active. A passive pixel has a photodiode, a capacitor, and a transistor. The transistor acts as a charge gate and switches the contents of each pixel to the charge amplifier at the bottom of each column of pixels. An active pixel has a photodiode, a capacitor, and an amplifier. The technique of reading the image out of array is the same for both the passive and the active pixel arrays. Each of the row selectors are sequentially clocked, which activates the switching transistor or charge amplifier for that row of pixels and transfers each pixel's charge to the column outputs. A readout register then serially transfers the column output values to an analogue to digital converter.

As each pixel, apart from having the light-sensitive area, must accommodate additional electronic components, its fill factor is considerably reduced. CMOS sensors are relatively cheap and have lower power consumption. Other physical characteristics include (1) random access, which allows fast readout of a small area of interest and (2) physical layout, which enables active electronic components to be located on each pixel and prevent blooming. CMOS sensors also have the following disadvantages: (1) lower sensitivity due to smaller fill factor, (2) higher temporal noise, (3) higher pattern noise, (4) higher dark current, and (5) nonlinear characteristic curve. Due to random access to individual pixels, it is possible to readout a region of interest (ROI) of the whole sensor, thereby high frame rates can be achieved for small ROIs.

The recording of dynamic processes, e.g., the measurement of time-dependent deformations with electronic speckle pattern interferometry (ESPI), requires cameras with a suitable frame rate and the possibility of a simultaneous exposure of all pixels. Therefore, only IT and FT sensors are suitable in CCD cameras and only integrating sensors with a global shutter in the case of CMOS cameras. CMOS sensors are interesting alternatives to the CCD sensors in optical metrology.

3.7.2 Time-Delay and Integration (TDI) Mode of Operation

Conventional CCD cameras are restricted to working with stationary objects. Object motion during exposure blurs the image. Time-delay and integration (TDI) is a special mode of CCD cameras that provide a solution to the blurring problem. In TDI mode, the charges collected from each row of detectors are shifted to the neighboring sites at a fixed time interval. As an example, consider four detectors operating in TDI mode as shown in Figure 3.10. The object is in motion and at time t_1, its image is formed on the first detector, which creates a charge packet. At time t_2, the image

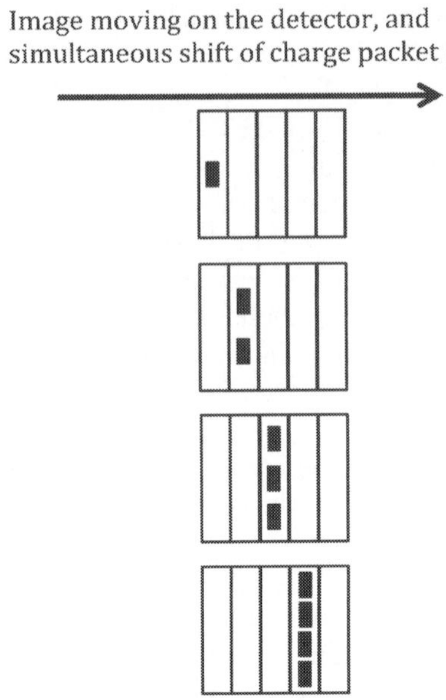

FIGURE 3.10 TDI mode of operation.

moves to the second detector. Simultaneously, the pixel clock moves the charge packet to the well under the second detector. Here, the image creates additional charge, which is added to the charge shifted from the first detector. Similarly, at time t_3, the image moves to the third detector and creates a charge. Simultaneously, charge from the second detector is moved to the well of the third detector. The charge thus increases linearly with the number N of the detectors in TDI. Therefore, the signal increases with N. The noise also increases but as $\sqrt{N}$; hence, SNR increases as $\sqrt{N}$. The well capacity limits the maximum number of TDI elements that can be used. The charge packet must always be in synchronism with the image for the camera to work in TDI mode. The mismatch between the image scan rate and clock rate can adversely smear the output.

3.8 SPATIAL LIGHT MODULATORS

Like array CCDs, SLMs are both one-dimensional (1D) and two-dimensional (2D) devices that can modulate either the amplitude, or the phase or the polarization of an incident wave as a function of spatial coordinates and time. In this section, we will discuss only 2D devices that can be addressed either electrically or optically.

There are several technologies for realizing SLMs, but we will discuss only those SLMs that are based on liquid crystal technology. Liquid crystals share the properties of both the solids and the liquids. The molecules of the liquid crystals are ellipsoidal with a single long axis about which there is a circular symmetry at any plane normal to the axis. There are three phases of the liquid crystals: (1) nematic, (2) smectic, and (3) cholesteric. Of these, nematic and a special class of smectic liquid crystals are used in SLMs. In nematic liquid crystals (NLCs), the molecules favor a parallel orientation, but their centers are randomly located. In smectic liquid crystals, the molecules also favor a parallel orientation, but their centers lie in one plane, but their spatial locations are random.

A liquid crystal device usually consists of two glass plates with a separation of 5 μm to 10 μm. Inner sides of the glass plate have a coating of a soft material (a polyamide) that is scratched with parallel grooves to place a mechanical constraint on the molecules, which align along the direction of grooves. To electrically interrogate the liquid crystal, transparent metal electrodes are deposited on the plates. We now consider the NLC molecules placed between the plates that have grooves perpendicular to each other. Because the nematic crystal molecules will align along the grooves, there will be a twist of the molecules from one plate to other. The situation is somewhat different for smectic liquid molecules of C* phase, which are called ferroelectric liquid crystals (FLCs). Due to the constraint that the centers of molecules lie in a plane, they form the layers. Therefore, the angle of the molecules within a single layer is constrained to lie at a specific declination angle with respect to the layer normal, and thus there is a cone of possible orientations for any given layer. The directions of orientation between the layers form a helical spiral. Further, the FLCs have a permanent dipole moment perpendicular to the long axis of the molecules. On the other hand, NLCs have only the induced dipole moment (i.e., when an electric field is applied) when the charges are induced at the ends of the long axis of the molecule. Therefore, the applied field will exert a torque on the molecules. Thus, the molecules, which are not in close proximity of the alignment layers, will rotate freely and align their long axis along the applied field. Further, if the polarity of the applied field is changed, the direction of dipole moment also reverses. Therefore, the direction of torque exerted by the field is independent of polarity of the applied voltage. This means that the molecules of NLCs align in the same direction with respect to the applied field, regardless of polarity. In order not to cause permanent chemical changes to NLC, the NLC cells are usually driven by AC voltage of about 5 V and the frequency in the range of 1 kHz to10 kHz. Typical time constants to align with the electric field are in the range of 100 μs, and it takes about 20 ms for the molecules to relax back to the original state.

As FLCs have a permanent dipole moment, the molecules orient at an angle, θ or $-\theta$, depending on the polarity, and retain the current state even after the applied electric field is removed. The FLC cell is thus bistable and has a memory. DC fields of opposite polarity must be applied to switch the FLC between states. The FLC cells are typically 1 μm to 2 μm thick, applied voltages are in the range of 5 V to 10 V, and the switching time is approximately 50 μs.

Due to the fact that molecules are ellipsoidal, which favors parallel orientation, but spatial positions are completely random, the NLC medium is inhomogeneous and anisotropic. It acts as a uniaxial crystal with optic axis parallel to the elongated direction of the molecules. The refractive index μ_e along the elongated direction is the extraordinary refractive index, which has a higher value compared with the value of ordinary refractive index μ_o. One of the remarkable properties of NLCs is that the difference between the refractive indices $(\mu_e - \mu_o)$ is very large being in the range of 0.2. Now consider an NLC cell in which the molecules are constrained as shown in

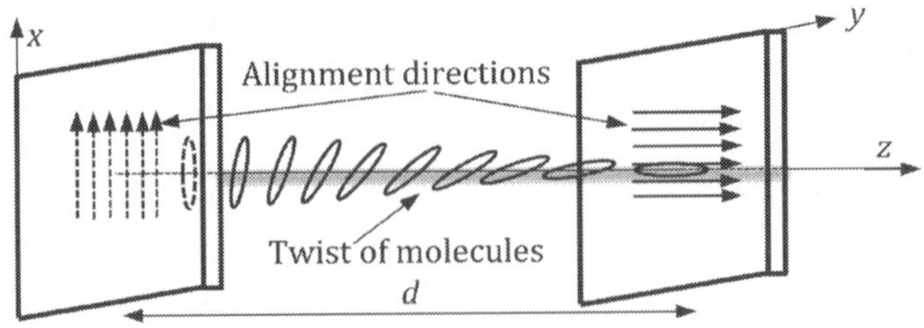

FIGURE 3.11 Alignment of nematic liquid crystal molecules under constraint.

Figure 3.11. The separation between the plates is d. At the inside of the plate the molecules are aligned along the x-axis and their orientation becomes horizontal at the other plate. Therefore, due to alignment constraints, the molecules undergo a twist. The angle of twist θ depends linearly on distance z and is expressed as

$$\theta = \alpha\, z$$

where α is the twist coefficient.

The direction of the optic axis is along the x-axis inside the first plate. At any z plane, the optic axis lies in x–y plane but makes an angle θ with the x-axis. To study the propagation of the polarized beam in the liquid crystal medium, we divide the medium into thin layers perpendicular to the axis of twist. Each of these layers acts as a uniaxial crystal, with the optic axis rotating slowly in a helical fashion. Each layer of thickness Δz rotates the plane of polarization by $\alpha\Delta z$ and introduces a phase difference $\beta\Delta z$, where $\beta = 2\pi\left(\mu_e - \mu_o\right)/\lambda_0$ and λ_0 is the vacuum wavelength. The cumulative effect of these layers on the transmitted wave is then determined considering both factors. Let us consider a case in which the incident wave is linearly polarized along the x-direction at $z = 0$ plane, which is the inner face of the first plate. It may be noted that $\beta \gg \alpha$, hence the linearly polarized wave maintains its linearly polarized state, but the plane of polarization rotates in alignment with the molecular twist. An NLC cell in a configuration shown in Figure 3.11 rotates the plane of polarization by $\pi/2$. In other words, it acts as a polarization rotator.

When an electric field is applied in the z-direction, which is parallel to the axis of twist, the induced dipole moment will align the molecules along the direction of field. The molecules lose their twisted character, except adjacent to the glass surface. Therefore, the polarization rotatory power of the nematic crystal molecules is deactivated. When the electric field is switched off, the orientations of the layers near the glass surfaces dominate, causing the molecules to return to their original twisted state and regaining the polarization rotatory power. Therefore, NLC can be used to rotate the polarization by withdrawing the applied electric field.

The situation with the FLCs is somewhat more complicated. Suffice to say that the FLC can also be used as a polarization rotator when the incident linearly polarized wave has its polarization aligned at a specified angle.

We can now discuss the construction and functioning of an SLM. The Hughes SLM is considered for this purpose. It is an optically addressable SLM, and there

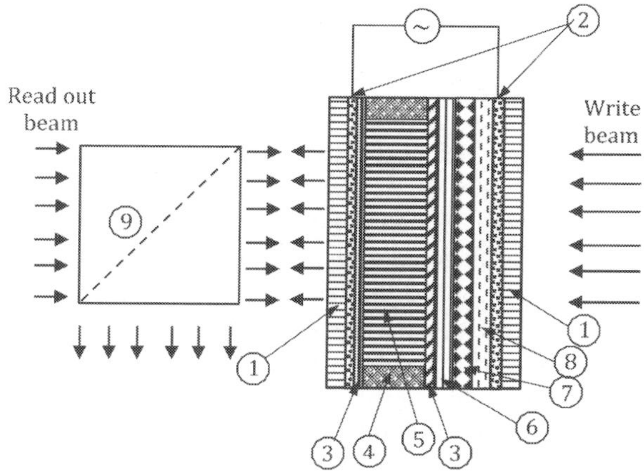

FIGURE 3.12 Hughes SLM. (1) Glass plates or fiber-optic faceplates. (2) Transparent conductive electrodes. (3) Alignment layers. (4) Spacer. (5) NLC. (6) Dielectric mirror. (7) Light-blocking layer. (8) Photoconducting layer. (9) Polarizing beam splitter.

are no pixels. It has two parts, write and readout, which are isolated. A schematic of the Hughes SLM is shown in Figure 3.12. The device can be written with either coherent or incoherent light beam of any polarization and is read out by a polarized coherent light beam.

The device consists of two glass/quartz or fiber optic faceplates. Fiber-optic faceplates are preferred to preserve resolution on passage through the thickness of the plates. Transparent metal electrodes (indium tin oxide) are deposited on the plates as a thin layer over which an alignment layer is deposited. The other alignment layer is on the dielectric mirror. The alignment layers have grooves/scratches at 45° with each other thus imparting a twist of 45°. The separation between these layers is between 5 μm and 10 μm, which is maintained constant by the spacers. This space is filled with an NLC material. After the dielectric mirror, there is a layer of cadmium telluride (CdTe), which blocks the write beam. Then there is a layer of photosensor, normally cadmium sulfide (CdS). Finally, the second plate with electrode layer deposited completes the constructional details of the device. The photoconductor should have the highest resistivity in the absence of light incident on it and the lowest possible resistivity when very strong light beam is incident on it. Within this range, the resistivity is proportional to the irradiance of the light beam. The dielectric mirror has a very high resistivity and thus prevents the flow of DC and hence protects the NLC from any permanent damage. An audio-frequency alternating field of root-mean-square (rms) voltage of 5 V to 10 V is applied to the electrodes. It may be seen that the device has two parts, a write part, which ends at the light blocking layer, and a readout part, which ends at the dielectric mirror.

When an image is projected on the right side of the device either from a cathode ray tube (CRT) or a projector, it would create a spatially varying electric field in the NLC layer. This image is the write beam. The spatially varying beam will untwist the molecules of the NLC and introduce birefringence; the untwist is proportional to the electric field and hence the irradiance of the write beam. In order to read the untwist, a linearly polarized beam with its polarization parallel to the elongated direction of the molecule is sent from the left side. In the absence of an electric field, the plane of polarization of the beam will rotate by 45° on reaching the dielectric mirror, which would reflect the beam, essentially changing the handedness. This beam again passes through the twisted nematic crystal medium and will emerge in the same state of polarization it had before entering the device. An analyzer with its polarization direction orthogonal to the polarizer in the incident beam will completely block the beam and hence the field will appear dark. However, when the electric field is turned on and it is strong enough so that the molecules align along the electric field, there is no twist and no birefringence, and the beam returns in the same state of polarization and is blocked by the analyzer. Between these two extremes, there is a birefringence and twist and hence the reflected beam is elliptically polarized. Thus, the analyzer will transmit the light. Hence the spatially varying electric field distribution gets converted to an amplitude variation. It in fact forms a negative coherent image.

Over the years there have been great technological improvements from a format of 1024 × 768 pixels that are 36 μm in size in 2012 to the SLMs with a picture format of 4160 × 2464 pixels, which are 3.74 μm in size and are commercially available.

There are also electrically addressable SLMs. One of the typical examples is liquid crystal television (LCTV). The LCTV is made by arranging micro-size (~50 μm × 50 μm) liquid crystal cells in a matrix form. They contain NLCs usually with either a 90° or a 270° twist. There may be 100 to 200 pixels in both the horizontal and the vertical directions. Their optical quality is poor, hence, they are used only in rudimentary demonstration works instead of for serious research work.

QUESTIONS

3.1 Derive Equation (3.5) using Equation (3.4).

3.2 Consider a spherical ball with a 10 cm radius at a temperature of 1000 K. Assuming it to be a blackbody, calculate the total emissive power, total radiation emitted in 100s, and the spectral emittance at a 3 μm wavelength.

3.3 External quantum efficiency $\eta(\lambda)$ of a photodetector is given by

$$\eta(\lambda) = \zeta\,(1-R)\left[1-e^{-\alpha(\lambda)d}\right]$$

where ζ is the fraction of electron-hole pairs that contribute to the photocurrent, R is the reflectivity of the interface, $\alpha(\lambda)$ is the absorption coefficient, and d is the thickness of the active layer. Taking $\zeta = 1$ and $R = 0.3$, calculate the quantum efficiency for germanium photodetectors for (1) $\alpha(\lambda) = 10^4$ cm^{-1} and $d = 1$ μm, and (2) $\alpha(\lambda) = 10^5$ cm^{-1} and $d = 0.1$ μm.

3.4 A low-contrast object is imaged on a photographic film. The irradiance distribution incident on the film can be expressed as

$$I(x,y) = I_0 + \Delta I(x,y):\ \Delta I(x,y) << I_0$$

where I_0 is the constant irradiance. Assuming that the exposure lies in the linear region of the H&D curve, show that the contrast distribution transmitted by the film when illuminated with constant irradiance is linearly related to the incident contrast distribution $\Delta I(x,y)/I_0$.

3.5 For an object of 2-cm dimension, what is the minimum spatial resolution (in centimeters) when the object is digitized into an array of 512 samples?
How many harmonics will be present in the Fourier transform of the digitized object?

What is the lowest (but not DC) spatial frequency involved in the transform of the object described earlier?

3.6 An object wave is incident normally on a photographic plate, whereas the collimated reference wave makes an angle of 15° with the normal to the plate. What are the resolution requirements when either (1) a transmission hologram, or (2) a reflection hologram is recorded?

4 Interferometry

4.1 INTRODUCTION

The phenomenon of interference has been known since the time of Newton. It accounts for the colors observed on oil slicks and soap bubbles. Interference is not peculiar to light, but it is observed with all kinds of waves. When two or more waves are superposed, irradiance variations as a result of interference are observed in the region of superposition. The phenomenon provides a very high magnification and hence is a powerful tool for the measurement of several parameters. To observe interference effect, certain conditions have to be satisfied. These conditions are both source and observer (recording material) dependent and hence have undergone considerable changes over the years. However, to observe a stationary interference pattern, the two interfering waves must be coherent to each other, have the same wavelength, and they should not be orthogonally polarized. Interference over a large path difference can be observed only when the waves are monochromatic or nearly monochromatic.

4.2 EARLY HISTORY

Interference colors observed in thin films and soap bubbles when illuminated with sunlight were the forerunner for early interference experiments. A notable experiment is "Newton's rings," in which a thin air film is created by placing a long radius of curvature surface on a plane surface and the interference takes place between waves reflected from the surfaces that enclose the air wedge of varying thickness. A white light source suffices for this purpose and hence a system of colored fringes (rings) is observed. A central spot corresponding to the point of contact is black, signifying destructive interference due to a phase change of π taking place when the reflection is from an air-glass interface. The rings are seen beautifully with a sodium lamp or any other low vapor pressure lamp like cadmium, mercury, etc. The fringes are the loci of constant path difference and in this case are loci of constant thickness, so they are called fringes of constant thickness. The method immediately suggests that this could be utilized to measure flatness of the surface provided a reference flat surface is available. Fortunately, a reference plane surface is naturally provided by the surface of a quiet liquid. With reference to this, a glass surface, often called a flat, can be realized in a workshop that acts as a reference flat surface. An interferometer for testing the flat surfaces was devised by Fizeau in 1876. Figure 4.1(a) shows a schematic of the interferometer. The test surface is kept slightly away from the reference surface; the test surface and the reference surface are nominally parallel to each other. The interference pattern is observed between the light beams reflected from the reference and the test surfaces. Michelson invented an interferometer in 1881 that was more versatile than that invented by Fizeau, which is shown in Figure 4.1(b).

DOI: 10.1201/9781003531845-4

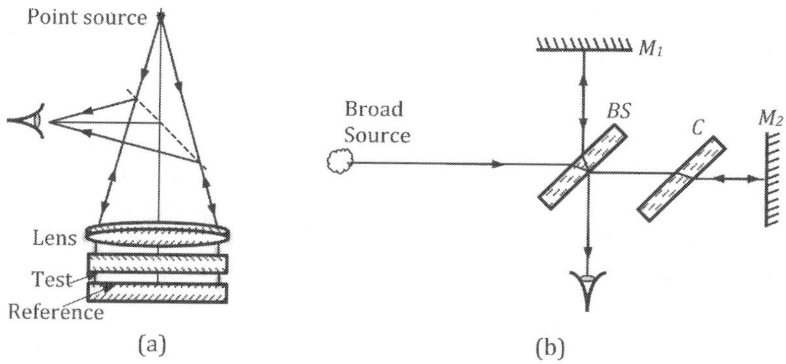

FIGURE 4.1 Schematic of (a) Fizeau interferometer and (b) Michelson interferometer.

It uses a broad quasi-monochromatic source. In order to use it with a white light source a compensating plate *C* was introduced in the path. One of the mirrors could be moved over large distances. Michelson used this interferometer for the calibration of the meter. It is frequently used for several other applications in the laboratory like measuring refractive index, thickness, wavelength, etc. It is interesting to note that the interferometer could be used to obtain spectral distribution of the source and is often used in Fourier transform spectroscopy.

This interferometer was not suitable to test optical components in the workshop. Twyman and Green modified the Michelson interferometer in 1916. The modified interferometer, often called the T-G interferometer, was immensely suited to test varieties of optical components, and it became a very useful testing tool in the workshop environment. Because it did not have compensating plate, it required a low-pressure mercury lamp as a source. The beam was collimated and the fringes were unlocalized. Other interferometers where the pathlengths between the object and the reference waves were nearly same, were developed for optical testing. The coherence requirement was met by the ingenious experimental setups. Michelson also provided a very ingenious experimental setup, known as the Michelson stellar interferometer, to measure angular diameters of distant objects and angular separation of binaries. Interferometry during this period was also used in spectroscopy as it provided extremely good resolving power. Fabry–Perot interferometer, Lummer–Gehrcke plate, etc., are some examples. Indeed, a subtle application of the phenomenon can be traced to Abbe (1873) who developed the theory of imaging in the microscope. Its direct consequence was the invention of the phase contrast microscope by Zernike in 1935.

4.2.1 Arrival of Laser

With the advent of the laser, a long-coherence source, the interferometry received a boost in applications. It was now possible to observe interference from widely separated surfaces. The laser unequal path interferometer (LUPI) was developed in 1967 and could be used to measure large radii of curvature, surface figure of mirrors, etc., using a short-length reference beam. Shear interferometry became very popular as shear could be realized with a plane parallel plate. This was also the time when the computers became available to optical physicists and were utilized more and more for the analysis of the interferograms. With the availability of enormous computation power and diode/charge-coupled device (CCD) arrays, an entirely new approach for interferogram evaluation and hence for optical testing evolved. Now it is quite often that the experimenter does not see an interference pattern but gets the final result. The basic principle of interference between two waves and the visibility of fringes have been discussed in Chapter 1.

4.3 GENERATION OF COHERENT WAVES/SOURCES

To observe a stable interference pattern, the interfering beams are derived from the parent beam either by (1) the wave front division or (2) the amplitude division. There can be two or more interfering beams. As an example, one can have three beams, four beams, etc., for observing interference. Experimental setups utilizing more than two beams for interference will be discussed under multiple-beam interferometry in sections 4.3.1 and 4.3.2. Table 4.1 lists various arrangements/interferometers utilizing two-beam and multiple-beam interference.

It is not intended to discuss all these interferometers. Attempt will, however, be made to highlight some of the important concepts.

TABLE 4.1
Two-Beam and Multiple-Beam Interferometers

Wave Front Division		Amplitude Division	
Two-Beam Interference	**Multiple-Beam Interference**	**Two-Beam Interference**	**Multiple-Beam Interference**
Fresnel biprism	Ronchi grating	Newton's ring	Fabry–Perot
Fresnel bimirror	Grating	Fizeau	Lummer–Gehrcke
Llyod's mirror	Talbot/Lau	Jamin	
Split-lens	Zone-plate	Michelson	
Young's double-slit		Mach–Zehnder	
Michelson stellar		Twyman–Green	
Rayleigh		Sagnac	
Point diffraction		Cyclic	
		Polarization	
		Scatter plate	
		Dyson	
		Mirau	
		Kösters	

4.3.1 Wave Front Division: Double-Slit Experiment

One of the very well-known examples of wave front division is the famous double-slit experiment by Young. A pair of slits samples a wave front at two spatial locations. These sampled wave fronts are diffracted by the slits thereby expanding their spatial extent so that they superpose at a plane some distance away from the slit plane. Alternately, the pair of slits could be considered as a pair of secondary sources excited by the parent wave.

Figure 4.2 shows the experimental arrangement of the double-slit experiment. The double-slit lies at $z = 0$ plane and O is the origin of the Cartesian coordinate system. The phase difference between the two waves at any point $P(x,y,z)$ on the x–y plane distance z from the slit plane, under first-order approximation, is

$$\delta = k\ d \sin\theta \approx k \frac{x\ d}{z} \tag{4.1}$$

where $k = 2\pi / \lambda$; d is the separation between the slits; and θ is the angle of diffraction, which is assumed to be small. The irradiance distribution in the interference pattern, however, is given by

$$I(\delta) = 2I_0\ \text{sinc}^2\left(\frac{2b\ x}{\lambda\ z}\right)\left[1 + \cos\delta\right] \tag{4.2}$$

where I_0 is the irradiance on the axis due to each slit, $\text{sinc}\left(\frac{2bx}{\lambda z}\right) = \frac{\sin\frac{2\pi bx}{\lambda z}}{\frac{2\pi bx}{\lambda z}}$ is the normalized amplitude distribution due to diffraction at the slit of width $2b$. The irradiance distribution consists of two parts: the first part is due to diffraction at the single slit and the second part in the square bracket is due to the interference between the waves from the slits. It is said that the irradiance distribution is modulated by the diffraction term. When each slit is very narrow, the diffracted irradiance is distributed over a very wide angular range as the sinc function will have a very large width, i.e., the irradiance remains practically constant. This irradiance is modulated by the interference term. The fringe width in the interference pattern is given by

$$\bar{x} = \frac{\lambda\ z}{d} \tag{4.3}$$

The fringe width increases linearly with z and inversely with d. If the number of slits is increased keeping d constant, the fringe width remains unchanged, but the irradiance distribution changes remarkably; the fringes become narrower.

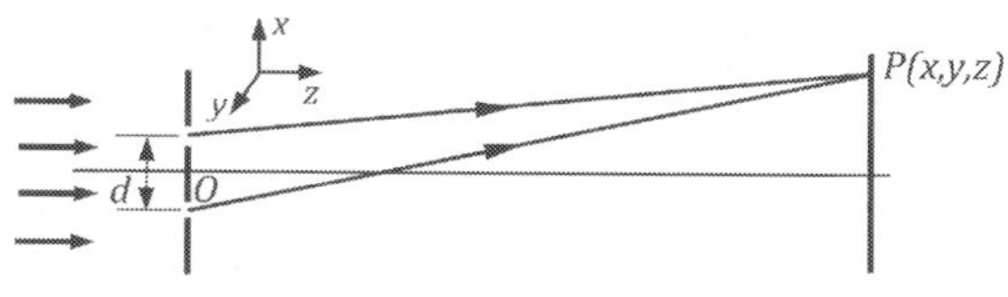

FIGURE 4.2 Schematic of the double-slit arrangement.

For a grating comprised of N slits with period d and slit width $2b$, which is illuminated by a plane beam normally, the number of interfering beams is N, and the phase difference between two consecutive beams is $\delta = k\ d\sin\theta$, where θ is the angle of diffraction. The irradiance distribution in the multiple beam interference pattern is given by

$$I(\delta) = I_0\ \text{sinc}^2(2b\sin\theta/\lambda)\frac{\sin^2(N\delta/2)}{\sin^2(\delta/2)} \tag{4.4}$$

where I_0 is the irradiance of the beam incident on the grating. Again, the first term is the diffraction term, and the second term is the interference term. Interference fringes are extremely narrow and are called the principal maxima. The principal maxima are formed wherever

$$d\sin\theta = m\lambda : m = 0, \pm1, \pm2, \ldots \tag{4.5}$$

is satisfied. Equation (4.5) is known as the grating equation. For the same width and period, the location of maxima in the interference pattern is the same whether two-beam, three-beam, or N-beam interference is observed. The irradiance distribution, however, gets drastically modified. For N-beam interference, the width of the maxima is very narrow. When a polychromatic beam of light is incident on the grating, different wavelengths get angularly separated in the same order m. Two closely separated wavelengths can be resolved if the principal maximum of one wavelength falls on the minimum of the second wavelength. The resolving power defined as $R = \lambda/\Delta\lambda$ is given by

$$\frac{\lambda}{\Delta\lambda} = m\ N \tag{4.6}$$

where $\Delta\lambda$ is the wavelength difference at λ, which is just resolved.

4.3.2 Amplitude Division: Plane Parallel Plate

One of the earliest instruments based on amplitude division is the Michelson interferometer. The path matching with dispersion compensation between the two arms of the interferometer is accomplished by introducing a compensator so that low coherence length sources could also be used. A plane parallel plate interferometer is another example of wave front division, but it requires moderate coherence length of the source. All polarization interferometers utilize amplitude division. The Fabry–Perot interferometer and thin film structures utilize multiple interference by amplitude division.

Consider a plane wave incident at an angle θ_i on a plane parallel plate of refractive index μ and thickness t as shown in Figure 4.3. An incident ray is reflected at the front face and is refracted and then reflected from the rear face and refracted again from the front face. Therefore, an incident ray generates two reflected rays; in actuality there are a large number of rays due to multiple reflections, but the amplitude of

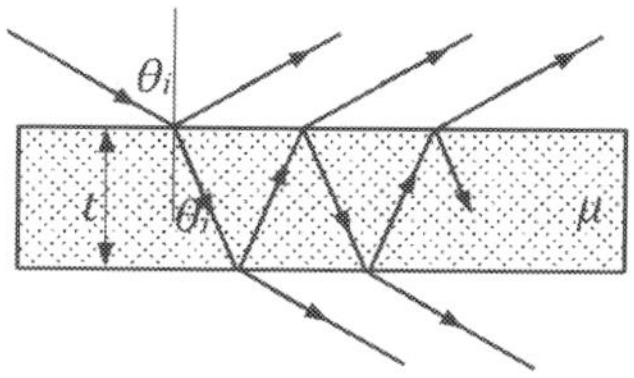

FIGURE 4.3 Amplitude division; plane parallel plate interferometer.

the successive rays falls off very rapidly. In practice we consider only two reflected rays. These rays are spatially separated and have a path difference Δ between them, which is given by

$$\Delta = 2\ t\ \mu\ \cos\theta_r \tag{4.7}$$

where θ_r is the angle of refraction in the plate. Because we are considering two plane waves, there is an overlap region in which an interference pattern is formed. As both t and θ_r are constant, the overlap region has uniform irradiance, which would change if θ_r is changed. On the other hand, if the illumination of the plate is with a divergent spherical wave, the interference pattern will have straight-line fringes.

If the surfaces of a plane parallel plate are coated for high reflectivity or two plates with one surface of each plate coated for high reflectivity and the other surface anti-reflection coated are aligned parallel to each other and illuminated with a lightly spherical wave normally, then an infinite number of beams participate in interference. The irradiance distribution in transmission is expressed as

$$I(\delta) = I_0 \frac{1}{1 + F\ \sin^2(\delta/2)} \tag{4.8}$$

where the parameter F, called the coefficient of finesse, is defined as

$$F = 4R/(1-R)^2$$

Here $\delta = k\ \Delta$ and R is the reflectivity of the surfaces of the plane parallel plate. The bright fringes are formed when

$$\delta = \frac{2\pi}{\lambda}\ 2\mu t \cos\theta_r + 2\phi = 2m\pi \Rightarrow 2\mu t \cos\theta_r = m\lambda + \frac{\phi}{\pi}\ \lambda \tag{4.9}$$

where ϕ is the phase change on internal reflection and m takes integral values. Because t is constant and θ_r varies, the fringes are loci of constant inclination: the fringe width decreasing outward from the center and fringes are very narrow. These are well-known Fabry–Perot fringes, and the arrangement is called the Fabry–Perot etalon. If the thickness t is varied, it becomes the Fabry–Perot interferometer.

When reflectivity R is small compared with unity, the value of parameter F is also small. The interference pattern then approximates that arising due to two-beam interference. However, when R approaches unity, the value of F tends to become very large, and the fringes become extremely fine. The angular width at half-height, ε, of the interference fringe is given by

$$\varepsilon = \frac{4}{\sqrt{F}} \tag{4.10a}$$

We now define finesse $\mathcal{F}$ as the ratio of the angular separation between two consecutive fringes to the angular fringe width, i.e.,

$$\mathcal{F} = \frac{2\pi}{\varepsilon} = \frac{\pi\sqrt{F}}{2} \tag{4.10b}$$

Using the Rayleigh criterion of resolution, it can be shown that the two wavelengths at λ and $\lambda + \Delta\lambda$ are just resolved when

$$\varepsilon = \frac{2.07\,\pi}{\mathcal{F}} \tag{4.10c}$$

Using Equation (4.10c) along with Equation (4.9), the resolving power of the Fabry–Perot etalon/interferometer is expressed as

$$\frac{\lambda}{\Delta\lambda} = 0.97\, m\mathcal{F} \approx \frac{2\,\mu\, t}{\lambda}\,\mathcal{F} \tag{4.11}$$

Comparing this with the resolving power of the grating with N periods given by Equation (4.6), $0.97\mathcal{F}$ may be called as the effective number of interfering beams. The resolving power of an etalon of fused quartz ($\mu = 1.54$) of thickness 2 mm and reflectivity R of 0.95 operating around 500 nm will be of the order of 7.545×10^5, which is much larger than that achievable with a large grating. However, the effective number of interfering beams is around 60. Here it may be noted that a grating has a large number of interfering beams but very small order, whereas the Fabry–Perot etalon has very large number of order and moderate number of effective interfering beams. Another quantity of interest is the free spectral range, $(\Delta\lambda)_{SR}$, which is the wavelength difference corresponding to a displacement of one order, i.e., $(\Delta\lambda)_{SR} = \lambda / m = \lambda^2 / 2\mu t$ for normal incidence. Higher resolving power results in a smaller free spectral range.

4.4 FRINGE PATTERNS

4.4.1 Interference between Two Plane Waves

Let us consider two plane waves, one propagating along the z-direction and other inclined at an angle θ with z-axis and lying in the x–z plane. The phases of these waves can be written as

$$\delta_1 = \delta_0 \tag{4.12a}$$

and

$$\delta_2 = k\ x\ \sin\theta \tag{4.12b}$$

Ignoring the constant phase δ_0, the phase difference between these two waves is given by

$$\delta = \delta_2 - \delta_1 = k\ x\ \sin\theta \tag{4.13}$$

The fringes run parallel to the y-axis and the fringe spacing is $\bar{x} = \lambda / \sin\theta$. The fringe pattern is a system of straight lines running parallel to the y-axis and have a constant spacing. The irradiance distribution in the pattern is co-sinusoidal. If the plane waves were incident symmetrically about the z-axis and enclosed a θ between them, the fringe spacing will be $\bar{x} = \lambda / 2\sin(\theta/2)$. The fringes will run parallel to the y-axis. For small θ, the fringe width is almost equal in both cases.

Let us now consider the interference between a plane wave propagating along the z-axis and a spherical wave emanating from a point source on the origin. The phase difference is given by

$$\delta = \delta_0 + k\frac{x^2 + y^2}{2z} \tag{4.14}$$

where δ_0 is a constant phase difference and z is the distance between the point source and the observation plane. This expression is valid under paraxial approximation.

The bright fringes are formed when

$$x^2 + y^2 = 2m\lambda z\text{: } m = 0, \pm1, \pm2, \pm3\ldots\ldots \tag{4.15}$$

The fringe pattern consists of circular fringes with radii proportional to the square root of the natural number and the brightness of the zeroth-order fringe depends on δ_0. As the observation plane is shifted parallel to itself, the fringe pattern still contains circular fringes but with different diameters. The surfaces of constant phase therefore are surfaces of cones. If the plane wave is added at an angle the fringes will appear as arcs of the circles.

When the interference pattern is observed in space between a counter-propagating plane wave and a spherical wave, the surfaces of constant phase are spherical. They become plane surfaces when both interfering waves are plane waves counter propagating along z-direction.

4.4.2 Interference between Two Plane Waves of Slightly Different Frequencies: Two-Wavelength Interferometry (TWI)

Let us consider two waves of frequencies ω_1 and ω_2, respectively and traveling along the z-axis. Assuming these waves are the same amplitude, they are described by

$$u_1(z;t) = u_0\cos(\omega_1 t - k_1\ z) \tag{4.16a}$$

and

$$u_2(z;t) = u_0\cos(\omega_2 t - k_2\ z) \tag{4.16b}$$

where k_1 and k_2 are the propagation vectors of the respective waves.

The resultant amplitude $u(z;t)$ is

$$u(z;t) = u_1 + u_2 = u_0\cos(\omega_1 t - k_1\ z) + u_0\cos(\omega_2 t - k_2\ z) \tag{4.17a}$$

$$\begin{aligned} u(z;t) &= 2u_0\cos\left(\frac{\omega_1+\omega_2}{2}t - \frac{k_1+k_2}{2}\ z\right)\cos\left(\frac{\omega_1-\omega_2}{2}t - \frac{k_1-k_2}{2}\ z\right) \\ u(z;t) &= 2u_0\cos(\bar{\omega}t - \bar{k}\ z)\ \cos\left(\frac{\omega_1-\omega_2}{2}t - \frac{k_1-k_2}{2}\ z\right) \end{aligned} \tag{4.17b}$$

This represents a wave of an average frequency $\bar{\omega}$ modulated by a low frequency component. The irradiance distribution is obtained as the time average of $|u(z;t)|^2$ and is given by

$$I(z) = 2u_0^2\ \cos^2\left(\frac{\omega_1-\omega_2}{2}t - \frac{k_1-k_2}{2}\ z\right) \tag{4.18a}$$

or

$$I(z) = u_0^2\ \{1+\cos[(\omega_1-\omega_2)t - (k_1-k_2)\ z]\} \tag{4.18b}$$

The irradiance in the interference pattern is not constant but varies with time at any plane of observation. Alternately, we can also obtain the instantaneous irradiance distribution using Equation (4.17a) as

$$I(z,t) = u_0^2\cos^2(\omega_1 t - k_1\ z) + u_0^2\cos^2(\omega_2 t - k_2\ z) + 2u_0^2\cos(\omega_1 t - k_1\ z)\cos(\omega_2 t - k_2\ z) \tag{4.19a}$$

or

$$\begin{aligned} I(z,t) &= u_0^2\cos^2(\omega_1 t - k_1\ z) + u_0^2\cos^2(\omega_2 t - k_2\ z) \\ &\quad + u_0^2\{\cos[(\omega_1+\omega_2)t - (k_1+k_2)\ z] + \cos[(\omega_1-\omega_2)t - (k_1-k_2)\ z]\} \end{aligned} \tag{4.19b}$$

Taking the time average over the observation time, which is very large compared with the period of oscillation, we obtain

$$\begin{aligned} I(z,t) &= \frac{1}{2}u_0^2 + \frac{1}{2}u_0^2 + 0 + u_0^2\cos[(\omega_1-\omega_2)t - (k_1-k_2)\ z] \\ I(z,t) &= u_0^2\{1+\cos[(\omega_1-\omega_2)t - (k_1-k_2)\ z]\} \end{aligned} \tag{4.20}$$

It is seen that the irradiance in the interference pattern at any point varies with time or at any instant of time it varies with the position. It is like a traveling wave of frequency ($\nu_1 - \nu_2$) and wavelength Λ $\left[\Lambda = (\lambda_2\lambda_1)/(\lambda_2 - \lambda_1)\right] = \lambda_{av}^2 / \Delta\lambda$. The wavelength Λ could be very large when the two wavelengths are close: this is often called the synthetic wavelength or the effective wavelength. This is two-wavelength interferometry (TWI), which can provide a very long effective wavelength and thus helps resolve the ambiguity problem encountered in single-wavelength interferometry. Two-frequency interferometry has been utilized for the measurement of length.

4.5 SOME MORE INTERFEROMETERS

4.5.1 Two-Frequency Interferometer

An interferometer utilizing two frequencies for length measurement is developed by Hewlett-Packard. It uses a Zeeman-split He-Ne laser that outputs two orthogonally polarized waves with frequencies ν_1 and ν_2 with roughly 2 MHz difference between the two. A beam splitter reflects these waves, and their polarizations are then oriented so that these waves interfere on a photodetector that outputs a signal with frequency ($\nu_1 - \nu_2$). The transmitted waves are incident on a polarization beam splitter (PBS), which separates these two waves. One wave travels toward a stationary cube corner and is returned to the PBS. The other wave travels to a moving cube corner and hence the return wave is Doppler-shifted. These two waves after their linear polarizations are oriented in the same direction, say using a polarizer, interfere on another photodetector that generates a signal with frequency [$\nu_1 - (\nu_2 \pm \Delta\nu)$], where $\Delta\nu$ is the Doppler shift due to the motion of the cube corner. If the measurement is made over a time interval T, then the counts from the first photodetector are N_1 where $N_1 = \int_0^T (\mathrm{v}_1 - \mathrm{v}_2)\,dt$ and from the other photodetector are N_2 where $N_2 = \int_0^T \left[\mathrm{v}_1 - (\mathrm{v}_2 \pm \Delta\mathrm{v})\right]dt$. The difference between the two counts $|N_2 - N_1|$ is given by

$$|N_2 - N_1| = \int_0^T \Delta\mathrm{v}\,dt = \int_0^T \frac{2V(t)\mathrm{v}}{c}\,dt = 2\frac{\mathrm{v}}{c}\int_0^T V(t)\,dt = 2\frac{\mathrm{v}}{c}L = \frac{2L}{\lambda} \tag{4.21}$$

where $V(t)$ is the instantaneous velocity of the cube corner and L is the traverse of the cube corner over the time T. The length L is, therefore, obtained from the difference between the two counts.

4.5.2 Doppler Interferometer: Velocity Interferometer System for Any Reflector (VISAR)

The Doppler interferometer is basically a Michelson interferometer in which the path difference between the two arms is kept fixed. The interferometer is illuminated by the light from a moving target, i.e., it is illuminated by a Doppler-shifted light. If the velocity of the target is changing, then the wavelength of light reflected from the target is also changing with time. The phase difference between the two interfering waves is given by $\delta = (2\pi / \lambda)\Delta$, where Δ is the path difference between the two arms, which has a constant value. The condition of interference maxima is $\delta = 2m\pi$ or $\Delta = m\ \lambda$. Because none of the components in the interferometer is translated, Δ is fixed and has a constant value; hence, the fringe formation is due to the change in wavelength. The change in wavelength results in the change in fringe order, i.e., $|dm| = |d\lambda| \frac{\Delta}{\lambda^2}$. To get an appreciable change in the fringe order, the path difference Δ between the arms of the interferometer must be very large. This interferometer, therefore, requires very large coherence length and one of the beams should be considerably delayed. Because the interferometer is used with any object, reflecting or diffuse, a 4-f arrangement is used so that the pupils in both the arms of the interferometer are superposed resulting in high-contrast fringes. When the beam from the moving object is fed into the interferometer, the change in wavelength is related to the velocity v by $\Delta\lambda = 2(v / c)\lambda$. Therefore, the change in fringe order is $\Delta m = (2v / \lambda c)\Delta$. The velocity is obtained as $v = (\lambda c / 2\Delta)\Delta m$. The sensitivity of the interferometer is defined as the velocity that changes the fringe order by one, i. e., $\left|\frac{v}{\Delta m}\right|_{\Delta m=1} = \frac{\lambda}{2}\frac{c}{\Delta}$. The interferometer can sense changes in velocity if the path difference Δ is very large. This interferometer therefore provides velocity history of the moving object and is used extensively in monitoring projectiles and conducting ballistic research.

4.5.3 Cyclic Interferometer

This may also be considered a variation of the Michelson interferometer in which the compensating plate has been dispensed with, and the mirrors are tilted so that both the beams travel identical triangular paths. The beam splitter performs the roles of beam splitting and recombining. This is a common path interferometer and fully path-compensated. It is insensitive to vibrations and temperature fluctuations. Apart from applications in optical testing, it is used for testing beam quality of ultrashort pulses. But one of the most common applications has been in gyroscopes. The cyclic interferometer is used in laser ring gyroscopes based on the Sagnac effect. The Sagnac frequency $\Delta\nu$ is given by $\Delta\nu = S\Omega$, where S is the scale factor and Ω is the angular frequency.

4.5.4 Shear Interferometer

In shear interferometers, the interference is observed between a wave front and its sheared version; there is no need to have a reference wave front. The sheared replica is obtained either by a linear shear or a rotational shear or a radial shear. Though shear can be introduced in a Mach–Zehnder interferometer with a tilt to one of the mirrors or by a shift of a mirror in a cyclic interferometer or by some other means, true impact of shear was realized after the advent of the laser. Though shear interferometry was carried out using Mach–Zehnder interferometer, the first simplest shear device is a plane parallel plate, which is illuminated by a collimated beam; the beams reflected from the front and back surfaces of the plate are linearly sheared. This required a light beam with considerable coherence. A variety of configurations exist to introduce linear shear, rotational shear, radial shear, inversion shear, and folding shear. The shear interferometry has been used for optical testing, in speckle metrology, etc.

4.6 PHASE SHIFTING

Although the earlier reference to the phase-shifting technique dates back to 1966, it was actually implemented in the early 1970s when the table-top-computer power and array detectors became available. This data collection and analysis method is now an integral part of many interferometers. It is applicable to both monochromatic and white light interferometry where wavelength-independent phase-shift methods are available. Phase shift can be introduced in succession or simultaneously. Therefore, two separate phase-shift methods, namely temporal and spatial, are currently in use. Transient phenomena are studied using spatial phase-shift techniques.

4.6.1 TEMPORAL PHASE SHIFTING

The irradiance distribution at any point (x,y) of a monochromatic interferogram is given by

$$I(x,y) = I_r(x,y) + I_t(x,y) + 2\sqrt{I_r(x,y)\; I_t(x,y)}\;\cos\left[\delta(x,y) + \delta_c\right] \quad (4.22a)$$

where $I_r(x,y)$ is the reference beam irradiance, $I_t(x,y)$ is the test beam irradiance, $\delta(x,y)$ is the unknown phase to be determined, and δ_c is the constant phase. This equation can be rewritten as

$$I(x,y) = I_0(x,y)\left\{1 + V(x,y)\cos\left[\delta(x,y) + \delta_c\right]\right\} \quad (4.22b)$$

where $I_0(x,y)\left[= I_r(x,y) + I_t(x,y)\right]$ is the total irradiance and $V(x,y)$ $\left[= 2\sqrt{I_r(x,y)\; I_t(x,y)} / I_0(x,y)\right]$ is the visibility of the fringes.

Equation (4.22b) has three unknowns, namely $I_0(x,y)$, $V(x,y)$, and $\delta(x,y)$. The quantity of interest is $\delta(x,y)$, which would lead us to the desired parameter, such as height variation, refractive variation, etc. The value of $\delta(x,y)$is obtained by setting up a minimum of three equations by changing the constant phase in steps or continuously over a fixed range and solving them for $\delta(x,y)$. Phase δ_c is changed, for example, by shifting the reference beam mirror by known amounts. For this purpose, the mirror is mounted on a piezoelectric translator to translate it by the precise amounts. Phase-shifted interferograms are recorded on a CCD. The phase at each pixel is then calculated using phase-shift algorithms. We demonstrate this by using a four-step algorithm in which the phase of the reference beam is shifted by $\pi/2$. That is, four interferograms are captured at $\delta_c = 0, \pi/2, \pi, 3\pi/2$, and phase $\delta(x,y)$ is obtained from

$$\delta(x,y) = \arctan\frac{I_4 - I_2}{I_1 - I_3} \quad (4.23)$$

where I_1, I_2, I_3, and I_4 are the irradiance values as given by Equation (4.22b) when $\delta_c = 0$, $\pi/2$, π, and $3\pi/2$, respectively. The phase is thus obtained at each pixel. This phase is discontinuous as the function arctan is defined over the limited range of angles, namely between $-\pi/2$ to $\pi/2$. The first step is to extend this range from 0 to 2π. This is easily achieved, as enough information is available in the recorded interferograms.

TABLE 4.2
Phase-Shift Algorithms

Some Phase-Shift Algorithms

	Steps	Algorithms
1	$-\alpha, 0, \alpha$	$\delta(x,y)=\arctan\left[\frac{1-\cos\alpha}{\sin\alpha}\left(\frac{I_1-I_3}{2I_2-I_1-I_3}\right)\right]$
		$\delta(x,y)=\arctan\left(\frac{I_1-I_3}{2I_2-I_1-I_3}\right)$ for $\alpha=\pi/2$
2	$0, \pi/2, \pi, 3\pi/2$	$\delta(x,y)=\arctan\left(\frac{I_2-I_4}{I_3-I_1}\right)$
3	$-3\alpha/2, -\alpha/2, \alpha/2, 3\alpha/2$	$\tan\left[\delta(x,y)+3\alpha/2\right]=\tan(\alpha/2)\left(\frac{I_3-I_1+I_4-I_2}{I_1-I_3+I_4-I_2}\right)$
		with
		$\tan\left[\frac{\alpha(x,y)}{2}\right]=\sqrt{\frac{3(I_2-I_3)-(I_1-I_4)}{(I_2-I_3)+(I_1-I_4)}}$
4	$-2\alpha, -\alpha, 0, \alpha, 2\alpha$	$\delta(x,y)=\arctan\left[2\sin\alpha\left(\frac{I_2-I_4}{2I_3-I_5-I_1}\right)\right]$
		with
		$\alpha(x,y)=\arccos\left[\frac{1}{2}\left(\frac{I_5-I_1}{I_4-I_2}\right)\right]$

Thus, the phase is available modulo 2π. Finally, phase-unwrapping algorithms convert the modulo 2π phase to the continuous phase. From this phase information, the wave front shape is calculated.

There are several algorithms available in the literature and are compiled in the third edition of the book *Optical Shop Testing*. A few of these are given in Table 4.2. The irradiance values I_1, I_2, I_3, and I_4 correspond to phase δ_c as indicated in the column Steps.

It may be noted that the error in the calculated phase decreases with the number of steps at which measurements are made. In many applications, a five-step phase algorithm is found to provide reasonably high accuracy.

4.6.2 Spatial Phase Shifting

For transient events, phase-shifted interferograms are captured at the same time. One way is to use several CCD cameras; each CCD camera captures a phase-shifted interferogram. A novel method using a single CCD camera is the pixelated approach. A polarization phase shifter is used and the four interferograms, phase-shifted by $\pi/2$, are gathered simultaneously. Another method is the Fourier transform method.

In order to use this method, a carrier frequency is introduced. The irradiance distribution in the interference pattern can now be expressed as

$$I(x,y)=I_1(x,y)+I_2(x,y)+2\sqrt{I_1(x,y)\,I_2(x,y)}\,\cos\left[\delta(x,y)+\frac{2\pi}{\lambda}x\,\sin\theta\right] \quad (4.24a)$$

where $\sin\theta/\lambda$ is the carrier frequency.

This can also be written as

$$I(x,y)=A(x,y)+B(x,y)\,\cos\left[\delta(x,y)+\frac{2\pi}{\lambda}x\,\sin\theta\right] \quad (4.24b)$$

where $A(x,y)=I_1(x,y)+I_2(x,y)$ and $B(x,y)=\frac{2\sqrt{I_1(x,y)\,I_2(x,y)}}{I_1(x,y)+I_2(x,y)}$. Equation (4.24b) can also be expressed as

$$I(x,y)=A(x,y)+C(x,y)\,e^{i(2\pi/\lambda)\,x\,\sin\theta}+C^*(x,y)\,e^{-i(2\pi/\lambda)\,x\,\sin\theta} \quad (4.25)$$

where $C(x,y)=\frac{1}{2}\,B(x,y)\,e^{i\,\delta(x,y)}$ and $C^*(x,y)$ is its complex conjugate. Taking Fourier transform of this irradiance distribution, we obtain

$$\mathcal{I}(f_x,f_y)=\tilde{A}(f_x,f_y)+\tilde{C}(f_x-f_0,f_y)+\tilde{C}^*(f_x+f_0,f_y) \quad (4.26)$$

where $f_0\left[=\sin\theta/\lambda\right]$ is the carrier frequency and $\mathcal{I}$, $\tilde{A}$, $\tilde{C}$, and $\tilde{C}^*$ are the Fourier transforms of I, A, C, and C^*, respectively. If the carrier frequency is properly chosen, the terms $\tilde{A}$, $\tilde{C}$, and $\tilde{C}^*$ are separated. It may be noticed that term $\tilde{C}(f_x-f_0,f_y)$ contains the phase term $\delta(x,y)$. When this is shifted to the origin, it becomes $\tilde{C}(f_x,f_y)$. On inverse Fourier transformation, we obtain $C(x,y)\left[=\frac{1}{2}\,B(x,y)\,e^{i\,\delta(x,y)}\right]$. It may be seen that the imaginary part of its natural logarithmic leads to the phase as

$$\ln C(x,y)=\ln\left[\frac{1}{2}\,B(x,y)\right]+i\,\delta(x,y) \quad (4.27)$$

The phase at each point is calculated using the relation

$$\delta(x,y)=\arctan\frac{\operatorname{Im}\,C(x,y)}{\operatorname{Re}\,C(x,y)} \quad (4.28)$$

The signs of the numerator and denominator must be considered separately to obtain values in the range $-\pi$ to π. The ambiguities of 2π can be overcome by complementary information. This is the wrapped phase. To obtain true phase, the wrapped phase has to be unwrapped.

4.7 PHASE-UNWRAPPING PROCEDURES

The need for phase unwrapping arises in applications such as three-dimensional (3D) imaging where the 3D surface profile of an object is encoded in the phase of the light field that is either reflected from or transmitted through that object. For some

3D imaging methods, such as digital holography (DH), the recovered object phase is wrapped, meaning the phase is represented across an interval of $[-\pi, \pi]$ or $[0, 2\pi]$. To obtain the true phase, the wrapped phase must be unwrapped, thus the need for phase-unwrapping algorithms.

4.7.1 1D Phase Unwrapping in MATLAB

MATLAB offers some local functions for one-dimensional (1D) phase wrapping and unwrapping, which are briefly discussed here. Starting with an arbitrary parabolic phase that could represent the curvature of a convex lens in one dimension (Figure 4.4(a)), that phase can be wrapped on the interval of $[-\pi, \pi]$ using MATLAB's *wrapToPi* function (Figure 4.4(b)), or on the interval of $[0, 2\pi]$ using MATLAB's *wrapTo2Pi* function (Figure 4.4(c)).

The concept is easy to understand and implement in one dimension. As the phase increases moving toward the center of the lens from either edge, the *wrapToPi* function simply subtracts 2π from any value y in the range $\pi < y \leq 3\pi$, then subtracts 4π from any value y in the range $3\pi < y \leq 5\pi$, and so on. For the *wrapTo2Pi* function, 2π would be subtracted from any value y in the range $2\pi < y \leq 4\pi$, 4π would be subtracted from any value y in the range $4\pi < y \leq 6\pi$, and so on. The general form of these operations is easy to write down and implement with a few lines of code, a for loop, and some conditional statements. The MATLAB local function for phase unwrapping, called *unwrap*, applies these rules in reverse.

4.7.2 2D Phase Unwrapping Using PUMA

For two-dimensional (2D) phase unwrapping, various algorithms can be found in the literature, each with their own strengths and weaknesses. The same concept from before is applied to 2D phase unwrapping; that is, checking neighboring pixel values for absolute phase discontinuities of $> 2\pi$ and then correcting those discontinuities by adding or subtracting 2π. However, this only works well if the true phase being recovered has low noise and does not violate the Itoh condition, which is the

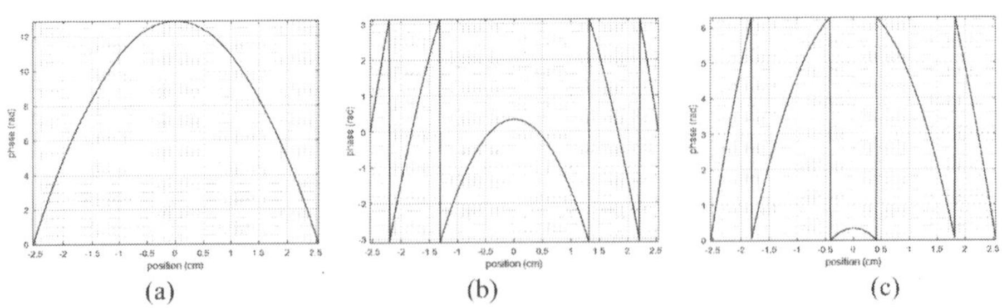

FIGURE 4.4 (a) True phase representing the curvature of a simulated convex lens with a diameter of 5.08 cm. (b) Phase shown in (a) wrapped to the interval $[-\pi, \pi]$ using MATLAB local function *wrapToPi*. (c) Phase shown in (a) wrapped to the interval $[0, 2\pi]$ using MATLAB local function *wrapTo2Pi*. (Courtesy Professor Partha Banerjee.)

assumption made by most phase-unwrapping algorithms that the maximum absolute phase difference between neighboring pixels of the true phase is $< \pi$. For true phase surfaces that are noisy and/or discontinuous, the problem of phase unwrapping becomes quite difficult.

In MATLAB, a 2D wrapped phase map can be generated in several ways, but only one will be shown here. First, the parabolic equation used to generate the 1D lens surface phase profile is extended to 2D. This 2D parabola represents the true phase of an arbitrary optical field and is shown in Figure 4.5(a). The amplitude of the simulated optical field is chosen as a Gaussian distribution as shown in Figure 4.5(b). The Gaussian amplitude (A_{Gauss}) and the parabolic phase (ϕ_{Para}) are combined to represent an optical field by $U_{\text{field}} = A_{\text{Gauss}} \exp(i\phi_{\text{Para}})$. The wrapped phase of this field then can be calculated using the MATLAB local function *angle*, which takes the arctangent of the ratio of imaginary and real components:

$$\phi = \tan^{-1}\left(\frac{A_{\text{Gauss}}\sin\phi_{\text{Para}}}{A_{\text{Gauss}}\cos\phi_{\text{Para}}}\right) = \tan^{-1}\left(\frac{\sin\phi_{\text{Para}}}{\cos\phi_{\text{Para}}}\right) \tag{4.29}$$

The wrapped phase generated in this way is shown in Figure 4.5(c).

The wrapped phase in Figure 4.5(c) can also be generated by using *wrapToPi* on the true phase in Figure 4.5(a), but the simulated optical field method is more representative of real-world cases producing 2D wrapped phase maps. For example, DH relies on the numerical propagation of optical fields to recover the phase at specific locations in space, resulting in wrapped phase maps.

The phase unwrapping via max-flow/min-cut (PUMA) algorithm was developed by Bioucas-Dias and Valadao. They introduced an energy minimization framework for phase unwrapping, and the minimization is accomplished through a series of max-flow/min-cut calculations. The suite of functions coded in MATLAB to implement the PUMA algorithm, which were also written by Bioucas-Dias and Valadão, are open source. Once downloaded and added to the MATLAB current folder, the main function can be called *[unwrapped_phase, iter, erglist]=puma_ho(wrapped_phase, p)*, where *unwrapped_phase* is the output unwrapped phase map (2D array), *iter* is the number of iterations the algorithm runs through (integer), *erglist* is the

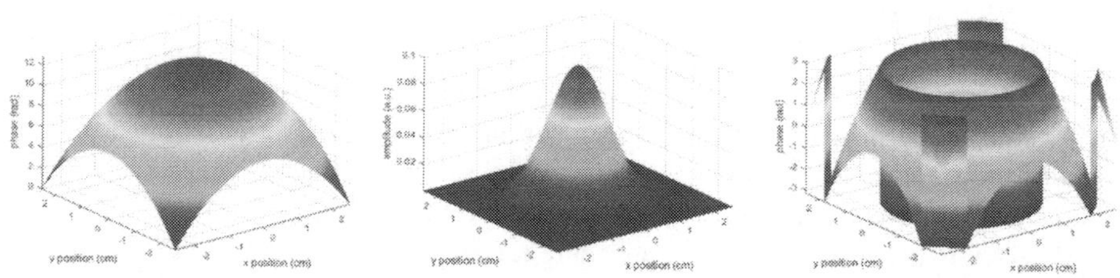

(a) (b) (c)

FIGURE 4.5 (a) Parabola from Figure 4.4(a) extended to 2D representing the true phase of the simulated optical field. (b) Gaussian amplitude of the simulated optical field. (c) Wrapped phase of the simulated optical field calculated using MATLAB local function *angle*. (Courtesy: Professor Partha Banerjee.)

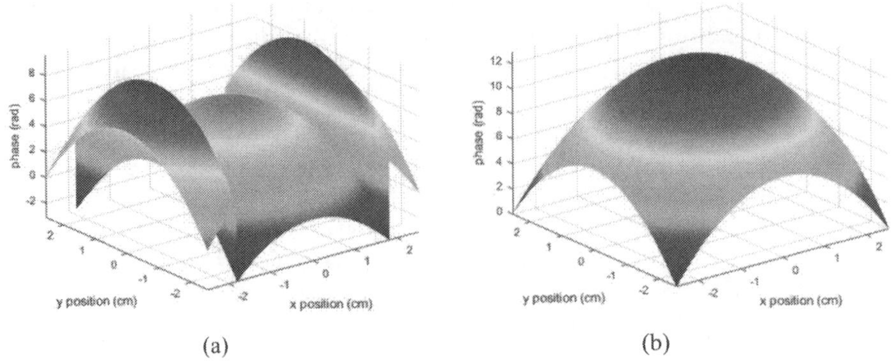

FIGURE 4.6 (a) True phase recovered using MATLAB local function *unwrap*. (b) True phase recovered using PUMA phase-unwrapping algorithm. (Courtesy Professor Partha Banerjee.)

sequence of energies of the unwrapped phases (1D array), *puma_ho* is the function name, *wrapped_phase* is the input wrapped phase map, and *p* defines the clique potential exponent (> 0). The *iter* and *erglist* outputs can be suppressed by replacing their names with "~" and the clique potential exponent used in examples shown here is $p = 1$. The line of code to call and implement the PUMA algorithm is now rewritten as *[unwrapped_phase, ~, ~] = puma_ho(wrapped_phase, 1).*

Figure 4.6 shows the result of unwrapping the wrapped phase in Figure 4.5(c) using MATLAB local function *unwrap* (Figure 4.6(a)) and PUMA (Figure 4.6(b)).

As seen in Figure 4.6(a), MATLAB local function *unwrap* fails to fully recover the true phase shown in Figure 4.5(a), even for this noiseless and continuous phase map, which is an ideal case. Comparing Figure 4.6(b) with Figure 4.5(a) clearly shows that PUMA is successful in recovering the true phase.

In the next example, PUMA is used to unwrap the phase map obtained from using multi-wavelength digital holography (MWDH) to recover 3D topograms of fingermarks coated with columnar thin films (CTFs). In MWDH, digital holograms of an object are recorded using a CCD (or complementary metal oxide semiconductor [CMOS]) camera at two different wavelengths. The recorded interference patterns are then numerically propagated back to the object plane where the phase of each object field can be extracted after separation from the DC and twin-image components. It is important to note that to correctly subtract the phases generated from the propagated holograms recorded at two different wavelengths, pixel-to-pixel matching must be ensured. Each image (recorded digital hologram) has the same number of pixels, but because they are recorded in different wavelengths, the physical size each pixel will represent at the object plane (after propagation) will differ. Thus, to ensure proper pixel-to-pixel matching, zero-padding of the hologram recorded at the longer wavelength is done prior to propagation. The difference between the phase maps from each wavelength is calculated to produce a phase map corresponding to the synthetic wavelength Λ, which is calculated as $\Lambda = \lambda_1\lambda_2 / |\lambda_1 - \lambda_2|$, where λ_1 and λ_2 are the two wavelengths used to record the holograms. This new phase map must be unwrapped to recover the true phase, after which the phase is converted to

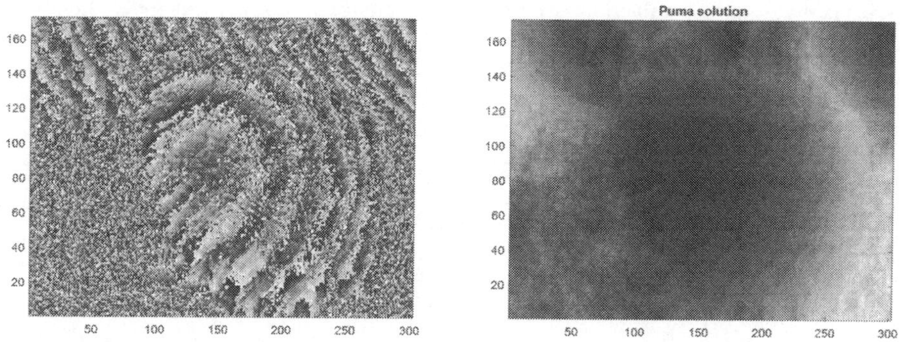

(a) (b)

FIGURE 4.7 (a) Wrapped phase map corresponding to a synthetic wavelength $\Lambda = 4.162\ \mu$m. (b) True phase recovered using the PUMA phase-unwrapping algorithm. (Courtesy Professor Partha Banerjee.)

depth using $d = \phi\Lambda / 4\pi$. Figure 4.7 shows a wrapped phase map corresponding to a synthetic wavelength of $\Lambda = 4.162\ \mu$m and the unwrapped true phase recovered using PUMA.

In the unwrapped phase map shown in Figure 4.7(b), some details of fingermark ridges are now visible. There is still some postprocessing to do before the 3D topogram is recovered, such as phase curvature removal and noise filtering; however, the phase-unwrapping step is crucial, and PUMA has been shown to be very robust and computationally fast.

4.7.3 Phase Unwrapping via the Transport of Irradiance

Another way to get the unwrapped phase of an object is by using the transport of irradiance equation (TIE). TIE is derived from the Helmholtz equation using the paraxial approximation. More specifically, it is the imaginary part of the paraxial wave equation and physically represents the conservation of energy for a traveling electromagnetic wave. TIE is a partial differential equation (PDE) of the form (see Appendix)

$$\nabla_\perp^2 \varphi(x, y; z) + k_0 \frac{1}{I(x, y; z)} \frac{\partial I(x, y; z)}{\partial z} \approx 0 \tag{4.30}$$

where $\varphi(x, y; z)$ is the phase, $I(x, y; z)$ is the optical field irradiance, k_0 is the wavenumber equal to $2\pi / \lambda$ with λ being the wavelength, $\nabla_\perp^2$ is the transverse Laplacian, and z denotes the propagation direction. Note that for TIE to take this form the assumption of uniformity of the irradiance near the image (or object) plane at $z = z_0$ is made, which is generally true for pure phase objects near the image plane. The advantage of making this assumption and having TIE in the form shown in Equation

(4.30) is that it can be solved by use of Fourier transforms. Doing this gives the phase at plane $z = z_0$ as

$$\varphi(x, y; z_0) \approx \mathcal{F}^{-1}\left\{\frac{1}{\left(k_x^2 + k_y^2\right)}\mathcal{F}\left\{\frac{k_0}{I(x, y; z_0)}\frac{\partial I(x, y; z_0)}{\partial z}\right\}\right\} \tag{4.31}$$

where $\mathcal{F}$ and $\mathcal{F}^{-1}$ denote the forward and inverse 2D Fourier transform operators, and k_x and k_y are spatial frequency variables in the x- and y-direction, respectively. The irradiance derivative at image plane $\partial I(x, y; z_0)/\partial z$ can be approximated by a finite difference method, i.e., $\partial I(x, y; z_0)/\partial z = \left[I(x, y; z_0 + \Delta z) - I(x, y; z_0 - \Delta z)\right]/2\Delta z$, where $I(x, y; z_0 + \Delta z)$ and $I(x, y; z_0 - \Delta z)$ are the optical field irradiances at two planes on opposite sides of the image plane at distances of $\pm\Delta z$. The irradiance at the image plane can also be taken as the average of the two defocused irradiances, or $I(x, y; z_0) \approx \left[I(x, y; z_0 + \Delta z) + I(x, y; z_0 - \Delta z)\right]/2$, meaning a minimum of two irradiance recordings of the object near the image plane is needed to solve for the phase of the object. The phase calculated via TIE is already unwrapped, so there is no need for PUMA if this approach is taken.

Another advantage of TIE is that it can be used in conjunction with digital holograms to act as a phase-unwrapping algorithm instead of using PUMA. This approach is generally referred to as DH + TIE, and it requires the numerical propagation of a recorded digital hologram from the recording plane to two planes near the object plane. To obtain the two defocused irradiance images needed to solve Equation (4.31), the DC and twin-image components must first be removed from the recorded digital hologram. This can be done by cropping the image of interest from the propagated holograms if an off-axis hologram is recorded, or by filtering in the frequency domain before propagation if an in-line hologram is recorded. In either case, numerical propagation and isolation of the images of interest can provide two defocused irradiance images of the object, which are then substituted into Equation (4.31) to get an unwrapped phase profile of the object, thereby avoiding explicit phase-unwrapping algorithms like PUMA.

To show that TIE provides an unwrapped phase, MATLAB is used to simulate an ideal phase object having a Gaussian phase distribution with a maximum phase of 60 rad, width of 50 μm, and a uniform amplitude. The uniform amplitude and Gaussian phase is combined into a simulated optical field by $U_{\text{field}} = A_{uni}e^{j\varphi_{\text{Gauss}}}$, where A_{uni} represents the uniform amplitude distribution, and φ_{Gauss} is the Gaussian phase distribution. This field is then propagated to two planes located at $z = \pm\Delta z$, where $\Delta z = 10$ μm, and the irradiance distribution at those planes is calculated as $I_{\Delta z} = \left|U_{\text{field}}^{\Delta z}\right|^2$ and $I_{-\Delta z} = \left|U_{\text{field}}^{-\Delta z}\right|^2$, where $U_{\text{field}}^{\Delta z}$ and $U_{\text{field}}^{-\Delta z}$ are the propagated fields at planes $z = \Delta z$ and $z = -\Delta z$. The object size is 1024×1024 pixels with a pixel pitch of 0.31 μm, and the wavelength used is $\lambda = 660$ nm, which is assumed to be coherent illumination. The simulated object is shown in Figure 4.8. Figure 4.9 shows the corresponding amplitude distributions at planes $z = \pm\,\Delta z$ and the recovered phase using TIE.

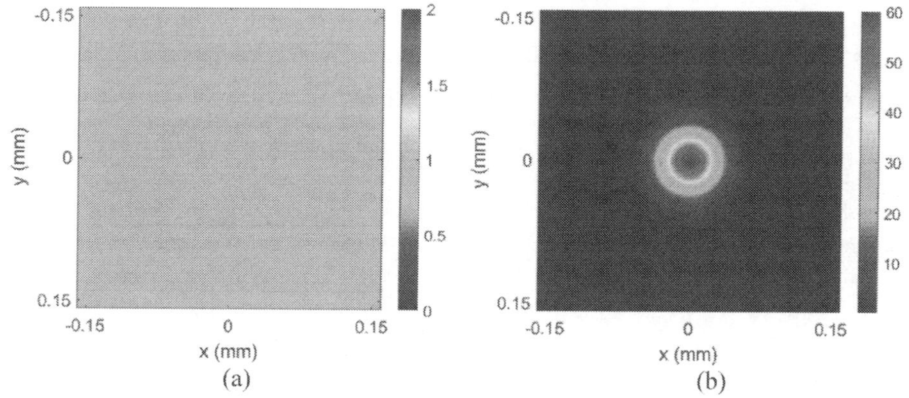

FIGURE 4.8 (a) Uniform amplitude and (b) Gaussian phase for the simulated ideal phase object. The maximum phase is 60 rad with a width of 50 μm. (Courtesy Professor Partha Banerjee.)

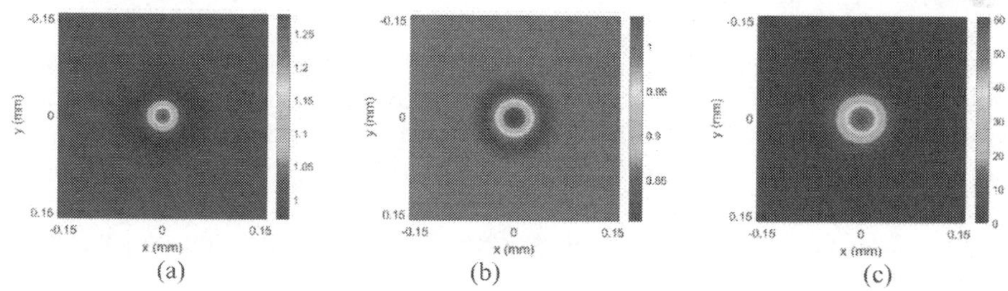

FIGURE 4.9 Numerical generated amplitude images at planes (a) $z = -\Delta z$ and (b) $z = \Delta z$. (c) Phase retrieval results using TIE for the Gaussian phase object shown in Figure 4.8. (Courtesy Professor Partha Banerjee.)

APPENDIX: DERIVATION OF THE TRANSPORT OF IRRADIANCE AND PHASE EQUATIONS

In optics, the starting point is the Helmholtz equation for the scalar optical field

$$\vec{\nabla}^2 E_p + k_0^2 E_p = 0 \tag{A.1}$$

where k_0 is the propagation constant, and

$$E_p(\vec{r}_\perp, z) = E_e(\vec{r}_\perp, z)\,\exp(-jk_0 z) \tag{A.2}$$

denotes the optical field phasor, with E_e denoting the envelope of the optical field. Also, $\vec{r}_\perp$ denotes the transverse dimensions, with z representing the nominal direction of propagation. E_p can be represented as

$$E_p = a\,\exp(-j\tilde{\varphi}) \tag{A.3}$$

where $a(\vec{r}_\perp, z) = \sqrt{I(\vec{r}_\perp, z)}$ denotes the amplitude with I denoting the irradiance. It is noted that $\tilde{\varphi}(\vec{r}_\perp, z)$ represents the phase of E_p and is related to the phase $\varphi(\vec{r}_\perp, z)$ of E_e through the relation $\tilde{\varphi} = \varphi + k_0 z$. Similar to Equation (A.3), E_e can be expressed as

$$E_e = \sqrt{I}\,\exp(-j\varphi) \tag{A.4}$$

Substituting Equation (A.3) into Equation (A.1) and separating the real and imaginary parts yields, after some straightforward algebra,

$$\vec{\nabla}\tilde{\varphi} \cdot \vec{\nabla}\tilde{\varphi} = k_0^2 + \frac{1}{a}\nabla^2 a \tag{A.5}$$

$$\vec{\nabla} \cdot \left(a^2 \vec{\nabla}\tilde{\varphi}\right) = 0 \tag{A.6}$$

respectively. In Equation (A.5), $\vec{\nabla}\tilde{\varphi}$ gives the direction of the spatial rate of change of $\tilde{\varphi}$, or the local direction of the "rays." Without the last term on the right-hand side, Equation (A.5) is the mathematical enunciation of the traditional eikonal equations in the absence of diffraction. The last term is the contribution from diffraction, which is absent if one uses the traditional ray approach to optical propagation. Equation (A.6) is a restatement of the conservation of energy, which can be seen if both sides are integrated over a closed volume and the divergence theorem is invoked. The integrand $a^2\nabla\tilde{\varphi}$ is indicative of the directed flux flowing out, and the integral is equal to zero because there are no sources assumed within the control volume. In summary, Equations (A.5) and (A.6) model optical propagation in the presence of diffraction while also satisfying the conservation of energy and must be simultaneously solved to determine the amplitude and phase of a propagating optical field.

If, instead, Equation (A.2) is substituted into the Helmholtz equation and E_e is assumed to be a slowly varying function of z, i.e.,

$$\frac{\left|\dfrac{\partial^2 E_e}{\partial z^2}\right|}{\left|\dfrac{\partial E_e}{\partial z}\right|} \ll k_0 \tag{A.7}$$

the paraxial wave equation

$$\left(\vec{\nabla}_\perp^2 - 2jk_0 \frac{\partial}{\partial z}\right) E_e = 0 \tag{A.8}$$

is obtained. Substituting Equation (A.4) into Equation (A.8) and separating into real and imaginary parts yields, after some straightforward algebra,

$$\vec{\nabla}_\perp\varphi \cdot \vec{\nabla}_\perp\varphi + 2k_0\left(\frac{\partial\varphi}{\partial z}\right) = \frac{\vec{\nabla}_\perp{}^2\sqrt{I}}{\sqrt{I}} \tag{A.9}$$

$$\vec{\nabla}_\perp.\left(I\vec{\nabla}_\perp\varphi\right) = -k_0\frac{\partial I}{\partial z} \tag{A.10}$$

respectively. Equation (A.9), like Equation (A.5), now describes paraxial propagation in the presence of diffraction, and is referred to as the *transport of phase equation* (TPE). Equation (A.10), like Equation (A.6), is again a restatement of the conservation of energy, and is also called the *transport of irradiance equation*. As in the case of the eikonal equation in the presence of diffraction and energy conservation discussed previously, both Equations (A.9) and (A.10) must be simultaneously satisfied for a propagating optical field. Under the assumption of near constancy of irradiance I (which is theoretically exactly true for an imaged phase object), Equation (A.10) reduces to Equation (4.30), which is the simplified form of TIE.

It can be readily verified that Equations (A.9) and (A.10) can be derived by starting from Equations (A.5) and (A.6) and using $\tilde{\varphi} = \varphi + k_0 z$ and the slowly varying approximation, as expected.

QUESTIONS

4.1 Two sinusoidal waves of equal amplitude are propagating along the positive z-direction. If the path difference between them is $\lambda/4$, what is the resultant amplitude? Assuming that the amplitudes of the two waves are unequal (1:3) and the phase difference between them is $3\pi/4$, what is the resultant amplitude? What is the resultant irradiance in the interference pattern?

4.2 In double-slit experiments shown in Figure 4.4, the path difference between the two waves at any point P is given by $\Delta = r_1 - r_2$. Show that r_2 is given by

$$r_2 = \left(r_1^2 - 2\, r_1 d\, \sin\theta + d^2\right)^{\frac{1}{2}}$$

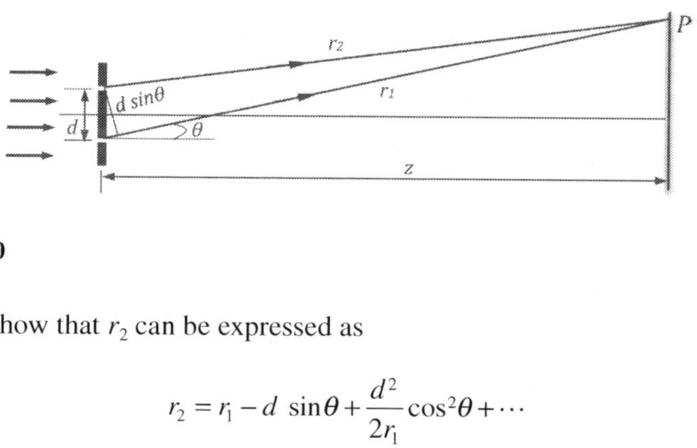

FIGURE 4.10

Further show that r_2 can be expressed as

$$r_2 = r_1 - d\, \sin\theta + \frac{d^2}{2r_1}\cos^2\theta + \cdots$$

If the path difference $(r_1 - r_2)$ is to be expressed as $d\sin\theta$, then show that $d^2/r_1 \ll \lambda$.

4.3 Light from a laser at 488 nm illuminates a pair of fine slits separated by 0.2 mm and the interference pattern is observed at a distance of 1.6 m from the plane of the slits. What is the separation between the fifth-order and zeroth-order fringes? If this experiment is conducted under water (μ = 4/3), what is the separation between the fifth-order and zeroth-order fringes?

4.4 Obtain the irradiance distribution in the interference pattern of N slits, which are randomly distributed parallel to each other. The width of each slit is $2b$.

4.5 A bi-convex lens of radius of curvature R_1 is placed on a spherical surface test plate of radius of curvature R_2. The arrangement is illuminated by a light beam of wavelength λ. Show that the radius r_m of the mth dark ring is given by

$$r_m = \sqrt{\frac{m\,\lambda\,(R_1 R_2)}{(R_2 - R_1)}}$$

4.6 A plano-convex cylindrical lens is placed on a plane surface with its cylindrical surface in contact. Show that the fringes obtained are straight lines and the location of mth straight dark fringe is

$$x_m = \sqrt{m\,\lambda\,R}$$

where R is the radius of curvature of the cylindrical surface and λ is the wavelength of light. The axis of the cylindrical surface is along the y-direction.

4.7 A thin film of refractive index 1.362 at wavelength 546 nm is coated on a glass plate of refractive index 1.515. What is the thickness of the film for which there is minimum reflection at this wavelength? What is the reflectivity of the plate in the absence of film? Estimate the reflection loss when 20 such interfaces are encountered by the light in propagation through them. This highlights the importance of anti-reflection coatings on multi-element imaging optics.

4.8 Compare the resolving power of a grating of 10 cm square with 3000 line/mm operating in first order with that of a Fabry–Perot etalon 20 mm thick made of fused quartz (μ = 1.552) and coated with a film to provide 85% reflectivity. Assume the wavelength around 600 nm.

4.9 The mirrors of a Fabry–Perot interferometer are coated to provide an amplitude reflection coefficient of 0.9488. Calculate the finesse, and angular full-width at half height of the fringes. What is the role of separation between the two mirrors?

5 Holography and Digital Holography

Optical metrology utilizes a large number of techniques for the measurement, and Chapters 5 to 9 introduce these techniques. These are whole-field techniques and include (1) holography and digital holography, (2) speckle photography and speckle interferometry, (3) moiré interferometry, (4) photoelasticity, and (5) microscopy. These techniques are extensively used for deformation studies, vibration analysis, and surface topography. We will discuss the basic principles of these techniques and their variations in the next few chapters and their applications will follow in later chapters.

5.1 HOLOGRAPHY

The birth of holography was due to a technological demand of improving the resolution of an electron microscope. The resolution of the electron microscope was extremely poor compared with the resolution limit given by the wave theory due to poor performance of the electron lenses, and any technological solution to improve their performance was not possible then. To solve this problem, Gabor invented a two-step process that involved the recording and the reconstruction steps. In order to understand this, let us examine the process of seeing, i.e., how we see the objects. An object is seen by the light diffracted/scattered by the object. This scattered light is manipulated by the imaging system or the eye lens to create an irradiance distribution similar to that existing on the object. This distribution is detected by the receiver/detector or the retina producing an image. Detectors in the optical regime are energy detectors, that is, they respond to absolute amplitude squared, hence the phase information is lost. On the other hand, the scattered light from the object, which is called the object wave, contains all the information about the object: its amplitude and phase are coded. The recording step in Gabor's two-step process records both the amplitude and the phase of the wave. The object wave is thus stored in the recording step and played back in the reconstruction step. Gabor's idea was to record the electron wave from the object and play it with visible light, thereby completely avoiding the role of electron lenses. Gabor's two-step process is known as holography, which literally means total recording, i.e., recording of both amplitude and phase of the wave. Because optical detectors respond to irradiance, the phase information is converted to irradiance information by the phenomenon of interference. Basically, holography involves the recording of an interference pattern between an object wave and a reference wave. The record, after processing, is called a hologram. Gabor demonstrated his two-step process using micro-objects in the form of transparencies and illuminating them with filtered radiation from nearly a point source. In order to carry out the

DOI: 10.1201/9781003531845-5

recording on the low-resolution film available at that time, he resorted to in-line geometry where both the reference and the object waves travel along the same direction. In fact, his choice of objects was such that they generated both reference and object waves.

After its invention in 1948, holography remained practically dormant until the arrival of the laser because a long coherence length source was needed to record a hologram of an object. Earlier recordings of three-dimensional (3D) objects using laser radiation were made on Kodak 649F plates and the results were very impressive. The holographic reconstructions were as real as the objects and provided the necessary parallax. Holography, therefore, came to be known as a 3D photography. It, however, is more than ordinary 3D photography as it provides 3D views with changing perspectives. A hologram is like a window with a memory. Different perspectives of the scene are seen through different portions of the hologram. Holography, apart from recording holograms of 3D objects and creating 3D reconstructions, and being an inspiration to artists and painters, found many scientific and commercial applications. One of the remarkable scientific applications of holography is holographic interferometry (HI).

HI emerged as a technique of unparalleled applications because it provided interferometric comparison of real objects or events separated in time and space. Various kinds of HI have been developed, for example, real-time, double-exposure, time-averaged, etc. Further, it could be performed with one reference wave, two reference waves, etc. These reference waves could be of the same wavelength or different wavelengths. The reference wave could come from the same side of the hologram as the object wave or from the other side of the hologram. HI could be performed with a continuous wave laser or a pulsed laser. The record could be made on photo-emulsions, thermoplastics, photopolymers, charge-coupled devices (CCDs), etc. Digital holography allows the comparison of objects situated at different locations. Small objects can be studied for their responses to external agency. The possibilities are endless and so are the applications.

5.1.1 Hologram Recording

An object is illuminated by a wave from a laser and the diffracted field is received on a recording plate lying in the x–y plane. A reference wave is added to this field at the recording plane as shown in Figure 5.1(a). The diffracted field from the object constitutes the object wave, which is represented by $\boldsymbol{O}(x,y) = O_0(x,y)\exp\left[i\delta_0(x,y)\right]$ where $O_0(x,y)$ is the amplitude of the object wave and $\delta_0(x,y)$ is its phase. The complex amplitude $\boldsymbol{R}(x,y)$ of the reference wave at the recording plane is expressed as $\boldsymbol{R}(x,y) = R_0\exp(2\pi i f_R y)$, where f_R is the spatial frequency of the wave. A plane wave incident at an angle of θ with the z-axis is taken as a reference wave, hence $f_R = \sin\theta/\lambda$. These waves are derived from the same wave (source) and hence are coherent to each other. The total amplitude at the recording plane is

$$\boldsymbol{A}(x,y) = \boldsymbol{O}(x,y) + \boldsymbol{R}(x,y) \tag{5.1}$$

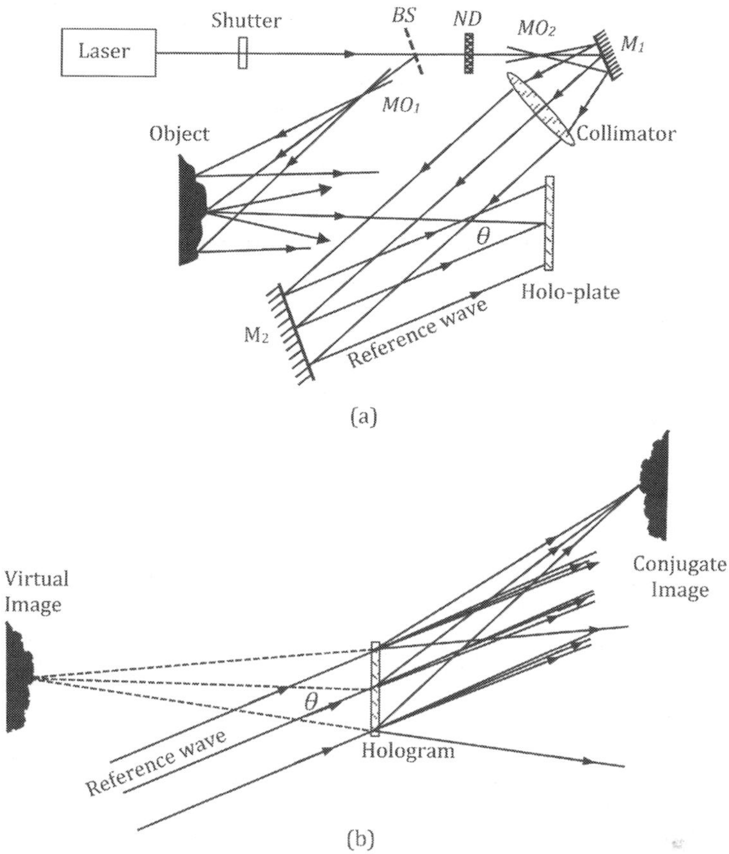

FIGURE 5.1 (a) Recording of a hologram and (b) its reconstruction.

Therefore, the irradiance distribution $I(x,y)$ on the recording plane is given by

$$I(x,y) = O_0^2(x,y) + R_0^2 + 2\,O_0(x,y)\,R_0\,\cos\left[2\pi f_R y - \delta_0(x,y)\right] \tag{5.2}$$

It is, thus, seen that both the amplitude variations $O_0(x,y)$ and the phase variations $\delta_0(x,y)$ of the object wave have been converted into irradiance variations to which the recording material responds. We assume that the recording material is a photo-emulsion, say, a holographic plate or a film. The irradiance is recorded over an appropriate time period T resulting in an exposure variation $E(x,y) = I(x,y)T$. After development, the plate/film is called a hologram. On the hologram, the exposure variation is converted into density variation, which leads to an amplitude transmittance variation. The amplitude transmittance is a complex function of position as the hologram introduces both the amplitude and the phase variations to the wave incident on it; the phase variations are a result of thickness variations. The phase variations can be eliminated if the hologram is placed in a liquid gate. Further, if the phase variations are shared between two wave fronts then they are not seen when these wave fronts interfere. We assume that the amplitude transmittance of the hologram is proportional to the exposure incident during recording. Under this assumption, the amplitude transmittance $t(x,y)$ of the hologram is expressed as follows:

$$t(x,y) = t_0 - \beta E(x,y) \tag{5.3}$$

where β is a constant dependent on processing parameters, exposure, etc., and t_0 is the bias transmittance.

5.1.2 Reconstruction

The hologram is placed back in the same position, which it held during recording and illuminated by the reference wave. The field just behind the hologram is given by

$$t(x,y)\ \boldsymbol{R}(x,y) = t_0'\ \boldsymbol{R}(x,y) - \beta T\boldsymbol{R}(x,y)\left\{O_0^2(x,y) + 2O_0(x,y)\ R_0\ \cos\left[2\pi f_R y - \delta(x,y)\right]\right\} \tag{5.4}$$

where $t_0' = t_0 - \beta T R_0^2$ is the modified DC transmittance. Equation (5.4) can also be written as follows:

$$t(x,y)\ \boldsymbol{R}(x,y) = t_0'\ \boldsymbol{R}(x,y) - \beta T\boldsymbol{R}(x,y)\left[O_0^2(x,y) + \boldsymbol{O}(x,y)\ \boldsymbol{R}^*(x,y) + \boldsymbol{O}^*(x,y)\ \boldsymbol{R}(x,y)\right] \tag{5.5}$$

where * signifies complex conjugate. There are four waves just behind the hologram of which the wave $t_0'\ \boldsymbol{R}(x,y)$ is the reference wave uniformly attenuated. The wave $-\beta T\boldsymbol{R}(x,y)O_0^2(x,y)$ also propagates in the direction of the reference wave. Due to the slow spatial variation of $O_0^2(x,y)$, there is a diffracted field around the direction of the reference wave. The third wave $-\beta T R_0^2\ \boldsymbol{O}(x,y)$ is the original object wave multiplied by a constant, $-\beta T R_0^2$. This wave propagates in the direction of the object wave and has all the attributes of the object wave except that its spatial dimensions are restricted by the size of the hologram. The negative sign signifies that a phase change of π has taken place due to the wet development process. The fourth wave $-\beta T\boldsymbol{R}^2(x,y)\ \boldsymbol{O}^*(x,y)$ represents a conjugate wave, which propagates in a different direction. This wave can also be written as $-\beta T R_0^2\ O_0(x,y)\exp\left[2\pi i 2 f_R y - \delta_0(x,y)\right]$; the phase of the reference wave is modulated by that of the object wave but it essentially travels in the direction ϕ where $\sin\phi = 2\sin\theta$. Figure 5.1(b) shows that several waves are generated during the reconstruction step.

The recording and reconstruction geometries just described are the off-axis holography geometries due to Leith and Upatnieks. However, the reference beam can be added axially to the object beam. This is in-line holography or Gabor holography. In-line holography has applications in particle size measurements, etc.

5.1.3 In-Line Holography or Gabor Holography

Figure 5.2(a) shows a schematic of in-line holography. A collimated beam illuminates a transparency, which has high transmittance. An example is a glass plate over

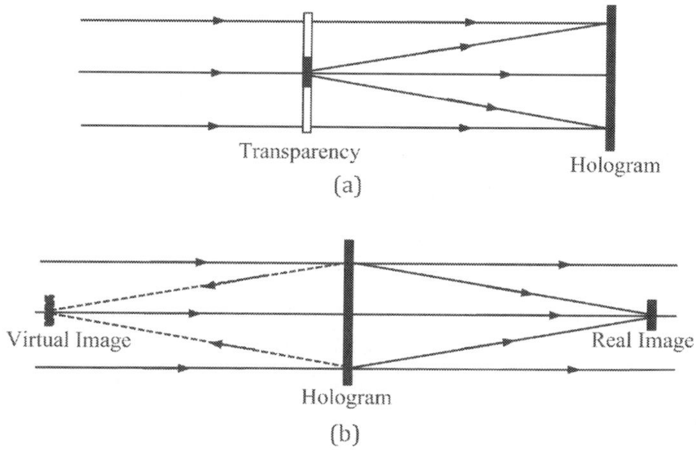

FIGURE 5.2 (a) Recording of an in-line hologram. (b) Reconstruction from in-line hologram.

which text is written. Incident beam scattered by the object is the object beam and the directly transmitted beam is the reference beam.

On reconstruction, as seen in Figure 5.2(b), one sees both the real and the virtual images in-line, one behind the other. One may be focused but the other will appear defocused. When the reference beam irradiance is much larger than that of the object beam, the images appear in a nearly uniform bright background. On the other hand, if the reference beam is weak, the images may not be clearly seen.

5.1.4 Off-Axis Holography

Gabor holography suffers from two serious drawbacks: (1) the images appear in-line, one focused and other defocused and (2) it is restricted to a class of objects that have high transmittance. These limitations are removed in off-axis holography. Initially it was also conducted with transparency as an object, but the reference beam was added at an angle. Figure 5.3 shows the schematic of the setup.

On reconstruction, the images are angularly separated. The off-axis reference wave produces the carrier frequency and hence requires a recording medium of higher resolution. Further, a source of longer coherence length is needed. These requirements are more than offset by the beautiful reconstruction. Off-axis holography also enabled recording of real 3D objects with full parallax.

It is also possible to record an object when the object wave and the reference wave lie on either side of the recording medium. The recording medium must be thick so that the fringes formed due to interference between these two waves are recorded inside the medium. The reconstruction is due to reflection from the Bragg-like planes inside the medium. Such holograms, also called reflection type or Denisyuk holograms, can be reconstructed with white light.

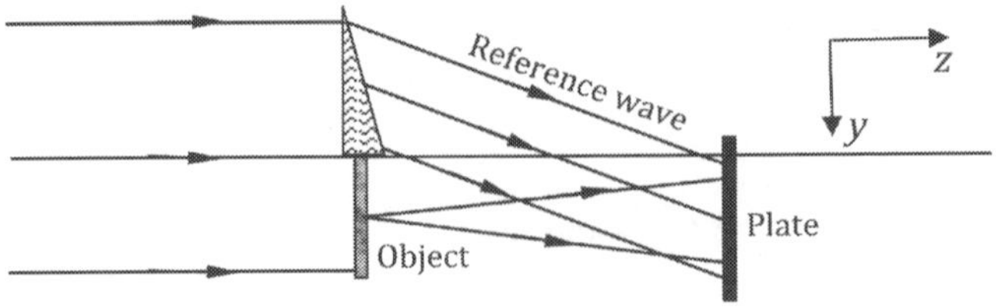

FIGURE 5.3 Recording of an off-axis hologram.

5.1.4.1 Choice of Angle of Reference Wave

One of the shortcomings of Gabor holography is that all the diffracted waves propagate in the same direction. These waves are angularly separated if the reference wave is added at an angle during the recording of a hologram, but how large should the angle be? In fact, the off-axis reference wave acts as a carrier of object information. It also results in very fine interference fringes into which object information is coded. The fringe frequency is given by $\sin\theta/\lambda$, where θ is the mean angle between the reference and the object waves. The recording medium should be capable of resolving this fringe frequency. Further, as has been described earlier, the various waves are angularly separated by nearly θ when θ is small. But a restriction on θ is placed by a simple condition that the spectra of the various waves should not overlap. If the bandwidth of the object wave is f_0, then it is easily seen from Figure 5.4 that the spectra will be just separated when $3f_0 = f_R$. This immediately gives the smallest angle θ_{min} as $\theta_{min} = \sin^{-1}(3\lambda f_0)$. For the angles smaller than θ_{min}, the various diffracted beams will overlap.

5.1.4.2 Choice of Irradiance of the Reference Wave

It is tacitly assumed that the amplitude transmittance of the hologram is proportional to the exposure. This is valid only when one operates over a very small portion of the *t-E* curve, which is shown in Figure 5.5. To meet this condition, the irradiance of the reference wave should be 3 to 10 times more than that of the object wave. If this condition is not met, then nonlinearities will begin to play their role and higher order images are produced.

5.1.5 Types of Holograms

Holograms are classified by a variety of ways, as summarized in Table 5.1.

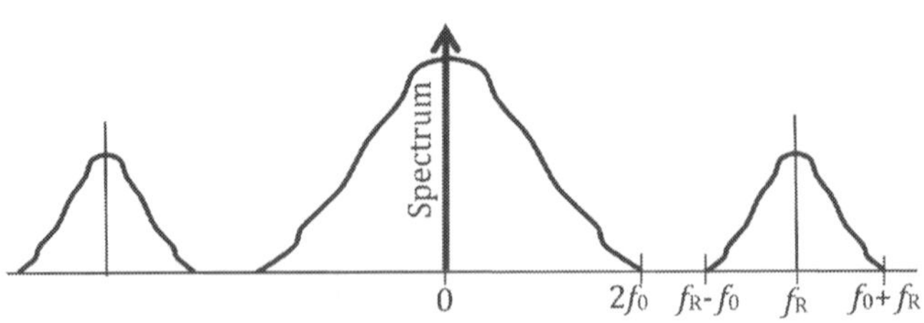

FIGURE 5.4 Fourier transform of the transmittance of a hologram.

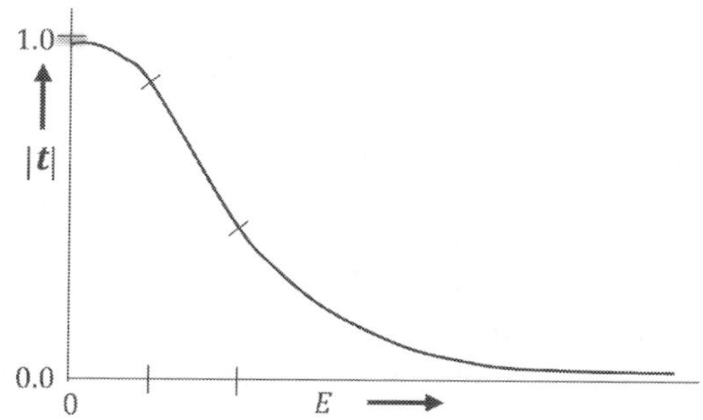

FIGURE 5.5 Amplitude transmittance versus exposure ($|t|$ vs. E) curve for a typical photo-emulsion.

TABLE 5.1
Classification of Hologram Types

Properties	Hologram Types
Transmission function	Amplitude hologram
	Phase hologram
Region of diffraction	Fresnel hologram
	Fraunhofer hologram
	Fourier hologram
	Image-plane hologram
Recording geometry	On-axis/Gabor/in-line hologram
	Off-axis/Leith–Upatnieks/carrier frequency hologram
	Reflection/Denisyuk hologram
	Benton hologram
Exposure	Single-exposure/real-time/live-fringe hologram
	Double-exposure/lapsed-time/frozen-fringe hologram
	Multiple-exposure hologram
	Time-Averaged Hologram
Emulsion thickness/Q-parameter	Thin hologram $Q \leq 1$: $Q = 2\pi\lambda d / \mu\Lambda^2$, where d-thickness, and μ-refractive index of the film and Λ-fringe width
	Thick hologram $Q \geq 10$
Reconstruction	Monochromatic light hologram
	White light hologram
Recording configuration	Transmission hologram
	Reflection hologram
	Rainbow hologram
	Digital hologram
Object type	Transmission objects/phase-object hologram
	Transmission objects-diffuse illumination hologram
	Opaque objects hologram

5.1.5.1 Diffraction Efficiency

Diffraction efficiency is a measure of the amount of light that goes into forming the useful image when the hologram is illuminated by the reference wave. It is calculated for the ideal case of interference between two plane waves. For a thin amplitude transmission hologram, the maximum diffraction efficiency is 6.25%. This value can be improved by bleaching, i.e., the amplitude hologram is converted into a phase hologram. It can then reach up to 33.6% for a thin-phase transmission hologram. The diffraction efficiency for a thick-phase hologram can approach 100%: essentially then, the entire incident light is utilized for image formation and there is only one diffraction order.

5.1.6 Experimental Arrangement

The experimental arrangement, as seen in Figure 5.1(a), consists of a laser, a beam splitter, beam-expanding optics, and the recording medium. We describe these in detail in the following sections.

5.1.6.1 Lasers

There are a large number of lasers to choose from, and some of these are described in the following:

1. *Ruby laser:* This is a pulsed laser with a pulse width of the order of 10 ns. Its coherence length could be more than 2 m when used with an etalon, and it can deliver more than 1 J in a single pulse. The laser for HI has a dual-pulse facility with variable pulse separation for dynamical studies.
2. *Argon-ion laser:* This is a continuous wave laser with output exceeding 5 W. It gives a multiline output, hence, a dispersion prism is mounted to select a spectral line. Further, an inter-cavity etalon increases the coherence length to a usable length for studying moderate-sized objects. It can be used with photoresist and thermoplastics as the recording media.
3. *He-Ne laser:* This is the most commonly used laser for holography and HI with power outputs in the range of 25 mW to 35 mW. The coherence length is more than 20 cm.
4. *Nd:YAG semiconductor pumped, first harmonic (green) laser:* This laser is finding acceptance for holography. It offers a very long coherence length with continuous wave operation. The output ranges from 40 mW to 150 mW.
5. *He-Cd laser:* It is a continuous wave laser with output in the range of 40 mW to 150 mW. It is suitable for recording on photoresist. Both blue and ultraviolet lines are usable for holography.
6. *Semiconductor or diode lasers (LDs):* These can be run both as continuous and pulsed. Continuous wave lasers are preferred for holography but are not really very common. They require temperature and current stability. They can output up to 500 mW continuous wave.

5.1.6.2 Beam Splitters

Usually, an unexpanded laser beam is split into two or more beams. For this, a glass plate (preferably a wedge plate) serves as a good beam splitter. However, when an expanded collimated beam is to be split, a cube beam splitter or a plane parallel plate splitter with one side anti-reflection coated is recommended. Pellicle beam splitters are also used. In some cases, polarization optics (Wollaston prism) is employed as beam splitter.

5.1.6.3 Beam Expanders

With high-power lasers, a suitable concave or negative lens is used for beam expansion; otherwise, a microscope objective of 5×, 10×, 20×, 40× (45×), or 60× (63×) could be used. Due to dust particles on the optical surfaces, the beam is usually dirty, i.e., it has circular rings, dark specks, etc. It can be cleaned by placing a pinhole at the focus of the microscope objective. This arrangement is known as a spatial filter. The pinhole size is to be matched with the Airy disk of the microscope objective. To expand the beam, an achromatic lens is placed in the beam such that the pinhole is at its focus. This arrangement in called a beam expander.

5.1.6.4 Object Illumination Beam

For small-sized objects, a collimated beam illumination is preferred as the propagation vector for the illumination beam is then constant over its whole surface. For large objects, a spherical diverging wave is used. If the object is cylindrical in shape, a cylindrical diverging wave is preferred. Objects must be located in the coherence volume. Special recording geometries, which relax the coherence requirement, may be used for large objects.

5.1.6.5 Reference Beam

Either a diverging spherical wave or a plane wave can be used as a reference wave. However, if the hologram is to be reconstructed at a later time, or at a different location, use of a plane wave is recommended. The beam is overexpanded to make it uniform over the hologram plane. The reference beam irradiance must be 3 to 10 times stronger than the object beam irradiance at the hologram plane for linear recording.

5.1.6.6 Angle between Object and Reference Beams

The object wave is a diffuse wave; therefore, the angle between the object wave and the reference wave (even when it is a plane wave) will vary over the recording plane. We, however, take the mean angle. The fringe spacing on the recording plane will be approximately $\sin\theta/\lambda$, where θ is the mean angle between the reference wave and the object wave incident nearly normally on the recording plane. Recording medium should be able to resolve this fringe structure. Too large an angle puts higher demand on the resolution. A small angle, on reconstruction, may not result in the separation of the beams. Therefore, the mean angle between the object wave and the reference wave is to be chosen judiciously.

5.1.7 Holographic Recording Materials

A variety of recording materials are available and have been described in Chapter 3. The choice of material is made based on the wavelength sensitivity, resolution, and other characteristics of the material.

5.2 HOLOGRAPHIC INTERFEROMETRY

HI is used to compare two waves from the real objects. These two waves usually are from the initial (unstressed) state and the final (stressed) state of the object. It is, however, assumed that the microstructure of the surface does not change as a result of loading and hence the two waves differ due to path difference changes rather than microstructural changes. HI can be performed in a variety of ways. The most common methods are (1) real-time HI, (2) double-exposure HI, and (3) time-averaged HI. Several new configurations have been developed depending on the application and also to exploit the strengths of HI. They are discussed in the following sections.

5.2.1 Real-Time HI

A hologram of an object is recorded, processed, and placed back in the experimental setup exactly at the same location that it occupied during recording. Reconstruction of the hologram by the reference wave generates a replica of the original object wave, which propagates in the direction of the original wave. This wave is, however, phase-shifted by π due to the wet photographic development process. Because the object wave is also present, it will suffer diffraction on passage through the hologram generating several waves: one wave transmitted by the DC transmittance of the hologram will propagate in the original direction of the object wave. Therefore, there are two object waves: one released from the hologram by the interaction of the reference wave and the other transmitted by the DC transmittance of the hologram. These waves are identical in all respects but for a phase change of π; hence, they produce a dark field on interference. If the object is now loaded, the object wave carries the deformation phase and hence an interference pattern is observed. This pattern changes in real time with the change in load. One can therefore monitor the response of the object to the external loading agency continuously until the fringes become too fine to be resolved. The technique is also known as single-exposure or live-fringe HI.

It can be shown that the irradiance distribution on the object, under the condition that the reference wave is much stronger than the object wave, can be expressed as

$$I_{\text{obj}}(x,y) = I_0(x,y)\left[1 + \frac{\left(t_0 - \beta T R_0^2\right)^2}{\left(\beta T R_0^2\right)^2} - 2\frac{\left(t_0 - \beta T R_0^2\right)}{\left(\beta T R_0^2\right)}\cos\delta(x,y)\right] \tag{5.6}$$

where $I_0(x,y)$ is proportional to the irradiance of the object wave, $\delta(x,y)$ is the deformation phase, and the other symbols have their usual meaning. Dark fringes are formed wherever $\delta(x,y) = 2m\pi$, with m being an integer. The contrast η of the fringes is given by

$$\eta = 2\frac{\left(\beta TR_0^2\right)\left(t_0 - \beta TR_0^2\right)}{\left(\beta TR_0^2\right)^2 + \left(t_0 - \beta TR_0^2\right)^2} \tag{5.7}$$

It may be seen that the contrast of the fringes can be controlled by the reference wave irradiance during reconstruction. In general, the contrast is less than 1 but by appropriate increase of the irradiance of the reference wave, it is possible to achieve unit contrast fringes. As an example, if the bias transmittance t_0 is equal to $2\beta TR_0^2$, the unit contrast fringe pattern results.

5.2.2 Double-Exposure HI

Here, both the exposures belonging to the initial and the final states of the object are recorded sequentially on the same photographic plate. The total exposure recorded can be expressed as

$$E(x,y) = T\left[I_1(x,y) + I_2(x,y)\right] \tag{5.8}$$

where $I_1(x,y)$ and $I_2(x,y)$ are the irradiances recorded during the first and second exposures. Inserting the expressions for the irradiances and assigning the amplitude transmittance to the double-exposure hologram, the irradiance distribution in the reconstructed object can be shown to be

$$I_{\text{obj}}(x,y) = 2\left(\beta TR_0^2\right)^2 O_0^2(x,y)\left[1 + \cos\delta(x,y)\right] = \mathrm{I}_0(x,y)\left[1 + \cos\delta(x,y)\right] \tag{5.9}$$

Bright fringes are formed wherever $\delta = 2m\pi$, with m being an integer. The technique is also known as frozen fringe or lapsed-time HI. It may be noted that only two states of the object are being compared by this technique. Figure 5.6 shows a double-exposure interferogram of a pipe with a defect. The pipe was loaded by an application of hydraulic pressure between the exposures. It is seen that the defect is a region of thinner wall thickness.

FIGURE 5.6 Double-exposure interferogram of a pipe with a defect.

5.2.3 Time-Averaged HI

Time-averaged HI is utilized for the study of vibrating bodies like musical instruments, objects subjected to impact, etc. As the name suggests, the recording is carried out over a time period that is great deal longer than the period of vibration. Because an object vibrating sinusoidally spends most of the time at the locations of maximum displacement, a recording of such a vibrating object is equivalent to a double-exposure record. However, the irradiance distribution on the reconstructed image is modified considerably due to the time of excursion between these extreme positions. To study this phenomenon, we can write for the instantaneous irradiance on the recording plane

$$I(x,y;t)=O_0^2(x,y;t)+R_0^2+\boldsymbol{O}^*(x,y;t)\boldsymbol{R}(x,y)+\boldsymbol{O}(x,y;t)\boldsymbol{R}^*(x,y) \quad (5.10)$$

where the object wave $\boldsymbol{O}(x,y;t)$ is expressed as

$$\boldsymbol{O}(x,y;t)=O_0e^{i\delta_0}\ e^{i\delta(x,y;t)} \quad (5.11)$$

The phase $\delta(x,y;t)$ represents the phase change introduced by the vibration and δ_0 is a constant phase. If the body is vibrating with an amplitude $A(x,y)$ at a frequency ω and is illuminated and observed along the directions making angles θ_1 and θ_2 with the local normal, then the phase difference δ can be written as

$$\delta(x,y;t)=\frac{2\pi}{\lambda}\ A(x,y)(\cos\theta_1+\cos\theta_2)\ \sin\omega t \quad (5.12)$$

The irradiance distribution expressed by Equation (5.10) is recorded over a period T much longer than the period of vibration, i.e., irradiance distribution is recorded over a large number of vibration cycles. The average irradiance recorded over the period T is

$$I(x,y)=\frac{1}{T}\int_0^T I(x,y;t)\ dt \quad (5.13)$$

This record, on development, is called a time-averaged hologram, and the procedure is known as time-averaged HI. The hologram is reconstructed with the reference wave. Because, on reconstruction, various waves generated are separable, we consider only the amplitude of the desired wave $a(x,y)$, which is given by

$$a(x,y)=-\beta TR_0^2\frac{1}{T}\int_0^T \boldsymbol{O}(x,y;t)\,dt=-\beta TR_0^2O_0(x,y)\frac{1}{T}\int_0^T e^{\frac{2\pi i}{\lambda}A(x,y)(\cos\theta_1+\cos\theta_2)\ \sin\omega t}\ dt \quad (5.14)$$

The time integral $\frac{1}{T}\int_0^T e^{\frac{2\pi i}{\lambda}A(x,y)(\cos\theta_1+\cos\theta_2)\ \sin\omega t}dt$ is called the characteristic function of the sinusoidal vibration, and it is represented by M_T. Thus, the irradiance distribution in the reconstructed object is given by

$$I_{\text{obj}}(x,y)=a(x,y)a^*(x,y)=\beta^2T^2R_0^4O_0^2|M_T|^2=I_0(x,y)|M_T|^2 \quad (5.15)$$

The characteristic function has been generalized for various motion types and $|M_T|^2$ has been tabulated in Table 5.2 for motions of various kinds. The characteristic function for the sinusoidal motion as included in Equation (5.14) can be analytically obtained. It is given by

$$M_T = \frac{1}{T}\int_0^T e^{\frac{2\pi i}{\lambda} A(x,y)(\cos\theta_1+\cos\theta_2)\sin\omega t}\,dt = J_0\left[\frac{2\pi}{\lambda} A(x,y)(\cos\theta_1+\cos\theta_2)\right] \tag{5.16}$$

where $J_0(x)$ is the Bessel function of zeroth order and first kind. The irradiance distribution in the reconstructed image will be

$$I_{\text{obj}}(x,y) = I_0(x,y)\, J_0^2\left[\frac{2\pi}{\lambda} A(x,y)(\cos\theta_1+\cos\theta_2)\right] \tag{5.17}$$

It may be seen that the reconstructed image is modulated by the $J_0^2(x)$ function. The variation of $J_0^2(x)$ with the argument x is shown in Figure 5.7. Thus, the reconstructed object is covered by the fringes. Figure 5.8 shows a time-averaged

TABLE 5.2
Characteristic Functions $|M_T|^2$

HI Type	Displacement	$\|M_T\|^2$
Real-time	Static (**L**)	$1+c^2-2c\cos(\mathbf{k}\cdot\mathbf{L})$
	Harmonic of amplitude **A**(x, y)	$1+c^2-2cJ_0(\mathbf{k}\cdot\mathbf{A})$
	Harmonic (time-averaged)	$1-J_0(\mathbf{k}\cdot\mathbf{A})$
Real-time with reference fringes	Harmonic	$1+c^2-2c\cos(\mathbf{k}\cdot\mathbf{L})J_0(\mathbf{k}\cdot\mathbf{A})$
Real-time stroboscopic	Harmonic: pulses at ωt = π/2 and 3π/2	$1+c^2-2c\cos(2\mathbf{k}\cdot\mathbf{A})$
Double-exposure	Static (**L**)	$\cos^2(\mathbf{k}\cdot\mathbf{L}/2)$
Double-exposure stroboscopic	Harmonic: pulses at ωt = π/2 and 3π/2	$\cos^2(\mathbf{k}\cdot\mathbf{A})$
Time-averaged	Harmonic of amplitude **A**(x, y)	$J_0^2(\mathbf{k}\cdot\mathbf{A})$
	Constant velocity $\mathbf{L}_r = \mathbf{v}T$	$\operatorname{sinc}^2(\mathbf{k}\cdot\mathbf{L}_r/2)$
	Constant acceleration from rest	$\dfrac{C^2(\sqrt{\mathbf{k}\cdot\mathbf{A}})+S^2(\sqrt{\mathbf{k}\cdot\mathbf{A}})}{(2/\pi)(\mathbf{k}\cdot\mathbf{A})}$
Time-averaged	Irrationally related modes	$J_0(\mathbf{k}\cdot\mathbf{A}_1)J_0(\mathbf{k}\cdot\mathbf{A}_2)$
Temporally frequency translated	Harmonic motion	$J_m^2(\mathbf{k}\cdot\mathbf{A})$
	Amplitude modulated reference wave $f_r(t)=e^{i(\omega t-\Delta)}$	$J_m^2(\mathbf{k}\cdot\mathbf{A})\cos^2\Delta$
	Phase modulated reference wave $f_r(t)=e^{iM_R\sin\omega t}$	$J_m^2(\mathbf{k}\cdot\mathbf{A}-M_R)$

Note: c is the contrast; **k** is the propagation vector, and C and S are Fresnel cosine and sine integrals.

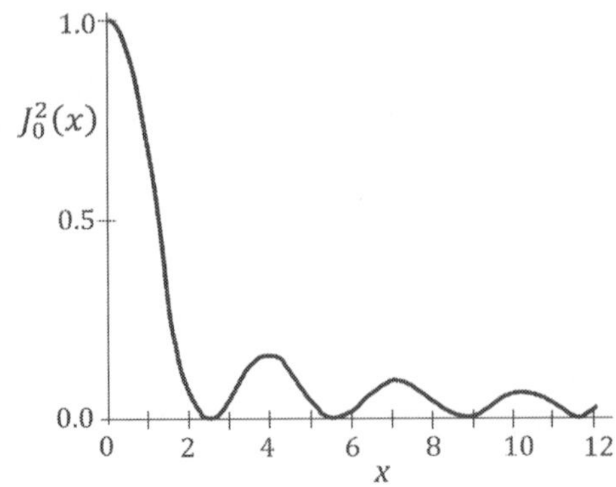

FIGURE 5.7 $J_0^2(x)$ distribution, which defines the irradiance distribution in time-averaged HI fringes for a sinusoidally vibrating object.

interferogram of an edge-clamped diaphragm vibrating in the second harmonic mode. The zero irradiance in the reconstructed object occurs at the zeros of the $J_0^2(x)$ function. It may also be seen that the irradiance is maximum at the stationary region [$A(x,y) = 0$] and decreases to zero when

$$\frac{2\pi}{\lambda} A(x,y)(\cos\theta_1 + \cos\theta_2) = 2.4048 \tag{5.18}$$

This gives the amplitude of vibration at the first zero of the Bessel function. Assuming illumination and observation directions along the normal to the surface of the object, the amplitudes at the first and the successive zeros of the Bessel function are given by

$$A(x,y) = \frac{\zeta\,\lambda}{4\pi} : \zeta = 2.4048,\ 5.5200,\ 8.6537,\ 11.7015,\ \ldots. \tag{5.19}$$

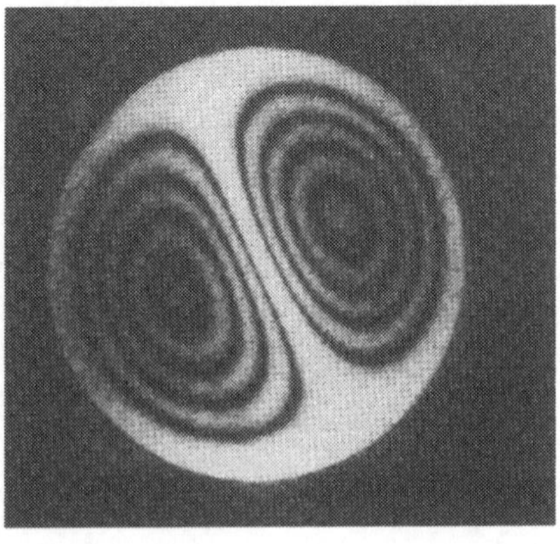

FIGURE 5.8 Time-averaged interferogram of an edge-clamped diaphragm vibrating in second harmonic mode.

Because the irradiance falls off rapidly with amplitude, high amplitudes of vibration are difficult to monitor by this technique. In brief, time-averaged HI gives the vibration map (phase of the vibration is lost) over a range of frequencies, i.e., the amplitude of the vibration can be measured.

5.2.4 Real-Time, Time-Averaged HI

In this technique, a single-exposure record of a stationary object is first made. After development, the hologram is precisely relocated. The object is now set in vibration and viewed through the hologram. Interference between the reconstructed wave from the hologram, and the directly transmitted wave from the vibrating object, will yield an irradiance distribution, which is proportional to $\left[1-J_0(x)\right]^2$, where $x=4\pi\ A(x,y)/\lambda$. Thus, the nodal points on the object appear dark. The contrast of this fringe pattern is very low. While observing the real-time pattern, the laser beam can be chopped at the frequency of vibration. The method is then equivalent to the real-time HI for static objects. The fringes correspond to the displacement between the initial position and the position when the laser beam illuminates the object. By varying the time at which the light pulse illuminates the object during its vibration cycle, the displacement at different phases of vibration can be mapped.

5.2.5 Stroboscopic Illumination/Stroboscopic HI

The object is illuminated by laser pulses whose duration is much shorter than the period of vibration: the object is thus frozen at any phase of vibration. The recorded hologram ends up like a double-exposure hologram. Let the first record be made when the object is in state 1 and the second record be made when the object is in state 2 of the vibration cycle as shown in Figure 5.9. The double-exposure hologram will display fringes corresponding to the displacement $A(x,y)$ between these two vibration

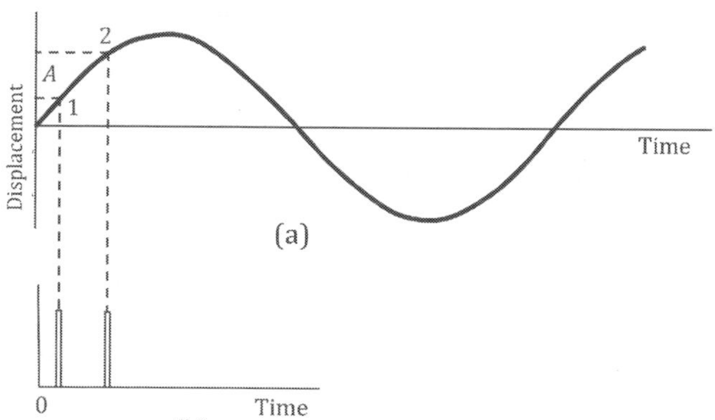

FIGURE 5.9 Stroboscopic HI of a vibrating object. (a) Displacement of a sinusoidally vibrating object and (b) pulse illumination.

states. By varying the pulse separation, the holograms for different phases of the vibration cycle and consequently for different displacements can be recorded.

Alternately, the pulse separation may be adjusted to correspond to one-quarter of the time period, and the double-exposure holograms may be recorded at various phases of the vibration cycle. The advantage of stroboscopic HI is that the fringes of unit contrast, similar to those in double-exposure HI, are formed. In addition, the use of pulse illumination does not require vibration isolation of the holographic setup.

5.3 SPECIAL TECHNIQUES IN HOLOGRAPHIC INTERFEROMETRY

5.3.1 Two-Reference Beam HI

While discussing the phase-shift method of evaluation of the phase in an interferogram, it was mentioned that an additional phase is to be introduced that is independent of the phase to be measured. In real-time HI, it was provided by shifting the phase of the reference wave during reconstruction. However, it is not possible to achieve this in double-exposure HI, as the reconstruction of both recorded waves is done with a single reference beam.

In order to have independent access to the two reconstructed waves, and introduce the desired phase difference between them, a holographic setup with two reference waves is required (Figure 5.10). The first exposure of a double-exposure hologram is made with the object in its initial state and with reference wave R_1 (reference wave R_2 is blocked). The object is loaded, and the second exposure is made with reference wave R_2 (reference wave R_1 is blocked). The reconstruction of the hologram is accomplished with both reference waves R_1 and R_2 simultaneously to generate the interference pattern between the initial and final states of the object: the interference pattern may be varied by changing one of the reference waves while leaving the other unaffected. Mathematically, the procedure is described as follows:

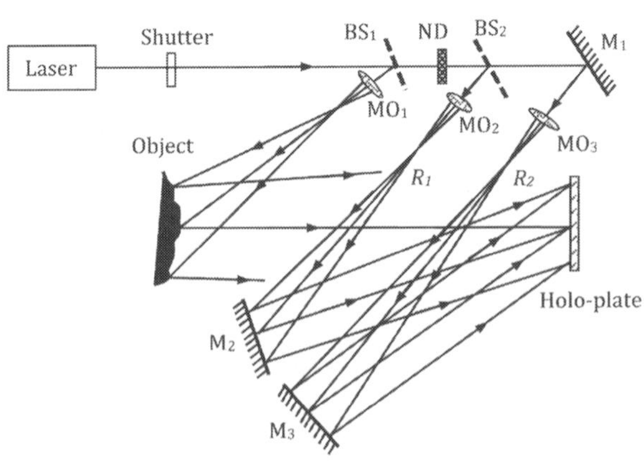

FIGURE 5.10 A schematic for two reference waves HI.

An irradiance distribution $I_1(x,y)$ with reference wave $\boldsymbol{R}_1$ is recorded in the first exposure where $I_1(x,y)$ is given by

$$I_1(x,y) = O_0^2 + R_{01}^2 + \boldsymbol{O}\ \boldsymbol{R}_1^* + \boldsymbol{O}^*\boldsymbol{R}_1 \tag{5.20}$$

The irradiance distribution $I_2(x,y)$ recorded in the second exposure with reference wave $\mathbf{R}_2$ is given by

$$I_2(x,y) = O_0^2 + R_{02}^2 + \boldsymbol{O}'\ \boldsymbol{R}_2^* + \boldsymbol{O}'^*\boldsymbol{R}_2 \tag{5.21}$$

where $\boldsymbol{O}'(x,y) = \boldsymbol{O}(x,y)e^{i\delta(x,y)}$, being the object wave from the deformed state of the object. The amplitude transmittance $t(x,y)$ of the double-exposure hologram is given by

$$t(x,y) = t_0 - \beta T(I_1 + I_2) \tag{5.22}$$

This hologram is illuminated by both reference waves. Therefore, the amplitude of the waves just behind the hologram is $(\boldsymbol{R}_1 + \boldsymbol{R}_2)\ t(x,y)$. This hologram generates a multiplicity of images that may or may not overlap. However, $\boldsymbol{R}_1\boldsymbol{R}_1^*\boldsymbol{O}$ and $\boldsymbol{R}_2\boldsymbol{R}_2^*\boldsymbol{O}'$ completely overlap and give rise to interference fringes. Other undesired images can be separated from this interference pattern if the angle between the two reference waves is made very large. Generally, two reference waves are taken on the same side of the object normal and have mutual angular separation much larger than the angular size of the object.

5.3.2 Sandwich HI

HI records phase difference irrespective of the source of the phase difference. However, when applying HI for studying strains in the object, the main interest is to measure local deformations on the object surface. Depending on the loading conditions, the object very often undergoes rigid body movements (translations and tilts) and these are commonly much greater in magnitude than the local deformations. These rigid body deformations might be too large to be measured by HI, or these may completely mask the local deformations of interest. Therefore, there is a considerable interest in exploring techniques that can compensate for the influence of rigid body movements. Fringe control techniques have been developed but they are applicable only with real-time HI. Sandwich HI offers an attractive alternative for compensating rigid body movements.

As the name suggests, the recording is done on two holographic plates that form a sandwich. A variety of possibilities have been described by Abramson for recording the sandwich holograms and later extracting the information from these sandwiches. The sandwich can be manipulated to introduce controlled path difference and hence can be used to compensate the influence of unwanted tilts and translations. In order to fully utilize the strength of sandwich HI, a reference surface is placed by the side of the actual object. The reference surface should be free from fringes on reconstruction. However, when the fringe pattern on the object is compensated, a linear fringe pattern appears on the reference surface, which is used to determine the tilt given to the sandwich hologram. Alternately, the reference surface is placed so that it is unaffected by loading but is influenced by the bodily tilts and translations. Then, on reconstruction, the reference surface will also carry a fringe pattern, which should be fully compensated in order to obtain the correct deformation map of the object.

5.3.3 Reflection HI

Another method to eliminate the contributions of rigid body movements and tilts to the measured phase difference is to clamp the holographic plate to the object under study as shown in Figure 5.11(a). The photographic plate with its emulsion side facing the object is illuminated by a plane wave at normal incidence. In the first pass through the plate, this wave acts as a reference wave, while the light scattered from the object forms the object wave. Therefore, a reflection hologram is recorded within the emulsion as interference planes between the incident wave and the one scattered by the object. The recording is done on a thick emulsion, say of about 10 μm thickness. Agfa-Gevaert 8E75 plates are suitable for recording with He-Ne lasers.

For interferometric comparison, two exposures with the object loaded in between exposures are made. The hologram is reconstructed with the laser light as shown in Figure 5.11(b). Due to emulsion shrinkage, the reconstruction is possible only at a suitable lower wavelength. Otherwise, the emulsion is swelled to its original thickness by soaking it in an aqueous solution of triethanolamine $(CH_2OHCH_2)_3N$ followed by slow and careful drying in air to observe the fringe pattern at the recording wavelength.

5.3.4 Heterodyne HI

It has been shown that the fringe separation usually corresponds to $\lambda/2$ in path difference. With phase-shifting techniques, deformations as small as $\lambda/30$ can be easily measured. If even smaller deformations are to be monitored, other means are to be explored. Heterodyne HI is one such technique as it is capable of giving resolution of $\lambda/1000$. This is, however, achieved at the expense of a more complicated experimental setup and slower data acquisition.

Figure 5.12 shows an experimental arrangement; it is a two-reference wave geometry. One of the reference waves passes through two acousto-optical

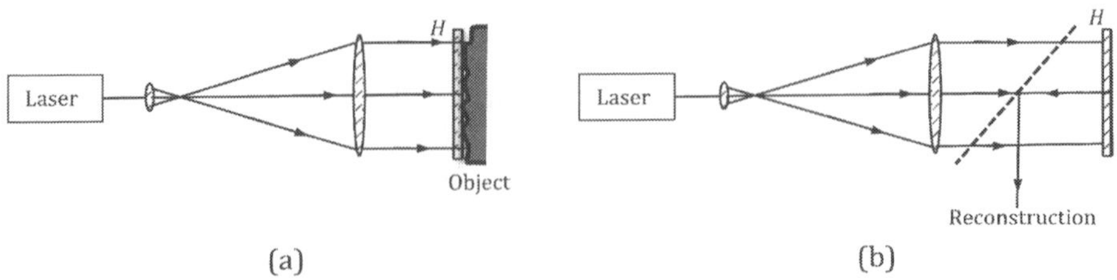

FIGURE 5.11 (a) Recording of a hologram in reflection HI and (b) reconstruction.

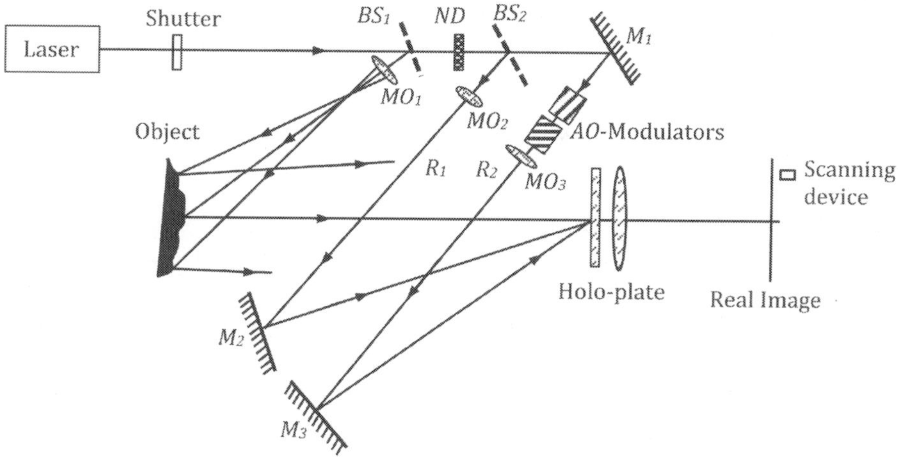

FIGURE 5.12 A schematic for the heterodyne hologram interferometry.

modulators. The first exposure is made with reference beam R_1 (R_2 blocked), and the second exposure, after the object has been loaded, is made with reference wave R_2 (R_1 blocked). Both these exposures are made at the same wavelength with the *AO* modulators turned off. The reconstruction is done with both reference waves; the *AO* modulators are turned on so that the wavelengths of the two reference waves are different. As an example, one *AO* modulator modulates the beam at 40 MHz. It means that the frequency of the light wave diffracted in the first order is shifted by 40 MHz. This then passes through the second *AO* modulator, which modulates it at 40.08 MHz. We now consider the diffraction in –1 diffraction order to cancel the deviation of the beam and obtain a light wave that is frequency-shifted by 80 kHz (= 40.08–40.0 MHz). Therefore, the two reference waves are now frequency-shifted by 80 kHz.

As has been pointed out earlier, there is a multiplicity of images that are produced in two-reference wave HI. We, however, consider only terms of interest in the amplitude transmittance of the hologram, i.e., $\boldsymbol{O}\boldsymbol{R}_1^* + \boldsymbol{O}'\boldsymbol{R}_2^*$, where reference waves $\boldsymbol{R}_1$ and $\boldsymbol{R}_2$ have the same frequency. During reconstruction, the frequency of reference wave $\boldsymbol{R}_2$ has been shifted. Considering the terms of interest, it can be shown that the irradiance distribution in the image of the object is given by

$$I(x,y) = a(x,y) + b(x,y)\cos\left[\delta(x,y) + \Delta\omega t\right] \tag{5.23}$$

where $a(x,y)$, $b(x,y)$ contain various terms of object and reference waves and $\Delta\omega$ is the frequency difference between the two reference waves during reconstruction. The irradiance at any point on the reconstructed image of the object varies sinusoidally with time. The deformation phase to be measured introduces phase shift in this signal. Unfortunately, no information about the deformation phase can be extracted from this signal. However, the phase difference between the deformation phases at any two points can be measured very accurately using an electronic phase-meter by measuring the phase shifts at the two points (x_1,y_1) and (x_2,y_2) as can be seen from

$$\Delta\delta(x_1,y_1;x_2,y_2)=\left[\delta(x_1,y_1)+\Delta\omega t\right]-\left[\delta(x_2,y_2)+\Delta\omega t\right]=\delta(x_1,y_1)-\delta(x_2,y_2) \tag{5.24}$$

There are two ways to perform the temporal heterodyne evaluation. In one approach, the phase difference is measured by keeping one detector fixed to a reference point and scanning the image with another detector. In this way, phase difference modulo 2π with respect to the phase subsisting at the reference point is measured. The interference phase difference can be summed up from point to point to yield the phase distribution along a line or a plane. In the second method, we scan the real image using a pair, triplet, quadruplet, or quintuplet of photodetectors with known separation, and then measure the phase differences. In practice, the detector consists of a photodiode to which a fiber is pigtailed. The other end of the fiber scans the real image.

Heterodyne HI cannot be implemented as a real-time HI because of the extremely high stability requirement of the experimental setup and also of the surrounding environment.

5.4 HOLOGRAPHIC CONTOURING/SHAPE MEASUREMENT

For complete stress analysis, it is necessary to know the shape of the object under study. The following are three techniques to obtain the shape of an object:

1. Change of wavelength between exposures: Dual-wavelength method
2. Change of refractive index of the medium surrounding the object between exposures: Dual-refractive index method.
3. Change of direction of illumination beam between exposures: Dual-illumination method.

5.4.1 Dual-Wavelength Method

The dual-wavelength technique requires that a holographic recording of the object be made on the same plate with two slightly different wavelengths, λ_1 and λ_2. Therefore, the first exposure is made when the object is illuminated with wavelength λ_1, and the reference wave is incident at an angle θ with the normal to the plate. Before making a second exposure with wavelength λ_2, the reference wave is adjusted such that it makes an angle ϕ with the normal to the plate where $k_2 \sin\phi = k_1 \sin\theta$. Here k_1 and k_2 are the propagation vectors of the two waves of wavelength λ_1 and λ_2, respectively. The processed plate, the hologram, is then illuminated with the reference wave at wavelength λ_1 so that the original object wave is reconstructed at λ_1 together with the distorted reconstruction of the object recorded at wavelength λ_2: both waves propagate in the same direction and interfere to produce the fringe pattern in which the fringes correspond to the loci of constant depth.

The irradiance distribution on the reconstructed image of the object is given by

$$I(x_0, y_0, z_0) = I_0(x_0, y_0, z_0)\left[1+\cos(k_2 - k_1)z_0\right] = I_0\left[1+\cos 2\pi \frac{|\lambda_2 - \lambda_1|}{\lambda_1\lambda_2} z_0\right] \quad (5.25)$$

where (x_0, y_0, z_0) are the coordinates of a point on the object and I_0 is the irradiance distribution on the object.

Bright fringes are formed wherever

$$z_0(x_0, y_0) = m\frac{\lambda_1\lambda_2}{|\lambda_2 - \lambda_1|} = m\ \lambda_s \quad (5.26)$$

The wavelength $\lambda_s$is called the synthetic wavelength. The contour interval is $\Delta z_0(x_0, y_0) = \frac{\lambda_1\lambda_2}{|\lambda_2-\lambda_1|}$. Essentially, the method produces interference planes perpendicular to the z-axis and with a separation of $\lambda_1\lambda_2/|\lambda_1 - \lambda_2|$. The sensitivity of the contour fringes can be varied from 1 μm to several millimeters using two suitable wavelengths. Figure 5.13 is an interferogram of an object taken with two wavelengths. The object is a spherical surface and hence the intersection by parallel planes yields circular contour lines.

A variant of this, called two-wavelength holography by some researchers, uses Gabor in-line geometry. It uses a collimated reference beam and applies to both transmission and opaque objects. As an example, consider a lens as an object on which a collimated beam is incident. A hologram is recorded at a plane between the focal plane and the lens by adding a collimated beam axially. The first exposure is done with light of wavelength λ_1, and the second exposure is done with light of wavelength λ_2. This double-exposure hologram when reconstructed with the light of either wavelength will generate fringes caused by aberrations of the lens. The fringes are due to the synthetic wavelength $\lambda_1\lambda_2/|\lambda_1 - \lambda_2|$.

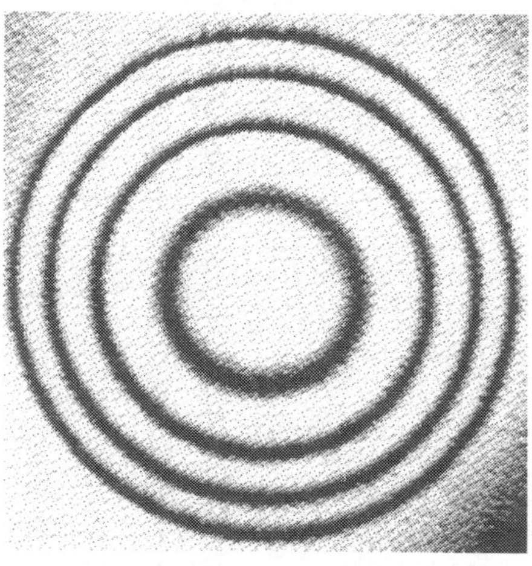

FIGURE 5.13 Fringes over a spherical surface obtained with dual-wavelength HI.

5.4.2 Dual-Refractive Index Method

A similar result is obtained when double-exposure HI is performed by changing the medium surrounding the object in between exposures. The object is placed in a medium of refractive index μ_1 in an enclosure with a transparent window and the first exposure is made. The medium surrounding the object is replaced with a medium of refractive index μ_2 and the second exposure is made on the same plate. One way to change the refractive index of the medium is to replace water used in the first exposure by a mixture of water and alcohol before the second exposure. The hologram, when viewed, displays the object covered with interference planes.

If the vacuum wavelength of the recording light is λ_0, then the wavelength in a medium of refractive index μ_1 is $\lambda_1 = \lambda_0/\mu_1$, and in the other medium of refractive index μ_2 is $\lambda_2 = \lambda_0/\mu_2$. Essentially, this is like a two-wavelength HI, and the contour interval should be given by

$$\Delta z_0\left(x_0, y_0\right) = \frac{\lambda_1\lambda_2}{\left|\lambda_2 - \lambda_1\right|} = \frac{\lambda_0^2 / \mu_1\mu_2}{\left|\dfrac{\lambda_0}{\mu_1} - \dfrac{\lambda_0}{\mu_2}\right|} = \frac{\lambda_0}{\left|\mu_2 - \mu_1\right|} \tag{5.27}$$

There is no need to correct for the angle of the reference wave prior to the second exposure as is done in two-wavelength HI. The contour interval can be varied by a suitable choice of the media surrounding the object.

5.4.3 Dual-Illumination Method

The dual-illumination method is frequently used for contouring due to its simplicity. A collimated beam illuminates the object at an angle θ with the optical axis, and a real-time hologram is recorded in the usual way. On reconstruction, a dark field is observed. Now if the illumination beam is tilted by a small angle, $\Delta\theta$, a system of equi-spaced interference planes intersects the object. These planes run parallel to the bisector of the angle, $\Delta\theta$. The contour interval $\Delta z_0(x_0, y_0)$ is given by

$$\Delta z_0\left(x_0, y_0\right) = \frac{\lambda}{\sin\theta \sin\Delta\theta} \approx \frac{\lambda}{\sin\theta}\frac{1}{\Delta\theta} \tag{5.28}$$

Usually, the interference planes do not intersect the object perpendicular to the line of sight. Large objects are illuminated by a spherical wave from a point source, which is translated laterally to produce the interference surfaces. Double-exposure HI can also be performed for contouring by shifting the point source between exposures.

5.5 DIGITAL HOLOGRAPHY

5.5.1 Recording of Digital Holograms

Instead of using photo-emulsions or other recording media, a CCD is used as a recording medium and the information is stored electronically. The reconstruction is performed numerically on the electronically stored data. Advantages of recording on CCD are that the recording of the hologram is done at video frequency and no

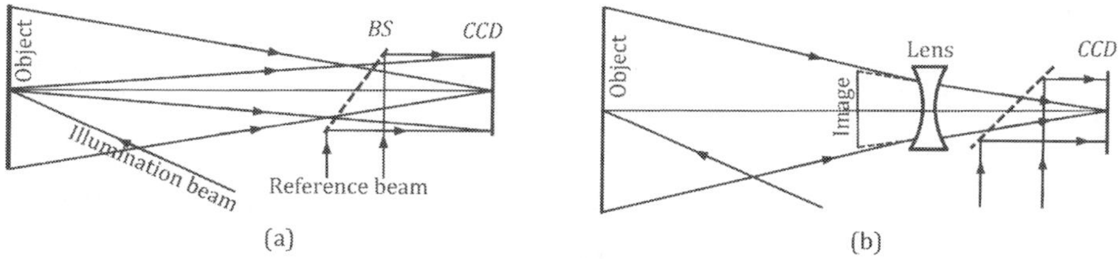

FIGURE 5.14 Schematic arrangements for recording a digital hologram (a) small object and (b) large object.

chemical or physical process for development is necessary. CCD cameras, however, have resolutions of ~100 l/mm, which is at least one order of magnitude smaller than that of photographic emulsions commonly used in holography. For this reason, the maximum admissible angle between the object and the reference waves is ~1°.

A CCD detector contains N × N pixels of sizes Δx and Δy along the x- and y-directions. Assuming that the adjacent pixels have no spacing in-between and no overlap, Δx and Δy are also pixel center distances. In digital holography, a Fresnel hologram is generated on the CCD target by the superposition of an object wave and a reference wave. The hologram is digitized, quantized, and stored in the memory of the image processing system. Figure 5.14(a) shows a schematic for recording a digital hologram. A plane reference wave is assumed for simplicity. The reconstruction is done numerically. However, according to the sampling theorem, only the spatial frequencies less than $\mu_{\max}$ along the x-direction and $\nu_{\max}$ along the y-direction can be reliably reconstructed where $\mu_{\max} = \frac{1}{2\,\Delta x}$ and $\nu_{\max} = \frac{1}{2\,\Delta y}$. This sets the maximum admissible limit on the angle between the object wave and the reference wave. The angle θ in the x–z plane can be written as $\theta = \sin^{-1}\left(\lambda \mu_{\max}\right)$. As a numerical example, consider a CCD detector of 1024 × 1024 pixels, each 6.8 μm × 6.8 μm, and the illumination of the object by a He-Ne laser beam at 633 nm. The maximum admissible angle is 2.67°. Therefore, an object placed at a distance of 1 m from the CCD must be smaller than 4.3 cm along the x-direction. Larger objects either have to be placed further away from the CCD plane or have their apparent sizes reduced by a lens as shown in Figure 5.14(b).

5.5.2 Reconstruction of Digital Holograms

Usually, the reconstruction is done by illuminating the hologram with a plane reference wave. The amplitude in the real image can be expressed as follows:

$$A(\xi,\eta) = \frac{i\boldsymbol{R}}{\lambda\, z}\, e^{-i\frac{\pi}{\lambda\, z}\left(\xi^2+\eta^2\right)} \int_{-\infty}^{\infty}\int_{-\infty}^{\infty} t(x,y)\, e^{-i\frac{\pi}{\lambda\, z}\left(x^2+y^2\right)} e^{\frac{2\pi i}{\lambda z}(x\xi+y\eta)}\, dx\, dy \qquad (5.29)$$

where $\boldsymbol{R}$ is the amplitude of the reference wave. This equation is valid under Fresnel approximation, which is

$$z^3 \gg \frac{\pi}{4\lambda}\left[\left(x-\xi\right)^2 + \left(y-\eta\right)^2\right]^2$$

where z is the distance between the object and the hologram, and hence between the hologram and the real image.

Because the hologram is recorded on a CCD detector, the data are available in digital form, so Equation (5.29) is written as follows:

$$\boldsymbol{A}(m,n) = \frac{\boldsymbol{iR}}{\lambda z} e^{-i\frac{\pi}{\lambda z}(m^2\Delta\xi^2 + n^2\Delta\eta^2)} \sum_{k=0}^{N-1}\sum_{l=0}^{N-1} t(k,l) e^{-i\frac{\pi}{\lambda z}(k^2\Delta x^2 + l^2\Delta y^2)} e^{\frac{2\pi i}{\lambda z}\left(\frac{k\ m\Delta x\Delta\xi}{N} + \frac{l\ n\Delta y\Delta\eta}{N}\right)} \tag{5.30}$$

where $m = 0, 1, 2, \ldots\ldots(N-1)$; $n = 0, 1, 2, \ldots.(N-1)$, and $t(k, l)$ is a matrix of $N \times N$ (pixels in CCD) data that describes the digitally sampled amplitude transmittance of the hologram, Δx, Δy, and $\Delta\xi$, $\Delta\eta$ are the pixel sizes in the hologram plane and the plane of the real image, respectively. The amplitude distribution in the real image is obtained as the inverse Fourier transform of the product consisting of the hologram transmittance $t(k, l)$ multiplied by an exponential factor that contains a quadratic phase factor. The Fourier transform is performed by a Fast Fourier Transform algorithm. The amplitude $\boldsymbol{A}(m,n)$ in the reconstructed image is a complex function and hence the irradiance and the phase are computed at each pixel. The irradiance $I(m, n)$ and the phase $\phi(m, n)$ are determined as follows:

$$I(m,n) = \left|\boldsymbol{A}(m,n)\right|^2 = \left\{Re\left[\boldsymbol{A}(m,n)\right]\right\}^2 + \left\{Im\left[\boldsymbol{A}(m,n)\right]\right\}^2 \tag{5.31a}$$

and

$$\phi(m,n) = \tan^{-1}\frac{Im\left[\boldsymbol{A}(m,n)\right]}{Re\left[\boldsymbol{A}(m,n)\right]} \tag{5.31b}$$

with values of ϕ lying between $-\pi$ and π.

As a consequence of surface roughness, the phase $\phi(m, n)$ varies randomly. In digital holography, only irradiance variation on the object is of interest; hence, several calculations like that of $iR/\lambda z$, phase $\phi(m, n)$ and the exponential term $e^{-i\frac{\pi}{\lambda z}(m^2\Delta\xi^2 + n^2\Delta\eta^2)}$ are not carried out.

5.6 DIGITAL HOLOGRAPHIC INTERFEROMETRY

Comparison of states of the object due to loading can be performed by digital HI in a manner similar to that of double-exposure HI. Two holograms with the amplitude transmittances $t_1(x, y)$ and $t_2(x, y)$ belonging to the two states of the object are digitally recorded and stored. During reconstruction, we have two procedures to obtain the interferogram representing the change in the state of the object. In the first procedure, the transmittances of both the holograms are added pixel by pixel and the Fresnel transform is taken. This will reconstruct the sum of the two amplitudes. When irradiance distribution is calculated, it would exhibit cosine variation characteristics of the interference between the two waves. This interference pattern can be evaluated by any one of the well-known procedures. In the second method, the holograms are individually reconstructed and their phases $\phi_1(m, n)$ and $\phi_2(m, n)$ are calculated pixel-wise. Though $\phi_1(m, n)$ and $\phi_2(m, n)$ are random functions, their difference $\delta(m,n)\{=[\phi_2(m,n)-\phi_1(m,n)]\}$ will be deterministic and will give the phase change due to loading. The phase change $\delta(m, n)$ is calculated as

$$\delta(m, n)=\phi_2(m, n)-\phi_1(m, n) \qquad \phi_2(m, n)\geq\phi_1(m, n) \tag{5.32a}$$

$$=\phi_2(m, n)-\phi_1(m, n)+2\pi \qquad \phi_2(m, n)<\phi_1(m, n) \tag{5.32b}$$

These two approaches are summarized in Table 5.3. Figures 5.15(a) and 5.15(b) show the numerically reconstructed phase of an undeformed and deformed object,

TABLE 5.3
Procedures for Obtaining Digital Holographic Interferograms

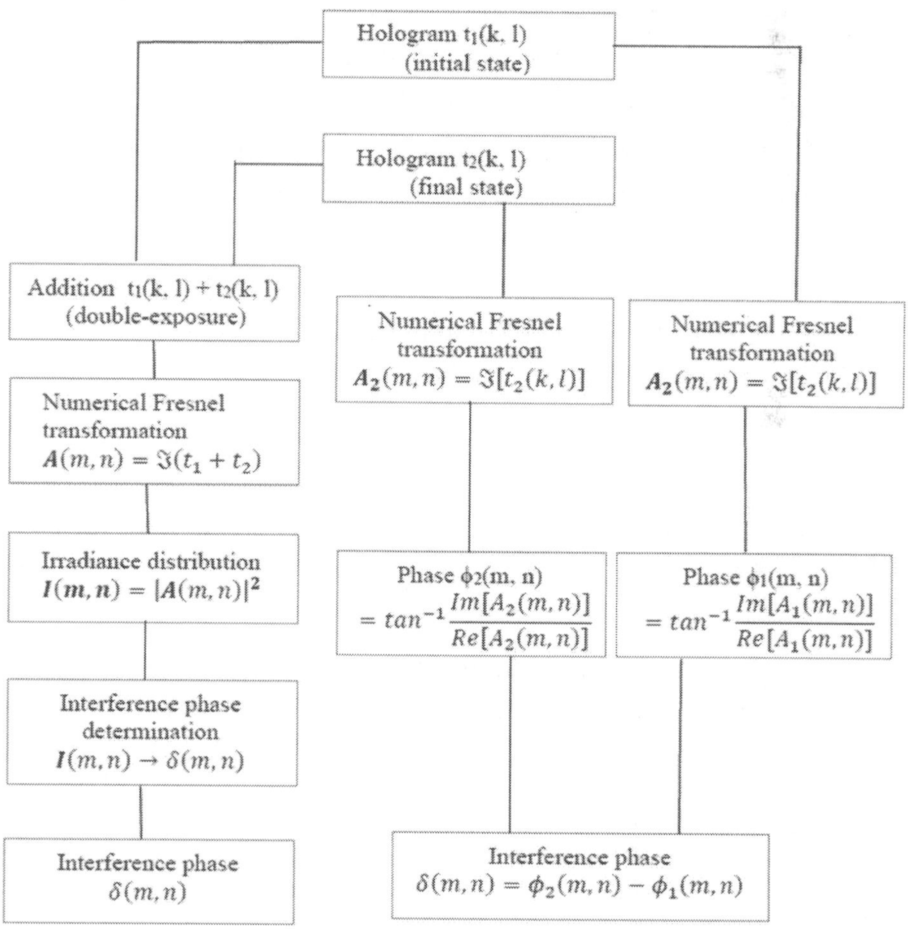

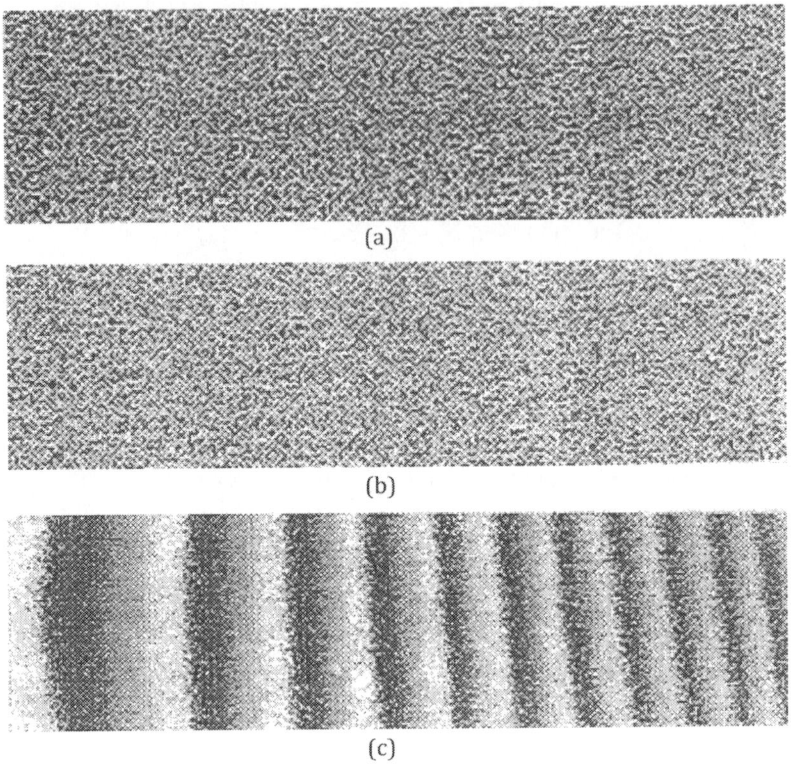

FIGURE 5.15 Numerically reconstructed phase of (a) an undeformed object, (b) deformed object, and (c) phase difference between (a) and (b) producing an interferogram.

respectively, and the difference between these two is shown in Figure 5.15(c) as an interferogram.

Because digital HI can handle only small objects, it has potential applications in the testing and the characterization of micro-systems. Comparison of remote objects and their responses to an external agency can be conveniently done with digital holography.

5.7 FRINGE FORMATION AND MEASUREMENT OF DISPLACEMENT VECTOR

We have mentioned that object deformation causes a phase change between the two waves, at least one of which is derived from the hologram. This phase change is responsible for the formation of the fringe pattern. In this section we address two issues:

1. How is the phase change related to the deformation vector?
2. Where are the fringes really localized?

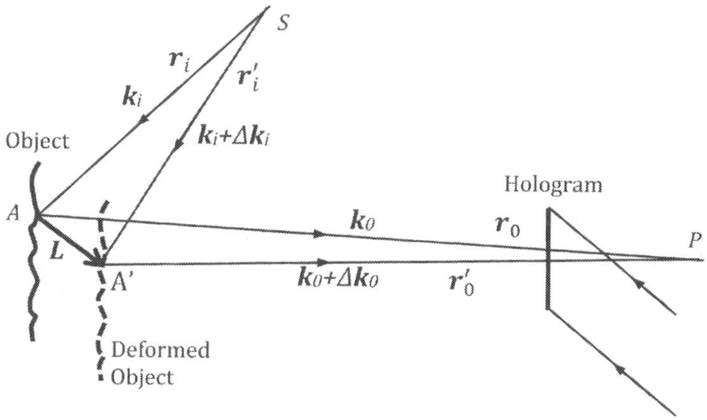

FIGURE 5.16 Calculation of the phase difference.

Let us consider a point A on the object surface. Due to loading, this point moves to a different location A'. The vector distance AA' is the deformation vector $\boldsymbol{L}$. We consider the geometry as shown in Figure 5.16 to calculate the phase difference δ introduced by the deformation. The phase difference δ at the observation point P is given by

$$\delta = \boldsymbol{k}_i \cdot \boldsymbol{r}_i + \boldsymbol{k}_0 \cdot \boldsymbol{r}_0 - (\boldsymbol{k}_i + \Delta\boldsymbol{k}_i) \cdot \boldsymbol{r}_i' - (\boldsymbol{k}_0 + \Delta\boldsymbol{k}_0) \cdot \boldsymbol{r}_0' \tag{5.33}$$

where $\mathbf{r}_i$ and $\mathbf{r}_0$ are the distances from the source point S to the image point P via a point A on the object and $\boldsymbol{r}_i'$ and $\boldsymbol{r}_0'$ are the distances after the object undergoes deformation. The deformation vector is expressed as $\boldsymbol{L} = u\hat{i} + v\hat{j} + w\hat{k}$, with u, v, and w as the components along the x-, y-, and z-directions. We can express $\boldsymbol{r}_i'$ and $\boldsymbol{r}_0'$ as $\boldsymbol{r}_i' = \boldsymbol{r}_i + \boldsymbol{L}$ and $\boldsymbol{r}_0' = \boldsymbol{r}_0 - \boldsymbol{L}$. Substituting for $\boldsymbol{r}_i'$ and $\boldsymbol{r}_0'$, we obtain

$$\delta = \boldsymbol{k}_i \cdot \boldsymbol{r}_i + \boldsymbol{k}_0 \cdot \boldsymbol{r}_0 - (\boldsymbol{k}_i + \Delta\boldsymbol{k}_i) \cdot (\boldsymbol{r}_i + \boldsymbol{L}) - (\boldsymbol{k}_0 + \Delta\boldsymbol{k}_0) \cdot (\boldsymbol{r}_0 - \boldsymbol{L}) \tag{5.34}$$

On simplification, we obtain

$$\delta = (\boldsymbol{k}_0 - \boldsymbol{k}_i) \cdot \boldsymbol{L} = \boldsymbol{k} \cdot \boldsymbol{L} \tag{5.35}$$

Here we have neglected terms $\Delta\boldsymbol{k}_0 \cdot \boldsymbol{L}$ and $\Delta\boldsymbol{k}_i \cdot \boldsymbol{L}$ being second-order terms and have set $\Delta\boldsymbol{k}_i \cdot \boldsymbol{r}_i = 0$ and $\Delta\boldsymbol{k}_0 \cdot \boldsymbol{r}_0 = 0$ as these vectors are orthogonal to each other in an experimental setup. Vector **k** is known as the sensitivity vector. Equation (5.35) is the governing equation for designing the experimental setup and carrying out calculation of deformation vector components.

Equation (5.35) tells us when and how the fringe pattern is formed; it does not tell us where the fringe pattern is actually formed. The situation is quite different from that encountered in classical interferometry. In HI, the fringes are normally not localized on the object surface but on a surface in space. This surface is known as the surface of localization. The knowledge of the location of the surface of localization helps in determining the deformation vector. In fact, the surface of localization depends on several factors, including the deformation vector. Several theories have been presented in the literature. Needless to say, a holographer should know about the fringe localization in detail before he carries out any quantitative evaluation.

5.8 LOADING OF THE OBJECT

Because HI sees only the change in the state of the object, the object must be subjected to an external agency to change its state. This can be performed by any one of the following methods:

1. Mechanical loading
2. Thermal loading
3. Pressure/vacuum loading
4. Vibratory or acoustical loading
5. Impact loading

The most commonly used method is mechanical loading, wherein the object is subjected to either a compressive or a tensile force. If the dynamic response needs to be studied, it is subjected to either a vibratory or an impact loading. In holographic nondestructive testing (HNDT), it is essential to use an appropriate kind of loading. For example, debonds are easily seen when a pressure or a vacuum stressing is applied.

QUESTIONS

5.1 A transmission hologram is recorded in a thick photo-emulsion with a He-Ne laser ($\lambda = 632.8$ nm). The object and the reference beams make an angle of $+45°$ and $-45°$ with the normal to the recording plane. What is the spatial frequency of fringes in the hologram? Take the refractive index of the photo-emulsion as 1.6.

5.2 A reflection hologram is recorded with a He-Ne laser in a thick photo-emulsion. The object beam is incident normally on the hologram plane. The reference beam makes an angle of 60° with the normal to the recording plane. What is the spatial frequency of the fringes in the holograms. Refractive index of the photo-emulsion is 1.6.

5.3 Show that the contrast of fringes in real-time HI is given by

$$\eta = 2\frac{(\beta T R_0^2)(t_0 - \beta T R_0^2)}{(\beta T R_0^2)^2 + (t_0 - \beta T R_0^2)^2}$$

where β is a constant, T is exposure time, t_0 is the DC transmittance of the hologram, and R_0 is the amplitude of the reference wave. Further, how is the contrast related to the amplitude of the reference wave?

5.4 A hologram of a flat circular diaphragm clamped at the edges is recorded using a He-Ne laser. The diaphragm is illuminated by a collimated beam incident at angle of 60° to the normal to the diaphragm. The hologram is processed in situ and the diaphragm is viewed along the normal to its surface (*single-exposure HI*). When the pressure inside the diaphragm is slightly changed, five dark circular fringes are seen on the diaphragm; the center of the fringe pattern is a dark spot. Calculate the deflection at the center of the diaphragm due to pressure change.

5.5 Show that the characteristic function M_T for double-exposure holography is given by

$$M_T = \cos^2(\mathbf{k}.\mathbf{L}/2)$$

where $\boldsymbol{k} = (\boldsymbol{k}_0 - \boldsymbol{k}_i)$ is the sensitivity vector and **L** is the deformation vector.

5.6 Derive the Equation (5.25) that gives the irradiance distribution on the reconstructed image of an object when the two-wavelength method is used for contouring an object.

5.7 Calculate the deflection profile from the interferogram given in Figure 5.8 along the line joining the centers of the two lobes. The diaphragm is clamped at the edges and is illuminated with a He-Ne laser ($\lambda = 632.8$ nm) at an angle of 30° and viewed normally.

6 Speckle Phenomenon, Speckle Photography, and Speckle Interferometry

6.1 SPECKLE PHENOMENON

Illumination of a diffuse object by coherent light produces a grainy structure in space. This grainy light distribution is known as a speckle pattern. It arises due to self-interference of numerous waves from the scattering centers on the surface of the diffuse object as shown in Figure 6.1. The amplitudes and the phases of these scattered waves are random variables. We assume that (1) the amplitude and the phase of each scattered wave are statistically independent variables and are independent of the amplitudes and the phases of all other waves, and (2) the phases of these waves are uniformly distributed between $-\pi$ and π. Such a speckle pattern is a fully developed speckle pattern. The resultant complex amplitude $\boldsymbol{u}(x,y) = u(x,y)\exp(i\phi)$ is given by

$$u(x,y)e^{i\phi} = \frac{1}{\sqrt{N}}\sum_{k=1}^{N} u_k = \frac{1}{\sqrt{N}}\sum_{k=1}^{N} a_k e^{i\phi_k} \tag{6.1}$$

where a_k and ϕ_k are the amplitude and the phase of the wave from the kth scatterer. For such a speckle pattern the resultant complex amplitude of $\boldsymbol{u}(x,y)$ obeys the Gaussian statistics. The irradiance in the speckle pattern is obtained as $I(x,y) = |\boldsymbol{u}(x,y)|^2$.

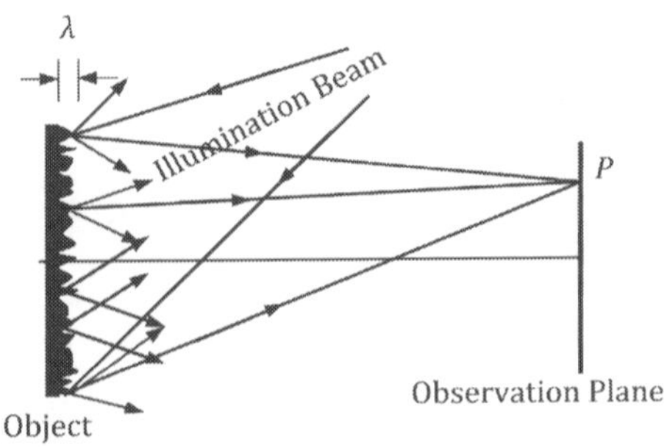

FIGURE 6.1 Formation of a speckle pattern.

DOI: 10.1201/9781003531845-6

The probability that the irradiance at a point in the speckle pattern will have a value I is given by the probability density function $p(I)$, which is expressed as

$$p(I) = \frac{1}{\overline{I}} e^{-I/\overline{I}} \tag{6.2}$$

where $\overline{I}$ is the average irradiance. The probability density function in the speckle pattern follows a negative exponential law. The most probable irradiance value is zero. A measure of the contrast c in the speckle pattern is the ratio $c = \sigma / \overline{I}$ where σ is the standard deviation of the speckle irradiance. The contrast in the fully developed linearly polarized speckle pattern is unity.

6.1.1 Average Speckle Size

The grains or the speckles in the pattern are not well defined in shape and size but have a structure. However, an average size is associated with a speckle. This is considered under the following two cases.

6.1.1.1 Objective Speckle Pattern

The speckle pattern formed in space due to free space propagation, when a diffuse object is illuminated by a coherent wave, is termed as an objective speckle pattern. The speckle size, s_{ob}, in an objective speckle pattern is given by

$$s_{ob} = \frac{\lambda z}{D} \tag{6.3}$$

where D is the size of the illuminated area of the object and z is the distance between the object and the observation plane as shown in Figure 6.2(a). The size is governed by the interference between the waves from the extreme scattering points on the object. The relationship is the same as that expected from the Young's double-slit experiment; the slits are at the extreme positions in the illuminated area. The speckle size increases linearly with the separation between the object and the observation plane.

6.1.1.2 Subjective Speckle Pattern

The speckle pattern in the image of a diffuse object illuminated with a coherent wave is called the subjective speckle pattern. This arises due to the interference of waves

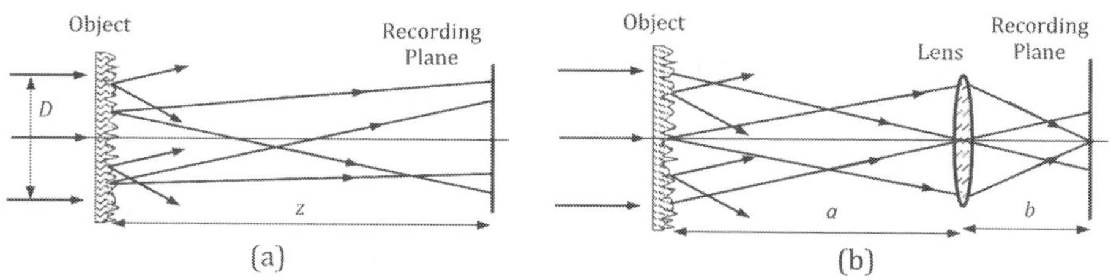

FIGURE 6.2 (a) Objective speckle pattern and (b) subjective speckle pattern.

from several scattering centers in the resolution element (area) of the imaging lens: the randomly dephased impulse response functions are added in the image of this resolution area, resulting in a speckle. Therefore, the speckle size is governed by the well-known Airy formula. The diameter of the Airy disk is approximately given by

$$s_{sub} \simeq \frac{\lambda\, b}{D'} \tag{6.4}$$

where s_{sub} is the subjective speckle size, D' is the diameter of the lens, and b is the image distance as seen in Figure 6.2(b). Here again, the speckle size is determined by the maximum aperture of the lens. The objective speckle pattern just in front of the lens is transmitted through it and appears on the other side of the lens. The speckles at the periphery of this pattern determine the speckle size at the image plane. By introducing the *f*-number $F\#$ $(= f/D')$ of the lens, the average speckle size is expressed as follows:

$$s_{sub} = (1+m)\lambda F\# \tag{6.5}$$

Here we have introduced magnification m $\{= -b/a = -(b-f)/f\}$ of the lens, a is the object distance, and f is the focal length of the lens. It is thus seen that the speckle size can be controlled by (1) the magnification m and (2) the $F\#$ of the lens. The control of speckle size by $F\#$ is often used in speckle metrology to match the speckle size with the pixel size of the charge-coupled device (CCD) array detector.

The speckle patterns can be added either on the amplitude basis or the irradiance basis. An example of the addition of speckle patterns on the amplitude basis is in shear speckle interferometry where the two speckle patterns are shifted and then superposed. In such a speckle pattern, the statistics of the resultant speckle pattern remains unchanged. However, when the speckle patterns are added on the irradiance basis, for example, two speckle records are made on the same plate, the speckle statistics is completely modified and is governed by the correlation coefficient.

In the explanation of a fully developed linearly polarized speckle pattern, it is stressed that the phases are uniformly distributed between $-\pi$ and π and there are a large number of scatterers participating in the speckle formation. This results in a unit contrast speckle pattern. However, if the surface is smoother than necessary to satisfy this condition, the speckle contrast decreases. The speckle contrast depends on the surface roughness and the coherence of light. In fact, the speckle contrast has been employed to measure the surface roughness over a large range using both monochromatic and polychromatic illuminations.

6.1.2 Relation between Object Displacement and Speckle Shift

The speckle phenomenon is used to measure the change in the object either brought about by bodily shift/tilt or by deformation due to an application of external forces. It is, therefore, instructive to know how the speckle pattern changes with changes in the object or its position.

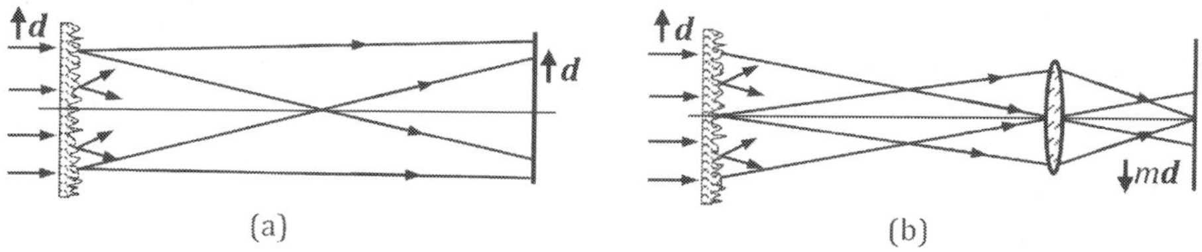

FIGURE 6.3 In-plane displacement (a) objective speckle pattern and (b) subjective speckle pattern.

6.1.2.1 In-Plane Displacement

Let us assume a translucent object, which is translated by a distance d in its own plane: the objective speckle pattern also translates by the same magnitude in the same direction as shown in Figure 6.3(a). However, the structure of the speckle pattern begins to change (i.e., decorrelation sets in) when some of the scattering centers go out of the illumination beam. For the subjective speckle pattern, the speckle motion is in the direction opposite to that of the object and its magnitude is md, where m is the magnification as seen in Figure 6.3(b).

6.1.2.2 Out-of-Plane Displacement

Let us consider that an objective speckle is formed at a position $P(r,0)$ as shown in Figure 6.4(a). This speckle is due to the superposition of all the waves from the object. When the object translates axially by a small distance ε, all these waves are phase-shifted nearly by the same amount. If this phase shift is a multiple of 2π, then a similar state of speckle will exist at point $P'(r - \Delta r, 0)$; that is, the speckle would have radially shifted by Δr. This gives

$$|\Delta r| = \varepsilon \frac{r}{z} \tag{6.6}$$

A similar situation prevails for a subjective speckle pattern as shown in Figure 6.4(b). We therefore obtain

$$|\Delta r| = \varepsilon \frac{r}{a} \tag{6.7}$$

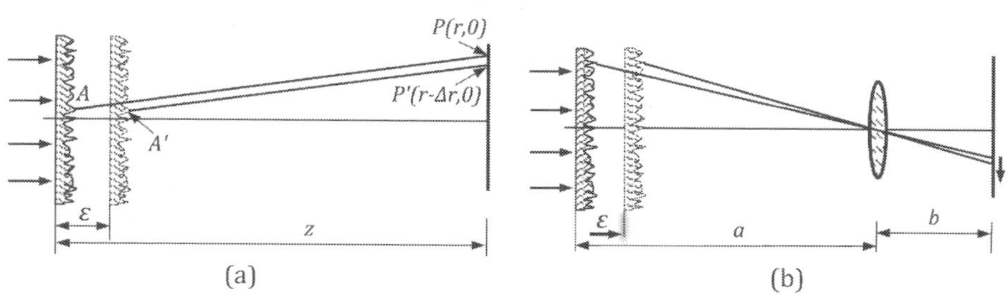

FIGURE 6.4 Out-of-plane displacement. (a) Objective speckle pattern and (b) subjective speckle pattern.

where a is the object distance. Thus, for axial translation of the object, the speckle pattern shifts radially; it either expands or contracts depending on the direction of the shift of the object. It may be noticed that the radial movement of the speckle pattern requires rather large out-of-plane displacement of the object due to r/z or r/a factor, which is very small.

6.1.2.3 Tilt of the Object

Let the object be illuminated in the direction making an angle α with the z-axis, which is also in the direction of the local surface normal, and the speckle pattern is observed at a point P in the direction making an angle β as shown in Figure 6.5(a). When the object is tilted by a small angle, $\Delta\phi$, the speckle pattern shifts to a new position that is angularly shifted by $\Delta\psi$, where $\Delta\psi = (1+\cos\alpha / \cos\beta)\Delta\phi$. Only for the case of normal illumination and observation directions (angles α and β are very small) is the angular shift of the speckle pattern twice the tilt of the object. The situation is quite different when we consider the subjective speckle pattern. In fact, there is no shift at the image plane due to the tilt of the object. Between the image plane and the focal plane (or FT plane), the speckle shift is due to both in-plane motion and tilt contributions. At the FT plane, the speckle pattern does not shift due to in-plane motion but shifts only due to tilt: the shift Δx_f is given by

$$\Delta x_f = f\ \Delta\phi$$

where $\Delta\phi$ is the tilt given to the object as seen in Figure 6.5(b). When the object is illuminated by a divergent wave, one can find a plane that is sensitive only to the in-plane motion and another plane sensitive to tilt alone.

It is, thus, seen that a speckle pattern undergoes changes when the object is either translated or tilted. However, for deformation measurement, we are interested in measuring changes/shifts at various points on the objects; hence, we employ only a subjective speckle pattern. This way, a correspondence between the object and the image points is maintained. In other words, the local changes in the object cause changes locally in the speckle pattern rather than over the whole speckle pattern plane. The changes in a speckle pattern due to the object deformation are (1) positional changes accompanied by irradiance changes and decorrelation and (2) phase

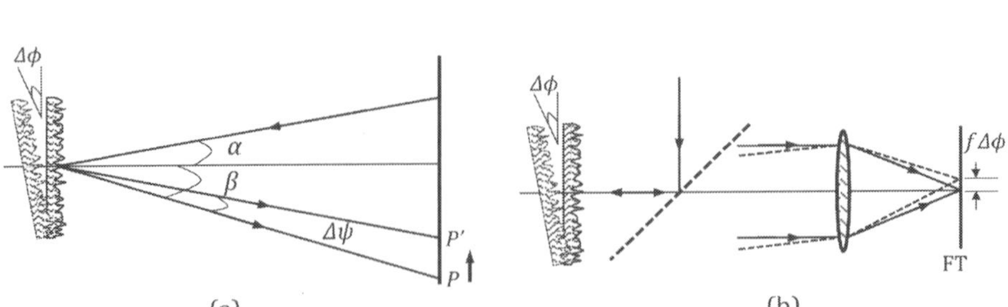

FIGURE 6.5 Motion of the speckle due to the tilt of the object (a) at a far plane (objective speckle pattern) and (b) at the FT plane.

changes that are made visible by adding a specular or a speckled reference wave at the image plane. Often, both these changes do occur simultaneously but one may dominate the other.

The speckle methods used for deformation measurement and vibration analysis can be placed under the following four categories:

1. Speckle photography
2. Speckle interferometry
3. Speckle shear interferometry
4. Electronic/digital speckle pattern interferometry (ESPI/DSPI) and shear ESPI/DSPI (shearography).

Methods 1 to 3 employ photographic medium, thermoplastics, photorefractive crystals, etc., for recording, and method 4 employs electronic detection. Correlation, as in speckle photography, also can be accomplished digitally.

6.2 SPECKLE PHOTOGRAPHY

In speckle photography the object is illuminated either obliquely or normally with the laser radiation as seen in Figure 6.6(a). It is imaged in its initial state on a photographic plate capable of resolving the speckles in the pattern. After the object is subjected to external forces, that is, after deformation, it is imaged again on the same plate. Deformation will result in translation of the speckles in the image of the object locally. After development, this double-exposure record is the specklegram. The specklegram, thus, contains two speckled images of the object: in one of the images the speckles are locally translated by $\boldsymbol{d}$. We need to find $\boldsymbol{d}$ at various locations on the plate and then generate the deformation map. It was pointed out earlier that speckle photography has poor sensitivity for the axial (out-of-plane) displacement component. Hence, speckle photography is used mostly to measure in-plane displacements and in-plane vibration amplitudes.

Let us first examine the specklegram (negative film or plate) realized by making a single exposure. The irradiance recorded is given by $I(x,y) = |\boldsymbol{u}(x,y)|^2$, where $\boldsymbol{u}(x,y)$ is the complex resultant amplitude. Assuming linear recording, the amplitude transmittance $t(x,y)$ of this negative (specklegram) is expressed by

$$t(x,y) = t_0 - \beta T\ I(x,y) \tag{6.8}$$

where t_0 is the bias transmittance, β is a constant, and T is the exposure time. Because the speckle pattern consists of a grainy structure, each grain is identified by a δ-function, and the irradiance $I(x,y)$ could also be expressed as follows:

$$I(x,y) = \iint I(x',y')\,\delta(x-x',y-y')\,dx'\,dy' \tag{6.9}$$

where $I(x',y')$ is the irradiance of a speckle at (x',y'). When this specklegram is placed in the setup shown in Figure 6.6(b), and illuminated by a collimated beam of

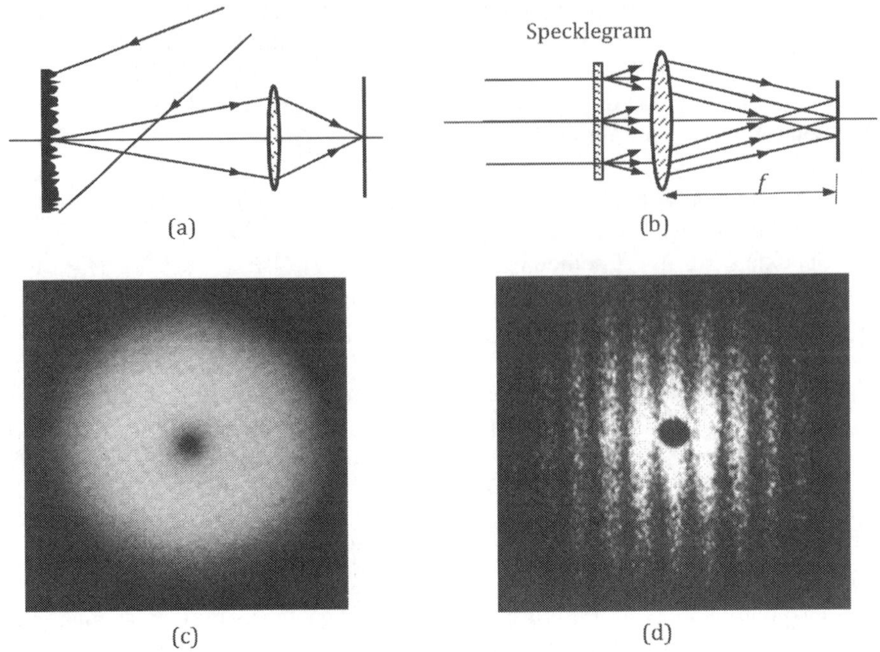

FIGURE 6.6 (a) Speckle photography setup to measure bodily in-plane displacement, (b) setup to observe diffracted field from the specklegram at the focal plane of a lens, (c) halo from a single-exposure specklegram, and (d) fringes in the halo from a double-exposure specklegram.

light, the amplitude transmitted is given by $\boldsymbol{u}_0(x,y)\, t(x,y)$, where $\boldsymbol{u}_0(x,y)$ is the amplitude of the illuminating plane wave. The specklegram will diffract light over a reasonably large cone depending on the speckle size.

We collect this diffracted light with a lens and make an observation at its back focal plane. Essentially, we are taking the Fourier transform of the amplitude transmittance $t(x,y)$ of the speckled image record. The amplitude at the FT plane, assuming unit amplitude illumination wave, is given by

$$\boldsymbol{U}(f_x, f_y) = \iint t(x,y)\, e^{2\pi i(f_x x + f_y y)}\, dx\, dy$$

$$\begin{aligned} \boldsymbol{U}(f_x, f_y) &= \iint t_0 e^{2\pi i(f_x x + f_y y)}\, dx\, dy - \beta T \iint I(x,y)\, e^{2\pi i(f_x x + f_y y)}\, dx\, dy \\ &= t_0 \delta(f_x, f_y) - \beta T \Im\left[I(x,y)\right] \end{aligned} \tag{6.10}$$

where $\Im[\ ]$ signifies Fourier transform. This definition of Fourier transform is consistent with the counterclockwise rotation of the phaser or the wave description in the form of $e^{i(\omega t - \boldsymbol{k}\cdot\boldsymbol{r})}$. The irradiance distribution at the FT plane, $\left|\boldsymbol{U}(f_x, f_y)\right|^2$, consists of a strong central peak and a light distribution around, which is called the halo. This is shown in Figure 6.6(c); the central portion is blocked while recording the halo. The halo contains a range of spatial frequencies between 0 and $s_{sub} / \lambda f$, where s_{sub} is the average size in the subjective speckle pattern. The diameter of the halo is $(2f\lambda/s_{sub})$. The halo distribution is given by the autocorrelation of the aperture function of the imaging lens. A physical insight into the formation of the halo can be obtained if we consider a specklegram as having a large number of sinusoidal gratings of continuously varying pitch and random orientations. When the specklegram is illuminated, these gratings diffract the beam in various directions forming the halo at the back focal plane of a lens.

Let us now consider a double-exposure specklegram. In the first exposure, an irradiance distribution $I_1(x,y)$ is recorded. The object is then deformed and the second exposure $I_2(x,y)$ is recorded on the same plate. Due to the deformation, the speckle pattern shifts locally. Therefore, the irradiance distributions $I_1(x,y)$ and $I_2(x,y)$ can be expressed as follows:

$$I_1(x,y) = \iint I(x',y')\delta(x-x',y-y')\,dx'dy' \tag{6.11a}$$

$$I_2(x,y) = \iint I(x',y')\delta\left(x+d_x-x',y+d_y-y'\right)\,dx'dy' \tag{6.11b}$$

where d_x and d_y are the components of $\boldsymbol{d}$ along the x- and y-directions, respectively. The total irradiance recorded is

$$I_t(x,y) = I_1(x,y) + I_2(x,y)$$

$$I_t(x,y) = \iint I(x',y')\left[\delta(x-x',y-y') + \delta\left(x+d_x-x',y+d_y-y'\right)\right]dx'dy' \tag{6.12}$$

When a collimated beam illuminates this double-exposure specklegram, then one obtains, at the focal plane of the lens, a central order and the superposition of halos belonging to the initial and the final states of the object. Mathematically, the amplitude transmittance of the double-exposure specklegram is

$$t(x,y) = t_0 - \beta T\left\{I_1(x,y) + I_2(x,y)\right\} \tag{6.13}$$

The amplitude at the FT plane is given by

$$\Im\left[t(x,y)\right] = t_0\delta\left(f_x,f_y\right) - \beta T\Im\left[I(x,y)\right] \tag{6.14}$$

For simplicity, we now confine ourselves to one dimension, and hence write the total irradiance as follows:

$$I_t(x) = \iint I(x')\left[\delta(x-x') + \delta\left(x+d_x-x'\right)\right]dx' \tag{6.15}$$

which is a convolution of $I(x)$ with $[\delta(x) + \delta(x + d_x)]$. Therefore

$$\Im[t_t(x)] = \Im[I_1(x) + I_2(x)] = \Im[I(x,y)]\ \Im[\delta(x) + \delta(x + d_x)] \quad (6.16)$$

The Fourier transform of $\delta(x)$ is a plane wave propagating on-axis, while that of $\delta(x + d_x)$ is also a plane wave propagating inclined with the axis, that is,

$$\Im[\delta(x) + \delta(x + d_x)] = \int \delta(x)e^{2\pi i f_x x}\,dx + \int \delta(x + d_x)e^{2\pi i f_x x}\,dx = c + ce^{2\pi i f_x d_x} \quad (6.17)$$

where c is a constant, being the Fourier transform of a δ-function. The spatial frequency f_x is defined as $f_x = x_f/f\lambda$, with x_f is the x-coordinate on the FT plane and f is the focal length of the lens. Suppressing the central order, we can write for the irradiance distribution in the FT plane (in the halo) as follows:

$$\mathcal{I}(f_x) \propto 2\beta^2 T^2 c^2 [1 + \cos(2\pi f_x d_x)] = \mathcal{I}_0(f_x)\cos^2(\pi f_x d_x) \quad (6.18)$$

It is therefore seen that the halo, given by $\mathcal{I}_0(f_x)$, is modulated by the $\cos^2(\pi f_x d_x)$ term. In other words, a fringe pattern appears in the halo. Figure 6.6(d) shows the fringe pattern in the halo. This fringe pattern is similar to the Young's double-slit fringe pattern. Therefore, the fringes are termed Young's fringes. In the $\cos^2(\pi f_x d_x)$ term, there are two variables, namely f_x, the coordinate along the x-axis, and d_x, which is a function of local coordinates on the specklegram if the deformation is not constant. For constant d_x, the interpretation is simple and straightforward: the halo distribution gets modulated by the co-sinusoidal fringes, whose spacing is $\bar{x}_f = \frac{f\lambda}{d_x}$. Therefore, the magnitude of d_x may be obtained from the fringe-width measurement. The direction of displacement d_x is always along the normal to the fringes. However, if d_x is not constant, then each value of d_x will produce its own fringe pattern, which on superposition may completely wash out this irradiance variation. In such a situation, the specklegram is interrogated with a narrow or unexpanded laser beam to extract the displacement information from each region of interrogation. The displacement in the region of interrogation is assumed to be constant.

6.2.1 Methods of Evaluation

A double-exposure specklegram contains two shifted speckle patterns and this shift is to be determined at a number of locations on the specklegram in order to generate the deformation map. It has been earlier remarked that the Young's type fringes are formed when a narrow beam illuminates the specklegram. From this fringe pattern, one obtains the direction and the magnitude of the displacement at a point. The process of extracting the information from a specklegram is called filtering. The filtering is done both at the plane of the specklegram and at its Fourier transform plane. These methods are called the (1) point-wise filtering method and the (2) whole-field filtering method.

One more method of filtering, usually applicable to an out-of-plane displacement measurement, is known as the Fourier filtering method. It can also be used in other cases; the fringes are generally localized on the specklegram.

6.2.1.1 Point-Wise Filtering Method

It was mentioned earlier that when the displacement of the speckles is nonuniform, no fringe pattern may be formed at the Fourier plane when the whole specklegram is illuminated by a collimated beam. However, it may be safe to assume that the speckle movement over a very small region of the image (specklegram) is constant. Therefore, if the double-exposure specklegram is illuminated by a narrow (unexpanded) beam and the observation is made at a plane sufficiently far away (far field) as shown in Figure 6.7(a), a system of Young's fringes will be formed. The fringes always run perpendicular to the direction of displacement and the fringe width $\overline{p}$ is inversely proportional to the displacement, i.e., $\overline{p} = \lambda z / |\boldsymbol{d}|$. Thus, both the direction and the magnitude of the displacement at each interrogation region on the specklegram are obtained. The sign ambiguity is still not resolved. The sign ambiguity can be resolved by giving a linear known displacement to the photographic plate before the second exposure is made. By obtaining the magnitude and the direction of the displacement at a large number of points on the specklegram, the displacement map on the specklegram is generated, which is then translated to the object surface through the magnification of the imaging system.

The contrast of the Young's fringes is influenced by several factors: the most important ones are nonuniform displacement and missing speckle pairs in the region of illumination. Further, it may be kept in mind that the halo distribution is nonuniform and background irradiance may not be constant. Therefore, Young's fringes do not have unit contrast and the positions of the maxima and minima of the fringes are shifted. This shift introduces errors in the calculation of the displacement. Methods have been developed to correct for this.

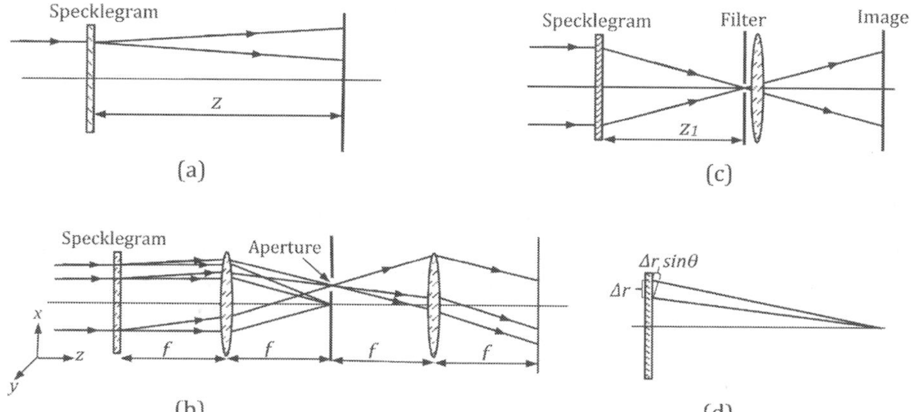

FIGURE 6.7 (a) Point-wise filtering arrangement. (b) Whole-field filtering arrangement. (c) Fourier filtering arrangement. (d) Path difference from a speckle pair.

6.2.1.2 Whole-Field Filtering Method

We consider an arrangement shown in Figure 6.7(b). This is a $4f$ configuration. The filtering is carried out at the FT plane by an aperture. The image of the specklegram is formed at unit magnification on the output plane by the light allowed through the filtering aperture. The image contains the fringes, which represent constant in-plane displacements in the direction of the filtering aperture. To understand the working of the whole-field filtering technique, let us place an aperture of adequate size at a position x_f along the x-direction. All the identical pair-scatterers (speckles) on the specklegram will diffract light in phase in the direction of the filtering aperture if their separation is such that the waves diffracted by these pair-scatterers have a path difference of integral multiples of λ, that is,

$$d_x \sin\theta = n\lambda : n = 0, \pm 1, \pm 2, \pm 3, \ldots \tag{6.19}$$

where d_x is the x-component of the displacement vector and $\sin\theta = x_f/f$. Thus, we obtain

$$d_x = n\lambda f / x_f \tag{6.20a}$$

These areas, therefore, appear bright in the image, forming bright fringes, which are loci of constant d_x. Similarly, when the filtering aperture is placed at y_f along the y-axis, we obtain d_y as follows:

$$d_y = n'\lambda f / y_f \tag{6.20b}$$

These fringes are loci of constant d_y. In other words, the fringes represent the contours of constant in-plane displacement components that are separated by incremental displacements Δd_x and Δd_y, where

$$\Delta d_x = \lambda f / x_f \tag{6.21}$$

and

$$\Delta d_y = \lambda f / y_f \tag{6.22}$$

In the whole-field filtering method the fringes are formed over the whole object surface. To obtain d_x and d_y, we need to know n and n', and hence regions on the object where no displacement has taken place. It is also seen that the sensitivity of the method is variable and is maximum when the filtering aperture is placed at the periphery of the diffraction halo; the increase in sensitivity follows the decrease in the available light for image formation.

6.2.1.3 Fourier Filtering Method: Measurement of Out-of-Plane Displacement

It has been shown that the longitudinal or the axial displacement ε of the object results in a radial displacement of the speckle pattern, i.e., $|\Delta r| = \varepsilon\, r / z$. The displacement magnitude ε is obtained by the Fourier filtering method. The specklegram is illuminated by a collimated beam as shown in Figure 6.7(c) and an aperture is placed on the optical axis for filtering. A lens placed just behind the aperture images the specklegram at the observation plane. All those point pairs (identical scatterers) will diffract the incident light in phase at the filtering plane if $\Delta r \sin\theta = n\lambda$, where $\sin\theta = r / z_1$ and λ is the wavelength of light used for filtering as shown in Figure 6.7(d). Thus

$$\Delta r \sin\theta = (\varepsilon\, r / z)(r / z_1) = n\lambda$$

or

$$(\varepsilon\, r^2 / z\, z_1) = n\,\lambda \tag{6.23}$$

This indicates that a circular fringe pattern is observed at the observation plane. The displacement ε is obtained by measuring the radii of circular fringes of different orders as follows:

$$\varepsilon = \frac{n'\,\lambda\, z\, z_1}{\left(r_{n+n'}^2 - r_n^2\right)} \tag{6.24}$$

where $r_{n+n'}$ and r_n are the radii of the $(n + n')$th- and nth-order circular fringes, respectively.

6.2.2 Speckle Photography with Vibrating Objects: In-Plane Vibration

Let us consider an object that is executing an in-plane vibration of amplitude $A(x,y)$. The object is imaged on the photographic plate. Due to in-plane vibration, a speckle stretches to a line of length $2A(x,y)\,m$, where m is the magnification. Because a sinusoidally vibrating object spends more time at the positions of maximum displacement, the speckle line has a nonuniform brightness distribution. However, if the recording were done on a high-contrast photographic plate, the speckle line would have a uniform density after development. We therefore assume that the speckles are uniformly elongated to a length $2A(x,y)\,m$. When such a specklegram (time-averaged recording) is point-wise filtered, the halo distribution gets modulated by $\text{sinc}^2\left[2f_x mA(x,y)\right]$. The zeros of this function occur wherever

$$2f_x\, mA = n \text{ for } n = 0, \pm1, \pm2, \pm3, \ldots$$

or

$$A(x,y) = n\,\lambda\, z / 2\, m\, x_{fn} \tag{6.25}$$

where x_{fn} is the position of the nth fringe. From this expression, the amplitude of vibration at any location on the specklegram can be found.

In fact, when an object vibrates, the speckles in the regions, which are moving, smear out resulting in the reduction of contrast. The nodal regions where there is no movement have a unit contrast. Therefore, one observes a low-contrast vibration mode pattern with the nodal lines appearing dark. A visual speckle interferometer was proposed by Archbold et al. and later modified by Stetson. This interferometer uses a reference beam to visualize out-of-plane vibration amplitudes.

6.2.3 White Light Speckle Photography

The object is illuminated with white light or incoherent light. Most objects are good candidates for white light speckle photography as the surface structure is quite adequate for this. However, this property is enhanced by coating the surface with a retro-reflective paint. The images of the embedded glass balls in retro-reflective paint act as speckles. Deformation studies can, therefore, be done using white light speckles based on principles similar to those applied to laser speckles.

6.2.4 Sensitivity of Speckle Photography

The halo size is governed by the speckle size which, in turn, is governed by the *F#* number of the imaging lens. For the speckle displacement to be measurable, at least one fringe should be formed within a halo. In other words, the fringe width must be equal to or less than the halo diameter. Indeed, one fringe will be formed within the halo when the speckle shifts by an amount equal to its average size. This probably is the lower limit of the displacement that speckle photography can sense. The upper limit is also set by the speckle size. When the fringe width becomes equal to the speckle size, the fringes will not be discernible. In practice, we choose a fringe width of approximately 10 times the speckle size for it to be discernible. This, therefore, sets an upper limit for speckle displacement measurement. For whole-field filtering, similar limits apply. The quality of fringes in whole-field filtering is considerably poorer than in point-wise filtering.

One of the serious drawbacks of speckle photography is that the positions of the maxima of the fringes in the fringe pattern get shifted due to nonuniform halo distribution. The same is also true for the minima positions when background irradiance is present. Errors are introduced by this effect on the measured values of the displacements. Methods are available to correct for these errors.

6.2.5 Speckle Shear Photography

It is obvious that speckle photography is a tool to essentially measure in-plane displacement or in-plane component of deformation as it has very low sensitivity to the out-of-plane component. It is also used to measure bodily tilts. A stress analyst is often interested in strain rather than displacement. Strain is obtained from the displacement data by numerical differentiation, which is error prone. However, strain can be obtained by optical differentiation, that is, by obtaining displacement values at two closely separated points. With a double-exposure specklegram, this can be achieved by interrogating it with two closely spaced parallel narrow beams. Each beam produces Young's type fringes in the diffraction halo and the superposition of the two patterns, from each illumination beam, will generate a moiré pattern, from which strain can be calculated. Alternatively, two states of the object can be recorded on two separate films/plates. During filtering, one plate is displaced with respect to the other to introduce shear. Again, a moiré pattern is formed from which strain is calculated. These techniques provide a variable shear by varying the beam separation or specklegram displacement.

When two narrow beams are used for illumination, the diffraction halos are spatially displaced. For low angle diffraction, the overlap region may be very small to provide a meaningful moiré pattern. On the other hand, when two double-exposure films/plates are used, the underlying assumption that the object has been identically loaded for each double-exposure record may be difficult to realize. It is therefore desirable to record a single double-exposure shear specklegram.

6.2.6 Particle Image Velocimetry (PIV)

Speckle photography of seeded flows carried out with pulsed lasers or with scanning laser beams provides information about the flow (spatial variation of the velocity of flow, in fact, the velocity of seeds). The technique is known as particle image velocimetry (PIV). First, a record is made with a short duration laser pulse so that the motion is frozen and then the second exposure is made a short time later. The seeds (particles) would have moved during this time to new locations depending on their velocities. The double-exposure record is point-wise filtered to generate the velocity map.

6.2.7 Digital Image Correlation (DIC)

Digital image correlation (DIC) is a powerful technique typically used in experimental mechanics that allows noncontact measurement of full-field displacement and strain fields on the surface of structures. DIC operates by comparing digital images of an object taken at different times or under different loading conditions with respect to a reference state. DIC works principally by identifying small subsets of pixels within an image and tracks the said pixels movement over time using image correlation algorithms. Strain fields can then be calculated through the relative displacement of neighboring subsets.

Over the years, DIC has undergone several improvements in terms of accuracy, speed, and robustness. For example, advances in camera technology have enabled image capturing at higher resolutions and frame rates, while developments in computer processing power have allowed for faster and more complex algorithms to be implemented. The use of random irradiance pattern (so-called "speckle patterns") has become more widespread, which has led to improved tracking algorithms and more accurate measurements of displacement and strain fields. In addition, the required hardware became cheaper and readily available, thus DIC became an increasingly widespread measurement technique.

The advantages of DIC are that it is a noncontact and noninvasive technique due to its optical imaging and that it has no temporal resolution limitation. The temporal resolution is solely determined by the sampling rate of the cameras used. This makes DIC suitable for time-lapse recordings of relatively slow movements that take place over hours to days. The spatial resolution of the method is highly dependent on the cameras and lenses used, as well as on the contrast of the surface examined, allowing characterization from nanoscales to macroscales.

There are two main types of DIC, namely 2D-DIC and 3D-DIC (also known as stereo-DIC). 2D-DIC is used when the surface for analysis has little or no curvature and is predominantly deformed in a plane. 3D-DIC is used for objects with a nonplanar surfaces and/or to analyze out-of-plane deformation. 2D-DIC uses images from a single camera whose sensor is aligned parallel to the sample surface and does not require calibration of the measurement field. In contrast, 3D-DIC analysis is based on stereo images captured by two synchronized cameras with identical sensors. These sensors are aligned at a stereo angle to the object of approximately 20 to 30°, thus capturing the surface from different viewing angles. Here, the measurement volume must be calibrated before the actual experiment. In both systems, the resolution of the camera is a crucial determinant of the resulting spatial resolution of the measurements. It should be noted that especially for 3D-DIC, the depth of focus is an important factor because data processing requires completely sharp images. This requirement impedes, to a certain extent, the analysis of small structures, which require high optical magnification, and/or spatially very complex curved structures.

As discussed previously, the choice of camera system depends on, among other factors, the nature of the sample surface (2D-DIC for flat surfaces and 3D-DIC for curved/structured surfaces). However, not all surfaces are suitable for stereo camera imaging. For 3D-DIC analysis, all areas of the surface must always be visible in both cameras, otherwise this will cause gaps in the subsequent surface detection. Therefore, if the sample is highly curved or has pronounced edges or protrusions, the surface or parts of the surface cannot be tracked and analyzed. Once a suitable sample has been identified, preparing the surface is the next crucial step in DIC experiments. To obtain reliable and accurate results, the surface must contain a random speckle pattern to calculate the correlation.

6.3 SPECKLE INTERFEROMETRY

The speckle phenomenon itself is essentially an interference phenomenon. However, when a reference beam is added to the speckle pattern to code its phase, the technique is then termed as speckle interferometry. Speckle interferometry was first applied to measure in-plane displacement by Leendertz. The sensitivity of the measurement could be increased over that of speckle photography, which was limited by the speckle size. The basic theory was borrowed from holographic interferometry (HI) as the phase difference δ introduced by deformation is governed by the same equation, i.e., $\delta = (\boldsymbol{k}_2 - \boldsymbol{k}_1) \cdot \boldsymbol{d}$, where $\boldsymbol{k}_2$ and $\boldsymbol{k}_1$ are the propagation vectors of observation and illumination beams, and $\boldsymbol{d}$ is the displacement vector. When the object is illuminated with two beams with directions symmetrical to the object normal (also the optical axis) and observation is made along the optical axis, the arrangement generates fringes that are contours of constant in-plane displacement. The fringes are called correlation fringes, a name given to these fringes that will become obvious in the later sections.

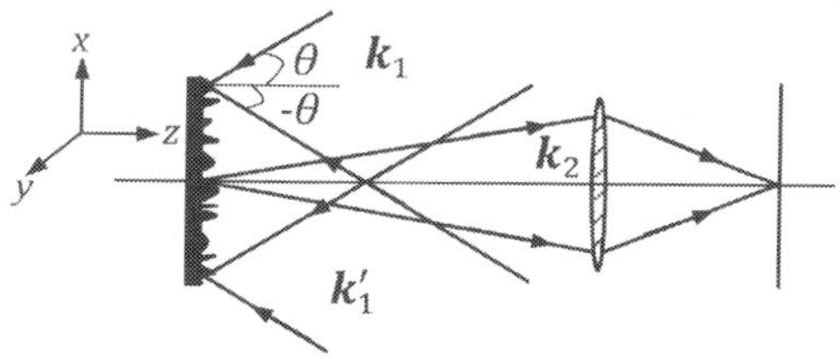

FIGURE 6.8 Configuration for in-plane displacement measurement due to Leendertz.

6.3.1 Leendertz Method

The schematic of the setup to measure in-plane displacement is shown in Figure 6.8. The object is illuminated by two-plane waves incident symmetrically at angles θ and $-\theta$ with respect to the optical axis. An image of the object is made on the recording material, for example, a photographic plate. The object is deformed, and a second exposure is made on the same plate. This double-exposure record, on filtering, generates fringes representing in-plane component along the x-direction. The whole process can be explained mathematically as follows.

Let us consider that the object is illuminated by unit amplitude plane waves, which are represented as $e^{i\,k\,x\sin\theta}$ and $e^{-i\,k\,x\sin\theta}$ at the plane of the object. The net amplitude of the waves at the plane of the object is $\left(e^{i\,k\,x\sin\theta}+e^{-i\,k\,x\sin\theta}\right)$. Let the scattering process from the object be represented by $\Re(x,y)$, which is a complex function and includes the surface characteristics. Thus, the field on the object surface just after scattering is

$$\Re(x,y)\left(e^{i\,k\,x\sin\theta}+e^{-i\,k\,x\sin\theta}\right)$$

The amplitude at the image plane is obtained through the superposition integral, that is,

$$\begin{aligned} u(x_i,y_i) &= \iint \Re(x,y)\left(e^{i\,k\,x\sin\theta}+e^{-i\,k\,x\sin\theta}\right) h(x_i-mx,y_i-my)\,dx\,dy \\ &= a_1(x_i,y_i)\,e^{i\phi_1(x_i,y_i)}+a_2(x_i,y_i)\,e^{i\phi_2(x_i,y_i)} \end{aligned} \tag{6.26}$$

where $h(x,y)$ is the impulse function and m is the magnification of the imaging lens. $a_1(x_i,y_i)$, $\phi_1(x_i,y_i)$, $a_2(x_i,y_i)$, $\phi_2(x_i,y_i)$ are the amplitudes and phases of the speckled waves at the image plane due to each of the illuminating waves. When the object is displaced by d_x in the x-direction, the amplitude in the image may be written as follows:

$$\begin{aligned} u'(x_i,y_i) &= \iint \Re(x+d_x,y)\left[e^{i\,k\,(x+d_x)\sin\theta}+e^{-i\,k\,(x+d_x)\sin\theta}\right]h(x_i-mx,y_i-my)\,dx\,dy \\ &= a_1(x_i,y_i)\,e^{i\phi_1(x_i,y_i)}e^{i\,k\,d_x\sin\theta}+a_2(x_i,y_i)\,e^{i\phi_2(x_i,y_i)}e^{-i\,k\,d_x\sin\theta} \end{aligned} \tag{6.27}$$

This equation has been arrived at by a tacit assumption that $\Re(x+d_x, y) = \Re(x, y)$. This assumption implies that the surface characteristics do not change over a distance d_x. Now, we can write the irradiances recorded in both the exposures as follows:

$$I_1(x_i, y_i) = a_1^2 + a_2^2 + 2a_1a_2\cos(\phi_1 - \phi_2) = a_1^2 + a_2^2 + 2a_1a_2\cos\phi \quad (6.28a)$$

$$I_2(x_i, y_i) = a_1^2 + a_2^2 + 2a_1a_2\cos(\phi + 2k\ d_x \sin\theta) \quad (6.28b)$$

For subsequent analyses of the various techniques, we will be writing for the irradiance distribution in the second exposure as follows:

$$I_2(x_i, y_i) = a_1^2 + a_2^2 + 2a_1a_2\cos(\phi + \delta) \quad (6.29)$$

where δ is the phase introduced by deformation. In this case, $\delta = 2k\ d_x \sin\theta = \frac{4\pi}{\lambda}\ d_x \sin\theta$. Later, we will use different arguments to prove that δ is indeed equal to $2k\ d_x \sin\theta$ for the experimental configuration of Figure 6.8. For the present, we may mention that ϕ_1, ϕ_2, a_1, and a_2 are random variables because $\Re(x, y)$ is a random variable. Therefore, $\phi = (\phi_1 - \phi_2)$ is also a random variable.

The total irradiance recorded will be

$$I_T(x_i, y_i) = I_1 + I_2 = 2a_1^2 + 2a_2^2 + 4a_1a_2\cos(\phi + \delta/2)\cos(\delta/2) \quad (6.30)$$

In fact, an examination of Equations (6.28a) and (6.29) for I_1 and I_2 reveals that when $\delta = 2m\pi$, both equations are identical, i.e., the speckles in both exposures are fully correlated. When $\delta = (2m + 1)\pi$, the speckles are uncorrelated. This correlation does provide the basis for fringe formation.

It should be noted that the contrast of the fringes is very poor in this arrangement, mainly due to the very strong granular bias term $2(a_1^2 + a_2^2)$. We will discuss other arrangements in which this bias term has been isolated thereby improving the contrast of the fringes. First, let us prove that the phase change δ due to in-plane displacement d_x is indeed given by $2k\ d_x \sin\theta$. For this purpose, we express the phase change δ due to displacement $\boldsymbol{d}$ as

$$\delta = (\delta_2 - \delta_1) = (\boldsymbol{k}_2 - \boldsymbol{k}_1)\cdot\boldsymbol{d} - (\boldsymbol{k}_2 - \boldsymbol{k}_1')\cdot\boldsymbol{d} = (\boldsymbol{k}_1' - \boldsymbol{k}_1)\cdot\boldsymbol{d} \quad (6.31)$$

where $\boldsymbol{d} = d_x\hat{i} + d_y\hat{j} + z\hat{k}$, and $\boldsymbol{k}_1'$, $\boldsymbol{k}_1$ are the propagation vectors of the illumination beams and $\boldsymbol{k}_2$ is the propagation vector along the direction of observation. The equation of phase difference $\delta = (\boldsymbol{k}_2 - \boldsymbol{k}_1)\cdot\boldsymbol{d}$ was derived earlier (see Equation 5.35) and is also applicable to speckle interferometry. From the geometry of the experimental configuration of Figure 6.8, $\boldsymbol{k}_1'$ and $\boldsymbol{k}_1$ are expressed as follows:

$$\boldsymbol{k}_1' = \frac{2\pi}{\lambda}\left(\sin\theta\ \hat{i} - \cos\theta\ \hat{k}\right) \quad (6.32a)$$

$$\boldsymbol{k}_1 = \frac{2\pi}{\lambda}\left(-\sin\theta\ \hat{i} - \cos\theta\ \hat{k}\right) \quad (6.32b)$$

Thus, the vector $(\boldsymbol{k}_1' - \boldsymbol{k}_1)\left(= \frac{2\pi}{\lambda} 2\sin\theta\ \hat{i}\right)$ lies in the plane of the object. Hence the phase change δ introduced by deformation can be expressed as

$$\delta = (\boldsymbol{k}_1' - \boldsymbol{k}_1)\cdot \boldsymbol{d} = \left(\frac{2\pi}{\lambda} 2\sin\theta\ \hat{i}\right)\cdot\left(d_x\hat{i} + d_y\hat{j} + z\hat{k}\right) = \frac{2\pi}{\lambda} 2d_x \sin\theta$$

The phase change δ depends only on the in-plane component d_x of the displacement vector $\boldsymbol{d}$. Bright fringes are formed when

$$\frac{2\pi}{\lambda} 2d_x \sin\theta = 2m\pi$$

Therefore

$$d_x = \frac{m\ \lambda}{2\sin\theta} \tag{6.33}$$

The arrangement is sensitive only to the in-plane component of displacement lying in the plane of illumination beams. Consecutive in-plane fringes differ by $\lambda/2\sin\theta$. The sensitivity can be varied over a very wide range from zero to $\lambda/2$ per fringe by varying the inter-beam angle. Furthermore, this particular arrangement has the following features:

- As seen from the theoretical analysis, the contribution of the out-of-plane displacement component d_z has been fully compensated for collimated illumination. This is because both waves suffer equal phase changes arising from out-of-plane displacement and hence the net phase change is zero.
- The arrangement is not sensitive to the y-component, i.e., d_y. However, the component d_y can be measured by rotating the experimental arrangement through 90°.
- It offers variable sensitivity.

The disadvantage is that the fringes obtained are of low contrast.

6.3.2 Correlation Coefficient in Speckle Interferometry

The correlation coefficient of two random variables X and Y is defined by

$$\rho_{XY} = \frac{\langle XY\rangle - \langle X\rangle\langle Y\rangle}{\sigma_X\sigma_Y} \tag{6.34}$$

where $\sigma_X^2 = \langle X^2\rangle - \langle X\rangle^2$ and $\sigma_Y^2 = \langle Y^2\rangle - \langle Y\rangle^2$, and $\langle\ \rangle$ stands for the ensemble average. If X and Y are uncorrelated, then $\langle XY\rangle = \langle X\rangle\langle Y\rangle$ and the correlation coefficient is zero as expected. The correlation coefficient of the random variables $I_1(x_i,y_i)$ and $I_2(x_i,y_i)$, the irradiances recorded in the first and second exposures, is given by

$$\rho(\delta)=\frac{\langle I_1 I_2\rangle-\langle I_1\rangle\langle I_2\rangle}{\left[<I_1^2>-<I_1>^2\right]^{1/2}\left[<I_2^2>-<I_2>^2\right]^{1/2}} \tag{6.35}$$

where the irradiances I_1 and I_2 are expressed as follows:

$$I_1(x,y)=a_1^2+a_2^2+2a_1a_2\cos\phi=i_1+i_2+2\sqrt{i_1 i_2}\cos\phi \tag{6.36a}$$

and

$$I_2(x,y)=i_1+i_2+2\sqrt{i_1 i_2}\cos(\phi+\delta) \tag{6.36b}$$

In this expression, the irradiances i_1, and i_2 refer to the irradiances at the image plane due to individual illuminating beams. Equation (6.35) for the correlation coefficient is evaluated by noting the following

- The irradiances i_1, i_2, and the phase difference ϕ are independent random variables and hence can be averaged separately.
- $<\cos\phi> = <\cos(\phi+\delta)> = 0$, and
- $<i_1^2> = 2<i_1>^2$ and $<i_2^2> = 2<i_2>^2$

When the irradiance values I_1 and I_2 are substituted in Equation (6.35) for correlation coefficient and the averages taken, we obtain

$$\rho(\delta)=\frac{<i_1^2>+<i_2^2>+2<i_1><i_2>\cos\delta}{(<i_1>+<i_2>)^2} \tag{6.37}$$

The correlation coefficient depends on the irradiances of the beams and the phase introduced by the deformation. If we consider that $<i_1> = r<i_2>$, i.e., one beam is r times stronger than the other beam, the correlation coefficient takes a simpler form as

$$\rho(\delta)=\frac{1+r^2+2r\cos\delta}{(1+r)^2} \tag{6.38}$$

This has a maximum value of unity when $\delta = 2m\pi$ and minimum value of $(1-r)^2/(1+r)^2$ when $\delta = (2m + 1)\pi$. The minimum value will be zero if $r = 1$, i.e., when the average irradiances of both the beams are equal. The correlation coefficient then would vary between 0 and 1 as the value of δ varies over the record. This situation is completely at variance with HI where the reference beam is taken as being stronger than the object beam.

6.3.3 Out-of-Plane Speckle Interferometer

An interferometer for measuring the out-of-plane displacement is shown in Figure 6.9. It is the Michelson interferometer in which one of the mirrors has been

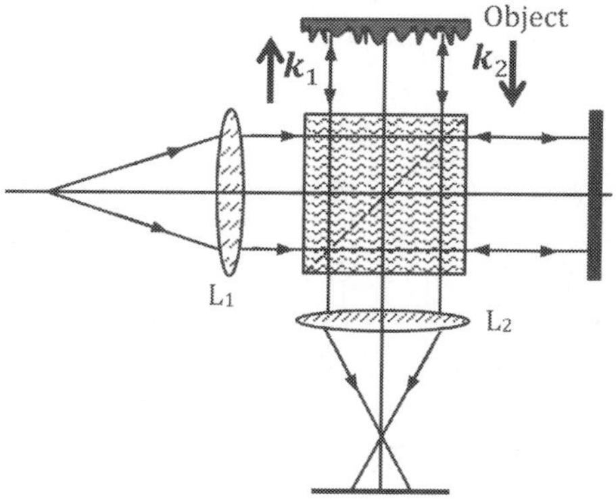

FIGURE 6.9 Configuration for out-of-plane displacement measurement.

replaced by an object under study. It uses either a speckled or a specular reference wave: here it is a specular or a smooth reference wave. The lens L_2 makes an image of the object at the recording plane. The record consists of an interference pattern between the smooth reference wave and the speckle field in the image of the object. The second exposure is recorded on the same plate after the object is loaded. This double-exposure specklegram (interferogram), when filtered, yields fringes, which are contours of constant out-of-plane displacement. As before, the irradiance distribution in the first exposure is

$$I_1(x_i, y_i) = a_1^2 + r_0^2 + 2a_1 r_0 \cos(\phi_1 - \phi_r) = a_1^2 + r_0^2 + 2a_1 r_0 \cos\phi \quad (6.39a)$$

where a_1, ϕ_1, and ϕ are random variables, and r_0 and ϕ_r are the amplitude and phase of the reference wave. The second exposure records the irradiance distribution given by

$$I_2(x_i, y_i) = a_1^2 + r_0^2 + 2a_1 r_0 \cos(\phi + \delta) \quad (6.39b)$$

where the phase difference δ, introduced by deformation, is given by

$$\delta = (\boldsymbol{k}_2 - \boldsymbol{k}_1) \cdot \boldsymbol{d} = \frac{2\pi}{\lambda} 2d_z$$

The total irradiance distribution $I_t(x_i, y_i)$ recorded is given by

$$I_t(x_i, y_i) = I_1(x_i, y_i) + I_2(x_i, y_i) = 2a_1^2 + 2r_0^2 + 4a_1 r_0 \cos\left(\phi + \frac{\delta}{2}\right)\cos\frac{\delta}{2} \quad (6.40)$$

The irradiance distribution is modulated by the term $\cos\frac{\delta}{2}$. The irradiance distributions I_1 and I_2 are correlated when $\delta = 2m\pi$. Thus, these areas on the specklegram will appear brighter. Substituting for δ, the bright fringes are formed wherever

$$\frac{2\pi}{\lambda} 2d_z = 2m\pi$$

or

$$d_z = m\lambda / 2 \tag{6.41}$$

The arrangement is sensitive to out-of-plane displacement component d_z.

Thus, the consecutive fringes are separated by out-of-plane displacement of $\lambda/2$. Again, the fringes are of low contrast. However, it may be noted that due to imaging geometry and customized configurations, only one of the components of deformation is sensed, unlike in HI where the phase difference δ introduced by deformation is invariably dependent on all the three components. Further, the correlation fringes are localized at the plane of the specklegram, unlike in HI where the fringe pattern is localized in space in general.

6.3.4 In-Plane Measurement: Duffy's Method

In the Leendertz method described earlier, the object is illuminated symmetrically with respect to the surface normal and observation is made along the bisector of the angle enclosed by the illuminating beams. This method, which is also called a dual-illumination and single-observation direction method, has very high sensitivity but poor contrast fringes.

It is also possible to illuminate the object along one direction and make an observation along two different directions symmetrical to the optical axis, taken along the local normal to the object. This method is due to Duffy and was developed in the context of moiré gauging. The method is also known as the single-illumination and dual-observation direction method. Figure 6.10(a) shows a schematic of the experimental arrangement. The object is illuminated at an angle θ, and a two-aperture mask is placed in front of the lens. The apertures enclose an angle of 2α at the object

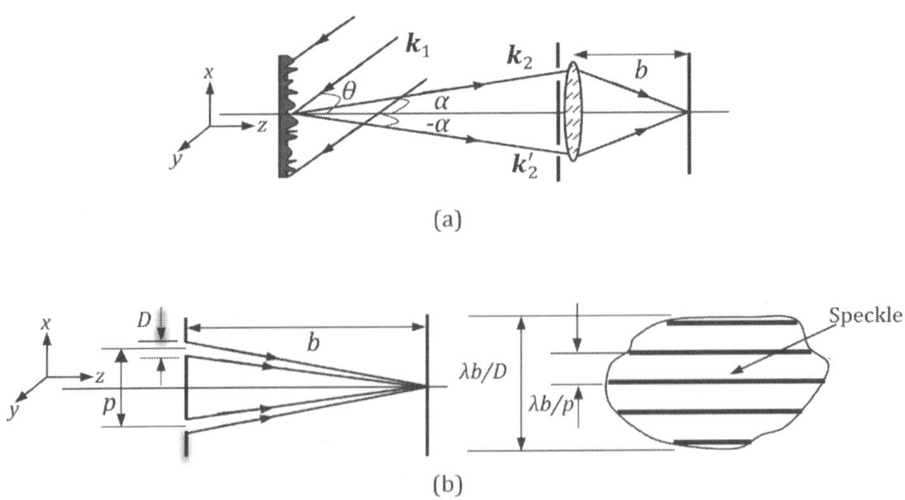

FIGURE 6.10 (a) Duffy's two-aperture arrangement and (b) modulation of a speckle.

distance. The lens makes an image of the object via each aperture, and these images are perfectly superposed.

Each wave passing through the aperture generates a speckled image with a speckle size $\lambda b/D$, where D is the aperture size. These waves are obliquely superposed and hence each speckle is modulated by a fringe pattern. The fringe spacing in the speckle is $\lambda b/p$, where p is the inter-separation between the two apertures. This is shown in Figure 6.10(b). When the object is deformed, these fringes shift in the speckle. When the deformation is such that a fringe moves by one period or a multiple thereof, it is then exactly superposed on the earlier recorded position, and the fringe contrast is high. The region will diffract strongly on filtering and hence these areas would appear bright. On the other hand, if the deformation is such that the fringe pattern moves by half a fringe width or an odd multiple of a half-period, the new pattern will fall midway in the earlier recorded pattern resulting in an almost complete washout of the fringe pattern. These regions will not diffract or will diffract poorly and appear dark on filtering. Therefore, bright and dark fringes correspond to regions where the displacement is an integral multiple of $\lambda b/p$ ($= \lambda/2\sin\alpha$) and odd integral multiple of half $\lambda b/p$, respectively.

Mathematically, we write the amplitudes of the waves via each aperture at the image plane as

$$a_{11} = a_0 e^{i\phi_{11}} \tag{6.42a}$$

$$a_{12} = a_0 e^{i(\phi_{12} + 2\pi\mu x_i)} \tag{6.42b}$$

where μ ($= p/\lambda b$) is the spatial frequency of the fringe pattern in the speckle pattern. The irradiance recorded in the first exposure is

$$I_1(x_i, y_i) = |a_{11} + a_{12}|^2 = a_0^2 + a_0^2 + 2a_0^2 \cos(\phi_1 + 2\pi\mu x_i) : \phi_1 = \phi_{12} - \phi_{11} \tag{6.43}$$

Similarly, when the object is loaded, the waves acquire additional phase changes δ_1 and δ_2, respectively, i.e., they can be expressed as

$$a_{21} = a_0 e^{i(\phi_{11} + \delta_1)} \tag{6.44a}$$

$$a_{22} = a_0 e^{i(\phi_{12} + 2\pi\mu x_i + \delta_2)} \tag{6.44b}$$

The irradiance distribution recorded in the second exposure is then

$$I_2(x_i, y_i) = |a_{21} + a_{22}|^2 = a_0^2 + a_0^2 + 2a_0^2 \cos(\phi_1 + 2\pi\mu x_i + \delta) \tag{6.45}$$

where δ ($= \delta_2 - \delta_1$) is the phase difference introduced by the deformation. The double-exposure record after processing (specklegram) is ascribed an amplitude transmittance $t(x_i, y_i)$, which is given by

$$t(x_i, y_i) = t_0 - \beta T (I_1 + I_2) \tag{6.46}$$

where β is a constant and T is the exposure time. Information about the deformation is extracted by filtering.

6.3.4.1 Filtering

The specklegram is placed in the setup shown in Figure 6.11 and is illuminated by a collimated beam, say, a unit amplitude wave. The field at the FT plane is the Fourier transform of the transmittance, i.e., $\Im\left[t(x,y)\right]$.

Due to the grating-like structure in each speckle, the halo distribution gets modified: it has a central halo (zero order) and plus and minus first-order halos. The zero order arises because of the terms $a_0^2 + a_0^2$, which do not carry any information. Because the speckle size is now larger due to the smaller apertures used for imaging, the halo size (zero-order halo) shrinks. The filtering is done by choosing any one of the first-order halos. A fringe pattern of almost unit contrast is obtained because the halo (zero order), which carries no information, has been isolated.

6.3.4.2 Fringe Formation

The phase differences δ_2 and δ_1, due to the deformation, experienced by the waves passing through the apertures can be expressed as:

$$\delta_2 = \left(\boldsymbol{k}_2' - \boldsymbol{k}_1\right)\cdot \boldsymbol{d} \text{ and } \delta_1 = \left(\boldsymbol{k}_2 - \boldsymbol{k}_1\right)\cdot \boldsymbol{d}$$

Hence the phase difference δ (= $\delta_2 - \delta_1$) is given by

$$\delta = \left(\boldsymbol{k}_2' - \boldsymbol{k}_2\right)\cdot \boldsymbol{d} \tag{6.47}$$

This phase difference generates an interference pattern. The illumination and observation beams lie in the x–z plane. Writing the wave vectors $\boldsymbol{k}_2'$ and $\boldsymbol{k}_2$, and displacement vector **d** in their components, the phase difference can be expressed as

$$\delta = \left(\boldsymbol{k}_2' - \boldsymbol{k}_2\right)\cdot \boldsymbol{d} = \frac{2\pi}{\lambda} 2d_x \sin\alpha$$

Bright fringes are formed wherever

$$\frac{2\pi}{\lambda} 2d_x \sin\alpha = 2m\pi$$

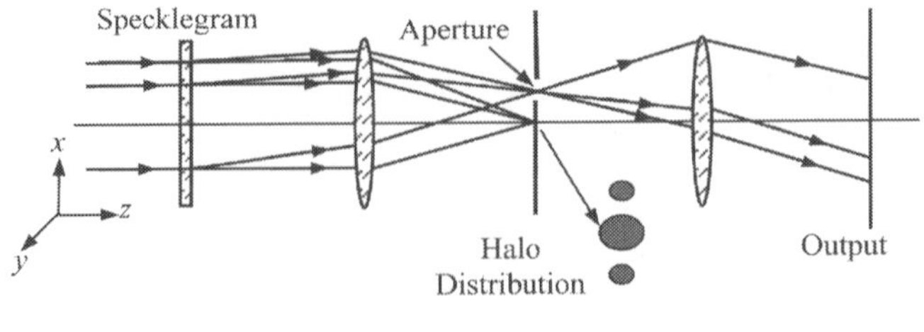

FIGURE 6.11 Whole-field filtering of a specklegram recorded with Duffy's arrangement.

This gives

$$d_x = m\lambda / 2\sin\alpha \tag{6.48}$$

This result is similar to that obtained earlier for the Leendertz method (Equation 6.33) except that the angle θ is replaced by the angle α. Obviously, α cannot take very large values as the magnitude of α is determined by the lens aperture or by the *F#* of the lens. Therefore, the method has intrinsically poor sensitivity. At the same time, the speckle size is very large compared with that in the Leendertz method; hence, the range of in-plane displacement measurement is large. The method yields high-contrast fringes due to the removal of the unwanted DC speckled field by the grating-like structure formed during recording.

It is indeed very simple and easy to extend Duffy's method to measure both the components of in-plane displacement simultaneously. Instead of a two-aperture mask in front of the lens, if a mask containing three or four apertures is used, then the magnitudes of both the in-plane components can be obtained by filtering from the appropriate halos.

Leendertz and Duffy methods can be combined into one: the object is illuminated by symmetric collimated beams and the specklegram is recorded using an apertured lens. This combination extends the range and sensitivity of the in-plane displacement measurement from a small value dictated by the Leendertz method to a large value governed by the Duffy method. On filtering the double-exposure specklegram, both systems of fringes are observed simultaneously.

Using a diffuser on one of the two apertures placed in front of the lens and illuminating it with a narrow beam, a diffuse reference beam can be generated. The other aperture is used for imaging. With this arrangement, the out-of-plane component of deformation can also be measured. In fact, by the judicious choice of number of apertures and their arrangement, it is possible to measure all the three components of the deformation vector simultaneously. In addition to obtaining all the components of the deformation vector from a single double-exposure specklegram, the aperturing of the lens could be used to multiplex the information record, i.e., several states of the object could be stored, and the information retrieved from the specklegram. The multiplexing can be done by two means: (1) frequency modulation where the apertures are laterally shifted on the lens after each exposure and (2) θ-multiplexing (modulation) where the apertures are angularly shifted. Both these methods can also be combined. The amount of information that can be recorded and retrieved depends on the lens size, the aperture size, the spatial frequency content in the object, and the dynamic range of the recording material. In some experiments, in addition to the measurement of displacement components, the slope information is also recorded and retrieved later.

6.3.4.3 Duffy's Arrangement: Enhanced Sensitivity

The sensitivity of the Duffy's arrangement is limited by the *F#* of the imaging lens. This limitation, however, can be overcome by the modification of the recording setup as shown in Figure 6.12. The object is illuminated normally and is observed along two symmetric directions: the beams are folded and directed to a pair of mirrors and

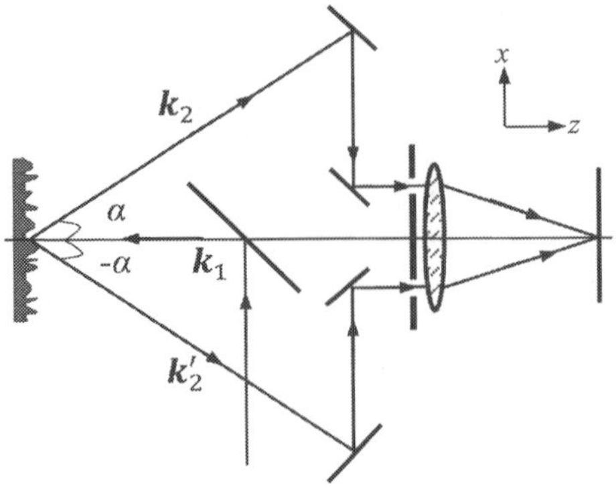

FIGURE 6.12 A configuration for in-plane measurement with enhanced sensitivity.

then onto a two-aperture mask in front of the lens. The image of the object is thus made via these folded paths. In this arrangement, the speckle size is governed by the aperture diameter as is the case with the other methods; the fringe frequency is determined by the aperture separation; and the sensitivity is governed by the angle α, which can be varied over a large range, and is not restricted by the *F#* of the lens. It can be shown that the phase difference due to deformation is given by

$$\delta = \delta_1 - \delta_2 = (\boldsymbol{k}_2 - \boldsymbol{k}_1)\cdot\boldsymbol{d} - (\boldsymbol{k}_2' - \boldsymbol{k}_1)\cdot\boldsymbol{d} = (\boldsymbol{k}_2 - \boldsymbol{k}_2')\cdot\boldsymbol{d} = \frac{2\pi}{\lambda}2d_x\sin\alpha$$

The in-plane fringes are extracted by filtering using one of the first-order halos. The technique introduces perspective errors when bigger objects are studied at larger angles. The perspective error, however, can be reduced using a pair of prisms to decrease the convergence. Obviously, the arrangement is not light efficient.

A twofold increase in sensitivity can be obtained if the object is coated with a retro-reflecting paint. In this situation, the directions of illumination and of observation are anti-parallel as shown in Figure 6.13(a). The object is illuminated and viewed in symmetrical directions. The phase change due to deformation is written as

$$\delta = \delta_1 - \delta_2 = 2(\boldsymbol{k}_2 - \boldsymbol{k}_2')\cdot\boldsymbol{d} = \frac{2\pi}{\lambda}4d_x\sin\alpha \tag{6.49}$$

Again, the in-plane fringes are obtained by filtering via one of the first-order halos. Figure 6.13(b) is a photograph showing twice the number of fringes in one half of a circular plate, which is coated with a retro-reflecting paint and remaining half with an ordinary white paint; the circular plate is tilted in between the exposures.

6.4 SPECKLE SHEAR INTERFEROMETRY

So far, we have discussed the techniques that measure the displacement components only. As mentioned earlier, a stress analyst is usually interested in strains rather than the displacements. The strain is obtained by fitting the displacement data

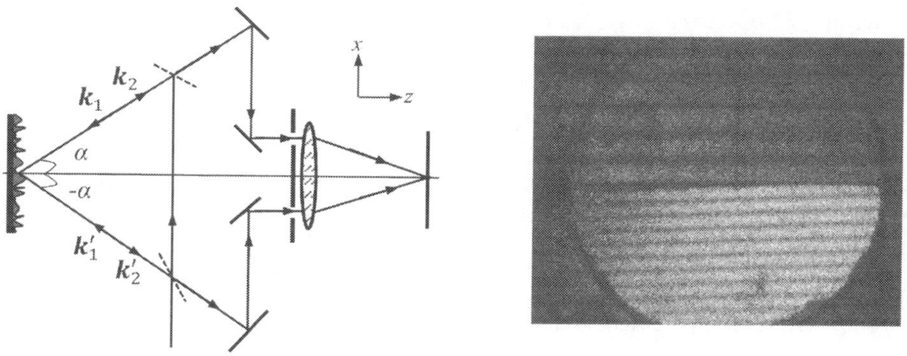

(a) (b)

FIGURE 6.13 (a) A configuration for in-plane measurement with two-fold increase in sensitivity. The object is coated with retro-reflective paint. (b) Photograph showing fringe with double the sensitivity in one half. (Courtesy Dr. N. Krishna Mohan, IIT Madras.)

numerically and then differentiating it. This procedure could lead to large errors. Therefore, methods are investigated that can yield fringe patterns that represent the derivatives of the displacement. This is achieved with speckle shear interferometry. Because all speckle techniques for displacement measurement use subjective speckles (i.e., image plane recordings), we restrict ourselves to shear at the image plane. The shear methods are grouped under the five categories and are listed in Table 6.1.

6.4.1 The Meaning of Shear

Shear has several dictionary meanings. We, however, use it to define the relative positioning and orientation of two beams/images derived from the same parent beam/object. When an object is imaged via two identical paths like in two-aperture arrangement, the images are perfectly superposed; there is no shear even though there are two images. Because the imaging is via two independent paths, the two images can be independently manipulated. In linear shear, one image is laterally shifted in any desired direction by an appropriate amount. In speckle shear interferometry, one image acts as a reference to the other; hence, there is no need to supply

TABLE 6.1
Shear Methods Used in Speckle Interferometry

Shear Types	Phase Difference Leading to Fringe Formation
Lateral shear or linear shear	$\delta(x + \Delta x, y + \Delta y) - \delta(x,y)$
Rotational shear	$\delta(r, \theta + \Delta\theta) - \delta(r,\theta)$
Radial shear	$\delta(r \pm \Delta r, \theta) - \delta(r,\theta)$
Inversion shear	$\delta(x,y) - \delta(-x,-y)$
Folding shear	$\delta(x,y) - \delta(-x,y)$: folding about the y-axis
	$\delta(x,y) - \delta(x,-y)$: folding about the x-axis

an additional reference wave. When using speckle shear interferometry, we compare the response of an object to the external agency at any point to that of a shifted point.

In rotational shear, one image is rotated, usually about the optical axis, by a small angle with respect to the other image. In radial shear, one image is either radially contracted or expanded with respect to the other image. The inversion shear allows a point at (x,y) to be compared with another point at $(-x,-y)$. This is equivalent to a rotational shear of π. In folding shear, a point is compared with its mirror image: the folding can be either about the y-axis or the x-axis.

6.4.2 Methods of Shearing

One of the most used methods of shearing employs the Michelson interferometer where the object is seen via two independent paths as shown in Figure 6.14(a). The object is seen via two paths *OABAD* and *OACAD*. When the mirrors M_1 and M_2 are normal to each other, and are at equal distances from A, the two images are perfectly superposed. Tilting one of the mirrors, as shown, displaces one image in the direction of the arrow, and the images are linearly separated. A detailed analysis of this arrangement reveals that there is a large spherical aberration introduced due to the use of a beam-splitting cube as the beams travel in a glass medium a distance of three times the size of the cube.

All the shear methods described in Table 6.1 can be conveniently introduced when an aperture mask is placed in front of an imaging lens. For linear shear, a pair of plane parallel plates, wedges as shown in Figure 6.14(b), or a biprism have been used. Shearing has also been done with gratings. A lens cut into two halves in which one half can either be translated in its own plane or along the optical axis makes an excellent shearing device: the device performs both the functions of shearing and imaging. In fact, a diffractive optical element can be designed that performs both the functions of imaging and shearing.

6.4.3 Theory of Speckle Shear Interferometry

It was pointed out earlier that in shear interferometry a point on the object is either imaged as two points or two points on the object are imaged as a single point.

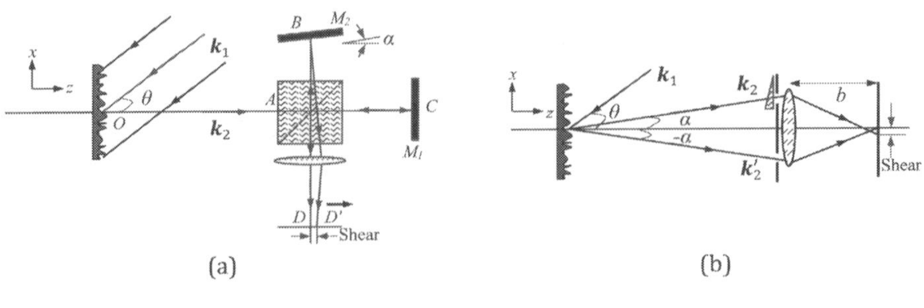

FIGURE 6.14 Shearing with (a) the Michelson interferometer and (b) Duffy's arrangement with a wedge.

One therefore obtains either an object plane shear or an image plane shear: these are related through the magnification of the imaging lens.

Let a_1 and a_2 be the amplitudes of the waves at any point on the image plane from the two points (x_0,y_0) and $(x_0 + \Delta x_0, y_0 + \Delta y_0)$ at the object plane, and let ϕ_1 and ϕ_2 be their phases. The irradiance distribution recorded at the image plane is given by

$$I_1(x_i, y_i) = a_1^2 + a_2^2 + 2a_1a_2 \cos(\phi_1 - \phi_2) = a_1^2 + a_2^2 + 2a_1a_2 \cos\phi \tag{6.50a}$$

After the object is loaded, the deformation vectors at the two points are represented by $\boldsymbol{d}(x_0,y_0)$ and $\boldsymbol{d}(x_0 + \Delta x_0, y_0 + \Delta y_0)$, respectively. Therefore, the waves from these two points arrive at the image point with a phase difference δ (= $\delta_2 - \delta_1$). The irradiance distribution in the second exposure can therefore be expressed as

$$I_2(x_i, y_i) = a_1^2 + a_2^2 + 2a_1a_2 \cos(\phi + \delta) \tag{6.50b}$$

The total irradiance recorded is $I_1(x,y) + I_2(x,y)$, and hence the amplitude transmittance of the double-exposure specklegram is

$$t(x_i, y_i) = t_0 - \beta T(I_1 + I_2) \tag{6.51}$$

On filtering, one obtains a fringe pattern representing the derivatives of the displacement components as will be shown later.

6.4.4 Fringe Formation

6.4.4.1 The Michelson Interferometer

The phase difference δ can be expressed, assuming shear only along the x-direction [Figure 6.14(a)], as follows:

$$\delta = (\boldsymbol{k}_2 - \boldsymbol{k}_1)\cdot \boldsymbol{d}(x_0 + \Delta x_0, y_0) - (\boldsymbol{k}_2 - \boldsymbol{k}_1)\cdot \boldsymbol{d}(x_0, y_0) = (\boldsymbol{k}_2 - \boldsymbol{k}_1)\cdot \frac{\partial \boldsymbol{d}}{\partial x_0}\Delta x_0 \tag{6.52a}$$

Writing $\boldsymbol{k}_2$, $\boldsymbol{k}_1$, and $\boldsymbol{d}$ in their components, Equation (6.52a) for the phase difference δ can be written as

$$\delta \cong \frac{2\pi}{\lambda}\left[\sin\theta \frac{\partial d_x}{\partial x_0} + (1+\cos\theta)\frac{\partial d_z}{\partial x_0}\right]\Delta x_0 \tag{6.52b}$$

The bright fringes are formed wherever

$$\left[\sin\theta \frac{\partial d_x}{\partial x_0} + (1+\cos\theta)\frac{\partial d_z}{\partial x_0}\right] = \frac{m\lambda}{\Delta x_0} : m = 0, \pm 1, \pm 2, \pm 3, \ldots. \tag{6.53}$$

The fringe pattern has contributions from both the strain $\left(\frac{\partial d_x}{\partial x_0}\right)$ and the slope $\left(\frac{\partial d_z}{\partial x_0}\right)$. However, when the object is illuminated normally, i.e., $\theta = 0$, the fringe pattern represents a partial x-slope pattern only, that is,

$$\frac{\partial d_z}{\partial x_0} = \frac{m\lambda}{2\Delta x_0} \tag{6.54}$$

The fringe pattern corresponding to partial y-slope $\left(\frac{\partial d_z}{\partial y_0}\right)$ is obtained when the shear is applied along the y-direction.

6.4.4.2 The Apertured Lens Arrangement

The phase difference δ can again be expressed, assuming shear only along x-direction [Figure 6.14(b)] as:

$$\delta = (\boldsymbol{k}_2 - \boldsymbol{k}_1)\cdot \boldsymbol{d}(x_0 + \Delta x_0, y_0) - (\boldsymbol{k}_2' - \boldsymbol{k}_1)\cdot \boldsymbol{d}(x_0, y_0)$$

This is rewritten as

$$\delta = (\boldsymbol{k}_2 - \boldsymbol{k}_1)\cdot \boldsymbol{d}(x_0, y_0) + (\boldsymbol{k}_2 - \boldsymbol{k}_1)\cdot \frac{\partial \boldsymbol{d}}{\partial x_0}\Delta x_0 - (\boldsymbol{k}_2' - \boldsymbol{k}_1)\cdot \boldsymbol{d}(x_0, y_0)$$

$$\begin{aligned} \delta &\cong (\boldsymbol{k}_2 - \boldsymbol{k}_2')\cdot \boldsymbol{d}(x_0, y_0) + (\boldsymbol{k}_2 - \boldsymbol{k}_1)\cdot \frac{\partial \boldsymbol{d}}{\partial x_0}\Delta x_0 \\ &= \frac{2\pi}{\lambda}2d_x \sin\alpha + \frac{2\pi}{\lambda}\left[\sin\theta\frac{\partial d_x}{\partial x_0} + (1+\cos\theta)\frac{\partial d_z}{\partial x_0}\right]\Delta x_0 \end{aligned} \tag{6.55}$$

A comparison with Equation (6.52b) for the Michelson interferometer reveals that there is an in-plane component dependent term in addition to the usual expression. This term arises due to the two apertures separated by a distance, an arrangement that has been shown to be inherently sensitive to the in-plane component. An interesting aspect of aperturing of a lens, as was pointed out earlier, is its ability to measure simultaneously in-plane, out-of-plane displacement components and their derivatives. Figure 6.15(a–c) shows photographs of the out-of-plane displacement, the partial x-slope, and the partial y-slope fringe patterns of a defective pipe obtained from the same double-exposure specklegram.

6.4.5 Speckle Shear Interferometry without the Influence of In-Plane Component

It was shown earlier that shear interferometry performed with an aperture mask in front of the imaging lens always yields a fringe pattern that is due to the combined effect of the in-plane displacement component and the derivatives of the displacement. At the same time, it is desirable to have an aperture mask for obtaining high-contrast fringe patterns. In order to retain this advantage and eliminate the in-plane

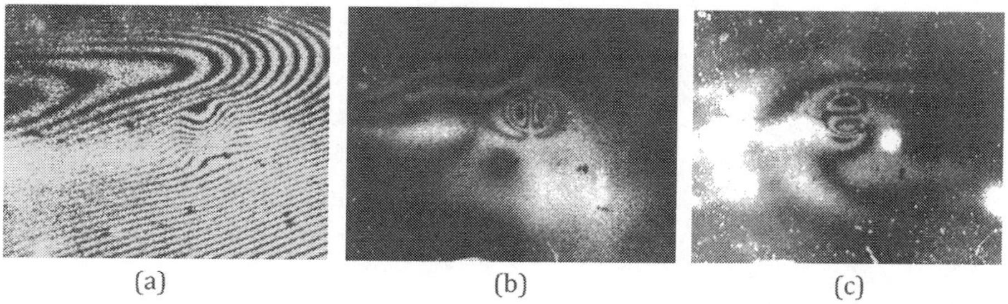

FIGURE 6.15 Interferograms from a double-exposure specklegram. (a) Out-of-plane displacement fringe pattern. (b) Partial x-slope fringe pattern. (c) Partial y-slope fringe pattern.

displacement component sensitivity, the object should be illuminated normally so that $\sin\theta = 0$ and is viewed axially. Therefore, one can evolve a configuration that is not sensitive to in-plane displacement components and only presents fringes that are due to slope.

6.5 ELECTRONIC SPECKLE PATTERN INTERFEROMETRY (ESPI)

The speckle size in speckle interferometry can be controlled by the *F#* of the imaging lens. Further, in a two-aperture arrangement, the size can be doubled by adding a reference beam axially. It is thus possible to use electronic detectors, which have limited resolution, for recording instead of photographic emulsions. Use of electronic detectors avoids the messy wet development process. Further, the processing is done at video rates making the technique almost real time. The photo-emulsions, as mentioned earlier, integrate the light irradiance falling on them. In speckle techniques with photographic recording, the two exposures were added in succession, and then techniques were developed to remove the undesired DC component. With electronic detection, the two exposures are independently handled, and the subtraction removes the DC component. Phase-shifting techniques are easily incorporated and hence deformation maps can be presented almost in real time. When recording is done on a CCD or a complementary metal oxide semiconductor (CMOS) and digital manipulations are carried out, the technique is known as DSPI. Both ESPI and DSPI are an excellent alternative to HI for studying the response of objects to external forces.

6.5.1 Out-of-Plane Displacement Measurement

Figure 6.16 shows one of the several configurations used for measuring the out-of-plane component of deformation when there is no in-plane component. Otherwise, the object should be illuminated axially, i.e., $\theta = 0$. The reference beam is added axially such that it appears to emerge from the center of the exit pupil of the imaging lens. The speckle size is matched with the pixel size by controlling the lens aperture. The irradiances of both the object and the reference beams at the CCD plane are adjusted to be equal. The first frame is stored in the frame grabber, and the second

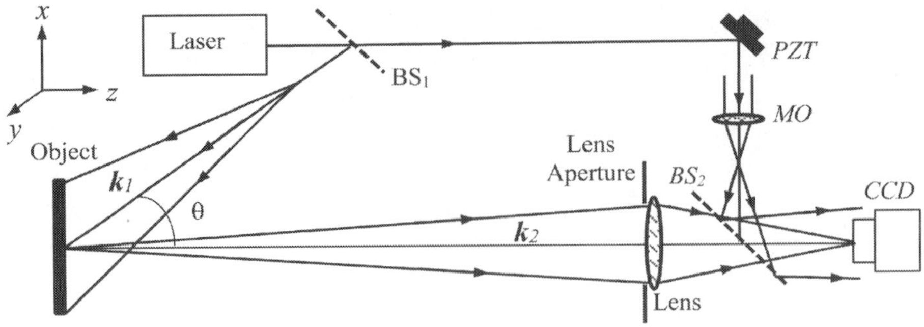

FIGURE 6.16 A configuration for ESPI.

frame, captured after loading the object, is subtracted pixel by pixel. The difference signal is rectified and then sent to the monitor to display the fringe pattern. This procedure can be mathematically described as

The irradiance recorded in the first frame is given by

$$I_1(x_i, y_i) = a_1^2(x_i, y_i) + a_2^2(x_i, y_i) + 2a_1(x_i, y_i)a_2(x_i, y_i)\cos\phi\text{: } \phi = \phi_2 - \phi_1 \quad (6.56a)$$

where a_1 and a_2 are the amplitudes of the object and reference waves and ϕ_1 and ϕ_2 are their phases at the detector plane. The output of the detector is assumed to be proportional to the irradiance incident on it.

The irradiance recorded in the second frame is

$$I_2(x_i, y_i) = a_1^2(x_i, y_i) + a_2^2(x_i, y_i) + 2a_1(x_i, y_i)a_2(x_i, y_i)\cos(\phi + \delta) \quad (6.56b)$$

where the phase difference δ introduced by the deformation is given by $\delta = (\boldsymbol{k}_2 - \boldsymbol{k}_1)\cdot \boldsymbol{d}$.

The subtracted signal $(I_2 - I_1)$ will generate a voltage signal ΔV as

$$\begin{aligned} \Delta V \propto I_2 - I_1 &= 2a_1(x_i, y_i)a_2(x_i, y_i)\left[\cos(\phi + \delta) - \cos\phi\right] \\ &= 4a_1(x_i, y_i)a_2(x_i, y_i)\sin\left(\phi + \frac{\delta}{2}\right)\sin\frac{\delta}{2} \end{aligned} \quad (6.57)$$

The brightness B on the monitor will be proportional to the voltage signal ΔV (difference signal) from the detector and hence

$$B = 4\wp a_1(x_i, y_i)a_2(x_i, y_i)\sin\left(\phi + \frac{\delta}{2}\right)\sin\frac{\delta}{2} \quad (6.58)$$

where $\wp$ is a constant of proportionality. As δ varies, $\sin(\delta/2)$ will vary between –1 and 1. The negative values of $\sin(\delta/2)$ will appear dark on the monitor resulting in the loss of signal. This loss is avoided by rectifying the signal before it is sent to the monitor. The brightness B is thus given by

$$B = 4\wp a_1(x_i, y_i) a_2(x_i, y_i) \left| \sin\left(\phi + \frac{\delta}{2}\right) \sin\frac{\delta}{2} \right| \tag{6.59a}$$

Alternately, the difference signal is squared before being sent to the monitor. The brightness B' on the monitor will now be given by

$$B' = 16\wp^2 a_1^2(x_i, y_i) a_2^2(x_i, y_i) \sin^2\left(\phi + \frac{\delta}{2}\right) \sin^2\frac{\delta}{2} \tag{6.59b}$$

Brightness will be zero when $\delta/2 = m\pi$: or $\delta = 2m\pi$; $m = 0, \pm 1, \pm 2,$. This means that the speckle regions in the speckle pattern, which are correlated, would appear dark. This is due to the difference operation. Also, due to this operation, undesirable terms are eliminated. Phase shifting is easily incorporated by reflecting the reference wave from a lead zirconate titanate (PZT)-mounted mirror. Because the irradiance distributions $I_1(x_i, y_i)$ and $I_2(x_i, y_i)$ are handled independently, it is possible to add them resulting in additive DSPI or multiply them resulting in multiplicative DSPI. The DC speckle terms are removed by low-pass filtering.

6.5.2 In-Plane Displacement Measurement

In-plane displacement can be measured using the experimental configuration due either to Leendertz or Duffy. The sensitivity can be varied by changing the inter-beam angle as has been explained in Section 6.3. Special configuration has been invented to measure in-plane displacement in cylindrical pipes.

6.5.3 Vibration Analysis

ESPI is an excellent tool for studying the vibration modes of an object. It can also be used to measure extremely small, moderate, and large vibration amplitudes. The arrangement used is the one suited for out-of-plane displacement measurement. The object is excited acoustically or by directly attaching a PZT that is run through a function generator, thereby scanning a large frequency range over which the response of the object could be studied. Because the video rates are much slower compared with the resonance frequencies, the pattern observed on the monitor represents time-averaged fringes. The irradiance distribution is given by $J_0^2\left[\frac{4\pi}{\lambda} A(x, y)\right]$, where $A(x,y)$ is the amplitude of vibration. The zeroth-order fringe occurs where the amplitude of vibration is zero, i.e., at the clamped portions of the object, and it would have the highest irradiance. The dark fringes occur wherever the argument of Bessel function is zero. Therefore, the amplitudes of vibration can be obtained over the whole surface. However, when the reference wave is also modulated at the frequency of object excitation, the irradiance distribution in the fringe pattern can be expressed as

$$I(x, y) \propto J_0^2\left\{\frac{4\pi}{\lambda}\left[A^2(x, y) + a_r^2 - 2A(x, y)\, a_r \cos(\phi - \phi_r)\right]^{1/2}\right\} \tag{6.60}$$

where a_r and ϕ_r are the amplitude and phase of the reference beam reflected from the vibrating mirror. When the object and the reference mirror vibrate in phase, the irradiance distribution can be expressed as $J_0^2\left\{\frac{4\pi}{\lambda}\left[A(x,y)-a_r\right]\right\}$. The zeroth-order fringe now occurs where $A(x,y) = a_r$. Therefore, large amplitudes of vibration can be measured. However, if very small vibration amplitudes are to be measured, the frequency of reference wave modulation is taken slightly different to that of the object vibration but still within the video frequency. Because of this, the phase of the reference wave varies with time. The irradiance distribution would now be proportional to

$$J_0^2\left\{\frac{4\pi}{\lambda}\left[A^2(x,y)+a_r^2-2A(x,y)\,a_r\cos(\phi-\phi_r(t))\right]^{1/2}\right\} \tag{6.61}$$

Because the phase $\phi_r(t)$ varies with time, the argument of the Bessel function varies between $\{A(x,y) + a_r\}$ and $\{A(x,y) - a_r\}$; hence, the irradiance on the monitor will fluctuate. However, if $A(x,y) = 0$, then the argument of the Bessel function remains constant and there is no fluctuation or flicker. Only at those locations where the flicker occurs will the amplitude of vibration be nonzero, thereby allowing very small vibration amplitudes to be detected.

6.5.4 Measurement on Small Objects

EPSI has been used for studying the performance of a variety of objects ranging in size from large to small. However, there is considerable interest in evaluating the performance of small-size objects particularly micro-electro-mechanical systems (MEMS) in real time. MEMS are the result of the integration of mechanical elements, sensors, actuators, and electronics on a common silicon substrate through microfabrication technology. They are used in varieties of fields such as telecommunication, computers, aerospace, automobiles, biomedical, micro-optics, etc. ESPI for the inspection and characterization of MEMS should not alter the integrity and the mechanical behavior of the device. Because MEMS have an overall size up to a few millimeters, a high spatial resolution measuring system is required, that is, a long working distance microscope with a combination of different magnification objective lenses is incorporated into ESPI. A schematic of microscopic ESPI is shown in Figure 6.17(a). Instead of a normal camera lens for imaging, a long working distance microscope is used for imaging on the CCD array. Phase shifting is accomplished by a PZT-driven mirror. The MEMS device chosen for study is a pressure transducer. The diaphragm is normally etched out in silicon; the deflection of the diaphragm due to an application of pressure is measured using Wheatstone circuitry. However, the deflection profile can be measured using ESPI as well. In fact, ESPI is used to calibrate the pressure transducer. Figure 6.17(b–d) show the results of measurement when some pressure is applied to the sensor in between two frames captured by CCD array camera. Figure 6.17(e) gives the deflection profile of the pressure sensor.

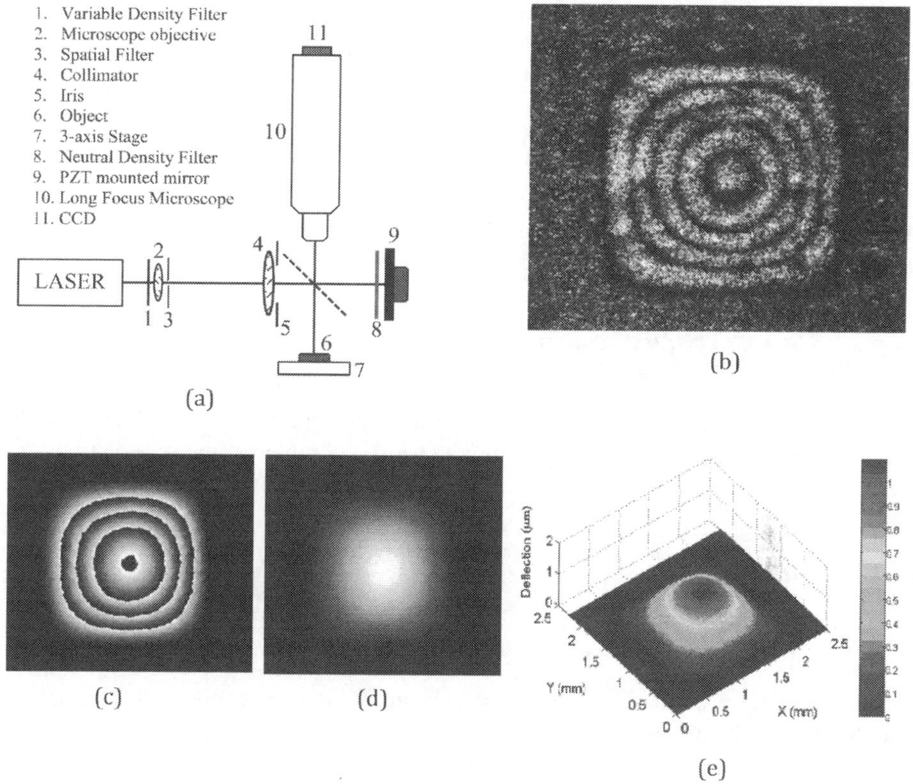

FIGURE 6.17 (a) Schematic of an ESPI system for studying small objects. (b) Correlation fringes. (c) Unwrapped fringes. (d) Wrapped phase. (e) Deflection profile of the sensor. (Courtesy Dr. N K Mohan, IIT Madras.)

6.6 CONTOURING WITH ESPI: SHAPE MEASUREMENT

The following methods are available for contouring using speckle interferometry and ESPI:

1. Change of direction of the illumination beam between exposures
2. Change of wavelength between exposures: dual-wavelength technique
3. Change of refractive index of the surrounding medium between exposures: dual-refractive index technique
4. Rotation of the object between exposures in an in-plane sensitive configuration.

6.6.1 Change of Direction of Illumination

The object is illuminated by a diverging wave from a point source and a usual ESPI setup is used. The first frame is grabbed and stored. The illumination point source is slightly shifted laterally to change the direction of illumination. The second frame captured now is subtracted from the stored frame and the contour fringes displayed on the monitor. The contour interval is given by

$$\frac{\lambda}{2\sin\theta}\frac{L}{\Delta s}=\frac{\lambda}{2\sin\theta\ \Delta\phi} \tag{6.62}$$

where Δs is the lateral shift of the point source, which causes an angular shift of $\Delta\phi$ and L is the distance between the point source and the object plane.

6.6.2 Change of Wavelength

The dual-wavelength method requires two sources of slightly different wavelengths or a single source that can be tuned. Two frames are grabbed and subtracted, each frame with one of the wavelengths of light. The true depth contours separated by Δz are generated by this method where separation Δz is given by

$$\Delta z=\frac{\lambda_{eff}}{2}=\frac{\lambda_1\ \lambda_2}{2|\lambda_1-\lambda_2|} \tag{6.63}$$

6.6.3 Change of Medium Surrounding the Object

Here, the medium surrounding the object is changed between exposures. Subtraction yields true depth contours. In fact, the method is equivalent to the dual-wavelength method as the wavelength in the medium changes when its refractive index is changed. We can thus arrive at the same result by writing the wavelength in terms of the refractive index and vacuum wavelength. The contour interval Δz is given by

$$\Delta z=\frac{\lambda}{2|\mu_1-\mu_2|}=\frac{\lambda}{2\ \Delta\mu} \tag{6.64}$$

Here, $\Delta\mu$ is the change in the refractive index when one medium is replaced by the other in between the exposures.

6.6.4 Tilt of the Object

Tilt of the object is a new contouring method and is applicable to speckle interferometry only. In this method, one uses an in-plane sensitive configuration, i.e., the Leendertz configuration. The object is rotated by a small amount between exposures. This converts the depth information, due to rotation, into an in-plane displacement to which the setup is sensitive. The depth contour interval Δz is given by

$$\Delta z=\frac{\lambda}{2\sin\theta\sin\Delta\phi}\cong\frac{\lambda}{2\sin\theta\ \Delta\phi} \tag{6.65}$$

where 2θ is the inter-beam angle of the illumination waves and $\Delta\phi$ is the angle of rotation. Several modifications of this technique have been published.

6.7 SHEAROGRAPHY

Shearography is a displacement gradient sensitive full-field optical technique that is resilient to the environmental disturbances and vibrations. Its importance arises because material scientists, stress analysts, quality assurance personnel, product developers, and many others are interested in strain values when the objects are deformed. Strain measurements are the basis for predicting how long an object will perform without failure when subjected to certain external influences. Shearography is one of the optical techniques that has evolved both for qualitative and quantitative strain (difference quotient) measurements.

Section 6.5 presents the theory of speckle shear interferometry when the recording is done on photo-emulsion. The same theory applies when the recording is electronic except that the two frames are now independently handled. Speckle shear interferometry, when recording is done on a CCD or CMOS, is known as shearography. Due to its importance as a nondestructive testing technique, it is explained in detail.

A typical experimental setup of shearography is shown in Figure 6.18. It consists of a light source, mostly a laser; an imaging system; and a shearing device placed either between the object and the lens or between the lens and the CCD. The object is illuminated by the laser beam and its image is formed on the detector via the shear device. In between the exposures, the object is subjected to an external loading that deforms the object. Due to shearing, a point on the object is imaged as two points. Alternately, two points on the object are imaged as a single point.

The theoretical framework of shearography is based on the same phase difference equation that is valid for both hologram interferometry and speckle pattern interferometry. The optical phase difference equation is given by

$$\phi = \left(\boldsymbol{k}_0 - \boldsymbol{k}_i\right) \cdot \boldsymbol{L} = \boldsymbol{k} \cdot \boldsymbol{L} \tag{6.66}$$

where $\mathbf{k}_0$ and $\mathbf{k}_i$ are the propagation vectors in the directions of observation and illumination, and $\boldsymbol{L}$ is the deformation vector at a point on the object. The vector $\mathbf{k}\left[=\left(\boldsymbol{k}_0 - \boldsymbol{k}_i\right)\right]$ is called the sensitivity vector. The deformation vector is expressed as $\boldsymbol{L} = u\hat{\boldsymbol{i}} + v\hat{\boldsymbol{j}} + w\hat{\boldsymbol{k}}$ with components u, v, and w along the x-, y-, and z-axes.

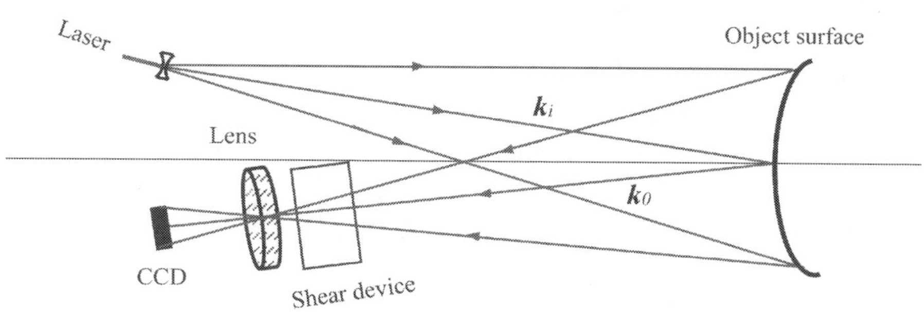

FIGURE 6.18 A schematic of a speckle pattern shear interferometer.

In shearography, there are two points on the object under consideration: point P and its sheared version Q. On loading the object, these two points undergo different displacements as shown in Figure 6.19. Points $P(x,y,z)$ and $Q(x+\Delta x,y,z)$ are very close to each other on the surface with lateral shear Δx along the x-axis. When the object is deformed, these points are displaced to their new locations $P'(x+u,y+v,z+w)$ and $Q(x+\Delta x+u+\Delta u,y+v+\Delta v,z+w+\Delta w)$. Point P undergoes a displacement $\boldsymbol{L}=u\hat{\boldsymbol{i}}+v\hat{\boldsymbol{j}}+w\hat{\boldsymbol{k}}$ and point Q undergoes a displacement $\boldsymbol{L}'=(u+\Delta u)\hat{\boldsymbol{i}}+(v+\Delta v)\hat{\boldsymbol{j}}+(w+\Delta w)\hat{\boldsymbol{k}}$.

The phase difference between the rays when point P displaces to point P'' is given by

$$\phi_P=(\boldsymbol{k}_0-\boldsymbol{k}_i)\cdot\left(u\hat{\boldsymbol{i}}+v\hat{\boldsymbol{j}}+w\hat{k}\right)=k_xu+k_yv+k_zw \tag{6.67}$$

where k_x, k_y, and k_z are the components of vector $\boldsymbol{k}=(\boldsymbol{k}_0-\boldsymbol{k}_i)$, i.e., $\boldsymbol{k}=k_x\hat{\boldsymbol{i}}+k_y\hat{\boldsymbol{j}}+k_z\hat{\boldsymbol{k}}$.

Similarly, the phase difference between the rays when point Q moves to point Q' is given by

$$\phi_Q=\boldsymbol{k}\cdot\left[(u+\Delta u)\hat{\boldsymbol{i}}+(v+\Delta v)\hat{\boldsymbol{j}}+(w+\Delta w)\hat{k}\right]=k_x(u+\Delta u)+k_y(v+\Delta v)+k_z(w+\Delta w) \tag{6.68}$$

Calculations of the phases ϕ_P and ϕ_Q in terms of coordinates of the source and a point on the detector, and their respective distances can be found in the books and in references.

Experimentally, the first exposure is made when the object is in its undeformed state. Because the wave front from the sheared point acts as a reference wave front to

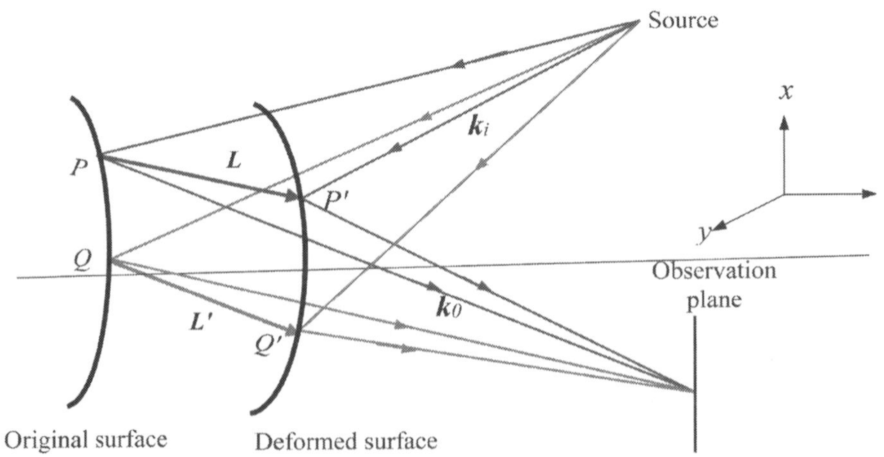

FIGURE 6.19 Original and deformed states of the object showing the location of sheared points on deformation.

the one from the other point or vice versa, there is no need for a reference wave front. The irradiance distribution in the first exposure can be expressed as

$$I_1(x_i, y_i) = 2a_0^2\left[1+\cos\left(\theta_Q-\theta_P\right)\right] = 2a_0^2\left(1+\cos\phi\right);\ \phi = \left(\theta_Q-\theta_P\right) \quad (6.69)$$

where a_0 is the amplitude of each wave, θ_P and θ_Q are the phases of the waves from point P and point Q, respectively, which are random phases due to the surface roughness of the real object surface. The amplitudes a_0 are random and hence a_0^2 provides the speckle pattern. Because the two points are very close to each other, the amplitudes of the waves from these points are taken equal. If the amplitudes of the two waves from points P and Q are unequal, the irradiance distribution is expressed as

$$I_1(x_i, y_i) = I_0\left(1+\gamma\cos\phi\right) \quad (6.70)$$

where I_0 is the total irradiance (sum of irradiances of both the waves) and γ is the modulation.

On loading, the surface deforms, and point P moves to point P' and point Q moves to point Q'. The waves from points P' and Q' acquire additional phases ϕ_P and ϕ_Q, respectively.

The irradiance distribution recorded now is given by

$$I_2(x_i, y_i) = 2a_0^2\left[1+\cos\left\{\left(\theta_Q+\phi_Q\right)-\left(\theta_P+\phi_P\right)\right\}\right] = I_0\left[1+\cos\left(\phi+\Delta\phi_{QP}\right)\right] \quad (6.71)$$

where $\Delta\phi_{QP} = \left(\phi_Q-\phi_P\right)$ is the phase difference between the two waves from points Q and P at a point on the recording plane. Rewriting the phase difference, we have

$$k_x\left(u+\Delta u\right)+k_y\left(v+\Delta v\right)+k_z\left(w+\Delta w\right)-k_x u+k_y v+k_z w = k_x\Delta u+k_y\Delta v+k_z\Delta w \quad (6.72)$$

Assuming the magnitude of the shear to be very small, the phase difference is expressed as

$$\Delta\phi_{QP} = \left(\phi_Q-\phi_P\right) = \left(k_x\frac{\partial u}{\partial x}+k_y\frac{\partial v}{\partial x}+k_z\frac{\partial w}{\partial x}\right)\Delta x = \left(\boldsymbol{k}_o-\boldsymbol{k}_i\right)\cdot\frac{\partial \boldsymbol{L}}{\partial x}\Delta x = \boldsymbol{k}\cdot\frac{\partial \boldsymbol{L}}{\partial x}\Delta x \quad (6.73)$$

where $\boldsymbol{L}'-\boldsymbol{L} = \Delta\boldsymbol{L}$. The phase difference depends on the derivatives of the components of the deformation vector if exceedingly small lateral shear is used. In practice, what is measured is not the derivative but the difference quotient. If the magnitude of the shear is made very small, the technique loses its sensitivity to a difference quotient. Thus far the discussion has been using the lateral shear, but it is possible to employ other shear types and obtain radial and theta derivatives.

In shearography the two exposures corresponding to $I_1(x_i, y_i)$ and $I_2(x_i, y_i)$ are independently handled. In one of the processing methods, the exposure corresponding to $I_2(x_i, y_i)$ is subtracted pixel by pixel from $I_1(x_i, y_i)$. The voltage output of the detector is proportional to $[I_1(x_i, y_i) - I_2(x_i, y_i)]$. The brightness on the monitor is proportional to the voltage output and hence it can be expressed as

$$B \propto \left[I_1(x_i, y_i) - I_2(x_i, y_i)\right] = I_0\left[\cos\phi - \cos\left(\phi + \Delta\phi_{QP}\right)\right] = 2I_0 \sin\left(\phi + \frac{\Delta\phi_{QP}}{2}\right)\sin\frac{\Delta\phi_{QP}}{2} \tag{6.74}$$

To avoid the loss of signal, the output signal is squared before being displayed and hence the brightness B on the monitor is expressed as

$$B \propto 4\ I_0^2 \sin^2\left(\phi + \frac{\Delta\phi_{QP}}{2}\right)\sin^2\left(\frac{\Delta\phi_{QP}}{2}\right) \tag{6.75}$$

The term $\sin^2\left(\phi + \frac{\Delta\phi_{QP}}{2}\right)$ contains the random phase and hence it represents the speckled portion of the brightness distribution on the monitor, which is modulated by the term $\sin^2\left(\frac{\Delta\phi_{QP}}{2}\right)$. There would be dark regions, called the fringes, wherever $\frac{\phi_{QP}}{2} = m\pi$; m being an integer taking values 0, ±1, ±2,. Substituting for $\Delta\phi_{QP}$, we obtain the condition for the formation of dark fringes as

$$\Delta\phi_{QP} = \left(k_x\frac{\partial u}{\partial x} + k_y\frac{\partial v}{\partial x} + k_z\frac{\partial w}{\partial x}\right)\Delta x = \left(\boldsymbol{k}_0 - \boldsymbol{k}_i\right)\cdot\frac{\partial \boldsymbol{L}}{\partial x}\Delta x = \boldsymbol{k}\cdot\frac{\partial \boldsymbol{L}}{\partial x}\Delta x = 2m\pi \tag{6.76}$$

The fringe pattern arises due to the gradient of the deformation vector. The fringe visibility depends on the correlation between the two waves.

It is seen from Equation (6.74) that the DC term is eliminated due to the subtraction process. In another scheme of recording, two speckle patterns with a phase difference of $\pi / 2$ are simultaneously obtained using polarized beam illumination for the deformed and the undeformed state of the object. Because the phase difference is fixed, a fast Fourier transform can be used to eliminate the DC terms of the speckle patterns.

An important feature of shearography is that the experimental setup could be configured to measure each derivative, $\frac{\partial u}{\partial x}$, $\frac{\partial v}{\partial x}$ or $\frac{\partial w}{\partial x}$ independently. For measuring in-plane derivatives, either symmetric illumination directions or symmetric observation directions with respect to local normal are used and the object is illuminated normally. For out-of-plane derivative measurement, the object is illuminated and observed normally. However, if three directions of illumination are used, i.e., three sources placed appropriately, all three derivatives can be obtained. Such a system is called multicomponent measuring system. For such a system, we can write

$$\Delta\phi_{QP1} = \left(k_{x1}\frac{\partial u}{\partial x} + k_{y1}\frac{\partial v}{\partial x} + k_{z1}\frac{\partial w}{\partial x}\right)\Delta x$$

$$\Delta\phi_{QP2} = \left(k_{x2}\frac{\partial u}{\partial x} + k_{y2}\frac{\partial v}{\partial x} + k_{z2}\frac{\partial w}{\partial x}\right)\Delta x$$

$$\Delta\phi_{QP3} = \left(k_{x3}\frac{\partial u}{\partial x} + k_{y3}\frac{\partial v}{\partial x} + k_{z3}\frac{\partial w}{\partial x}\right)\Delta x$$

where $\Delta\phi_{QP1}$, $\Delta\phi_{QP2}$, and $\Delta\phi_{QP3}$ are the phase differences recorded when the object is illuminated by three sources placed appropriately. These equations can be written in the matrix form as

$$\begin{bmatrix} \Delta\phi_{QP1} \\ \Delta\phi_{QP2} \\ \Delta\phi_{QP3} \end{bmatrix} = \begin{bmatrix} k_{x1} & k_{y1} & k_{z1} \\ k_{x2} & k_{y2} & k_{z2} \\ k_{x3} & k_{y3} & k_{z3} \end{bmatrix} \begin{bmatrix} \dfrac{\partial u}{\partial x} \\ \dfrac{\partial v}{\partial x} \\ \dfrac{\partial w}{\partial x} \end{bmatrix} \Delta x \tag{6.77}$$

and the displacement derivatives are obtained as

$$\begin{bmatrix} \dfrac{\partial u}{\partial x} \\ \dfrac{\partial v}{\partial x} \\ \dfrac{\partial w}{\partial x} \end{bmatrix} = \begin{bmatrix} k_{x1} & k_{y1} & k_{z1} \\ k_{x2} & k_{y2} & k_{z2} \\ k_{x3} & k_{y3} & k_{z3} \end{bmatrix}^{-1} \begin{bmatrix} \Delta\phi_{QP1} \\ \Delta\phi_{QP2} \\ \Delta\phi_{QP3} \end{bmatrix} \Delta x^{-1} \tag{6.78}$$

Similar equations are set up when the shear is along the y-axis.

It should be noted that the strain tensor S is given by

$$S = \begin{bmatrix} \varepsilon_x & \gamma_{xy} & \gamma_{xz} \\ \gamma_{yx} & \varepsilon_y & \gamma_{yz} \\ \gamma_{zx} & \gamma_{zy} & \varepsilon_z \end{bmatrix} = \begin{bmatrix} \dfrac{\partial \boldsymbol{u}}{\partial \boldsymbol{x}} & \dfrac{1}{2}\left(\dfrac{\partial \boldsymbol{u}}{\partial \boldsymbol{y}} + \dfrac{\partial \boldsymbol{v}}{\partial \boldsymbol{x}}\right) & \dfrac{1}{2}\left(\dfrac{\partial u}{\partial z} + \dfrac{\partial \boldsymbol{w}}{\partial \boldsymbol{x}}\right) \\ \dfrac{1}{2}\left(\dfrac{\partial \boldsymbol{v}}{\partial \boldsymbol{x}} + \dfrac{\partial \boldsymbol{u}}{\partial \boldsymbol{y}}\right) & \dfrac{\partial \boldsymbol{v}}{\partial \boldsymbol{y}} & \dfrac{1}{2}\left(\dfrac{\partial v}{\partial z} + \dfrac{\partial \boldsymbol{w}}{\partial \boldsymbol{y}}\right) \\ \dfrac{1}{2}\left(\dfrac{\partial \boldsymbol{w}}{\partial \boldsymbol{x}} + \dfrac{\partial u}{\partial z}\right) & \dfrac{1}{2}\left(\dfrac{\partial \boldsymbol{w}}{\partial \boldsymbol{y}} + \dfrac{\partial v}{\partial z}\right) & \dfrac{\partial w}{\partial z} \end{bmatrix} \tag{6.79}$$

where ε_x, ε_y, ε_z are the normal strain components and γ_{xy}, γ_{xz}, γ_{yx}, γ_{yz}, γ_{zx}, γ_{zy} are the shear strain components. Note that all elements of the shear tensor cannot be measured by shearography: only those with bold face can be measured.

Phase shifting is employed when a quantitative data about the strain field or the defects is required. Present day shearographic instruments are equipped with phase shifting. Most common phase-shifting algorithms are (1) three-step, (2) four-step, and (3) five-step phase algorithms. In the three-step phase-shifting algorithm, the phase of one of the beams is stepped by $2\pi/3$ in the interval 0 to 2π and the phase is obtained from,

$$\phi_{QP} = \tan^{-1}\frac{\sqrt{3}\left(I_3 - I_2\right)}{\left(2I_1 - I_2 - I_3\right)} \tag{6.80}$$

where I_1, I_2, and I_3 are the irradiance distributions captured at 0, $2\pi/3$, and $4\pi/3$ phase-step. In the four-step phase-shifting algorithm, the phase of one of the beams is stepped by $\pi/2$ in the interval 0 to 2π and the phase is obtained from

$$\phi_{QP} = \tan^{-1}\frac{(I_4 - I_2)}{(I_1 - I_3)} \tag{6.81}$$

where I_1, I_2, I_3, and I_4 are the irradiance distributions captured at 0, $\pi/2$, π, and $3\pi/2$ phase-step. The five-step algorithm is robust to noise and is insensitive to calibration errors. It captures five irradiance values at five equally spaced steps in the interval 0 to 2π. The value of the phase step as well as of the phase is obtained from the five irradiances captured during the phase shifting. The phase ϕ_{QP} is a wrapped phase and is to be unwrapped. The unwrapping poses no problem when the shear is small.

To study time-dependent deformations, the information is captured in a single frame and the Fourier transform method is used to obtain derivative values. The method, however, requires a carrier frequency so that various Fourier spectra are separated. Several methods to provide carrier frequency in the shearogram are available.

Though the Michelson interferometric arrangement is mostly used for shearing, many other methods have also been investigated. Further, both thermal and pressure/vacuum stressing methods are frequently used in between the exposures.

QUESTIONS

6.1 A diffuse object illuminated with a He-Ne laser is imaged by a 100-mm focal length lens at a magnification of 0.25. If the average size of the speckle in its image is 10 μm, what is the lens aperture?

6.2 Speckle photography is used to measure in-plane displacement of a diffuse object. A double-exposure specklegram is recorded by imaging the object at a magnification of 0.25. The double-exposure specklegram is interrogated with an unexpanded beam from a He-Ne laser (632.8 nm). At a plane distant 1.0 m from the specklegram, the measured fringe width is 3 mm. How much is the in-plane displacement of the object?

6.3 It is desired to introduce lateral and folding shears using a right-angle prism. Design such configurations.

6.4 In the in-plane sensitive configuration using an object coated with retro-reflective paint, the shear is introduced by the tilt of one of the mirrors in an arm. Show that the phase change δ introduced by deformation is given by

$$\delta = \frac{2\pi}{\lambda}4d_x \sin\theta + \frac{2\pi}{\lambda}2\left[\sin\theta\,\frac{\partial d_x}{\partial x_0} + \cos\theta\,\frac{\partial d_z}{\partial x_0}\right]\Delta x_0$$

where θ and $-\theta$ are the angles that the illumination beams make with the local normal, which is along the z-direction.

6.5 In plane displacement measurements are carried out using the configurations shown in Figures 6.12 and 6.13(a). On filtering, a fringe pattern shown in Figure 6.13(b) is obtained: the top portion of the interferogram is obtained using the configuration of Figure 6.12 and bottom portion using the configuration of Figure 6.13(a). In both configurations, the angle α is 30° and a He-Ne laser (632.8 nm) is used both for recording and filtering. Calculate the in-plane displacement.

6.6 For measuring a large amplitude of vibration using time-averaged ESPI, the frequency of the reference wave is shifted by n times the excitation frequency with which the object is excited. Show that the irradiance distribution on the monitor is given by $J_n^2\left[\frac{4\pi}{\lambda} A(x,y)\right]$. It is assumed that the directions of observation and illumination are along the local normal.

7 The Moiré Phenomenon

7.1 FORMATION OF MOIRÉ PATTERN

A moiré pattern is formed when two periodic or quasi-periodic objects are superposed. As an example, consider two coarse gratings with equal opaque and transparent regions placed over each other. When viewed in transmission, a moiré pattern is seen that changes on rotation or translation of one of the gratings. Formation of a moiré pattern is not restricted to linear gratings. Any periodic or quasi-periodic pattern of dots, lines, circles, arcs, spirals, etc., can generate the pattern. A moiré pattern is quite pleasing visually and can leave a lasting impression. The moiré phenomenon is a mechanical effect, and the formation of moiré fringes is best explained by mechanical superposition of the gratings. However, when the period in these periodic structures is very fine, diffraction effects play a very significant role.

In earlier chapters, we have discussed fringe formation by calculating the phase difference. Formation of the fringes can also be explained on the basis of the moiré phenomenon. As an example, both hologram interferometry (HI) and speckle interferometry record the initial and final states of the object through the phenomenon of interference. The resultant fringe pattern can be explained as the moiré between the interference structures that belong to the initial and the final states of the object. Two-wavelength interferometric fringes are also moiré fringes. In general, the moiré pattern can be regarded as a mathematical solution to the interference of two periodic functions. The holodiagram, a tool developed by Abramson to deal with several issues in HI, is a beautiful device to study fringe formation and fringe control using the moiré phenomenon.

We can explain the moiré phenomenon using the indicial equation or by the superposition of two sinusoidal gratings. We will follow both approaches, beginning with the indicial equation.

7.1.1 The Moiré Fringe Pattern between Two Linear Gratings

Let us consider a line grating with lines running parallel to the y-axis and having a period b as seen in Figure 7.1(a). This grating is described by

$$x = m\,b \tag{7.1}$$

where the various lines in the grating are identified by index m, which takes values 0, ±1, ±2, ±3,... The second grating of period a is inclined at an angle θ with the y-axis as shown in Figure 7.1(b). The lines in this grating are represented by

$$y = x\,\cot\theta - n\,a/\sin\theta \tag{7.2}$$

DOI: 10.1201/9781003531845-7

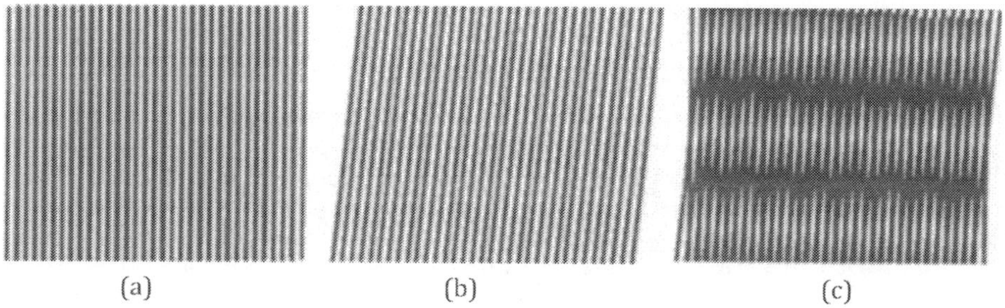

FIGURE 7.1 (a) Vertical grating. (b) Inclined grating. (c) Moiré pattern as a result of superposition.

where $a/\sin\theta$ is the intercept with the y-axis and the index n takes values 0, ±1, ±2, ±3,…. and hence identifies various lines in the grating. Formation of the moiré fringe pattern, on superposition of the two-line gratings, is governed by the indicial equation

$$m \pm n = p \tag{7.3}$$

where p is another integer that takes the values 0, ±1, ±2, ±3,…. The plus sign in the indicial equation generates sum moiré fringes that are of high frequency, while the minus sign generates the difference moiré fringes, which are the frequently used and observed moiré patterns. We will be using the *difference moiré pattern* ($m - n = p$) unless mentioned otherwise. We obtain the equation of the moiré fringes by eliminating m and n from Equations (7.1), (7.2), and (7.3) as

$$y = x\,\frac{b\cos\theta - a}{b\sin\theta} + \frac{p\,a}{\sin\theta} \tag{7.4}$$

The moiré fringe pattern is shown in Figure 7.1(c). Equation (7.4) can be written in a more familiar form like Equation (7.2) as

$$y = x\,\cot\phi + p\,d/\sin\phi \tag{7.5}$$

This implies that the moiré pattern is a grating of period d, which is inclined at an angle ϕ with the y-axis, where

$$\cot\phi = \frac{b\cos\theta - a}{b\sin\theta} \Rightarrow \sin\phi = \sin\theta\frac{b}{\sqrt{a^2 + b^2 - 2ab\cos\theta}} \tag{7.6a}$$

and

$$d = \frac{a\,b}{\sqrt{a^2 + b^2 - 2ab\cos\theta}} \tag{7.6b}$$

It is interesting to study the formation of moiré patterns for the following two situations.

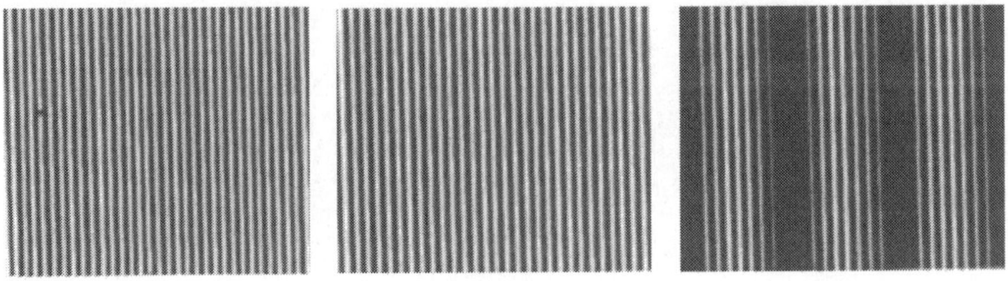

(a) (b) (c)

FIGURE 7.2 (a) Vertical grating. (b) Another vertical grating with a different pitch. (c) The moiré pattern as a result of pitch mismatch.

7.1.1.1 $a \neq b$ but $\theta = 0$: Pitch Mismatch

In this situation the grating lines are parallel to each other, but their periods are different. This is a well-known situation of pitch mismatch. The moiré fringes run parallel to the grating lines and the spacing d is $d = ab/|a - b|$. Here $(a - b)$ represents the pitch mismatch. When the gratings are of nearly the same pitch, $a \approx b$, then $d = a^2/|a - b|$. Figures 7.2(a) and (b) show the gratings of pitches a and b, respectively, with grating elements parallel to the y-axis. The moiré fringes due to pitch mismatch are shown in Figure 7.2(c). Physically, the moiré spacing is the distance over which pitch mismatch accumulates to the pitch of the grating itself. When the gratings are of equal period, the moiré spacing is infinite. This arrangement is therefore called as an infinite fringe mode.

7.1.1.2 $a = b$ but $\theta \neq 0$: Angular Mismatch

This is referred to as an angular mismatch between two identical gratings. It results in moiré pattern formation with a period $d = a/2\sin(\theta/2)$ and its orientation ϕ with the y-axis is $\pi/2 + \theta/2$. In fact, the moiré fringes run parallel to the bisector of the larger enclosed angle between the gratings.

Moiré fringe formation is easily appreciated when we work in the Fourier domain. In the Fourier domain, a sinusoidal grating of finite size generates three spectra (spots): the spots lie on a line, which passes through the origin, and is perpendicular to the grating elements. This is because the spectrum of a real grating (irradiance grating) is centro-symmetric. We can therefore consider only one half of the spectrum. The spots lie on a line, which is perpendicular to the grating elements (lines), and the distance between two consecutive spots is proportional to the frequency of the grating. The second grating also generates its spectrum, which is rotated by an angle equal to the angle between the two gratings. When two gratings are superposed, the difference Δr between the two spots (Figure 7.3) generates a grating. If Δr lies within the visibility circle (the spectrum lying in this circle will generate a grating that is seen by an unaided eye), the moiré pattern is formed: the pitch of the moiré fringes is inversely proportional to the length Δr, and the orientation is normal to it. When two gratings of equal period are superposed with an angular mismatch, a moiré fringe pattern running parallel to the bisector of these gratings, and the

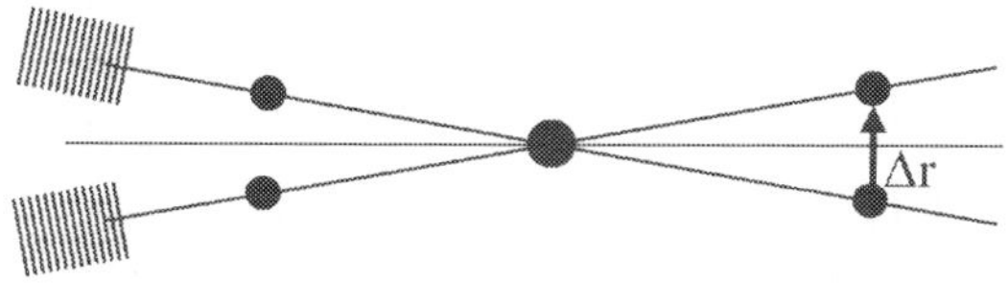

FIGURE 7.3 Spectra of two inclined sinusoidal gratings.

spacing inversely proportional to the angular mismatch, is formed. Alternately, if $\overline{g}_1$ and $\overline{g}_2$ are the grating vectors for the two gratings then $\overline{\Delta r}$ is the vector for the moiré fringes that run perpendicular to it.

7.1.1.3 Moiré Fringe Pattern between a Linear Grating and a Circular Grating

As mentioned earlier, moiré fringes are obtained by the superposition of periodic structures. The superposition of a linear grating with a circular grating is of some academic interest; we consider moiré formation by these two gratings. We take a linear grating of period b with its elements running parallel to the y-axis. It is represented, again, by

$$x = b\ m \text{ for } m = 0, \pm 1, \pm 2, \pm 3, \ldots \tag{7.7}$$

The circular grating of period a is centered at the origin of the coordinate system, hence it is represented by the equation of equi-spaced circles as

$$x^2 + y^2 = a^2 n^2 \text{ for } n = 0, \pm 1, \pm 2, \pm 3, \ldots \tag{7.8}$$

Using the indicial equation $m - n = p$, the moiré pattern is obtained as

$$\frac{x}{b} - \frac{\sqrt{x^2 + y^2}}{a} = p \Rightarrow \frac{x^2 + y^2}{a^2} = p^2 + \frac{x^2}{b^2} - \frac{2x\,p}{b} \tag{7.9}$$

This expression can be rewritten as follows:

$$x^2\left(\frac{1}{a^2} - \frac{1}{b^2}\right) + \frac{y^2}{a^2} + 2\frac{x}{b}p - p^2 = 0 \tag{7.10}$$

Equation (7.10) represents hyperbolas, ellipses, or parabolas depending on the relative grating periods.

For $a = b$, the moiré pattern is given by

$$\frac{y^2}{b^2} + 2\frac{x}{b}p - p^2 = 0 \tag{7.11}$$

which represents parabolic moiré fringes (see Figure 7.4). The Fourier transform (FT) of a circular grating lies on a circle; hence an overlap of a linear grating with a circular grating produces a moiré that has a very wide FT spectrum.

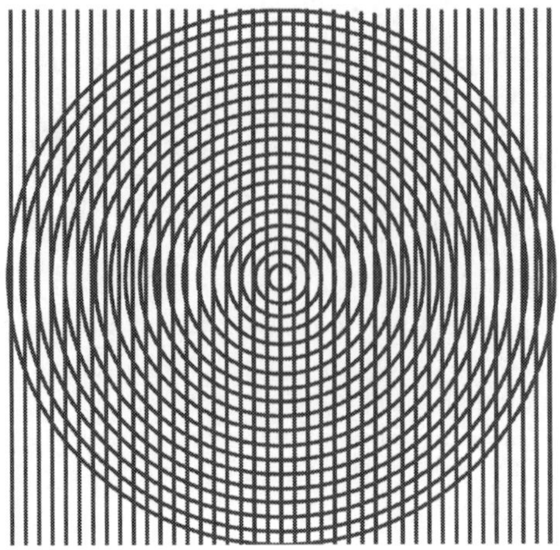

FIGURE 7.4 Moiré pattern between a linear grating and a circular grating.

7.1.1.4 Moiré Pattern between Sinusoidal Gratings

So far we have investigated the process of moiré formation using line gratings. These gratings are seldom used in practice. One normally uses gratings of lines of finite width, which are binary, because they are very easy to produce. However, when high-frequency gratings are used, these are generally fabricated interferometrically and hence have a sinusoidal profile. Even binary gratings can be Fourier decomposed into sinusoidal components. It is, therefore, instructive to see how the moiré pattern of sinusoidal gratings is formed.

The sinusoidal gratings can be recorded either on the same film or on two separate films, which are then superposed as was done with line gratings. We will examine these two situations separately. Let us consider a grating defined by the transmittance function $t_1(x)$:

$$t_1(x) = t_0\left[1 - M\cos(2\pi\ x/b)\right] = t_0\left[1 - M\cos(2\pi f_0 x)\right] \tag{7.12}$$

where t_0 is the bias transmission, M is the modulation, b is the period, and f_0 is the spatial frequency of the grating. The grating elements run parallel to the y-axis. When $M = 1$, the grating has unit contrast, and its transmission function lies between 0 and $2t_0$. Let us now take another sinusoidal grating that is inclined with respect to the first grating. Its transmission function $t_2(x,y)$ is expressed as

$$t_2(x,y) = t_0\left[1 - M\ \cos\frac{2\pi}{a}(-x\ \cos\theta + y\ \sin\theta)\right] = t_0\left[1 - M\ \cos 2\pi\left(-f_x x + f_y y\right)\right] \tag{7.13}$$

The grating is inclined at an angle θ with the y-axis and has a period of a. Its spatial frequencies along the x- and the y-directions are represented by f_x and f_y, respectively, such that $\frac{1}{a^2} = f_x^2 + f_y^2$. Further, both gratings are assumed to have the same modulation.

When the gratings are recorded on the same film/plate sequentially, the transmission function of the positive transparency, with proper processing, can be taken as being proportional to the sum of the two transmission functions: that is,

$$t(x,y) \propto \left[t_1(x,y) + t_2(x,y)\right]$$
$$= 2t_0\left\{1 - M\ \cos\pi\left[(f_x + f_0)x - f_y y\right]\ \cos\pi\left[(-f_x + f_0)x + f_y y\right]\right\} \tag{7.14}$$

This transmission function corresponds to a grating, which is modulated by a low-frequency grating (i.e., the moiré pattern). Bright moiré fringes are formed when

$$\cos\pi\left[(-f_x + f_0)x + f_y y\right] = -1 \Rightarrow (-f_x + f_0)x + f_y y = 2m + 1 \tag{7.15a}$$

where m is an integer. Similarly, dark moiré fringes are formed when

$$\cos\pi\left[(-f_x + f_0)x + f_y y\right] = 1 \Rightarrow (-f_x + f_0)x + f_y y = 2m \tag{7.15b}$$

The moiré fringes are inclined at an angle ϕ with the y-axis such that

$$\cot\phi = \frac{f_x - f_0}{f_y} = \frac{b\ \cos\theta - a}{b\ \sin\theta} \tag{7.16a}$$

When $a = b$, we have

$$\cot\phi = \frac{\cos\theta - 1}{\sin\theta} = -\tan\frac{\theta}{2} = \cot\frac{\pi + \theta}{2} \tag{7.16b}$$

The period of the moiré pattern is also obtained as

$$d = \frac{1}{\sqrt{(f_x - f_0)^2 + f_y^2}} = \frac{a\ b}{\sqrt{a^2 + b^2 - 2ab\ \cos\theta}} \tag{7.17}$$

These are the same formulae that were obtained for line gratings using indicial equation.

When the gratings are recorded on separate films and the moiré pattern due to their overlap is observed, the transmission function is obtained by the multiplication of their respective transmission functions, i.e., $t(x,y) = t_1(x)\ t_2(x,y)$. The moiré pattern is then obtained following the same procedure as explained here.

7.2 MOIRÉ PATTERN BETWEEN REFERENCE AND DEFORMED GRATINGS

When the moiré phenomenon is used for metrology, one of the gratings is mounted on the object, which is subjected to deformation. Therefore, one then observes the moiré between the deformed grating and the reference grating (undeformed grating). One can also obtain a moiré pattern between two deformed gratings (i.e., when two deformed states of the object are compared). It is thus instructive to study the moiré formation between deformed and undeformed gratings and learn how to extract the information about the deformation from the moiré pattern.

7.2.1 Reference and Deformed Gratings Oriented along Y-Axis

The deformation is represented by a function *f*(*x*,*y*) called distortion function, which is assumed to be slowly varying. The deformed grating can be expressed as

$$x + f(x,\ y) = m\ b \text{ for } m = 0, \pm1, \pm2, \pm3, \pm4, \ldots \tag{7.18a}$$

This grating is superposed on a reference grating represented by

$$x = n\ b \text{ for } n = 0, \pm1, \pm2, \pm3, \pm4, \ldots \tag{7.18b}$$

This results in a moiré pattern as

$$f(x,\ y) = p\ b \text{ for } p = 0, \pm1, \pm2, \pm3, \pm4, \ldots \tag{7.19}$$

The moiré fringes represent a contour map of *f*(*x*,*y*) with period *b*. Here, the elements in both the gratings run parallel to the *y*-axis with the deformed grating exhibiting slow variation.

7.2.2 Reference Grating Inclined

We can also obtain moiré between the deformed grating and a reference grating that is oriented at an angle θ with the *y*-axis. That is, the gratings are expressed as

$$x + f(x,\ y) = m\ b \tag{7.18a}$$

and

$$y = x\ \cot\theta - n\ b/\sin\theta \tag{7.20}$$

The moiré pattern, when the grating is inclined by a small angle such that $\cos\theta \sim 1$ and $\sin\theta \sim \theta$ is given by

$$y + \frac{f(x,y)}{\theta} = p\frac{b}{\theta} \tag{7.21}$$

This describes a moiré grating with the period and the distortion function magnified by a factor ($1/\theta$).

When two distorted gratings are superposed, the moiré pattern gives the difference between the two distortion functions. This difference can also be magnified when the finite fringe mode of moiré formation is used.

7.2.3 Gratings with Different Periods

Pitch mismatch between the gratings can also be used to magnify the effect of distortion. As an example, we consider a distorted grating and a reference grating given by

$$x + f(x, y) = m\ b \tag{7.18a}$$

and

$$x = n\ a \tag{7.22}$$

The moiré pattern is given by

$$x + \frac{a}{|a-b|} f(x,y) = \frac{a\ b}{|a-b|}\ p \tag{7.23}$$

The period of the moiré pattern is $ab/|a-b|$ and the distortion function has been magnified by $a/|a-b|$.

7.3 DERIVATIVE OF DISTORTION FUNCTION

The derivative of the distortion function is obtained by observing the moiré between the deformed grating and its displaced replica. For example, let us consider the deformed grating represented by

$$x + f(x, y) = m\ b \tag{7.18a}$$

Its replica has been displaced along the x-direction by Δx and hence it is represented by

$$x + \Delta x + f(x + \Delta x,\ y) = n\ b \tag{7.24}$$

When these gratings are superposed, the moiré pattern is given by

$$\Delta x + f(x + \Delta x,\ y) - f(x,\ y) = p\ b \tag{7.25}$$

In the limit, when the lateral shift Δx is small, we obtain

$$\Delta x + \frac{\partial f(x,y)}{\partial x} \Delta x = p\ b \tag{7.26}$$

The first term is a constant and just represents a shift of the moiré pattern. The moiré pattern thus displays the partial x-derivative of distortion function $f(x,y)$. As mentioned earlier, the moiré effect can be magnified by $(1/\theta)$ using finite fringe mode, i.e., when one of the gratings is inclined by θ.

7.4 MOIRÉ PATTERN WITH DEFORMED SINUSOIDAL GRATING

The transmission function $t_1(x,y)$ of a deformed sinusoidal grating is given by

$$t_1(x,y) = A_0 + A_1 \cos\frac{2\pi}{b}\left[x - f(x,y)\right] \tag{7.27}$$

where A_0 and A_1 are the constants specifying bias transmission and the modulation of the grating and $f(x,y)$ represents the distortion of the grating. The reference grating, oriented at an angle θ, is represented by

$$t_2(x,y) = B_0 + B_1 \cos\frac{2\pi}{a}(x\cos\theta - y\sin\theta) \tag{7.28}$$

Here, B_0 and B_1 are constants. The moiré fringe formation is described in the following sections.

7.4.1 MULTIPLICATIVE MOIRÉ PATTERN

Moiré fringes are formed when the angle θ is small, the deformation $f(x,y)$ varies slowly in space, and the periods of the two gratings are nearly equal; in general, $a = N\,b$, where N is an integer. The transmission function for multiplicative moiré is the product of their individual transmission functions, that is,

$$\begin{aligned} t(x,y) = t_1(x,y)t_2(x,y) &= A_0B_0 + A_1B_0\cos\frac{2\pi}{b}\left[x - f(x,y)\right] \\ &+ A_0B_1\cos\frac{2\pi}{a}(x\cos\theta - y\sin\theta) \\ &+ A_1B_1\cos\frac{2\pi}{b}\left[x - f(x,y)\right]\cos\frac{2\pi}{a}(x\cos\theta - y\sin\theta) \end{aligned} \tag{7.29}$$

Assuming the contrast of both gratings to be the same, i. e., $A_0 = B_0$ and $A_1 = B_1$, we obtain

$$\begin{aligned} t(x,y) &= A_0^2 + A_1A_0\cos\frac{2\pi}{b}\left[x - f(x,y)\right] + A_0A_1\cos\frac{2\pi}{a}(x\cos\theta - y\sin\theta) \\ &+ \frac{A_1^2}{2}\left\{\cos 2\pi\left[x\left(\frac{1}{b} + \frac{\cos\theta}{a}\right) - y\frac{\sin\theta}{a} - \frac{f(x,y)}{b}\right]\right. \\ &\left.+ \cos 2\pi\left[x\left(\frac{1}{b} - \frac{\cos\theta}{a}\right) + y\frac{\sin\theta}{a} - \frac{f(x,y)}{b}\right]\right\} \end{aligned} \tag{7.30}$$

In Equation (7.30), the first term is a DC term; the second, third, and fourth terms are the carriers; and the fifth term represents the moiré pattern. Moiré fringes are formed wherever

$$x\left(\frac{1}{b} - \frac{\cos\theta}{a}\right) + y\frac{sin\theta}{a} - \frac{f(x,y)}{b} = p \tag{7.31}$$

where p is an integer. The moiré fringes are deformed straight lines due to $f(x,y)$, the local deformation being $f(x,y)/\theta$, when θ is small and $a = b$.

7.4.2 Additive Moiré Pattern

An additive moiré pattern is obtained when the transmission functions of the individual gratings are added. Therefore, the transmission function, assuming gratings of the same modulation, is given by

$$\begin{aligned} t(x,y) = t_1(x,y) + t_2(x,y) = 2A_0 + A_1 \cos\frac{2\pi}{b}\left[x - f(x,y)\right] \\ + A_1 \cos\frac{2\pi}{a}(x\cos\theta - y\sin\theta) \end{aligned} \quad (7.32)$$

This can be written as follows:

$$\begin{aligned} t(x,y) = 2A_0 + A_1 \cos\pi\left[x\left(\frac{1}{b}+\frac{\cos\theta}{a}\right) - y\frac{\sin\theta}{a} - \frac{f(x,y)}{b}\right] \\ \cos\pi\left[x\left(\frac{1}{b}-\frac{\cos\theta}{a}\right) + y\frac{\sin\theta}{a} - \frac{f(x,y)}{b}\right] \end{aligned} \quad (7.33)$$

The second cosine term represents the moiré pattern, which modulates the carrier grating. Due to cosine variation of the moiré term, the phase of the carrier changes intrinsically at the crossover point. The visibility of the moiré pattern is generally poor, and imaging optics is required to resolve the carrier grating. The contrast can be improved by filtering the necessary information using the FT processor. For the purpose of filtering, the transparency $t(x,y)$ is placed at the input of the FT processor, and zero-order and two first-order diffraction spots are observed at the filter/frequency plane. Either of the first orders is filtered out and used for imaging. The irradiance distribution at the output plane is then proportional to

$$\begin{aligned} I(x,y) \propto A_1^2\cos^2\pi\left[x\left(\frac{1}{b}-\frac{\cos\theta}{a}\right) + y\frac{\sin\theta}{a} - \frac{f(x,y)}{b}\right] \\ = \frac{1}{2}A_1^2\left\{1+\cos2\pi\left[x\left(\frac{1}{b}-\frac{\cos\theta}{a}\right) + y\frac{\sin\theta}{a} - \frac{f(x,y)}{b}\right]\right\} \end{aligned} \quad (7.34)$$

This represents a unit contrast moiré pattern.

7.5 TALBOT PHENOMENON

When a periodic object is illuminated by a coherent monochromatic beam, its image is formed at specific planes called the self-image planes or the Talbot planes. The effect was first observed by Talbot in 1836, and its theory was worked out by Rayleigh in 1881. Self-imaging is due to diffraction and can be observed with periodic objects that satisfy the Montgomery condition. A linear (one-dimensional [1D]) grating is one such object. For a 1D grating of spatial frequency f_x, illuminated by a collimated beam of wavelength λ, the self-image planes are equi-distant and are located at distances $N / f_x^2\lambda$ from the object where $N = 1, 2, 3, \ldots$. gives the order of the Talbot planes. In other words, the transverse periodicity of the object manifests itself as the longitudinal periodicity. The imaging is called self-imaging because no imaging devices are used. A two-dimensional (2D) cross-grating with the same spatial frequency f_x in both directions also self-images at the planes located at $N / f_x^2\lambda$ from the grating.

7.5.1 Talbot Effect in Collimated Illumination

To explain the Talbot imaging, let us consider a 1D grating whose transmittance is given by

$$t(x) = ½\left[1 + \cos(2\pi f_x x)\right]$$

This grating is placed at the $z = 0$ plane and is illuminated by a collimated beam of amplitude A, as shown in Figure 7.5. The amplitude $u(x,0)$ of the wave just behind the grating ($z = 0$ plane) will be given by

$$u(x,0) = ½A\left[1 + \cos(2\pi f_x x)\right] \quad (7.35)$$

Using the Fresnel diffraction approach, the amplitude at any plane z is obtained as

$$u(x_1,z) = \frac{A}{2}\, e^{ikz}\left[1 + e^{-i\pi f_x^2 \lambda z}\cos(2\pi f_x x_1)\right] \quad (7.36)$$

The amplitude distribution at any plane z will be identical to the grating transmittance function except for a constant multiplicative phase factor, if $e^{-i\pi f_x^2 \lambda z} = 1$. This condition is satisfied when

$$\pi f_x^2 \lambda\, z = 2N\pi \text{ for } N = 0, 1, 2, 3, \ldots \quad (7.37)$$

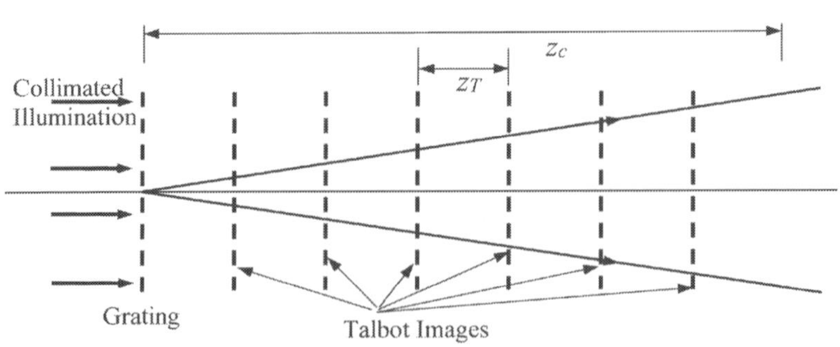

FIGURE 7.5 Formation of Talbot images in collimated illumination.

The planes at which this condition is satisfied are the Talbot planes. However, when N takes half integer values, we still obtain the transmittance function of a sinusoidal grating, but it is phase-shifted by π. Thus, the Talbot planes are separated by $z_T = 1/f_x^2\lambda$. In the case of collimated illumination, the Talbot planes are equi-spaced.

7.5.2 Cutoff Distance

The Talbot images are formed due to constructive interference among the diffracted waves at successive Talbot planes. For an infinite grating illuminated by an infinitely large beam, the various diffracted waves will continue overlapping irrespective of the distance, thus producing infinite number of Talbot images. In a practical situation, both the grating and the beam are of finite dimensions. Therefore, the diffracted waves will not overlap after a certain distance; consequently, no Talbot images are formed after this distance. Let us consider a grating of linear dimension D and spatial frequency f_x illuminated by a beam of size D. The cutoff distance z_c is defined as the distance from the grating over which the first-order diffracted beam deviates by half the beam size. This is given by $z_c = D/2 f_x \lambda$.

7.5.3 Talbot Effect in Noncollimated Illumination

Let us consider that a point source is placed at (0, 0) and the grating is at the $z = R$ plane. The grating is illuminated by a divergent spherical wave from the point source. The amplitude of the wave just behind the grating is

$$u(x,R) = \frac{A}{R} e^{ikR} e^{i\frac{k}{2R}x^2} \left(1 + \cos 2\pi f_x x\right) \tag{7.38}$$

where A is the amplitude of the divergent spherical wave at unit distance from the point source. Equation (7.38) is valid under paraxial approximation. Using Fresnel diffraction approximation, the amplitude at a plane distant z from the grating is

$$u(x_1, R+z) = \frac{iA}{\lambda(R+z)} e^{ik(R+z)} e^{i\frac{k}{2(R+z)}x_1^2} \left[1 + e^{-i\pi\frac{Rz}{R+z}f_x^2\lambda} \cos\left(2\pi f_x \frac{R}{R+z} x_1\right)\right] \tag{7.39}$$

If $e^{-i\pi\frac{Rz}{R+z}f_x^2\lambda} = 1$, then the Equation (7.39) represents a transmittance function of a grating of spatial frequency $f_x' = f_x \frac{R}{R+z}$ multiplied by a complex constant. Therefore, a grating with its spatial frequency dependent on distance z is formed. This yields the self-image planes distances $(z_T)_s$ as follows:

$$\left(z_T\right)_s = \frac{2N}{f_x^2\lambda - \frac{2N}{R}} \quad \text{for } N = 1, 2, 3, \ldots \tag{7.40}$$

The spacing between the successive Talbot planes increases with the order N. The period of the grating also increases as if it is geometrically projected. Similarly, when the grating is illuminated by a convergent spherical wave, the successive Talbot planes come closer and the spatial frequency increases. This is valid until some distance from the point of convergence.

QUESTIONS

7.1 An index grating moves over a scale/main grating in a length measuring system. Both the gratings have the same pitch of 10 μm and enclose an angle of 0.5°. What is the pitch d of the moiré fringes? If the pitch d of the moiré fringes is digitized into eight pulses, what is the least count of the length measuring system?

7.2 Artwork prepared by a computer is described by equation

$$a\left[x^3 + x\left(3y^2 + b\right)\right] = m$$

where a and b are constants and m is an integer (m = 1,2,3,4, ….). Two such examples of artwork displaced by $\pm d$ along the x-direction are superposed over each other. Show that the resultant moiré is a zone plate. What is its primary focal length in terms of a, b, and d?

7.3 Two circular gratings of different periods centered on the origin of a Cartesian coordinate system are superposed. Obtain the equation of moiré fringes. What is the pitch of the moiré fringes?

7.4 Two circular gratings with their canters displaced by $\pm d$ along the x-axis from the origin of the Cartesian coordinate system are superposed. Obtain the moiré pattern.

7.5 Two patterns are described by equations

$$6r^2 + 4x = m$$

$$-6r^4 + 4x + 2xy = n$$

where m and n are integers, and $r^2 = x^2 + y^2$. Obtain an equation of moiré when these two patterns are superposed.

7.6 A moiré pattern as a result of superposition of two patterns A and B is described by

$$4r^4 + 6r^2 - 2xy + 2x = p$$

where p is an integer. The pattern A is described by equation

$$6r^4 + 6r^2 - 2xy - 2x = m$$

where m is an integer. Obtain the equation of pattern B.

7.7 A circular grating is described by equation

$$x^2 + y^2 = m^2$$

When this circular grating is superposed on another circular grating, a moiré pattern described by the following equation is formed

$$x^2 + y^2 = 4p^2$$

Obtain the equation that describes the second circular equation.

8 Photoelasticity

Light is a very tiny part of the electromagnetic spectrum. The electromagnetic waves are transverse waves; the electric and the magnetic field vectors are orthogonal and vibrate perpendicular to the direction of propagation in free space or in an isotropic medium. In fact, the electric field vector, the magnetic field vector, and the propagation vector form an orthogonal triplet. The photo effects, like seeing or the recording on a photographic emulsion, etc., are attributed to the electric field vector; hence, while dealing with light, we are only concerned with the electric field vector (**E** vector).

Let us consider a plane wave propagating along the z-direction and the electric vector confined to the y–z plane. The tip of the electric vector describes a line in the y–z plane as the wave propagates. Such a wave is called a plane-polarized wave. The light emitted by an incandescent lamp, or a fluorescent tube, is not plane polarized as the waves emitted by the source, though plane polarized, are randomly oriented. Such a wave is called unpolarized or natural light. We can obtain polarized light from unpolarized light. This will be discussed in a later section.

8.1 SUPERPOSITION OF TWO PLANE POLARIZED WAVES

Let us consider two orthogonally polarized plane waves propagating in the z-direction: in one wave the **E** vector is vibrating along the y-direction, while in the other wave the **E** vector vibrating along the x-direction. These waves are described by

$$E_y(z;t) = E_{0y}\cos\left(\omega t - \kappa z + \delta_y\right) \tag{8.1a}$$

$$E_x(z;t) = E_{0x}\cos\left(\omega t - \kappa z + \delta_x\right) \tag{8.1b}$$

where δ_y and δ_x are the phases of these waves. These waves satisfy the wave equation. The sum of these waves, due to superposition principle, should also satisfy the wave equation. In general, a wave will have both the x- and y-components and can be written as

$$\boldsymbol{E}(z;t) = \hat{i}\ E_x(z;t) + \hat{j}\ E_y(z;t) \tag{8.2}$$

We wish to find out what track the tip of the electric vector traces when the waves are superposed. To do this, we introduce a variable $\tau = (\omega t - kz)$ and express the plane waves as

$$E_y(z;t) = E_{0y}\left(\cos\tau\ \cos\delta_y - \sin\tau\ \sin\delta_y\right)$$

$$E_x(z;t) = E_{0x}\left(\cos\tau\ \cos\delta_x - \sin\tau\ \sin\delta_x\right)$$

 DOI: 10.1201/9781003531845-8

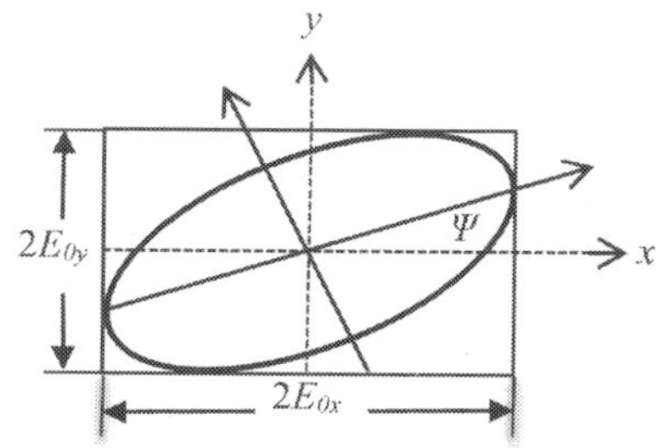

FIGURE 8.1 Elliptical polarization.

Eliminating τ from these equations, we obtain

$$\left(\frac{E_y}{E_{0y}}\right)^2+\left(\frac{E_x}{E_{0x}}\right)^2-2\frac{E_y}{E_{0y}}\frac{E_x}{E_{0x}}\cos\delta=\sin^2\delta \tag{8.3}$$

where $\delta = \delta_x - \delta_y$. This equation represents an ellipse. The ellipse is inscribed inside a rectangle of sides $2E_{0y}$ and $2E_{0x}$, which are parallel to the y- and the x-axes as shown in Figure 8.1. Hence, in the general case of the propagation of a monochromatic wave, the tip of its electric vector traces out an ellipse in any z-plane. Such a wave is called elliptically polarized. As it is a propagating wave, the tip of the ***E*** vector traces out a spiral. The tip of the ***E*** vector can rotate either clockwise or anticlockwise in the plane. These are termed as right-handed polarization (the ***E*** vector rotates clockwise when facing the source) and left-handed polarization (the ***E*** vector rotates anticlockwise when facing the source of light).

It is shown that the rotation direction is governed by the sign of the phase difference, δ. Let us consider a moment of time t_0 when $(\omega t_0 - kz + \delta_y) = 0$. At this moment, $E_y = E_{0y}$ and $E_x = E_{0x}\cos(\delta_x - \delta_y) = E_{0x}\cos\delta$ and $dE_x/dt = -\omega E_{0x}\sin\delta$. The rate of change of E_x, i.e., dE_x/dt is negative when $0 < \delta < \pi$ and positive when $\pi < \delta < 2\pi$. The former case corresponds to the right-handed polarized wave while the latter corresponds to the left-handed polarized wave.

There are two special cases of the elliptical polarization: plane polarization or linear polarization and circular polarization. Various polarization states of the light are shown in Figure 8.2.

8.1.1 Linear Polarization

When the phase difference between the two waves is a multiple of π, we obtain a linearly polarized wave. When both waves are in phase, i.e., $\delta = 2m\pi$: $m = 0, 1, 2,$ the ***E*** vector traces a line in the first and the third quadrants. When they are in anti-phase, i.e., $\delta = (2m + 1)\pi$, the ***E*** vector traces a line in the second and the fourth quadrants.

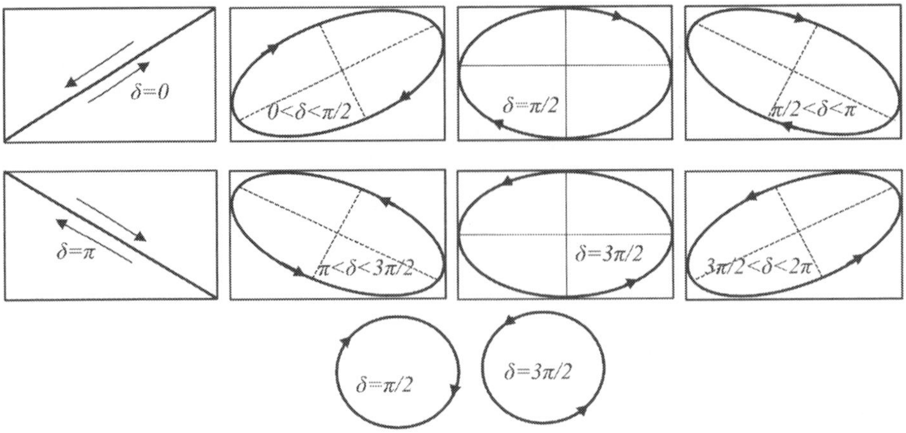

FIGURE 8.2 Polarization states of a sinusoidal electromagnetic wave: linear, elliptical with increasing phase difference δ, and circular when $E_{0x} = E_{0y}$ and $\delta = (2m + 1)\pi/2$.

8.1.2 Circular Polarization

When the phase difference between the two waves is an odd multiple of $\pi/2$, i.e., $\delta = (2m + 1)\pi/2$, the wave is an elliptically polarized wave, but its major and minor axes are now aligned parallel to the x- and the y-axes. However, if the amplitudes of the two waves are also equal, then it becomes a circularly polarized wave. It is right-handed circularly polarized if the phase difference is $\pi/2$, $5\pi/2$, $9\pi/2$, … and left-handed circularly polarized when $\delta = 3\pi/2$, $7\pi/2$, $11\pi/2$, …. Therefore, two conditions are to be met to obtain circularly polarized light: the phase difference must be an odd multiple of $\pi/2$ and the amplitudes of both waves must be equal.

8.2 PRODUCTION OF POLARIZED LIGHT

Natural light is unpolarized with the $\boldsymbol{E}$ vector taking all possible orientations randomly. We can, however, resolve the $\boldsymbol{E}$ vector orientations into two components, one oscillating in the plane of incidence or p-polarized while the other orthogonal to this or s-polarized. The amplitudes of these components at any instant are equal. It is also seen that we can, in general, obtain an elliptically polarized wave by introducing a phase difference between the orthogonally polarized components. However, most often we require a linearly polarized wave. Fortunately, there are a number of ways to obtain linearly polarized wave from the natural light. These methods are based on the following:

1. Reflection at a dielectric interface
2. Refraction at a dielectric interface
3. Double refraction
4. Dichroism
5. Scattering

Of these, the first four methods are used for making practical polarizers, which are devices that produce linearly polarized light wave from natural light.

8.2.1 Reflection

When a beam of light is incident at the air-dielectric interface at an incidence angle θ_B (called the Brewster angle), the reflected beam is linearly polarized and the $\boldsymbol{E}$ vector oscillates in the plane perpendicular to the plane of incidence. This is also called s-polarized beam. The transmitted beam is partially polarized. The Brewster angle θ_B is governed by the refractive index μ of the dielectric medium. For reflection at the air-dielectric interface, the Brewster angle is given by the relation

$$\tan\theta_B = \mu \tag{8.4}$$

It may be noted that the reflection at Brewster angle at the air-dielectric interface fixes the direction of vibration of the electric vector in the reflected beam and hence is used for calibration of polarizers, and so on.

8.2.2 Refraction

As mentioned earlier, the reflected beam is s-polarized, and the transmitted beam is partially polarized when the angle of incidence is the Brewster angle. The transmitted beam has less of an *s*-component as some of it has been removed by reflection at the Brewster angle. However, if several successive reflections are allowed on a number of plane parallel plates aligned at the Brewster angle, most of the s-polarized light will be removed by reflection and the transmitted beam is then p-polarized. This kind of polarizer is known as a pile-of-plates polarizer. This polarizer is often used with high-power lasers.

8.2.3 Double Refraction

There is a class of crystals in which an incident beam decomposes into two linearly orthogonally polarized beams inside the crystal. The structure of the crystal supports two orthogonally polarized beams. These are called anisotropic crystals. In these crystals, there is a direction along which there is no decomposition. This is known as the optic axis. Some crystals have only one optic axis and are called uniaxial crystals, whereas others have two optic axes and are called biaxial crystals. From the point of view of polarizers or other polarization components, the uniaxial crystals are of most importance, hence we would only discuss them. Two well-known examples of uniaxial crystals are calcite and quartz.

Let us consider a plate of a uniaxial crystal on which a beam of light is incident obliquely. This beam decomposes into two orthogonally polarized beams inside the plate. One beam obeys Snell's law of refraction and is called the ordinary beam or the o-beam, whereas the other beam does not obey this law and is called the extraordinary or the e-beam. The refractive index μ_0 for the o-beam is independent of the direction of propagation, whereas the refractive index for the e-beam varies with the direction, taking extreme value μ_e in a direction orthogonal to the optic axis. If $\mu_0 > \mu_e$, the crystal is a negative uniaxial, such as calcite ($\mu_0 = 1.658$ and $\mu_e = 1.486$ for the yellow sodium wavelength). The quartz crystal is a positive uniaxial crystal as $\mu_e > \mu_0$ ($\mu_0 = 1.544$ and $\mu_e = 1.553$). The optic axis is a slow axis in calcite, whereas the axis orthogonal to this is a fast axis. The ***E*** vector in the o-beam oscillates in a plane, which is perpendicular to the principal section of the crystal. The principal section contains the optic axis and the direction of propagation. The ***E*** vector of the e-beam lies in the principal plane.

8.3 OPTICAL ELEMENTS FROM CRYSTALS

8.3.1 Polarizers

As there are two linearly polarized beams inside the crystal, it is easy to obtain a linearly polarized beam by eliminating one of these beams. Fortunately, due to the angle-dependent refractive index of the e-beam and availability of media of refractive index intermediate to μ_0 and μ_e, it is possible to remove the o-beam by total internal reflection in the calcite crystal. One of the early devices based on this principle is a Nicol prism. Its more versatile companion is a Glan–Thompson prism, which is shown in Figure 8.3. It has two halves, which are cemented by Canada balsam. The direction of the optic axis (O.A.) in each half is shown in the Figure 8.3. The o-beam (p-polarized) is totally reflected at the cemented interface. In other words, the Glan–Thompson prism blocks the o-beam (p-polarized) or transmits only the e-beam (s-polarized). When an unpolarized beam is incident on the polarizer, the emergent light beam is linearly polarized (e-beam). However, if a linearly polarized beam is incident on it, the beam will be completely blocked if the ***E*** vector of the incident beam is perpendicular to the principal section. Therefore, a Glan–Thompson polarizer has a transmission axis when the ***E*** vector of the incident beam is in the principal plane. The polarizers obtained from anisotropic crystals are generally small in size but have very high extinction ratios. These are usually not used for photoelastic work, which requires larger polarizers as photoelastic models are usually moderately large in size.

8.3.2 Phase Plates

Along with obtaining polarizers from these crystals, we can also obtain phase plates. These plates provide a fixed but wavelength-dependent phase difference

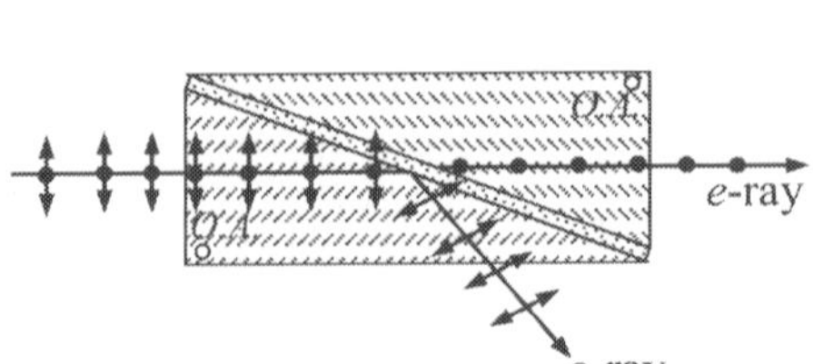

FIGURE 8.3 A Glan–Thompson polarizer.

between the two components. Let us consider a plane parallel plate of a uniaxial crystal with the optic axis lying in the surface of the plate. A linearly polarized beam is incident on this plate normally. This beam is decomposed into two beams that propagate with different velocities along the slow and fast axis of the plate. Let the refractive indices along these axes be μ_0 and μ_e, respectively. The plate will introduce a path difference $|(\mu_0 - \mu_e)|d$ between the two waves after they travel the plate of thickness d. Therefore, any path difference between the two waves can be introduced by an appropriate choice of the plate thickness of a given anisotropic crystal.

8.3.2.1 Quarter-Wave Plate

The plate thickness d is chosen to introduce a path difference of $\lambda/4$ or an odd multiple thereof, i.e., $(2m + 1)\lambda/4$, where m is an integer. In other words, such a plate introduces a path difference of a quarter-wavelength or its odd multiples. Therefore

$$d = \frac{(2m+1)}{|(\mu_0 - \mu_e)|}\frac{\lambda}{4} : m = 0, 1, 2, 3 \tag{8.5}$$

A quarter-wave plate is used to convert a linearly polarized beam into a circularly polarized beam. It is oriented such that the $\boldsymbol{E}$ vector of the incident beam makes an angle of 45° with either the fast or the slow axis of the quarter-wave plate. The components of the $\boldsymbol{E}$ vector inside the plate are then of equal amplitude and the plate introduces a path difference of $\lambda/4$ between these components. The emergent beam is thus circularly polarized. The handedness of circular polarization can be changed by rotating the plate by 90° about the axis of the optical beam.

8.3.2.2 Half-Wave Plate

A half-wave plate is obtained when its thickness d is such that it introduces a path difference of $\lambda/2$ or $(2m + 1)\lambda/2$. It rotates the plane of polarization of the linearly polarized beam. For example, if a beam of linearly polarized light with its azimuth 45° is incident on a half-wave plate, its azimuth is rotated by 90°. In other words, the beam emerges as linearly polarized but the orientation of $\boldsymbol{E}$ vector is rotated by 90°.

8.3.2.3 Compensators

The phase plates are devices that introduce fixed phase differences. Some applications require variable path difference while the others require compensation of the path difference. Compensators achieve these functions. There are two well-known compensators: the Babinet and the Soleil–Babinet compensators. The Babinet compensator consists of two wedge plates with the optic axes orthogonal to each other and parallel to the entrance face as shown in Figure 8.4(a). The role of o- and e-beams changes when the beams enter from one wedge to the other. The path difference introduced by the compensator is given by $|\mu_0 - \mu_e|[d_2(y) - d_1(y)]$, where $d_2(y)$ and $d_1(y)$ are the thicknesses of the two wedge plates at any position $(0,y)$. The path difference varies along the y-direction on the wedge plate. However, if we want a constant path difference between the two beams, the Soleil–Babinet compensator is used. This again consists of two elements: one is a plate and the other is a

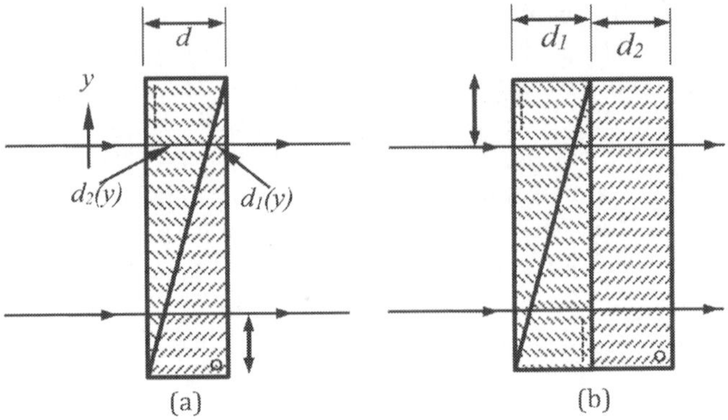

FIGURE 8.4 (a) Babinet compensator. (b) Soleil–Babinet compensator.

combination of two identical wedge plates forming a plane parallel plate. The optic axes in the plate and the wedge combination are orthogonal. By sliding one wedge over the other in a wedge combination, its thickness is varied. Thus, thickness difference $(d_2 - d_1)$ remains constant over the whole surface, where d_2 and d_1 are thicknesses of the plate and wedge combination, respectively. Figure 8.4(b) shows the schematic of a Soleil–Babinet compensator.

8.4 DICHROISM

There are anisotropic crystals that are characterized by different absorption coefficients with respect to the o- and e-beams. For example, a tourmaline crystal strongly absorbs an ordinary beam. Therefore, we can obtain an e-beam when a beam of natural light passes through a sufficiently thick plate of this crystal. Very large size polarizers based on selective absorption are available as sheets and are known as Polaroid or sheet polarizers. These are the ones often used in photoelastic work.

8.5 SCATTERING

The light scattered by particles is partially polarized. However, polarizers based on scattering are not practically used.

8.6 MALUS LAW

Consider a linearly polarized light beam of amplitude E_0 incident on a polarizer. The $\boldsymbol{E}$ vector of the beam makes an angle θ with the transmission axis of the polarizer. The beam is resolved into two components, one parallel to the transmission axis and the other perpendicular to it. The component perpendicular to the transmission axis is blocked. Therefore, the amplitude of the light transmitted by the polarizer is $E(\theta) = E_0 \cos\theta$. Hence, the irradiance of the transmitted light is given by

$I(\theta) = I_0 \cos^2\theta$, where I_0 is the irradiance of the incident beam. This is a statement of the Malus law. It may be seen that a polarizer could also be used as an attenuator in the linearly polarized beam.

8.7 THE STRESS-OPTIC LAW

The phenomenon of double refraction or optical anisotropy may also occur in certain isotropic materials, such as glass and plastics, when subjected to stress or strain. This condition is temporary and disappears when the stress is removed. This phenomenon was first observed by Brewster and forms the basis of photoelasticity. In photoelasticity, models of objects are cast or fabricated from isotropic materials, which are then subjected to stress. The stresses produce physical deformations, which completely alter the initial isotropic character of the material. We then characterize the material with three principal refractive indices, which are along the principal axes of the stresses.

The relationships between the principal indices of refraction μ_i of a temporary birefringent material and the principal stresses σ_i were formulated by Maxwell and are given by

$$\mu_1 - \mu_0 = C_1\sigma_1 + C_2(\sigma_2 + \sigma_3) \tag{8.6a}$$

$$\mu_2 - \mu_0 = C_1\sigma_2 + C_2(\sigma_3 + \sigma_1) \tag{8.6b}$$

$$\mu_3 - \mu_0 = C_1\sigma_3 + C_2(\sigma_1 + \sigma_2) \tag{8.6c}$$

where μ_0 is the refractive index of the unstressed (isotropic) material and C_1, C_2 are the constants depending on the material. Refractive indices μ_1, μ_2 and μ_3 of the stressed material are along the principal directions.

For materials under general triaxial stress, the stress-optic law is expressed as

$$\mu_1 - \mu_2 = C(\sigma_1 - \sigma_2) \tag{8.7a}$$

$$\mu_2 - \mu_3 = C(\sigma_2 - \sigma_3) \tag{8.7b}$$

$$\mu_1 - \mu_3 = C(\sigma_1 - \sigma_3) \tag{8.7c}$$

where $C = (C_1 - C_2)$ is the stress-optic coefficient of the photoelastic material.

Let us now consider a plate of an isotropic material of thickness d. This could be subjected to either (1) a uniaxial or (2) a biaxial state of stress. When the plate is subjected to a uniaxial state of stress, $\sigma_2 = \sigma_3 = 0$, and hence $\mu_2 = \mu_3$. The stress-optic law takes a very simple form

$$\mu_1 - \mu_2 = C\,\sigma_1 \tag{8.8}$$

The plate behaves like a uniaxial crystal. When the plate is subjected to a biaxial state of stress, i.e., $\sigma_3 = 0$, the stress-optic law takes the form

$$\mu_1 - \mu_2 = C(\sigma_1 - \sigma_2) \tag{8.9a}$$

$$\mu_2 - \mu_3 = C\,\sigma_2 \tag{8.9b}$$

$$\mu_1 - \mu_3 = C\sigma_1 \tag{8.9c}$$

The plate behaves like a biaxial crystal.

Now let us consider that a beam of linearly polarized light of wavelength λ is incident normally on a plate of thickness d of a photoelastic material. Within the plate, two linearly polarized beams, one vibrating in the x–z plane and the other in y–z plane are supported. These two waves acquire a phase difference on traverse in the plate and emerge at the exit surface with a phase difference δ, which is given by

$$\delta = \frac{2\pi}{\lambda}|\mu_1 - \mu_2|\,d = \frac{2\pi\,C}{\lambda}(\sigma_1 - \sigma_2)\,d \tag{8.10}$$

The phase change δ, depends linearly on the difference of the principal stresses, on the thickness of the plate, and inversely on the wavelength of light used. The bright fringes are formed wherever $\delta = 2m\pi$, i.e.,

$$\delta = \frac{2\pi\,C}{\lambda}(\sigma_1 - \sigma_2)\,d = 2m\pi \tag{8.11}$$

In photoelastic practice, it is more convenient to write Equation (8.11) in the form

$$(\sigma_1 - \sigma_2) = \frac{m\,f_\sigma}{d} \tag{8.12}$$

where m (= $\delta/2\pi$) is the fringe order and f_σ (= λ/C) is the material fringe value for a given wavelength of light. This relationship is known as the stress-optic law.

The principal stress difference ($\sigma_1 - \sigma_2$), in a two-dimensional (2D) model, can be determined by measuring the fringe order m, if the material fringe value f_σ of the material is known or is obtained by calibration. The fringe order at each point in the photoelastic model can be measured by observing the model in a polariscope.

At this juncture, it may be better to mention that a plate of thickness d and refractive index μ_0 introduce a phase delay of $k(\mu_0 - 1)d$; $k = 2\pi / \lambda$. When the plate is stressed, the linearly polarized components travel with different speeds and acquire the phase delays $k(\mu_1 - 1)d_1$ and $k(\mu_2 - 1)d_1$, where d_1 is the thickness of the stressed plate and is related to the thickness d of the unstressed plate by

$$d_1 = d\left[1 - \frac{\nu}{E}(\sigma_1 + \sigma_2)\right] \tag{8.13}$$

where E and ν are the Young's modulus and Poisson's ratio of the material of the plate. This change of thickness $(d_1 - d)$ is very important in interferometry and also in holo-photoelasticity.

8.8 THE STRAIN-OPTIC LAW

The stress-strain relationships for a material exhibiting perfectly linear elastic behavior under a 2D state of stress are given by

$$\varepsilon_1 = \frac{1}{E}\left(\sigma_1 - \nu\sigma_2\right) \tag{8.14a}$$

$$\varepsilon_2 = \frac{1}{E}\left(\sigma_2 - \nu\sigma_1\right) \tag{8.14b}$$

where ε_1 and ε_2 are the principal strains. From Equations (8.14a) and (8.14b), the difference between the principal stresses is given by

$$\left(\sigma_1 - \sigma_2\right) = \frac{E}{1+\nu}\left(\varepsilon_1 - \varepsilon_2\right) \tag{8.15}$$

Substituting this in the stress-optic law, we obtain

$$\left(\varepsilon_1 - \varepsilon_2\right) = \frac{m\ f_\varepsilon}{d} \tag{8.16}$$

where $f_\varepsilon \left[= f_\sigma (1+\nu)/E\right]$ is the material fringe value in terms of strain. The relationship given in Equation (8.16) is known as the strain-optic law in photoelasticity.

8.9 METHODS OF ANALYSIS

The optical system most often used for stress analysis is a polariscope. It takes a variety of forms depending on the end use. However, in general, a polariscope consists of a light source, a device to produce polarized light called a polarizer, a model, and a second polarizer called an analyzer. In addition, it may contain a set of lenses, quarter-wave plates, photographic or recording equipment, and loading equipment. We will discuss the optical system of a plane polariscope and a circular polariscope in the following sections.

8.9.1 THE PLANE POLARISCOPE

The plane polariscope consists of a light source, a light filter, collimating optics to provide a collimated beam, a polarizer, an analyzer, a lens, and photographic equipment as shown in Figure 8.5. The model is placed between the polarizer and the analyzer. The polarizer and the analyzer are crossed (their transmission axes are orthogonal), thus producing a dark field.

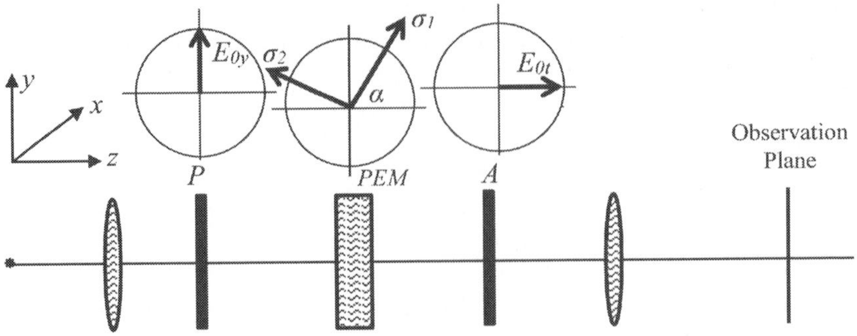

FIGURE 8.5 Schematic of a plane polariscope.

Let the transmission axis of the polarizer P be along the y-direction. The amplitude of the wave just behind the polarizer is expressed as follows:

$$E_y(z;t) = E_{0y}\cos(\omega t - kz) \quad (8.17)$$

where $k = 2\pi/\lambda$.

The field incident on the model is normally also given by this expression except that z refers to the plane of the model. Let us also assume that one of the principal stress directions makes an angle α with the transmission direction of the analyzer, i.e., x-axis. The incident field just at the entrance face of the model splits into two components, which are orthogonally polarized and vibrate in the planes of σ_1 and σ_2. The amplitudes of these components are E_{0y} sinα and E_{0y} cosα, respectively. The amplitudes of these components at the exit face of the model are

$$E_1(z;t) = E_{0y}\sin\alpha\cos\left[\omega t - kz - k(\mu_1 - 1)d\right] = E_{0y}\sin\alpha\cos(\omega t - kz - \delta_x) \quad (8.18a)$$

$$E_2(z;t) = E_{0y}\cos\alpha\cos\left[\omega t - kz - k(\mu_2 - 1)d\right] = E_{0y}\cos\alpha\cos(\omega t - kz + \delta - \delta_x) \quad (8.18b)$$

The model introduces a phase difference $\delta = \frac{2\pi}{\lambda}(\mu_1 - \mu_2)d = \frac{2\pi}{f_\sigma}(\sigma_1 - \sigma_2)d$ between these components where μ_1 and μ_2 are the refractive indices along σ_1 (slow) and σ_2 (fast) axes of the model of thickness d. The analyzer resolves these components further into components along and perpendicular to its direction of transmission, which is along the x-direction. The components along the direction of transmission are allowed through and produce a photoelastic pattern, whereas those in the orthogonal direction are blocked. The net transmitted amplitude is

$$\begin{aligned} E_1\cos\alpha - E_2\sin\alpha &= \frac{E_{0y}}{2}\sin 2\alpha\left[\cos(\omega t - kz - \delta_x) - cos(\omega t - kz + \delta - \delta_x)\right] \\ &= E_{0y}\sin 2\alpha\sin\frac{\delta}{2}\sin\left(\omega t - kz - \delta_x + \frac{\delta}{2}\right) \end{aligned} \quad (8.19)$$

This also represents a wave of amplitude $E_{0y}\sin 2\alpha \sin\frac{\delta}{2}$ propagating along the z-direction. The irradiance of this wave, just after the analyzer, therefore is given by

$$I = E_{0y}^2 \sin^2 2\alpha \, \sin^2\left(\frac{\delta}{2}\right) = I_0 \sin^2 2\alpha \, \sin^2\left[\frac{\pi(\sigma_1 - \sigma_2)}{f_\sigma}\right] \tag{8.20}$$

The propagation of the field through the plane polariscope can be followed using Jones calculus. Let the slow axis of the photoelastic material make an angle α with the x-axis and the phase difference introduced by the photoelastic material be δ, then

$$\mathbb{E}_{0t} = \mathbb{P}_x \mathbb{R}(-\alpha)\, \mathbb{J}_{PEM}(\delta)\, \mathbb{R}(\alpha) \mathbb{P}_y \mathbb{E}_{0y}$$

$$\begin{pmatrix} E_{0t} \\ 0 \end{pmatrix} = \begin{pmatrix} 1 & 0 \\ 0 & 0 \end{pmatrix} \begin{pmatrix} \cos\alpha & -\sin\alpha \\ \sin\alpha & \cos\alpha \end{pmatrix} \begin{pmatrix} e^{-i\frac{\delta}{2}} & 0 \\ 0 & e^{i\frac{\delta}{2}} \end{pmatrix} \begin{pmatrix} \cos\alpha & \sin\alpha \\ -\sin\alpha & \cos\alpha \end{pmatrix} \begin{pmatrix} 0 & 0 \\ 0 & 1 \end{pmatrix} \begin{pmatrix} 0 \\ E_{0y} \end{pmatrix} \tag{8.21}$$

Here $\mathbb{P}_y$ and $\mathbb{P}_x$ are the Jones matrices for the polarizer P and the analyzer A with transmission axes along y-and x-directions, respectively; $\mathbb{R}(\alpha)$ is the rotation matrix; and $\mathbb{J}_{PEM}(\delta)$ is the matrix that represents photoelastic material, which is assumed to be transparent. E_{0t} is the transmitted field.

This can be rewritten as

$$\begin{pmatrix} E_{0t} \\ 0 \end{pmatrix} = \begin{pmatrix} 1 & 0 \\ 0 & 0 \end{pmatrix} \begin{pmatrix} \cos^2\alpha\, e^{-i\frac{\delta}{2}} + \sin^2\alpha\, e^{i\frac{\delta}{2}} & -i\sin 2\alpha \sin\frac{\delta}{2} \\ -i\sin 2\alpha \sin\frac{\delta}{2} & \sin^2\alpha\, e^{-i\frac{\delta}{2}} + \cos^2\alpha\, e^{i\frac{\delta}{2}} \end{pmatrix} \begin{pmatrix} 0 & 0 \\ 0 & 1 \end{pmatrix} \begin{pmatrix} 0 \\ E_{0y} \end{pmatrix}$$

or

$$\begin{pmatrix} E_{0t} \\ 0 \end{pmatrix} = \begin{pmatrix} \cos^2\alpha\, e^{-i\frac{\delta}{2}} + \sin^2\alpha\, e^{i\frac{\delta}{2}} & -i\sin 2\alpha \sin\frac{\delta}{2} \\ 0 & 0 \end{pmatrix} \begin{pmatrix} 0 & 0 \\ 0 & 1 \end{pmatrix} \begin{pmatrix} 0 \\ E_{0y} \end{pmatrix}$$

$$= \begin{pmatrix} 0 & -i\sin 2\alpha \sin\frac{\delta}{2} \\ 0 & 0 \end{pmatrix} \begin{pmatrix} 0 \\ E_{0y} \end{pmatrix} = \begin{pmatrix} -i\sin 2\alpha \sin\frac{\delta}{2}\, E_{0y} \\ 0 \end{pmatrix} \tag{8.22}$$

The irradiance of the transmitted wave is

$$I_t(\delta) = \frac{|E_{0t}|^2}{2c\mu_0} = \frac{|E_{0y}|^2}{2c\mu_0} \sin^2 2\alpha \, \sin^2\frac{\delta}{2} = I_0 \sin^2 2\alpha \, \sin^2\frac{\delta}{2} = I_0 \sin^2 2\alpha \, \sin^2\left[\frac{\pi(\sigma_1 - \sigma_2)d}{f_\sigma}\right] \tag{8.23}$$

The irradiance of the transmitted beam is governed by α, the orientation of the principal stress direction with respect to the analyzer transmission axis, and the phase retardation δ. The transmitted irradiance will be zero when $\sin^2 2\alpha \ \sin^2\left(\frac{\delta}{2}\right) = 0$. In other words, the transmitted irradiance is zero when either $\sin 2\alpha = 0$ or $\sin\frac{\delta}{2} = 0$. When $\sin 2\alpha = 0$, the angle $\alpha = 0$ or $\alpha = \pi/2$. In either case, one of the principal stress directions is aligned with the polarizer transmission axis. Therefore, these dark fringes give the directions of the principal stresses at any point on the model and are known as isoclinics or isoclinic fringes.

When $\sin\frac{\delta}{2} = 0$, then $\delta = 2m\pi$ and hence $\frac{2\pi}{f_\sigma}(\sigma_1 - \sigma_2)d = 2m\pi$

or

$$\sigma_1 - \sigma_2 = \frac{m\ f_\sigma}{d} \tag{8.24}$$

The transmitted irradiance is zero when $(\sigma_1 - \sigma_2)$ is an integral multiple of f_σ / d. Therefore, the fringes are loci of constant $(\sigma_1 - \sigma_2)$ and are referred to as isochromatics. The adjacent isochromatics differ by f_σ / d. When white light is used for illumination of the model, these fringes are colored, and each color corresponds to a constant value of $(\sigma_1 - \sigma_2)$, hence the name isochromatics.

The isoclinics are the loci of points at which the directions of the principal stresses are parallel to the transmission axes of the polarizer and the analyzer. The isoclinic pattern is independent of the magnitude of the load applied to the model and the material fringe value. When white light is used for illumination, the isoclinics appear dark in contrast to the isochromatics, which, with the exception of the zeroth-order fringe, are colored. In regions where the directions of the principal stresses do not vary greatly from point to point, the isoclinics appear as wide diffuse bands. The isoclinics do not intersect each other except at an isotropic point, which is a point where the principal stresses are equal in magnitude, i.e., $(\sigma_1 - \sigma_2) = 0$. Further, at a point on a shear-free boundary where the stress parallel to the boundary has a maximum or a minimum value, the isoclinic intersects the boundary orthogonally.

It is seen that both the isoclinics and the isochromatics appear simultaneously in a plane polariscope. It is desirable to separate these fringe patterns. A circular polariscope does this function and provides only isochromatics.

8.9.2 The Circular Polariscope

It may be seen that the isoclinics appear because a linearly polarized light wave is incident on the model. These would disappear if the light incident on the model is circularly polarized. Therefore, a circular polarizer, which is a combination of a linear polarizer and a quarter-wave plate at 45° azimuth, is required. Further, to analyze this light, we also need a circular analyzer. Therefore, a circular polariscope consists of a light source, collimating optics, a polarizer, two quarter-wave plates, an analyzer, and the recording optics as shown in Figure 8.6.

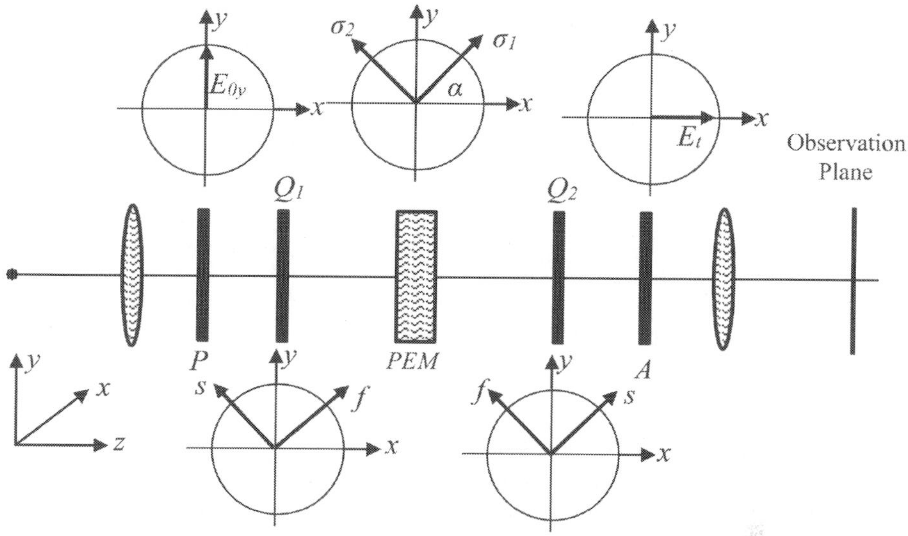

FIGURE 8.6 Schematic of a circular polariscope. *P*, polarizer; Q_1, Q_2, quarter-wave plates; *PEM*, model; and *A*, analyzer. For other symbols, see text.

The model is placed between the quarter-wave plates. As these plates are designated with slow and fast axes, they could be arranged in two ways, namely axes parallel and axes crossed. Similarly, the transmission axes of the polarizer and the analyzer can be parallel or crossed. Therefore, there are four ways of assembling a circular polariscope: two of these configurations give the dark field and the other two give the bright field at the output as given in Table 8.1.

We now consider a configuration in which the polarizer *P* has its transmission direction along the *y*-axis; the quarter-wave plate Q_1 is oriented with its slow axis making an angle of −45° with the *x*-axis. The model is placed between the quarter-wave plates and plate Q_2 has its slow axis at 45° with the *x*-axis and the analyzer *A* is crossed with the polarizer. Therefore, this is configuration 4 with polarizer and analyzer crossed and quarter-wave plates Q_1 and Q_2 also crossed resulting in a dark field output.

TABLE 8.1
Four Configurations of Circular Polariscope

Configuration	Polarizer and Analyzer Axes	Quarter-Wave Plates Axes	Field
1	Parallel	Parallel	Dark
2	Parallel	Crossed	Bright
3	Crossed	Parallel	Bright
4	Crossed	Crossed	Dark

The field transmitted by the polarizer is represented by Equation (8.17). This field is split into two components that propagate along the fast and the slow axes of the quarter-wave plate Q_1. The fields at the exit face of Q_1 along the axes are

$$E_{f1}(z;t) = \frac{E_{0y}}{\sqrt{2}}\cos\left[\omega t - kz - k(\mu_0 - 1)d'\right] = \frac{E_{0y}}{\sqrt{2}}\cos\left[\omega t - kz - \psi_1\right] \tag{8.25a}$$

$$E_{s1}(z;t) = \frac{E_{0y}}{\sqrt{2}}\cos\left[\omega t - kz - k(\mu_e - 1)d'\right] = \frac{E_{0y}}{\sqrt{2}}\cos\left[\omega t - kz - \frac{\pi}{2} - \psi_1\right]$$

$$= \frac{E_{0y}}{\sqrt{2}}\sin\left[\omega t - kz - \psi_1\right] \tag{8.25b}$$

where the phase difference $\pi/2$, introduced by Q_1 of thickness d', is expressed as $\frac{2\pi}{\lambda}(\mu_e - \mu_0)d' = \frac{\pi}{2}$ and $\psi_1 = \frac{2\pi}{\lambda}(\mu_0 - 1)d'$. This field is now incident on the model and hence gets further decomposed along the directions of the principal stresses. We assume that one of the principal stress directions makes an angle α with the analyzer transmission axis, i.e., with the x-axis. The fields at the entrance face of the model as decomposed along the σ_1 and σ_2 directions are now represented by

$$E_{\sigma1}(z;t) = E_{f1}(z;t)\cos\left(\frac{\pi}{4} - \alpha\right) - E_{s1}(z;t)\sin\left(\frac{\pi}{4} - \alpha\right) = \frac{E_{0y}}{\sqrt{2}}\cos\tau \tag{8.26a}$$

$$E_{\sigma2}(z;t) = E_{f1}(z;t)\sin\left(\frac{\pi}{4} - \alpha\right) + E_{s1}(z;t)\cos\left(\frac{\pi}{4} - \alpha\right) = \frac{E_{0y}}{\sqrt{2}}\sin\tau \tag{8.26b}$$

where $\tau = \omega t - kz - \psi_1 + \left(\frac{\pi}{4}\right) - \alpha$.

These field components travel along the σ_1 and σ_2 directions in the model of thickness d and acquire the phases. The fields at the exit face of the model along the σ_1 and σ_2 directions can be written as

$$E_{\sigma1e}(z;t) = \frac{E_{0y}}{\sqrt{2}}\cos(\tau - \psi_2) \tag{8.27a}$$

$$E_{\sigma2e}(z;t) = \frac{E_{0y}}{\sqrt{2}}\sin(\tau + \delta - \psi_2) \tag{8.27b}$$

where $\psi_2 = \frac{2\pi}{\lambda}(\mu_1 - 1)d$ and $\delta = \frac{2\pi}{\lambda}(\mu_1 - \mu_2)d$. Now, we decompose these fields along the axes of the second quarter-wave plate, which are inclined at 45° and –45° with the y-axis. These are expressed as follows:

$$E_{s2}(z;t) = E_{\sigma1e}(z;t)\cos\left(\frac{\pi}{4} - \alpha\right) + E_{\sigma2e}(z;t)\sin\left(\frac{\pi}{4} - \alpha\right)$$

$$= \frac{E_{0y}}{\sqrt{2}}\left[\cos\left(\frac{\pi}{4} - \alpha\right)\cos(\tau - \psi_2) + \sin\left(\frac{\pi}{4} - \alpha\right)\sin(\tau + \delta - \psi_2)\right] \tag{8.28a}$$

$$E_{f2}(z;t)= E_{\sigma 2e}(z;t)\cos\left(\frac{\pi}{4}-\alpha\right)-E_{\sigma 1e}(z;t)\sin\left(\frac{\pi}{4}-\alpha\right)$$

$$=\frac{E_{0y}}{\sqrt{2}}\left[\cos\left(\frac{\pi}{4}-\alpha\right)\sin(\tau+\delta-\psi_2)+\sin\left(\frac{\pi}{4}-\alpha\right)\cos(\tau-\psi_2)\right] \quad (8.28b)$$

The fields at the exit face of the quarter-wave plate Q_2 are obtained as

$$E_{s2}(z;t)=\frac{E_{0y}}{\sqrt{2}}\left[\cos\left(\frac{\pi}{4}-\alpha\right)\cos\left(\tau-\psi_2-\frac{\pi}{2}-\psi_1\right)\right.$$

$$\left.+\sin\left(\frac{\pi}{4}-\alpha\right)sin\left(\tau+\delta-\psi_2-\frac{\pi}{2}-\psi_1\right)\right]$$

$$=\frac{E_{0y}}{\sqrt{2}}\left[\cos\left(\frac{\pi}{4}-\alpha\right)\sin(\tau-\psi_2-\psi_1)+\sin\left(\frac{\pi}{4}-\alpha\right)\cos(\tau+\delta-\psi_2-\psi_1)\right]$$

(8.29a)

$$E_{f2}(z;t)=\frac{E_{0y}}{\sqrt{2}}\left[\cos\left(\frac{\pi}{4}-\alpha\right)\sin(\tau+\delta-\psi_2-\psi_1)+\sin\left(\frac{\pi}{4}-\alpha\right)\cos(\tau-\psi_2-\psi_1)\right]$$

(8.29b)

As the transmission direction of analyzer is along x-axis, the field transmitted by the analyzer is given as

$$E_t=\frac{E_{s2}(z;t)}{\sqrt{2}}-\frac{E_{f2}(z;t)}{\sqrt{2}}$$

$$=\frac{E_{0y}}{2}\left[\sin\left(\frac{\pi}{4}-\alpha\right)\cos(\tau+\delta-\psi_2-\psi_1)-\cos\left(\frac{\pi}{4}-\alpha\right)\sin(\tau+\delta-\psi_2-\psi_1)\right.$$

$$\left.+\cos\left(\frac{\pi}{4}-\alpha\right)\sin(\tau-\psi_2-\psi_1)-\sin\left(\frac{\pi}{4}-\alpha\right)\cos(\tau-\psi_2-\psi_1)\right]$$

$$=\frac{E_{0y}}{2}\left[-\sin\left(\tau+\delta-\psi_2-\psi_1+\frac{\pi}{4}-\alpha\right)+\sin\left(\tau-\psi_2-\psi_1+\frac{\pi}{4}-\alpha\right)\right]$$

$$=E_{0y}\ \sin\left(\frac{\delta}{2}\right)\left[cos\left(\tau-\psi_2-\psi_1+\frac{\pi}{4}-\alpha+\frac{\delta}{2}\right)\right]$$

(8.30)

This represents a wave of amplitude $E_{0y}\ \sin\left(\frac{\delta}{2}\right)$. The irradiance distribution is thus given by

$$I(\delta)=I_0\ \sin^2\frac{\delta}{2} \quad (8.31)$$

This represents a dark field configuration. Again, this analysis can be easily performed using Jones calculus. The progression of polarization states through various components can be followed with Jones calculus. It is given as

$$\mathbb{E}_{0t} = \mathbb{P}_x \mathbb{R}\left(\frac{-\pi}{4}\right) \mathbb{Q}_2\ \mathbb{R}\left(\frac{\pi}{4}\right)\mathbb{R}(-\alpha)\ \mathbb{J}_{PEM}(\delta)\ \mathbb{R}(\alpha)\mathbb{R}\left(\frac{\pi}{4}\right) \mathbb{Q}_1\ \mathbb{R}\left(\frac{-\pi}{4}\right)\mathbb{P}_y\ \mathbb{E}_{0y} \tag{8.32}$$

or

$$\mathbb{E}_t = \begin{pmatrix} 1 & 0 \\ 0 & 0 \end{pmatrix} \begin{pmatrix} \cos\frac{\pi}{4} & -\sin\frac{\pi}{4} \\ \sin\frac{\pi}{4} & \cos\frac{\pi}{4} \end{pmatrix} \begin{pmatrix} e^{\frac{-i\pi}{4}} & 0 \\ 0 & e^{\frac{i\pi}{4}} \end{pmatrix} \begin{pmatrix} \cos\frac{\pi}{4} & \sin\frac{\pi}{4} \\ -\sin\frac{\pi}{4} & \cos\frac{\pi}{4} \end{pmatrix} \begin{pmatrix} \cos\alpha & -\sin\alpha \\ \sin\alpha & \cos\alpha \end{pmatrix} \begin{pmatrix} e^{\frac{-i\delta}{2}} & 0 \\ 0 & e^{\frac{i\delta}{2}} \end{pmatrix}$$

$$\begin{pmatrix} \cos\alpha & \sin\alpha \\ -\sin\alpha & \cos\alpha \end{pmatrix} \begin{pmatrix} \cos\frac{\pi}{4} & \sin\frac{\pi}{4} \\ -\sin\frac{\pi}{4} & \cos\frac{\pi}{4} \end{pmatrix} \begin{pmatrix} e^{\frac{-i\pi}{4}} & 0 \\ 0 & e^{\frac{i\pi}{4}} \end{pmatrix} \begin{pmatrix} \cos\frac{\pi}{4} & -\sin\frac{\pi}{4} \\ \sin\frac{\pi}{4} & \cos\frac{\pi}{4} \end{pmatrix} \begin{pmatrix} 0 & 0 \\ 0 & 1 \end{pmatrix} \mathbb{E}_{0y} \tag{8.33}$$

This can be recast as

$$\begin{pmatrix} E_{0t} \\ 0 \end{pmatrix} = \begin{pmatrix} 1 & 0 \\ 0 & 0 \end{pmatrix} \frac{1}{\sqrt{2}} \begin{pmatrix} 1 & i \\ i & 1 \end{pmatrix} \begin{pmatrix} \cos^2\alpha\ e^{\frac{-i\delta}{2}} + \sin^2\alpha\ e^{\frac{i\delta}{2}} & -i\sin 2\alpha \sin\frac{\delta}{2} \\ -i\sin 2\alpha \sin\frac{\delta}{2} & \sin^2\alpha\ e^{\frac{-i\delta}{2}} + \cos^2\alpha\ e^{\frac{i\delta}{2}} \end{pmatrix}$$

$$\frac{1}{\sqrt{2}} \begin{pmatrix} 1 & -i \\ -i & 1 \end{pmatrix} \begin{pmatrix} 0 & 0 \\ 0 & 1 \end{pmatrix} \begin{pmatrix} 0 \\ E_{0y} \end{pmatrix}$$

On simplification, we obtain

$$\begin{pmatrix} E_{0t} \\ 0 \end{pmatrix} = \begin{pmatrix} 0 & (-\cos 2\alpha - i\sin 2\alpha)\sin\frac{\delta}{2} \\ 0 & 0 \end{pmatrix} \begin{pmatrix} 0 \\ E_{0y} \end{pmatrix} \tag{8.34}$$

The irradiance of the transmitted wave will be

$$I_t(\delta) = \frac{|E_{0t}|^2}{2c\mu_0} = \frac{|E_{0y}|^2}{2c\mu_0} (-\cos 2\alpha - i\sin 2\alpha)(-\cos 2\alpha + i\sin 2\alpha)\ \sin^2\frac{\delta}{2} = I_0\ \sin^2\frac{\delta}{2} \tag{8.35}$$

This is the dark field configuration, as the transmitted irradiance is zero when there is no stress.

On the other hand, if the analyzer is oriented such that its transmission axis is parallel to that of the polarizer, the transmitted field can be expressed as

$$\begin{pmatrix} 0 \\ E_{0t} \end{pmatrix} = \begin{pmatrix} 0 & 0 \\ 0 & 1 \end{pmatrix} \frac{1}{\sqrt{2}} \begin{pmatrix} 1 & i \\ i & 1 \end{pmatrix} \begin{pmatrix} \cos^2\alpha\, e^{\frac{-i\delta}{2}} + \sin^2\alpha\, e^{\frac{i\delta}{2}} & -i\sin 2\alpha \sin\frac{\delta}{2} \\ -i\sin 2\alpha \sin\frac{\delta}{2} & \sin^2\alpha\, e^{\frac{-i\delta}{2}} + \cos^2\alpha\, e^{\frac{i\delta}{2}} \end{pmatrix}$$

$$\frac{1}{\sqrt{2}} \begin{pmatrix} 1 & -i \\ -i & 1 \end{pmatrix} \begin{pmatrix} 0 & 0 \\ 0 & 1 \end{pmatrix} \begin{pmatrix} 0 \\ E_{0y} \end{pmatrix} \tag{8.36}$$

On simplification, we obtain

$$\begin{pmatrix} 0 \\ E_{0t} \end{pmatrix} = \begin{pmatrix} 0 & 0 \\ 0 & \sin^2\alpha \cos\frac{\delta}{2} + \cos^2\alpha \cos\frac{\delta}{2} \end{pmatrix} \begin{pmatrix} 0 \\ E_{0y} \end{pmatrix}$$

Thus

$$E_{0t} = E_{0y} cos\frac{\delta}{2} \Rightarrow I(\delta) = I_0 \cos^2\frac{\delta}{2} \tag{8.37}$$

When there is no stress distribution, $\delta = 0$, hence the transmitted irradiance is maximum and uniform. This therefore represents a bright field configuration.

Equations (8.31) and (8.37) show that the irradiance of light emerging from the analyzer in a circular polariscope is a function of δ only and hence the difference of the principal stresses $(\sigma_1 - \sigma_2)$ only. The isoclinics have been eliminated.

In a dark field configuration, the dark fringes occur wherever $\delta = 2m\pi$ ($m = 0, 1, 2, \ldots$), and they correspond to the integer isochromatic fringe order $m = 0, 1, 2, 3, \ldots$, respectively. An example of this fringe pattern is shown in Figure 8.7(a). However, for the bright field configuration, the dark fringes are obtained when $\delta = (2m + 1)\pi$. These corresponds to isochromatic fringes of half-order, i.e., $m = 1/2, 3/2, 5/2, \ldots$ An example of a light field fringe pattern is shown in Figure 8.7(b).

8.9.3 Evaluation Procedure

Directions of principal stresses at any point in the model are determined using a plane polariscope. The polarizer and analyzer are rotated about the optical axis until the isoclinic passes through the point of interest. The inclination of the transmission axis gives the principal stress direction. The principal stress difference $(\sigma_1 - \sigma_2)$ in the dark field is given by

$$(\sigma_1 - \sigma_2) = \frac{m\, f_\sigma}{d} \tag{8.38a}$$

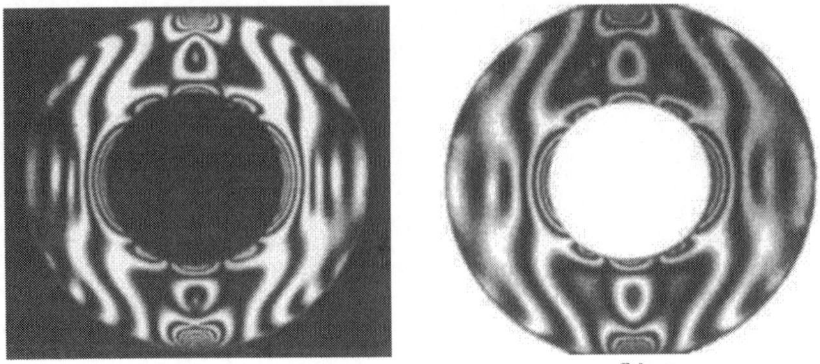

(a) (b)

FIGURE 8.7 Isochromatics in (a) dark field and (b) bright field.

and in the bright field by

$$(\sigma_1 - \sigma_2) = (m + ½)\frac{f_\sigma}{d} \tag{8.38b}$$

Therefore, the material fringe value, f_σ, needs to be known before $(\sigma_1 - \sigma_2)$ can be calculated. The material fringe value, f_σ, is obtained by calibration. A circular disc of the same photoelastic material and thickness is used as a model and diametrically loaded. The fringe order in the center of the disc is measured, and f_σ calculated using the formula,

$$f_\sigma = \frac{8F}{n\pi D} \tag{8.39}$$

where F is the applied force, D is the diameter of the disc, and n is the measured fringe order at the center of the disc. There are other calibration methods that use either tensile loading or bending.

The principal stress difference $(\sigma_1 - \sigma_2)$ in an arbitrary model is then found by using material fringe value, and the order m, of the isochromatics. However, the order m is to be counted from $m = 0$, which is normally not known. It can be easily found, if it exists in the model, by using white light illumination as $m = 0$ is an achromatic fringe while the higher-order fringes are colored. Therefore, a polariscope is usually equipped with both the white light and the monochromatic light sources: the white light source for locating the zero-order isochromatic and the monochromatic light source for counting the higher-order isochromatics. If the principal stress difference is desired at a point where neither a bright isochromatic nor a dark isochromatic passes, some means to measure fractional fringe order is to be implemented. There are methods to measure fractional fringe orders, which are discussed in the next section.

8.10 MEASUREMENT OF FRACTIONAL FRINGE ORDER

The methods described here assume that the directions of principal stresses are known. In one method, a Babinet or a Soleil–Babinet compensator is used. The principal axes of the compensator are aligned along the directions of principal stresses. An additional phase difference then can be introduced to shift the dark isochromatics to the point of interest and this additional phase shift is read from the compensator. Another method makes use of a quarter-wave plate for compensation and is known as Tardy's method.

8.10.1 The Tardy's Method

This makes use of a plane polariscope in dark field configuration where the irradiance distribution is given by $I = I_0 \sin^2 2\alpha \; \sin^2(\delta/2)$. The observation field contains both the isochromatics and the isoclinics. Let us now assume that we wish to measure the fractional isochromatics order at a point P as shown in Figure 8.8.

As it is a dark field configuration, the integral isochromatics orders correspond to dark fringes, and in order to work with the dark fringes we need to make the region at and around point *P* bright. For this purpose, the polarizer-analyzer combination is rotated by 45° so that its transmission axes make angles of 45° with the directions of the principal stresses. The irradiance distribution is now given by $I = I_0 \sin^2(\delta/2)$. A quarter-wave plate is now inserted between the model and the analyzer in such a way that its principal axes are parallel to the transmission axes of the polarizer and the analyzer. We can now shift the isochromatics by the rotation of the analyzer by an angle, which is related to the phase shift as shown in next paragraphs.

Let the transmission direction of polarizer make an angle of 45° with the principal stress direction σ_2 in the model: the principal stress directions in the model are taken along the *x*- and *y*-axes with σ_1 taken along *x*-axis. The arrangement is shown in Figure 8.9.

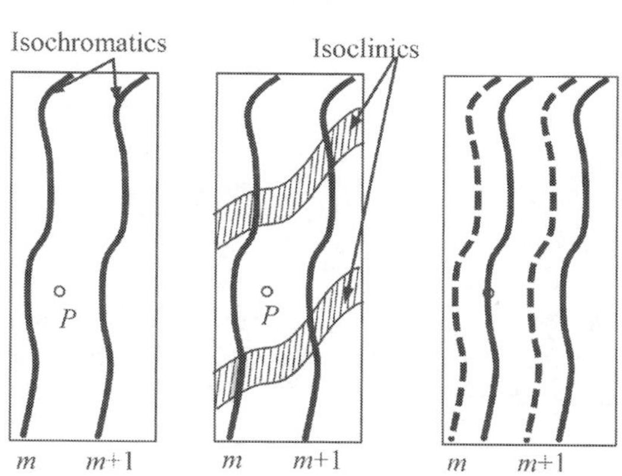

FIGURE 8.8 Tardy's method of compensation.

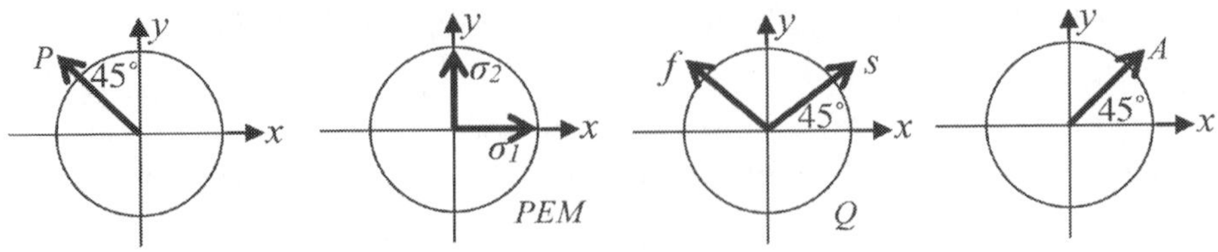

FIGURE 8.9 Arrangement of various components in Tardy's method.

The amplitude of the wave exiting from the polarizer is expressed as

$$E_p(z;t) = E_{0p}\cos(\omega t - kz) \tag{8.40}$$

The amplitude of the wave is resolved along the principal stress directions in the model. The amplitude components at the exit face of the model are given by

$$E_{\sigma 1e}(z;t) = -\frac{E_{0p}}{\sqrt{2}}\cos(\omega t - kz - \psi_2 - \delta) \tag{8.41a}$$

$$E_{\sigma 2e}(z;t) = \frac{E_{0p}}{\sqrt{2}}\cos(\omega t - kz - \psi_2) \tag{8.41b}$$

where $\psi_2 = \frac{2\pi}{\lambda}(\mu_1 - 1)d$ and $\delta = \frac{2\pi}{\lambda}(\mu_1 - \mu_2)d$: μ_1 and μ_2 are the refractive indices along the principal stress directions.

We now introduce a quarter-wave plate Q with its slow axis making an angle of 45° with the x-axis. The amplitudes exiting from the model are now to be resolved along the fast and slow axes of the quarter-wave plate. These amplitudes after passage through the quarter-wave plate are expressed as

$$E_{fe} = \frac{E_{0p}}{2}\left[\cos(\omega t - kz - \psi_2 - \psi_1) + \cos(\omega t - kz - \psi_2 - \delta - \psi_1)\right] \tag{8.42a}$$

$$\begin{aligned} E_{se} &= \frac{E_{0p}}{2}\left[\cos\left(\omega t - kz - \psi_2 - \psi_1 - \frac{\pi}{2}\right) - \cos\left(\omega t - kz - \psi_2 - \delta - \psi_1 - \frac{\pi}{2}\right)\right] \\ &= \frac{E_{0p}}{2}\left[\sin(\omega t - kz - \psi_2 - \psi_1) - \sin(\omega t - kz - \psi_2 - \delta - \psi_1)\right] \end{aligned} \tag{8.42b}$$

As the transmission axis of the analyzer is parallel to the slow axis of the quarter-wave plate, it will transmit only E_{se} component. Therefore, the amplitude of the wave transmitted by the analyzer is

$$E_{se} = E_{0p}\sin\left(\frac{\delta}{2}\right)\cos\left(\omega t - kz - \psi_2 - \frac{\delta}{2} - \psi_1\right) \tag{8.43}$$

This corresponds to the dark field configuration. Then we rotate the analyzer by χ. Now the analyzer will transmit components of both E_{fe} and E_{se}. The transmitted amplitude is given by

$$\begin{aligned} E_t &= E_{se}\cos\chi + E_{fe}\sin\chi \\ &= \frac{E_{op}}{2}\left[\cos\chi\sin\tau - \cos\chi\sin(\tau-\delta) + \sin\chi\cos\tau + \sin\chi\cos(\tau-\delta)\right] \\ &= \frac{E_{0p}}{2}\left[\sin(\chi+\tau) + \sin(\chi-\tau+\delta)\right] = E_{0p}\sin\left(\chi+\frac{\delta}{2}\right)\cos\left(\tau-\frac{\delta}{2}\right) \end{aligned} \quad (8.44)$$

where $\tau = \left(\omega t - kz - \psi_2 - \psi_1 - \frac{\pi}{2}\right)$.

This represents a wave with amplitude $E_{0p}\sin\left(\chi+\frac{\delta}{2}\right)$. Therefore, the irradiance of the wave is given by $I = I_0\sin^2\left(\chi+\frac{\delta}{2}\right)$. For the mth isochromatic, $\delta = 2m\pi$ and hence the irradiance at this location must be $I = I_0\sin^2\chi$, which is evidently zero when $\chi = 0$. Further, the irradiance at the same location will also be zero when $\chi = \pi$ but then, the mth isochromatic would have moved to $(m + 1)$th isochromatic. In other words, a rotation of analyzer by π shifts the isochromatics by one order. Therefore, if the analyzer is rotated by an angle χ_p to shift the isochromatics to point P, then the fractional order at that point must be χ_p/π.

8.11 PHASE SHIFTING

Phase shifting is a technique for the automatic evaluation of phase maps from the irradiance data. However, it assumes a special significance in photoelasticity as both the beams participating in interference travel along the same path and the phase of one cannot be changed independent of the other, as has been done for the other interferometric methods. We, therefore, present other methods of phase shifting in photoelasticity.

8.11.1 Isoclinics Computation

For this purpose, a bright field plane polariscope is used. The transmitted irradiance in this configuration is given by

$$\begin{aligned} I &= I_0 - I_0\sin^2 2\alpha\sin^2(\delta/2) \\ &= I_0\left[1 - \sin^2(\delta/2)\left(1-\cos^2 2\alpha\right)\right] \\ &= I_0\left[1 - \frac{1}{2}\sin^2(\delta/2)\left(1-\cos 4\alpha\right)\right] = I_B + V\cos 4\alpha \end{aligned} \quad (8.45)$$

where $I_B = I_0\left(1-\frac{1}{2}\sin^2(\delta/2)\right)$ and $V = \frac{1}{2}\sin^2(\delta/2)$. When the whole polariscope is rotated by β, the irradiance transmitted can be expressed by

$$I = I_B + V\cos 4(\alpha-\beta) \quad (8.46)$$

Both I_B and V depend on the value of the isochromatic parameter and consequently on the wavelength of light used. However, the isoclinic parameter does not depend on the wavelength of light used. The phase of the isoclinics is obtained by using a four-step algorithm with the irradiance data obtained at $\beta_i = (i-1)\pi/8; i = 1,2,3,\text{and } 4$ and the relation

$$\tan 4\alpha = \frac{I_4 - I_2}{I_3 - I_1} \tag{8.47}$$

The phase of the isoclinics is obtained from this relation except at regions where the modulation V is very small. Because V depends on δ, the low-modulation areas depend on the wavelength. If a monochromatic light is used there may be several areas where the value of δ makes the modulation unusable. This problem, however, can be overcome by using a white light source, as the low-modulation areas corresponding to a given wavelength will be high-modulation areas for another wavelength. Hence, the modulation is kept high enough for use, except at the zeroth-order fringe, where the modulation is obviously zero.

8.11.2 Computation of Isochromatics

As can be seen from the Tardy's method of compensation, the isochromatic fringe at a point can be shifted by the rotation of the analyzer from the crossed position: a rotation of analyzer by π shifts the isochromatic by one order. This presents a nice method of phase shifting but it suffers from the drawbacks: the principal axes should be known beforehand and the isochromatics are calculated for a fixed value of the isoclinic parameter. The phase of the isochromatics can be obtained pixel by pixel by taking irradiance data at four positions of the analyzer, say, at 0°, 45°, 90° and 135°.

We now present a method that is free from these shortcomings. It takes irradiance data at several orientations of a circular polariscope. In the following table we present the configurations along with the expressions for the transmitted irradiance:

No.	Polariscope Configuration	Transmitted Irradiance
1	$P_{\frac{\pi}{2}}\, Q_{\frac{\pi}{4}}\, Q_{\frac{\pi}{4}}\, A_{\frac{\pi}{4}}$	$I_1 = \frac{I_0}{2}(1 + \cos 2\alpha \sin\delta)$
2	$P_{\frac{\pi}{2}}\, Q_{\frac{\pi}{4}}\, Q_{\frac{-\pi}{4}}\, A_{\frac{\pi}{4}}$	$I_2 = \frac{I_0}{2}(1 - \cos 2\alpha \sin\delta)$
3	$P_{\frac{\pi}{2}}\, Q_{\frac{\pi}{4}}\, Q_{\frac{-\pi}{4}}\, A_0$	$I_3 = \frac{I_0}{2}(1 - \cos\delta)$
4	$P_{\frac{\pi}{2}}\, Q_{\frac{\pi}{4}}\, Q_{\frac{\pi}{4}}\, A_0$	$I_4 = \frac{I_0}{2}(1 + \cos\delta)$
5	$P_{\frac{-\pi}{4}}\, Q_{\frac{\pi}{2}}\, Q_{\frac{\pi}{2}}\, A_0$	$I_5 = \frac{I_0}{2}(1 + \sin 2\alpha \sin\delta)$
6	$P_{\frac{-\pi}{4}}\, Q_{\frac{\pi}{2}}\, Q_0\, A_{\frac{\pi}{2}}$	$I_6 = \frac{I_0}{2}(1 - \sin 2\alpha \sin\delta)$
7	$P_{\frac{-\pi}{4}}\, Q_{\frac{\pi}{2}}\, Q_0\, A_{\frac{\pi}{4}}$	$I_7 = \frac{I_0}{2}(1 - \cos\delta)$
8	$P_{\frac{-\pi}{4}}\, Q_{\frac{\pi}{2}}\, Q_{\frac{\pi}{2}}\, A_{\frac{\pi}{4}}$	$I_4 = \frac{I_0}{2}(1 + \cos\delta)$

From the eight transmitted irradiance data, the phase of the isochromatic pattern is computed at each pixel using the relation

$$\tan\delta = \frac{(I_1 - I_2)\cos 2\alpha + (I_5 - I_6)\sin 2\alpha}{\frac{1}{2}\left[(I_4 - I_3) + (I_8 - I_7)\right]} \tag{8.48}$$

It may be seen that I_3 and I_4 are theoretically equal to I_7 and I_8. However, in practice, due to polariscope imperfections, they may differ and hence all four values are used in the algorithm.

The Fourier transform method can also be used for phase evaluation. The carrier fringe pattern is introduced by a birefringent wedge plate of an appropriate wedge angle. Usually, a carrier frequency of three to five lines per millimeter is adequate. The plate is placed close to the model. The fringe pattern is captured and processed using Fourier transform method.

8.12 BIREFRINGENT COATING METHOD: REFLECTION POLARISCOPE

The use of a birefringent coating on the surface of an object extends the application of the photoelasticity to the measurement of surface strains on opaque objects and thus eliminates the need to make models. In this method, a thin layer of a birefringent material is bonded onto the surface of the object. Assuming the adhesion to be good, the displacements on the surface of the object on loading are transferred to the coating, which induce birefringence in the coating. The strain-induced birefringence is observed in reflection. In order to obtain good-reflected irradiance, either the surface of the object is polished to make it naturally reflecting or some reflective particles are added to the cement, which bonds the birefringent coating to the surface of the object. Figure 8.10 shows a schematic of a reflection polariscope used with birefringent coatings.

There are plane polariscopes and circular polariscopes. The isochromatics obtained with the circular polariscope give the difference between the principal stresses in the coating, i.e.,

$$(\sigma_1 - \sigma_2)_c = \frac{m\, f_{\sigma c}}{2d} \tag{8.49}$$

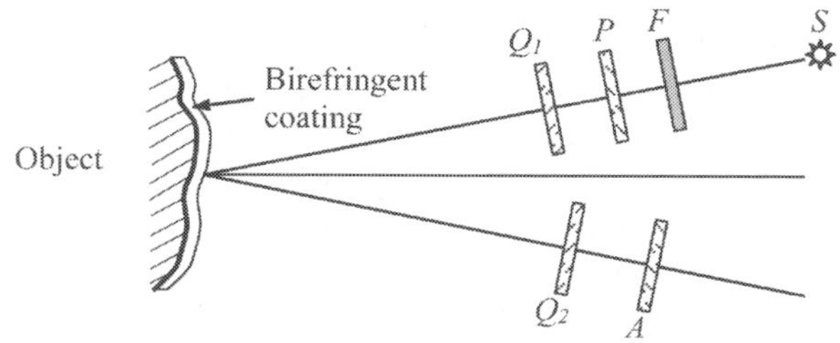

FIGURE 8.10 Schematic of a reflection polariscope; F, filter; P, polarizer; Q_1, Q_2, quarter-wave plates; A, analyzer.

where d is the thickness and $f_{\sigma c}$ is the fringe value of the coating. Subscript c stands for coating. Because the light travels through nearly the same region twice, the effective thickness is $2d$. The principal strains are related to the principal stresses through Hooke's law. We thus obtain the difference of the principal strains as

$$\varepsilon_1 - \varepsilon_2 = \frac{1+\nu_c}{E_c}(\sigma_1 - \sigma_2)_c \tag{8.50}$$

where E_c and ν_c are the elastic constants of the birefringent coating material. Similarly, we could express the difference of principal strains at the surface of the object as

$$\varepsilon_1 - \varepsilon_2 = \frac{1+\nu_o}{E_o}(\sigma_1 - \sigma_2)_o \tag{8.51}$$

where subscript o stands for object. Assuming that the strains in the coating and at the surface of the object are same, we have

$$(\sigma_1 - \sigma_2)_0 = \frac{E_o}{E_c}\frac{1+\nu_c}{1+\nu_o}(\sigma_1 - \sigma_2)_c \tag{8.52}$$

The separation of stresses in the coating is accomplished by the oblique incidence method. Hence, the principal strains in the coating are calculated. Having obtained the principal strains in the coating, the principal stresses at the surface of the object are obtained from the following equations:

$$\sigma_{10} = \frac{E_o}{1-\nu_o^2}(\varepsilon_1 + \nu_o\varepsilon_2) \tag{8.53a}$$

$$\sigma_{20} = \frac{E_o}{1-\nu_o^2}(\varepsilon_2 + \nu_o\varepsilon_1) \tag{8.53b}$$

The analysis is based on the assumption that the strains in the coating and at the surface of the object are the same.

8.13 HOLO-PHOTOELASTICITY

The separation of stresses requires that either σ_1, or σ_2, or $\sigma_1 + \sigma_2$ are known in addition to $(\sigma_1 - \sigma_2)$ obtained from photoelasticity. The sum of principal stresses is obtained interferometrically, say, by using a Mach–Zehnder interferometer. On the other hand, holo-photoelasticity provides fringe patterns belonging to $(\sigma_1 - \sigma_2)$ and $(\sigma_1 + \sigma_2)$ simultaneously, thereby effecting easy separation of stresses. However, the method requires coherent light for illumination. Here, we use holography to record the waves transmitted through the model and later reconstruct this record to extract the information. We can indeed use the technique in two ways. In one method, we obtain only isochromatics and hence the method is equivalent to a circular polariscope; it also provides the flexibility of leisurely evaluation of the fringe pattern. In the other method, both the isochromatics and isopachics fringe patterns are obtained. The method, which requires two exposures, is termed double-exposure holo-photoelasticity, while the former is a single-exposure method.

8.13.1 Single-Exposure Holo-Photoelasticity

The experimental arrangement for conducting a single-exposure holo-photoelastic experiment is shown in Figure 8.11. The light from a laser, generally polarized with its $\boldsymbol{E}$ vector vibrating in the vertical plane when the beam propagates in the horizontal plane, is used. The beam is expanded and collimated. In case the laser output is randomly polarized, a polarizer is used, followed by a quarter-wave plate oriented at 45°. In brief, the model is illuminated by a circularly polarized wave. The model is already stressed and hence is birefringent. The reference wave is also circularly polarized and of the same handedness so that both the components of electric field are interferometrically recorded.

The components of the wave just after the model are expressed as

$$E_{\sigma_1} = \frac{E_{0y}}{\sqrt{2}}\cos(\omega t - \kappa z + \psi_2) = \frac{E_{0y}}{\sqrt{2}}\cos(\tau + \psi_2) \tag{8.54a}$$

$$E_{\sigma_2} = -\frac{E_{0y}}{\sqrt{2}}\sin(\omega t - \kappa z + \psi_2) = -\frac{E_{0y}}{\sqrt{2}}\sin(\tau + \psi_2 + \delta) \tag{8.54b}$$

with $\psi_2 = -\frac{2\pi}{\lambda}(\mu_1 - 1)d_1$, $\delta = \frac{2\pi}{\lambda}(\mu_1 - \mu_2)d_1$ and d_1 is the thickness of the stressed model. Consistent with the treatment presented in the Chapter 5, we write these components with exponential functions as

$$E_{\sigma_1}(z;t) = Re\left[\frac{E_{0y}}{\sqrt{2}}\, e^{i(\tau+\psi_2)}\right] \tag{8.55a}$$

$$E_{\sigma_2}(z;t) = Re\left[\frac{E_{0y}}{\sqrt{2}}\, e^{i\left(\tau+\psi_2+\frac{\pi}{2}+\delta\right)}\right] \tag{8.55b}$$

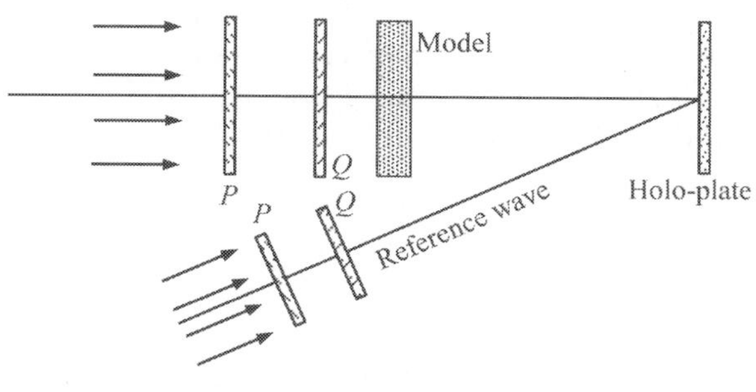

FIGURE 8.11 Experimental arrangement for single-exposure holo-photoelasticity.

Here τ_1 does not have the time dependence, which has been ignored because a monochromatic wave is used for illumination. Further, $Re\{\ldots\}$ refers to the real part. Similarly, the reference wave components are written as

$$E_{r_1} = Re\left[a_r\ e^{i\phi_r}\right] \tag{8.56a}$$

$$E_{r_2} = Re\left[a_r\ e^{i\left(\phi_r + \frac{\pi}{2}\right)}\right] \tag{8.56b}$$

Because these components are orthogonally polarized, they will interfere with the respective components; essentially, we record two holograms. The recorded irradiance is given by

$$I = \left|E_{\sigma_1} + E_{r_1}\right|^2 + \left|E_{\sigma_2} + E_{r_2}\right|^2 \tag{8.57}$$

This record, on processing, is a hologram. Assuming linear recording and illumination with the reference beam that releases the two beams, these beams interfere and generate an irradiance distribution of the type

$$I = I_0' \left| e^{i(\tau+\psi_2)} + e^{i(\tau+\psi_2+\delta)}\right|^2 = I_0\left(1+\cos\delta\right) \tag{8.58}$$

This is the irradiance distribution as obtained in a bright field circular polariscope. It may be noted that the quarter-wave–plate-analyzer combination is not placed behind the model during recording. The state of polarization in the reference wave serves the function of this assembly. If the state of polarization in the reference wave is orthogonal to that in the object wave from the model, i.e., the reference wave is of opposite handedness, the isochromatics pattern corresponding to the dark field circular polariscope will be obtained.

8.13.2 Double-Exposure Holo-Photoelasticity

The experimental arrangement for conducting double-exposure holo-photoelastic studies is similar to that shown in Figure 8.11. The model is illuminated by a circularly polarized wave, and another circularly polarized wave of the same handedness is used as a reference wave. The first exposure is made with the model unstressed and the second exposure with the model stressed. During the first exposure, the model is isotropic. However, to be consistent with our earlier treatment, we write the amplitudes of the object and reference waves recorded during the first exposure as

$$E_1(z;t) = Re\left[\frac{E_{0y}}{\sqrt{2}}\ e^{i(\tau+\psi_0)}\right] \tag{8.59a}$$

$$E_2(z;t) = Re\left[\frac{E_{0y}}{\sqrt{2}}\ e^{i\left(\tau+\psi_0+\frac{\pi}{2}\right)}\right] \tag{8.59b}$$

with $\psi_0 = -\frac{2\pi}{\lambda}(\mu_0 - 1)d$ and

$$E_{r_1} = Re\left[a_r\ e^{i\phi_r}\right] \tag{8.60a}$$

$$E_{r_2} = Re\left[a_r\ e^{i\left(\phi_r + \frac{\pi}{2}\right)}\right] \tag{8.60b}$$

In the second exposure, we record two waves from the stressed model. These waves are represented as

$$E_{\sigma_1}(z;t) = Re\left[\frac{E_{0y}}{\sqrt{2}}\ e^{i(\tau+\psi_2)}\right] \tag{8.61a}$$

$$E_{\sigma_2}(z;t) = Re\left[\frac{E_{0y}}{\sqrt{2}}\ e^{i\left(\tau+\psi_2+\frac{\pi}{2}+\delta\right)}\right] \tag{8.61b}$$

with $\psi_2 = -\frac{2\pi}{\lambda}(\mu_1 - 1)d_1$ and $\delta = \frac{2\pi}{\lambda}(\mu_1 - \mu_2)d_1$. The reference waves are the same as those used in the first exposure. As explained earlier, we record two holograms in the second exposure; the irradiance recorded is given by Equation (8.57). The total irradiance recorded can be written as

$$I = \left|E_1 + E_{r_1}\right|^2 + \left|E_2 + E_{r_2}\right|^2 + \left|E_{\sigma_1} + E_{r_1}\right|^2 + \left|E_{\sigma_2} + E_{r_2}\right|^2 \tag{8.62}$$

The amplitudes of the waves of interest, on reconstruction of the double-exposure hologram, are proportional to

$$2\ e^{i(\tau+\psi_0)} + e^{i(\tau+\psi_2)} + e^{i(\tau+\psi_2+\delta)} \tag{8.63}$$

These three waves interfere to generate a system of isochromatics $(\sigma_1 - \sigma_2)$ and isopachics $(\sigma_1 + \sigma_2)$ fringe patterns. The irradiance distribution in the interferogram is given by

$$I = I_0\left[1 + 2\cos\left(\frac{2\psi_2 + \delta - 2\psi_0}{2}\right)\cos\frac{\delta}{2} + \cos^2\frac{\delta}{2}\right] \tag{8.64}$$

Before proceeding further, we need to know what the term $(2\psi_2 + \delta - 2\psi_0)$ represents. Substituting for ψ_2, δ, and ψ_0 we obtain

$$\begin{aligned}
2\psi_2 + \delta - 2\psi_0 &= -\frac{2\pi}{\lambda}\left[2(\mu_1 - 1)d_1 - (\mu_1 - \mu_2)d_1 - 2(\mu_0 - 1)d\right] \\
&= -\frac{2\pi}{\lambda}\left[(\mu_1 + \mu_2)d_1 - 2\mu_0 d - 2\Delta d\right] \\
&= -\frac{2\pi}{\lambda}\left[(\mu_1 - \mu_0)d + (\mu_2 - \mu_0)d + (\mu_1 + \mu_2)\Delta d - 2\Delta d\right]
\end{aligned} \tag{8.65}$$

where $\Delta d = d_1 - d$. Assuming the birefringence to be small so that $(\mu_1 + \mu_2)$ can be replaced by $2\mu_0$, and then substituting for $(\mu_1 - \mu_0)$ and $(\mu_2 - \mu_0)$ from Equations (8.6a) and (8.6b), respectively, and using Equation (8.13) we obtain,

$$\begin{aligned} 2\psi_2 + \delta - 2\psi_0 &= -\frac{2\pi}{\lambda}\left[(C_1 + C_2)(\sigma_1 + \sigma_2)d - 2(\mu_0 - 1)\frac{\nu}{E}(\sigma_1 + \sigma_2)d\right] \\ &= -\frac{2\pi}{\lambda}\left\{\left[(C_1 + C_2) - 2(\mu_0 - 1)\frac{\nu}{E}\right](\sigma_1 + \sigma_2)d\right\} \\ &= \frac{2\pi}{\lambda}\left[(C_1' + C_2')(\sigma_1 + \sigma_2)d\right] = \frac{2\pi}{\lambda}C'(\sigma_1 + \sigma_2)d \end{aligned} \tag{8.66}$$

It is thus seen that the argument of the cos function in the second term in Equation (8.64) depends only on the sum of the principal stresses and hence generates the isopachic fringe pattern. We can rewrite Equation (8.64) as

$$I = I_0\left\{1 + 2\cos\left[\frac{\pi}{\lambda}C'(\sigma_1 + \sigma_2)d\right]\cos\left[\frac{\pi}{\lambda}C(\sigma_1 - \sigma_2)d\right] + \cos^2\left[\frac{\pi}{\lambda}C(\sigma_1 - \sigma_2)d\right]\right\} \tag{8.67}$$

It is seen that the second term in Equation (8.67) contains information about the isopachics, whereas the second and third terms contain information about the isochromatics. Figure 8.12 shows an interferogram depicting both types of fringes. We will now examine Equation (8.67) and study the formation of isochromatics and isopachics.

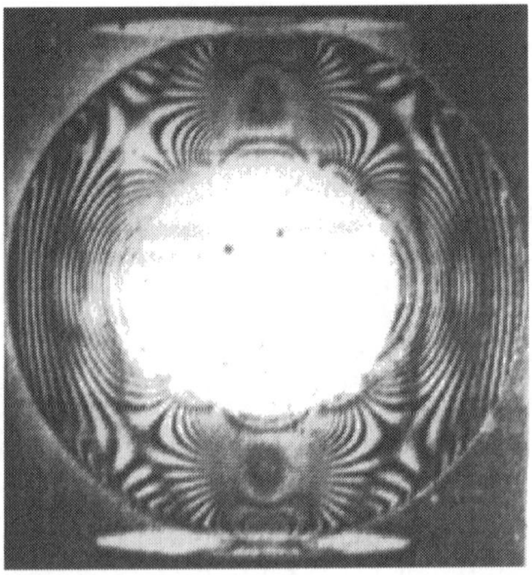

FIGURE 8.12 Interferogram showing both the isochromatics (broad fringes) and isopachics. Notice the phase shift of π when the isopachics cross an isochromatic.

Because we are using a bright field configuration, the dark isochromatics would occur when

$$\frac{\pi}{\lambda}C(\sigma_1-\sigma_2)d=(2n+1)\frac{\pi}{2}: n \text{ is an integer} \tag{8.68}$$

However, the irradiance of the dark isochromatics is not zero but I_0. The bright isochromatics occur when $\frac{\pi}{\lambda}C(\sigma_1-\sigma_2)d=n\pi$ and the irradiance in the bright isochromatics is given by

$$I=2I_0\left\{1+(-1)^n\ cos\left[\frac{\pi}{\lambda}C'(\sigma_1+\sigma_2)d\right]\right\} \tag{8.69}$$

The irradiance of the bright isochromatics is modulated by isopachics. Let us first consider a bright isochromatics of even order whose irradiance is given by

$$I=2I_0\left\{1+\ \cos\left[\frac{\pi}{\lambda}C'(\sigma_1+\sigma_2)d\right]\right\} \tag{8.70}$$

The irradiance in the bright isochromatics will be zero wherever

$$\frac{\pi}{\lambda}C'(\sigma_1+\sigma_2)d=(2K+1)\pi: K=0,\ 1,\ 2,\ 3,\ \ldots. \tag{8.71}$$

The integer K gives the order of isopachics. Wherever this condition is satisfied, the isochromatics would have zero irradiance. Therefore, the isopachics modulate the bright isochromatics. Let us now see what happens to the next order bright isochromatics. The irradiance distribution in this isochromatics will be

$$I=2I_0\left\{1-\ \cos\left[\frac{\pi}{\lambda}C'(\sigma_1+\sigma_2)d\right]\right\} \tag{8.72}$$

If we substitute the condition for the Kth dark isopachics in this equation, it gives a maximum irradiance of $4I_0$. This shows that the Kth isopachics has changed by one half-order in going over from one bright isochromatics to the next bright isochromatics. This interpretation is simple and holds valid when the two families of fringes are nearly perpendicular as shown in Figure 8.13. In the other extreme case, where the isochromatics and the isopachics run parallel to each other, this analysis breaks down. It is therefore advisable to use some method to separate out these two fringe patterns. The influence of birefringence can be eliminated by passing the beam twice through both the model and a Faraday rotator, thereby eliminating the isochromatics pattern. For the model, one can also use materials such as polymethyl methacrylate (PMMA), which has almost no or very little birefringence. For such a model, only the isopachic pattern would be observed. Holo-photoelasticity can also be performed in real time, which offers certain advantages.

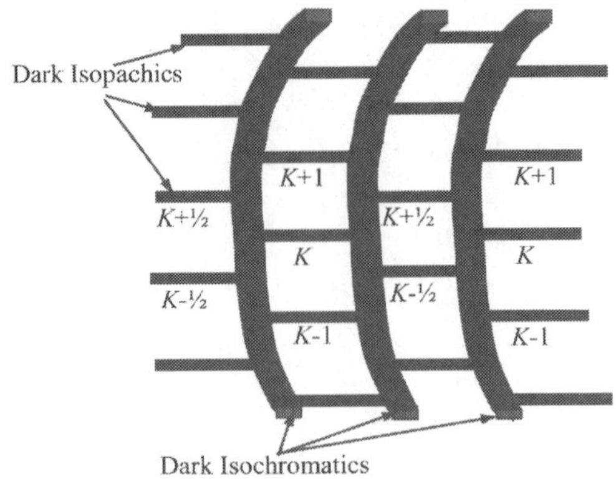

FIGURE 8.13 Simplified combined isochromatics and isopachics pattern (drawn).

8.14 THREE-DIMENSIONAL PHOTOELASTICITY

Photoelastic methods, thus far described, cannot be used for investigations of objects under a three-dimensional (3D) state of stress. When a polarized wave propagates through such an object, assuming it to be transparent, it integrates the polarization changes over the distance of travel. The integrated optical effect is so complex that it is impossible to analyze it or relate it to the stresses that produced it. There are, however, several methods for such investigations. Of these, we discuss two methods, namely the frozen stress method and the scattered light method. The frozen stress method is restricted in its application to static cases of loading by external forces.

8.14.1 The Frozen Stress Method

The frozen stress method is possibly the most powerful method of experimental stress analysis. It takes advantage of the multiphase nature of plastics used as model material. The procedure for stress freezing consists of heating the model to a temperature slightly above the critical temperature and then cooling it slowly to room temperature, typically at less than 2°C per hour under the desired loading condition. The load may be applied to the model either before or after reaching the critical temperature. Extreme care is taken to ensure that the model is subjected to correct loading as spurious stresses due to bending and gravitational load may be induced because of the low rigidity of the model material at the critical temperature.

After the model is cooled to the room temperature, the elastic deformation responsible for the optical anisotropy is permanently locked. The model is now cut into thin slices for examination under the polariscope. The optical anisotropy is normally not disturbed during slicing if the operation is carried out at high speeds and under coolant conditions.

8.14.2 Scattered Light Photoelasticity

When a beam of light passes through a medium containing fine particles dispersed in the volume, a part of the beam is scattered. The irradiance of the scattered light,

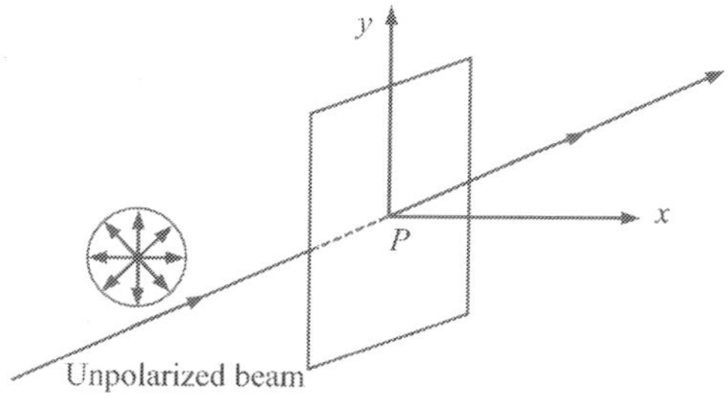

FIGURE 8.14 Scattering of unpolarized beam by a scatterer at *P*.

when the particles are much smaller than the wavelength of light, varies as ω^4, where ω is the angular frequency of the light waves. This phenomenon was investigated by Rayleigh in detail and is called Rayleigh scattering. The most beautiful observations of the red sunset and blue sky are due to scattering from the gaseous molecules in the atmosphere. Further, the light from the blue sky is partially linearly polarized. In some observation directions, the scattered light is linearly polarized.

Consider a scattering center located at a point *P* as shown in Figure 8.14. Let the incident light be unpolarized that can be resolved into two orthogonal linearly polarized components with random phases. The incident component vibrating in the y–z plane, when absorbed, will set the particle (rather the electrons in the particle) vibrating along the y-direction. The re-radiated wave will have zero amplitude along the y-direction. On the other hand, if the particle is oscillating along the x-direction, the re-radiated wave will have zero amplitude in that direction. Thus, when the observation direction lies along the y-direction in the x–y plane passing through point *P*, the scattered wave will be plane polarized. The particle acts as a polarizer.

Let us now consider that the incident wave is linearly polarized with the $\boldsymbol{E}$ vector vibrating in the y–z plane. The electrons in the particle will oscillate along the y-direction. The re-radiated wave will have zero amplitude when observed along the y-axis. The particle thus acts as an analyzer. This picture is equivalent to placing a polarizer and an analyzer anywhere in the model. Therefore, stress information can be obtained without freezing the stress and slicing the model. The scattered light method therefore provides a nondestructive means of optical slicing in three dimensions.

8.14.2.1 Examination of the Stressed Model in Scattered Light

8.14.2.1.1 Unpolarized Incident Light

Let us consider a stressed model in the path of a narrow beam of unpolarized light. We assume that there are a large number of scatterers in the model. Let us now consider the light scattered by a scatterer at point *P* inside the model when the observation direction is perpendicular to the incident beam. The scattered light is resolved into components along the directions of principal stresses σ_2 and σ_3 as shown in Figure 8.15. In traversing a distance *PQ* in the model, these two orthogonally polarized components acquire a phase difference. If an analyzer is placed in the observation direction, the transmitted irradiance will depend on the phase difference

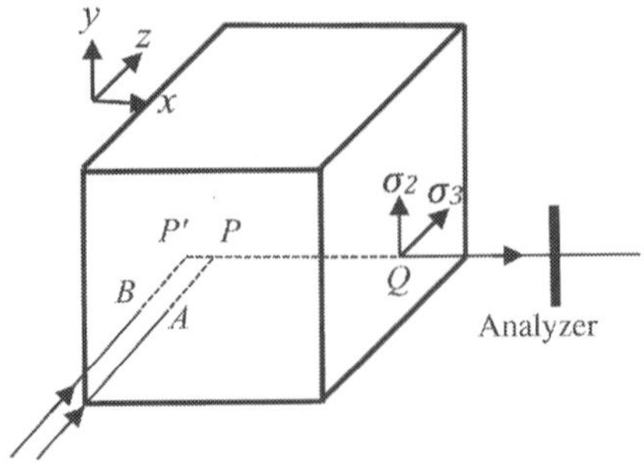

FIGURE 8.15 Stressed model. Illumination by an unpolarized beam and observation through an analyzer.

acquired. Because the incident beam is unpolarized, there is no influence of traverse *AP* in the model. Consider that the transmitted irradiance is zero for a certain location *P* of the scatterer; this occurs when the phase difference is a multiple of 2π. As the beam is moved to illuminate another scatterer at point *P*′ in the same plane along the line of sight, the transmitted irradiance will undergo cyclic variation between minima and maxima depending on the additional phase acquired when traversing the distance *PP*′. It is, however, assumed that the directions of principal stresses σ_2 and σ_3 do not change over the distance *PP*′.

Let m_1 and m_2 be the fringe orders when the light scattered from scatterers at points *P* and *P*′ is analyzed. Then

$$x_1(\sigma_2 - \sigma_3) = m_1 f \tag{8.73a}$$

and

$$x_2(\sigma_2 - \sigma_3) = m_2 f \tag{8.73b}$$

where $x_1 = PQ$ and $x_2 = P'Q$. Therefore, we obtain

$$(\sigma_2 - \sigma_3) = \frac{m_2 - m_1}{x_2 - x_1} f = \frac{dm}{dx} f \tag{8.74}$$

The principal stress difference at any point along the observation direction is proportional to the gradient of the fringe order.

8.14.2.1.2 *Linearly Polarized Incident Beam*

We now consider another situation of a linearly polarized beam incident on the model as shown in Figure 8.16. We assume, for the sake of simplicity, that the principal stress directions are along the *x*- and the *y*-axes. The transmission axis of the polarizer makes an angle α with the *x*-axis. The incident wave of amplitude E_0 is resolved along the *x*- and the *y*-directions. These linearly polarized components travel with

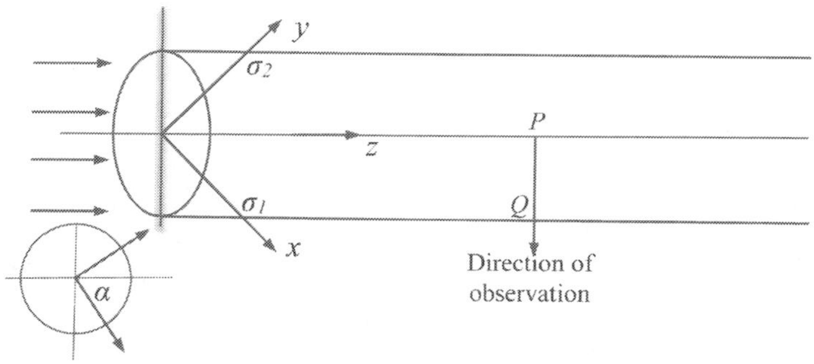

FIGURE 8.16 Stressed model. Illumination by a linearly polarized beam.

different velocities in the model and hence pick up a phase difference, δ. Thus, at any plane normal to the direction of propagation, the state of polarization of the wave in general will be elliptical. It is expressed by the equation

$$\frac{E_x^2}{E_0^2\cos^2\alpha} + \frac{E_y^2}{E_0^2\sin^2\alpha} - \frac{2\,E_x\,E_y}{E_0^2\cos\alpha\,\sin\alpha}\cos\delta = \sin^2\delta \tag{8.75}$$

where E_x and E_y are the components along the x- and the y-directions, respectively. The major axis of the ellipse makes an angle ψ with the x-axis where

$$\tan 2\psi = \tan 2\alpha \cos\delta \tag{8.76}$$

When $\delta = 2p\pi$, $p = 0, \pm1, \pm2, \pm3, \pm4, \ldots$, the state of polarization of the wave is linear with orientation $\psi = \pm\alpha$. For positive values of the integer p, the state of polarization of the wave at any plane is the same as that of the incident wave. In general, a scatterer at point P in any plane is excited by an elliptically polarized wave. In scattered light photoelasticity, we are looking in the model normal to the direction of propagation, i.e., the observation is confined to the plane of the elliptically polarized light. If the observation is made in the direction along the major axis of the ellipse, the amplitude of the re-radiated wave received by the observer will be proportional to the magnitude of the minor axis of the ellipse, and hence minimum. On the other hand, if the observation direction coincides with the minor axis, the irradiance will be a maximum.

As the beam propagates in the stressed model, the ellipse just described keeps on rotating as the phase difference changes. Therefore, the observation made in the scattered light normal to the direction of the incident beam will show variation in irradiance along the length of model in the direction of the incident beam. There is no influence of birefringence in the model on the traverse distance PQ. Figure 8.17(a) shows a scattered light stress pattern of a disc under diametral compression. The directions of incident beam and that of the scattered light are also shown in Figure 8.17(b).

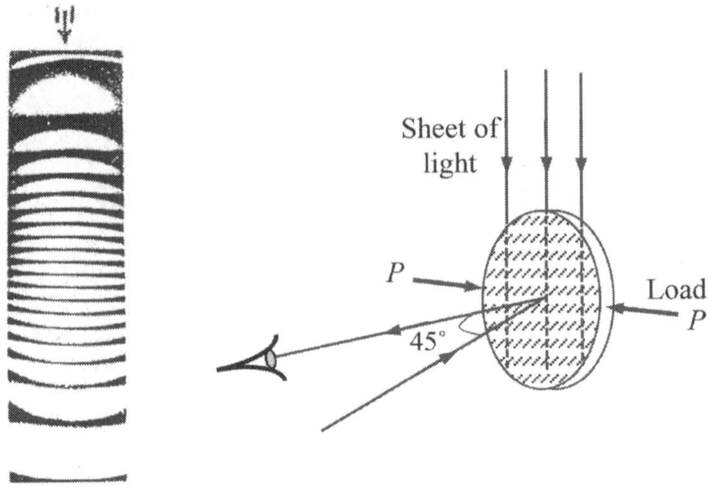

(a) (b)

FIGURE 8.17 (a) Scattered light pattern of a disc under diametral compression. (b) Schematic showing the directions of illumination and observation.

Due to the weak irradiance of the scattered light, an intense incident beam is used. This beam is realized from a high-pressure mercury lamp with suitable collimating optics. The laser is an attractive alternative as the beam can be used without optics. The model is placed in a tank containing an index-matching liquid to avoid refraction and polarization changes at the model surfaces. To facilitate proper adjustments, the model in the tank is mounted on a stage capable of translation and rotation motions.

QUESTIONS

8.1 With a monochromatic light source how would you distinguish isoclinics and isochromatics in a plane polariscope?

8.2 Why is polarized light used in photoelasticity experiments instead of unpolarized light, considering that stressed photoelastic models behave like crystals and can polarize light themselves?

8.3 Show that the irradiance distribution in a bright field plane polariscope is given by

$$I = I_0 - I_0 \sin^2 2\alpha \sin^2 (\delta / 2)$$

8.4 A circular polariscope is constructed using various optical arrangements of the elements. Establish using Jones calculus which of the given arrangements correspond to bright and dark fields. In the table, P is the orientation of the polarizer, Q_1 is the orientation of the slow axis of quarter-wave plate 1, Q_2 is the orientation of the slow axis of quarter-wave plate 2, and A is the orientation of the analyzer.

P	Q_1	Q_2	A
$\pi/2$	$3\pi/4$	$3\pi/4$	$\pi/4$
$\pi/2$	$3\pi/4$	$\pi/4$	0

8.5 Two discs, one made of epoxy ($f_\sigma = 12.5$ N / mm / fringe) and another made of polycarbonate ($f_\sigma = 8$ N / mm / fringe) are subjected to diametral compression in a photoelasticity experiment. The diameter and thickness of both discs are 60 mm and 6 mm, respectively. The load applied in both cases is 150 N. If the fringe order (m) at a particular point of the epoxy disc is 2, what would be the fringe order at the same point in the polycarbonate disc?

8.6 Explain the working principle of the Soleil–Babinet compensator and compare it against Tardy's method of compensation. Discuss the merits and demerits using an example.

8.7 In a photoelastic model, an isochromatic fringe order of 2.5 that is reasonably broad is passing through a specific point. As a first approximation, the fringe order at that point can be construed as 2.5. The material fringe value f_σ is 12 N/mm/fringe and the model's thickness is 4 mm. Using Tardy's compensation method, fringe order 2.0 moves to the point of interest precisely by rotating the analyzer by +100 degrees.

a. Calculate the initial principal stress difference $(\sigma_1 - \sigma_2)$ using the given fringe order and f_0.
b. Revise the fringe order at the point of interest by Tardy's method and calculate the principal stress difference.

8.8 Explain phase shifting in photoelasticity by comparing it with Tardy's method of compensation.

8.9 Select the correct statement(s) related to the photoelastic coating technique.

a. Surface stains are transmitted to the coating when the specimen is loaded
b. Surface stress is transmitted to the coating when the specimen is loaded
c. Material stress fringe value is same as material strain fringe value
d. Correction factors are required for the accurate measurement of specimen stresses

8.10 The photoelastic fringe patterns simulated using a virtual polariscope software, P_Scope are shown in Figure 8.18 for a 60-mm diameter and a 6-mm-thick circular epoxy disk (F_σ = 12 N/mm/fringe) subjected to diametral compression. Figure 8.18(a)–(c) depicts the fringe patterns in monochromatic light (589.60 nm) for three different diametral loads (P), namely, 282.74 N, 565.49 N, and 848.23 N, respectively. Further, Figure 8.18(d) represents the fringe pattern for the 848.23 N case under white light along with a color code correlating to the fringe order (N). Based on the given data, answer the following questions:

i. Label the fringe orders (N) for the monochromatic fringe patterns in Figure 8.18(a)–(c).
ii. Using the given fringe patterns, calculate the principal stress difference ($\sigma_1 - \sigma_2$) for all three load cases at the center point A (marked on Fig. 8.18a).

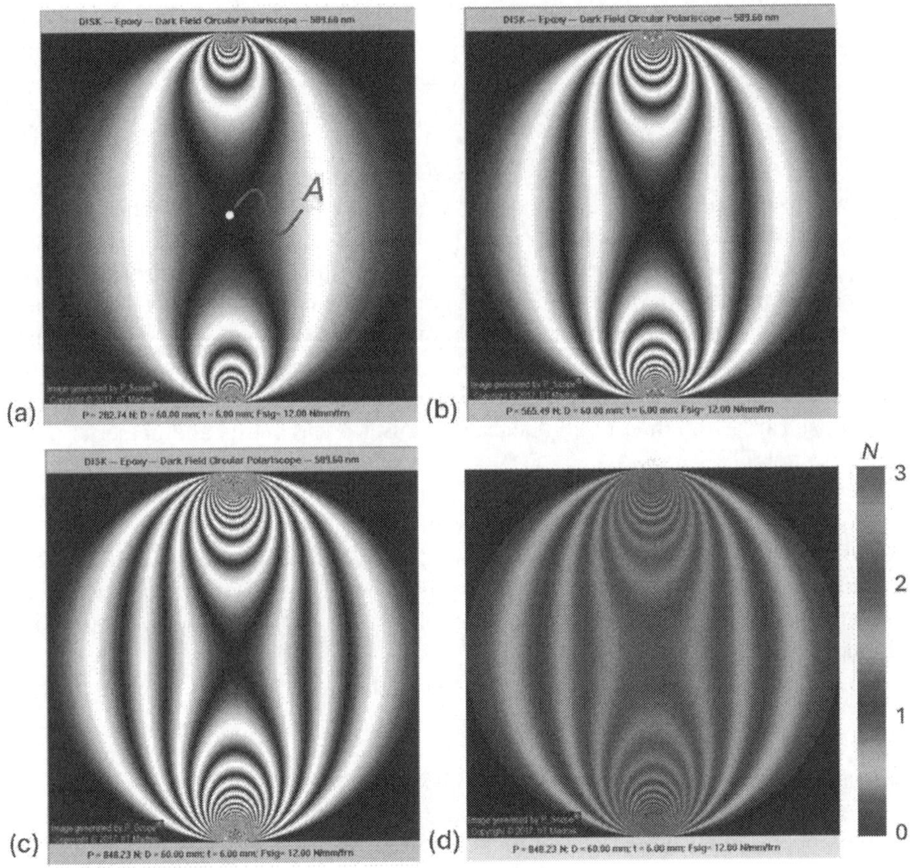

FIGURE 8.18 Photoelastic fringe patterns in monochromatic light (589.60 nm) for a circular disk under diametral compression for a diametral load P of (a) 282.74 N, (b) 565.49 N, (c) 848.23 N, and (d) colored isochromatic fringes under white light along with the corresponding color code for fringe order (N). (Courtesy K. Ramesh, "P_Scope®—A Virtual Polariscope," Photomechanics Lab, IIT Madras. [Online]. Available: https://home.iitm.ac.in/kramesh/p_scope.html.)

9 Microscopy

A microscope aids in seeing details. The resolving power of the eye is about 1 arcminute or 60 arcseconds (~2.9×10^{-4} rad). The minimum distance of the object from the eye, which can be imaged, is around 25 cm and is called the distance of distinct vision. Therefore, two points separated by about 0.1 mm can be distinctly seen by the eye. Objects closer than this have to be magnified before being presented to the eye. Magnifiers and microscopes are the devices that magnify the image.

Microscopes are also used to make dimensional measurements of micro-objects, study their birefringence, and examine surfaces. There are a variety of microscopes equipped with multiple objectives and measuring devices. The object can be either epi-illuminated or trans-illuminated.

9.1 MICROSCOPES

9.1.1 Simple Magnifier

Figure 9.1 shows a simple microscope, which essentially is a short focal length single positive lens. The object is placed close to the lens, within its focal distance. An erect image is formed at a distance of distinct vision from the eye, which is roughly 250 mm. Thus, the magnification M is $M = 1 + \frac{250}{f}$, where the focal length f is in millimeters. One cannot obtain high magnification using a single lens or a simple magnifier.

9.1.2 Compound Microscope

A compound microscope is a combination of two lenses, an objective and an eyepiece. The objective is a corrected short focal length compound lens that presents a magnified image. The eyepiece forms the image at infinity. A schematic of a compound microscope is shown in Figure 9.2.

The magnification M of the objective is given by $M = h'/h$. This can also be expressed as $M = b/a = f/z = z'/f_b$. Here f_b is the back focal length and z' is the

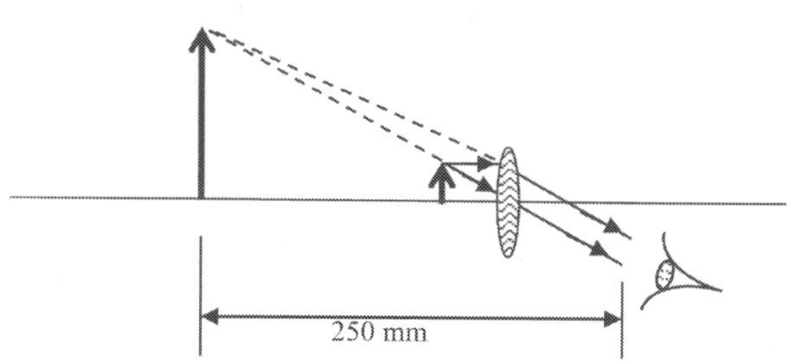

FIGURE 9.1 A simple microscope.

DOI: 10.1201/9781003531845-9

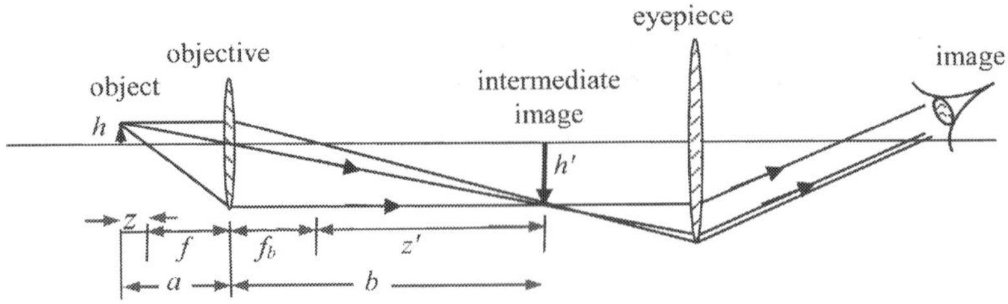

FIGURE 9.2 A compound microscope.

tube length. Tube length lies between 160 mm and 210 mm depending on the manufacturer. The magnification of the objective is to be multiplied by the magnification of the eyepiece to obtain the magnification of the compound microscope. The magnification of the eyepiece is given by $250/f_E$, where f_E is the focal length of the eyepiece. The schematic shown in Figure 9.2 is for the finite conjugate and the objective is corrected for finite conjugate positions. It may be seen that the introduction of any optical component in the convergent beam will introduce aberrations and degrade the image. Therefore, from 1980, infinity corrected objectives became available. The schematic of an infinity corrected compound microscope is shown in Figure 9.3. It has an additional lens called the tube lens after the objective.

The magnification of the objective is given by $M = h'/h = f_{\text{tube lens}}/f_{\text{objective}}$. The space between the objective and the tube lens is used to introduce other optical elements like polarizer, differential interference contrast (DIC) prism, etc.

A real microscope contains a trans-illumination system with filters, a condenser, a stage to carry the object, a set of objectives on a turret, and an eyepiece. A camera may replace the eyepiece, or both the eyepiece and the camera may be available. There is a set of diaphragms to control the illumination. Some microscopes may have both epi- and trans-illumination systems and an eyepiece with a measuring scale. In some microscopes the stage also carries a measurement scale for both x- and y-positions.

According to Abbe's theory of image formation in a microscope, diffraction takes place both at the object and at the objective of the microscope. Depending on the details in the object, the angle of diffraction could be large and hence the objective may not be able to collect the diffracted field from the object. A quantity that conveniently defines its ability to collect the light is the numerical aperture (NA) of the objective. The NA is defined as $NA = \mu \sin\theta$, where μ is the refractive index of

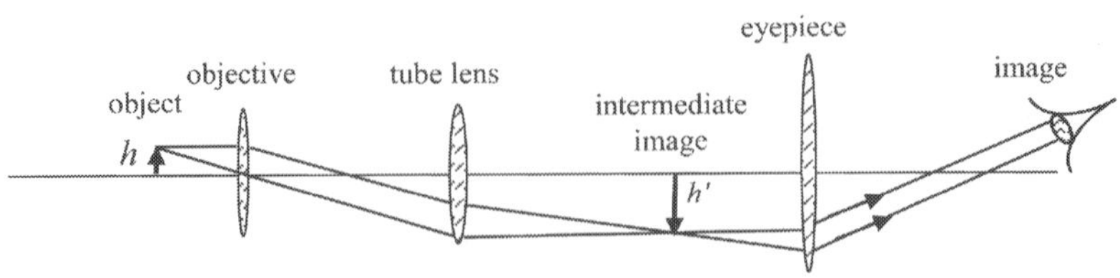

FIGURE 9.3 An infinity corrected compound microscope.

medium in front of the objective and θ is the semi-angle of the cone that the objective could capture. In theory the angle θ can take a value up to $\pi/2$. The objectives are designed to have very high value of *NA*.

Each microscope objective carries its specifications like value of *NA*, magnification, etc. The objectives usually are designed to work with a 0.17-mm thick glass plate of refractive index 1.515, called a coverslip. There are special objectives designed for polarized light, which carry abbreviations like *P*, *PO*, *POL*, or *SF* and all barrel engravings in red, *PH* for phase contrast and barrel engravings in green, DIC/ NIC (Nomarski interference contrast).

A microscope presents a well-resolved magnified image. The resolution is governed by diffraction as was pointed out in Chapter 1. If we consider that a point object is being imaged by the microscope objective, then its image will not be a point but the well-known Airy distribution. The irradiance distribution in the image of a point object is given by

$$I(r) = I_0 \left(\frac{2J_1(k\omega r)}{k\omega r} \right)^2 \tag{9.1}$$

where $k = 2\pi/\lambda$, ω is the sine of angle of diffraction, and r is the polar coordinate. The image consists of a central disk surrounded by rings with decreasing irradiance. Therefore, two incoherent points cannot be arbitrarily close to be seen distinctly. Following Rayleigh, the two-point objects are just resolved when the maximum of irradiance distribution of one object falls over the first minimum of the irradiance distribution of the second point. The distance, *d*, between the images of two points under Rayleigh criterion, is given by $d = \frac{1.22\,\lambda}{NA_{\text{obj}} + NA_{\text{condenser}}}$. When the numerical aperture NA of the condenser ($NA_{\text{condenser}}$) is matched with that of the objective, the distance d is then given by $d = \frac{0.61\,\lambda}{NA_{\text{obj}}}$.

The basic principle to be remembered is that the combination of direct and diffracted light (or manipulation of direct and diffracted light) is critically important in image formation. The key locations for such manipulation are the back focal plane of the objective and the front focal plane of the substage condenser. This principle is fundamental to most of the contrast-enhancing methods.

9.1.3 Köhler Illumination

Köhler in 1893 introduced a collector lens for the lamp and used it to focus the image of the lamp on the front aperture of the condenser. The field diaphragm was then focused on the specimen with the condenser focus control. The method provided bright, uniform illumination and fixed the positions of the focal planes of the microscope optics. All manufacturers of modern microscopes recommend this illumination as it produces uniform and glare-free specimen illumination. Correct alignment of microscope components ensures proper specimen illumination in order to achieve the best balance between the image contrast and the resolution. An understanding of the illumination and image-forming pathways is necessary in order to realize the full imaging potential of microscopes.

A microscope has two sets of conjugate planes: a set of four object or field planes and a set of four aperture or diffraction planes. The field planes are observed in normal viewing mode using an eyepiece. This mode is also called orthoscopic mode, and the image of the object is called an orthoscopic image. A Bertrand lens or an eyepiece with a telescope is required to view the aperture or the diffraction planes. Bertrand lens is focused on the back aperture of the objective. This mode is called the conoscopic mode. Figure 9.4 shows both the sets of conjugate planes.

The diameter of the view field in an eyepiece is expressed as a field of view number or field number (*FN*). Information about the *FN* of an eyepiece can yield the real diameter of the object view field using the formula

$$\text{Viewfield diameter}\left(\text{Obj}\right) = \frac{FN}{M_{\text{obj}} \times M_{\text{tube lens}}} \tag{9.2}$$

where *FN* is the field number of the eyepiece, M_{obj} and $M_{\text{tube lens}}$ are the magnifications of the objective and the tube lens, respectively.

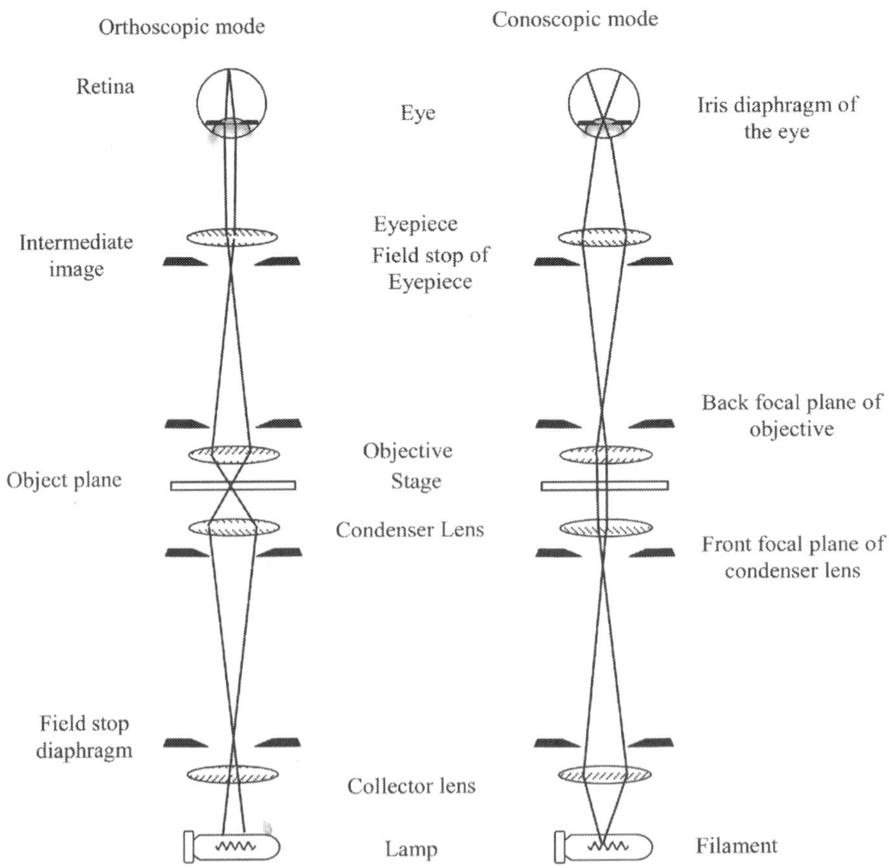

FIGURE 9.4 Conjugate planes in a microscope for Köhler illumination.

9.1.4 Empty Magnification

Earlier the eyepieces were available with magnifications ranging from 6.5× to 30× and sometimes even higher. When these eyepieces are used with objectives of high magnification, the problem of empty magnification may result. Today most eyepieces are in the magnification range of 10× to 20×.

NA of the objective/condenser system defines the range of useful magnification for an objective-eyepiece combination. There is a minimum magnification necessary for the details present in the image to be resolved, and this value is rather arbitrarily, set as 500 × NA. At the other end, the maximum useful magnification of an image is usually set at 1000 × NA. Exceeding this magnification does not yield any further information: the image suffers from the phenomenon of empty magnification. Table 9.1 lists the combinations of the objective and the eyepiece that satisfy the range of useful magnification.

9.1.5 Depth of Field

Depth of field is the distance along the optical axis over which image details are observed with acceptable clarity. Those factors that influence resolution also affect depth of field, but in the opposite direction. Therefore, a compromise must be reached between these two parameters, which becomes more difficult as the magnification increases. The depth of field, d_{axial}, can be estimated from:

$$d_{axial} = \frac{\lambda\sqrt{\mu^2 - NA^2}}{NA^2} = \frac{\lambda}{\mu \sin\theta}\cot\theta = \frac{\lambda}{NA}\cot\theta \tag{9.3}$$

where μ is the refractive index of the medium between the specimen and the objective ($n \sim 1.0$ for air), λ is the wavelength of light, and NA is the numerical aperture. This equation shows that depth of field increases as the NA decreases. Depth of field, in microscopy, is very small and usually measured in micrometers.

TABLE 9.1
Objective-Eyepiece Combinations

	Range of Useful Magnification {(500–1000) × NA_{obj}}				
Objective	**Eyepiece Magnification**				
Magnification/(NA)	**10×**	**12.5×**	**15×**	**20×**	**25×**
2.5× (0.08)	–	–	–	✓	✓
4× (0.12)	–	–	✓	✓	✓
10× (0.35)		✓	✓	✓	✓
25× (0.55)		✓	✓	✓	–
40× (0.70)	✓	✓	✓	–	–
60× (0.95)	✓	✓	✓	–	–
100× (1.42)	✓	✓	–	–	–

At high *NAs* of the microscope objectives, the depth of field is determined primarily by wave optics, whereas at lower *NAs*, the geometrical optical circle of confusion dominates the phenomenon. Using a variety of different criteria for determining when the image becomes unacceptably sharp, several authors have proposed different formulae to describe the depth of field in a microscope. The total depth of field is given by the sum of the wave and geometrical optical depths of fields as

$$d_{tot} = \frac{\lambda \mu}{NA^2} + \frac{\mu}{M \times NA} e \tag{9.4}$$

where d_{tot} represents the total depth of field, λ is the wavelength of illuminating light, μ is the refractive index of the medium (usually air [$\mu = 1.000$] or immersion oil [$\mu = 1.515$)] between the coverslip and the objective front lens element, and *NA* equals the objective *NA*. The variable *e* is the smallest distance that can be resolved by a detector that is placed in the image plane of the microscope objective whose lateral magnification is *M*.

9.1.6 Depth of Focus

The term depth of focus, which refers to image space, is often used interchangeably with the depth of field, which refers to object space. This interchange of terms could lead to confusion.

There are several formulae for the depth of focus in the literature. These are

$$d_{\text{depth of focus}} = \frac{1000}{7 \times NA \times M} + \frac{\lambda}{2\ NA^2} \tag{9.5a}$$

$$d_{\text{depth of focus}} = \frac{\lambda}{2NA^2} \tag{9.5b}$$

$$d_{\text{depth of focus}} = \frac{1000 M_{obj}}{7 \times NA \times M_{total}} + \frac{\lambda\ M_{obj}^2}{2\ NA^2} \tag{9.5c}$$

None of these seem to give the results, which compare well with experiments. Hence, the formula given below for the depth of focus is developed

$$d_{\text{depth of focus}} = \frac{\lambda}{4\ \mu \left(1 - \sqrt{1 - (NA/\mu)^2}\right)} \tag{9.5d}$$

Depth of focus varies with the *NA* and the magnification of the objective. Under some conditions, high *NA* systems (usually with higher magnification) have deeper focus depths than those systems with low *NA*, even though the depth of field is less.

9.2 CONTRAST-ENHANCING TECHNIQUES

Contrast is the measure of the ability of the objects seen distinctly against the background. It is defined as

$$\%\ \text{contrast} = \frac{I_b - I_s}{I_b} \times 100 \tag{9.6}$$

where I_b is the irradiance of the background and I_s is the irradiance of the scene/signal.

Some objects absorb light either completely or partially (amplitude objects) and hence can be easily seen in bright field microscopy. Others are naturally coloured or can be stained and hence are also seen in bright field microscopy. Many specimens particularly living, have very poor contrast: in many cases it is so poor that the specimen remains practically invisible. It is also important not to alter them by killing or by artificially staining or fixating them chemically. To overcome this drawback, many contrast-enhancing techniques have been developed.

9.2.1 Dark Field Microscopy

An object, when illuminated, diffracts light depending on the frequency contents in the object. For a faithful imaging, all transmitted and diffracted light should be collected by the objective. In dark field illumination, undiffracted (direct) light is eliminated, and only diffracted light is used for image formation. If the specimen has reflective edges or refractive index gradients, the reflected or refracted light, due to small deviations, will enter the objective and contribute to image formation.

Resolving power of the microscope is the same in dark field illumination as it is in bright field illumination, but the optical character of the image is not as faithfully reproduced. Dark field images are quite spectacular to see and the objects, which have very low contrast in bright field illumination, shine brilliantly in dark field illumination. Such illumination is quite good for revealing outlines, edges, and boundaries.

9.2.2 Rheinburg Illumination

Rheinburg illumination is a form of optical staining. The technique is a variation of low-to-medium power dark field illumination using colored gelatin or glass filters to provide rich color to both the specimen and the background. The central opaque dark field is replaced with a transparent, colored circular stop inserted into a transparent ring of a contrasting color. These stops are placed under the bottom lens of the condenser (front focal plane). It results in the specimen rendered in the color of the ring and the background having the color of the stop.

9.2.3 Hoffman Modulation Contrast

The Hoffman modulation contrast system is designed to increase visibility and contrast in unstained and living specimens by detecting optical gradients (or slopes) and converting them into variations of light irradiance. An optical amplitude filter, termed as modulator, is inserted at the back focal plane of the objective. The modulator has three zones: a small dark zone near the periphery of the back focal plane, which transmits only 1% of the light irradiance; a narrow gray zone, which transmits 15% of the light; and the remaining transparent zone covering most of the area at the back focal plane, which transmits 100% of the light. Light irradiance passing through it varies above and below the average value and thus is said to be modulated. When viewed under modulation contrast optics, the transparent objects that are essentially invisible in ordinary bright field microscopy take on an apparent three-dimensional (3D) appearance dictated by phase gradients. The Hoffman modulator does not alter the phase of the light passing through it but influences the principal zeroth-order maximum; higher-order maxima are unaffected.

Below the specimen stage, a condenser with a rotating turret is utilized to hold the remaining components of the Hoffman modulation contrast system. The turret condenser has a bright field opening for an aperture iris diaphragm for regular bright field microscopy and for alignment and establishing proper conditions for Köhler illumination. At each of the other turret openings, there is an off-center slit that is partially covered with a small rectangular polarizer. The size of the slit/polarizer combination is different for each objective of a different magnification, hence the need for a turret arrangement. When light passes through the off-axis slit, it is imaged at the back focal plane of the objective or at the Fourier plane, where the modulator has been installed. Like the central annulus and a phase ring in phase contrast microscopy, the front focal plane of the condenser containing the off-axis slit plate is optically conjugate to the modulator placed at the back focal plane of the objective. Image irradiance is proportional to the first derivative of the optical density in the specimen and is controlled by the zeroth-order of the phase gradient diffraction pattern.

Below the condenser, a circular polarizer is placed on the light exit port of the microscope: both polarizers (circular and slit) are below the specimen. Rotation of the polarizer controls the effective width of the slit. For example, if the two polarizers are crossed, the slit width would be minimum, and contrast in the image highest.

The slit is positioned at the periphery for high magnification objective to utilize fully its *NA*. The Hoffman modulation contrast system can work with objectives of all magnification.

9.3 PHASE CONTRAST MICROSCOPY

Phase objects do not attenuate the light irradiance and hence are not visible. In microscopy we come across objects that absorb light (amplitude objects), that change phase only (phase objects), and that change both amplitude and the phase (complex objects).

When light is incident on a phase object, the phase variations in the object diffract light. The transmitted light thus consists of both the direct light (undiffracted) and diffracted light. At the back focal plane, the undiffracted light is focused on-axis (zeroth order) and the diffracted light, depending on the angle, forms a halo. Both the diffracted and undiffracted light beams propagate and contribute to the image. Because diffracted and undiffracted beams are in quadrature, they do not interfere so the image does not appear visible.

Mathematically, under small phase approximation, we write for the phase object as

$$O(x,y) = e^{i\phi(x,y)} = 1 + i\,\phi(x,y) \tag{9.7}$$

where $\phi(x,y)$ is the phase introduced by the object. Unit amplitude wave incident on a phase object thus generates a transmitted wave of unit amplitude and a diffracted wave of amplitude $\phi(x,y)$, which is in phase quadrature with the transmitted wave (direct beam). The transmitted wave of unit amplitude is focused at the back focal plane of the objective, whereas the diffracted beam of amplitude $\phi(x,y)$ appears as a light distribution. In phase contrast microscopy the phase of the direct wave is either advanced or retarded by $\pi/2$. Because this beam is available for manipulation at the back focal plane, a phase plate is placed there. The phase plate either advances or retards the phase of direct beam by $\pi/2$. Therefore, at the back focal plane just after the phase plate, we obtain

$$O(x',y') = e^{\pm i\pi/2} + i\,\phi(x',y') = \pm i + i\,\phi(x',y') \quad (9.8)$$

These two beams are now either in phase or anti-phase and hence would interfere. This gives the irradiance distribution in the image as

$$I(x_i,y_i) = OO^* = \{\pm i + i\,\phi(x_i,y_i)\}\{\mp i - i\,\phi(x_i,y_i)\} = 1 \pm 2\,\phi(x_i,y_i) \quad (9.9)$$

If the direct beam is speeded up by $\pi/2$ ($e^{-i\pi/2}$), the two beams will interfere destructively. Thus, the areas of specimen with increasing phase would appear darker against the light background. This is called dark or positive phase contrast.

On the other hand, if the direct beam is retarded by $\pi/2$ ($e^{i\pi/2}$), the two beams interfere constructively resulting in areas of specimen with increasing phase as brighter areas. This is called bright or negative phase contrast.

9.4 INTERFERENCE MICROSCOPY

Phase objects can be seen under interference microscopy as the phase information can be converted into irradiance information. A special attachment, either a Michelson interferometer or a Mirau objective, is used for this purpose as seen in Figure 9.5(a–b). The Mirau objective is mostly used in metrology.

9.5 POLARIZATION MICROSCOPY

There are objects/crystals that are transparent but not isotropic; the refractive index varies with the direction of light propagation. These are called birefringent materials. On entering these objects the light beam decomposes into the ordinary and the extraordinary beams that propagate with different speeds. These are linearly polarized with their directions of vibration orthogonal to each other. There is, however, a direction(s) along which both these beams propagate with the same speed, and these are the optic axes. These objects are further classified as uniaxial and biaxial. In uniaxial crystal, there is only one axis, called optic axis, along which the refractive indices of both the beams are the same, whereas in biaxial crystals there are two such directions (two optic axes).

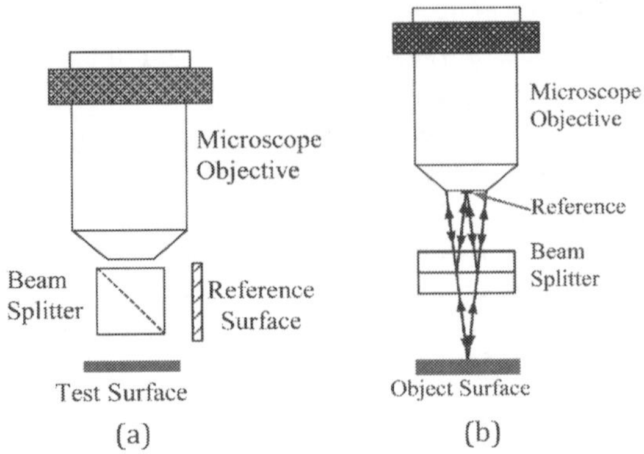

FIGURE 9.5 (a) Objective with Michelson interferometer. (b) Mirau objective.

The optical path introduced by the specimen (uniaxial) of thickness d is given by

$$\Delta = |\mu_0 - \mu_e| d \tag{9.10}$$

where μ_0 and μ_e are the refractive indices for the ordinary and the extraordinary beams. Polarized light is a contrast-enhancing agent that improves the quality of the image obtained with birefringent materials when compared with other techniques such as dark field and the bright field illumination, DIC, phase contrast, Hoffman modulation contrast, and fluorescence. Polarized light microscopes have a high degree of sensitivity and can be utilized for both quantitative and qualitative studies targeted at a wide range of anisotropic specimens.

Figure 9.6 shows a schematic of a polarizing microscope. A polarizer is used to linearly polarize the light from the source, which then illuminates the birefringent specimen through a condenser. In the specimen, two orthogonally polarized beams are generated that travel in slightly different directions. An analyzer brings the components of these beams in the same direction so that they interfere. The Bertrand lens projects the conoscopic image to the eyepiece.

For quantitative measurement of birefringence, a phase plate and other accessories are provided. The specimen table can be rotated. Polarization microscopy, however, requires strain-free optics (condenser, objective, eyepiece).

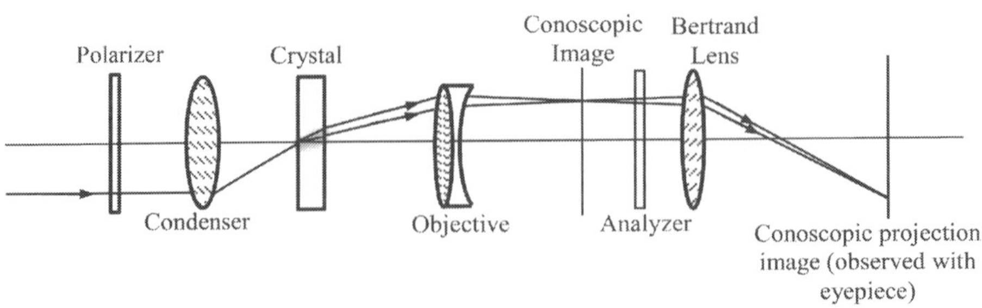

FIGURE 9.6 Schematic of a polarizing microscope.

9.6 DIFFERENTIAL INTERFERENCE MICROSCOPY (DIC)

Phase information can be converted into irradiance information by the phenomenon of interference. One beam passes through the specimen and carries its phase information. Another beam acts as a reference beam, which is added to the object beam. Interference between these beams reveals the object and can also be used to get quanitative information of the phase distribution in the object. However, if both the beams pass through the specimen, they would experience phase delays in their passage. The interference pattern would contain information about the phase difference between the two paths. However, if the beams are very close to each other, the interference pattern would contain information about the phase gradient. Mathematically, let a beam experience a phase delay $\phi(x,y)$ when it passes through it, while the second beam experience the phase delay $\phi(x - \Delta x,y)$. The interference pattern would contain information about the phase difference, i.e.,

$$\delta = \phi(x,y) - \phi(x - \Delta x, y) = \frac{\partial \phi}{\partial x} \Delta x \quad (9.11)$$

DIC makes use of two beams traveling very close to each other through the specimen. These beams are generated by the polarization optics.

In transmitted light DIC, light from the lamp passes through a polarizer located beneath the substage condenser as shown in Figure 9.7. It then passes through a modified Wollaston prism (modified by Nomarski) that splits it into two orthogonally polarized beams traveling in slightly different directions. The Wollaston prism is located at the front focal plane of the condenser. Light beams from the Wollaston prism vibrate at 90° relative to each other with a slight path difference. A different prism is needed for each objective of different magnification. A revolving turret on the condenser allows the microscopist to bring the appropriate prism into the light path when changing magnification (objective).

The two light beams travel very closely and pass through the specimen as parallel beams, and because they are orthogonally polarized they do not interfere. The distance between the beam, called the shear, is very minute, smaller than the Airy size. On passage through the specimen, their phases are altered dpending on the specimen's varying thickness, slopes, and refractive index. When the parallel beams enter the objective, they are focused at the back focal plane where they enter the second Wollaston prism that combines the two beams. This removes

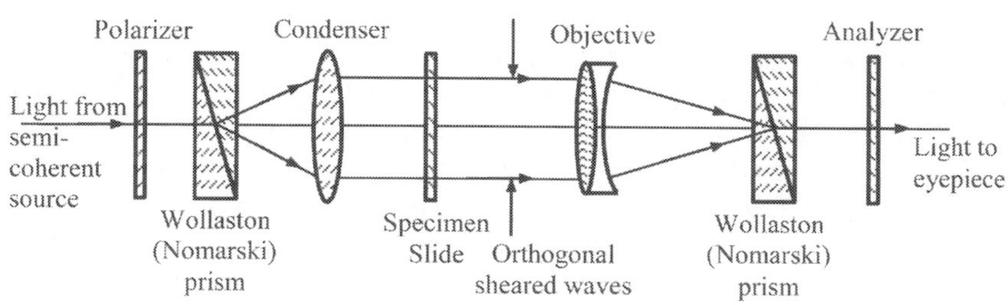

FIGURE 9.7 Schematic of a differential interference contrast microscope.

the shear and original path difference. Only the path difference due to traverse through the different areas of the specimen remains. In order for the beams to interfere, the vibrations of the beams must be brought into the same plane. This is accomplished by an analyzer placed after the Wollaston prism. Light then proceeds to the eyepiece where it can be observed as differences in irradiance and color. The design results in one side of a detail appearing bright (or possibly in color) while the otherside appears darker (or another color). This shadow effect gives a pseudo 3D appearance to the specimen.

The color and/or irradiance effects appearing in the image are related to the rate of change of the refractive index, specimen thickness, or both. Orientation of the specimen can have a pronounced effect on the relief-like appearance and often, rotation of the specimen by 180° changes the hill into a valley or vice versa. The 3D appearance is not representing the true geometric nature of the specimen; instead it is an exaggeration based on optical thickness. It is not suitable for accurate measurement of actual heights and depths.

DIC offers several advantages over phase contrast and Hoffman contrast modulation microscopy: it makes use of the full *NA* of the objective and hence achieves excellent resolution and provides optical staining (color). Use of a full objective aperture enables the microscopist to focus on a thin plane section of a thick specimen without confusing images from above or below the plane.

9.7 METROLOGICAL MICROSCOPE

The metrological microscope is used to make dimensional measurements on small objects with high accuracy. The measurement of dimensions always involves two measurements, and some sort of setting criterion at each measurement is to be established. The measurements are accomplished using one of the three methods:

1. Direct contact
2. Using x–y stage
3. Using eyepiece with a reticle

In the direct contact method, the object to be measured and the ruler are in direct contact. Both are magnified at the same time and the measurements made. This method is useful at low magnification and at a large field of view situation.

Use of an x–y measuring stage is quite common. The universal measuring microscope and the toolmaker's microscope are equipped with a precision x–y stage where linear measurement is made using objectives with magnification ranging from 1× to 8× and binocular eyepiece with a magnification of 10×. The setting is done using either a mechanical feeler or an optical feeler. Double image prism is also used for proper setting of the object. A measurement over 20 cm length and 10 cm width can be done with an accuracy of ±1 μm. In some microscopes, the eyepiece has a measuring graticule. To obtain accurate measurement of dimension, the magnification of the objective has to be precisely known.

9.8 CONFOCAL SCANNING OPTICAL MICROSCOPE (CSOM)

A conventional microscope has excellent resolution on the x–y plane but poor depth resolution and cannot be used for viewing thick objects. The confocal scanning optical microscope (CSOM) differs from a conventional optical microscope in that it illuminates and images the sample one point at a time through a pinhole and works on the dynamic focusing measurement principle based on a focus detection technique. Figure 9.8 shows the principle of a CSOM. A laser beam is focused on a pinhole P_1 by the lens L_1. The beam exiting from pinhole P_1 is partially reflected by the beam splitter BS and then is focused to a diffraction-limited spot (~0.5 μm) by lens L_2 at plane B. Lens L_2 images the diffraction-limited spot on pinhole P_2. Plane P_1 and plane B are conjugate planes. Plane B and plane P_2 are also conjugate planes. In such a situation, detector D receives the maximum light flux. However, if plane B is shifted a little from the correct focus position on either side, its image at plane P_2 becomes a patch and the output of the detector falls rapidly. Therefore, this arrangement can be used to determine displacement and step heights with a very high degree of accuracy. It has therefore been used for roughness measurement, measurement of surface undulations, and roundness of a specimen. It can also be used to make an image of an object by scanning and focus detection. Both two-dimensional (2D) and 3D objects can be imaged. In its various variants it has been used for imaging biological objects.

When the object is axially translated from the correct confocal position, the signal from detector D falls. The response curve of the confocal microscope can be generated by measuring the detector signal as a function of position of the object with respect to the confocal position. The distance d_{axial} between half-power points of the detector response is called the range resolution. For a reflection type CSOM as described earlier, the range resolution d_z is given by

$$d_{\text{axial}} = \frac{0.45\ \lambda}{1-\cos\theta} \tag{9.12}$$

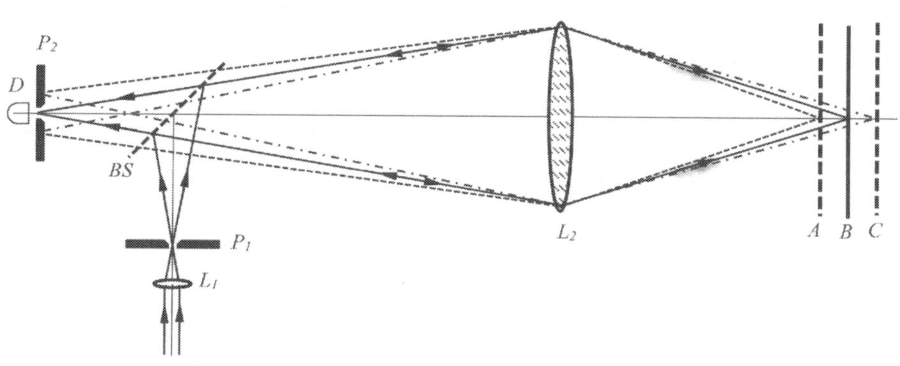

FIGURE 9.8 The principle of a confocal optical microscope.

where λ is the wavelength of laser radiation and θ is the half cone angle of the objective and related to the NA of the objective as $NA = \sin\theta$. The value of the range resolution for an objective (100× and $NA = 0.90$) when used with a He-Ne laser works out to be 0.5 μm. Obviously, high depth resolution is achievable with a confocal microscope.

Due to the narrower point spread function of the confocal microscope, lateral resolution is improved, and one can use the following formula

$$d_{\text{lateral}} = \frac{0.61\ \lambda}{\sqrt{2}\ NA} = \frac{0.4\ \lambda}{NA} \tag{9.13}$$

9.9 FLUORESCENCE MICROSCOPY

Fluorescence microscopy requires that the object is fluorescent; hence, the object is stained with fluorophore. Fluorescence is the emission of light when light of a shorter wavelength is absorbed by the fluorophore. The light emitted by fluorophore is used for imaging.

Figure 9.9 shows a schematic of an epi-fluorescence microscope.

The excitation filter allows the wavelength that would excite the fluorophore. The dichroic mirror reflects the excitation wavelength that passes through the objective and illuminates/excites the fluorophores in the object. The stained object emits light of a higher wavelength, i.e., Stokes shifted light. This light is used to image the object

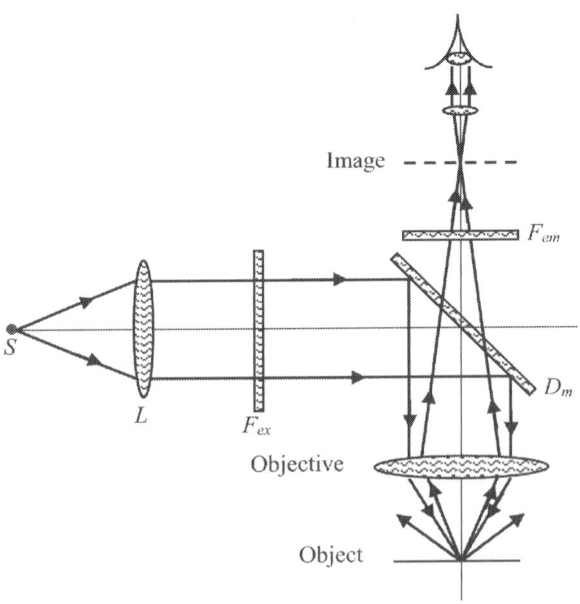

FIGURE 9.9 Schematic of a fluorescence microscope. S = source, F_{ex} = excitation filter, F_{em} = emission filter, and D_m = dichroic filter.

on the charge-coupled device (CCD). The light emitted by the fluorophores passes through the dichroic mirror and the filter F_{em}, which blocks the excitation light, if any, and unwanted fluorescent light. The resolution in epi-fluorescent microscopy is given by

$$d = \frac{1.22\ \lambda}{2\ NA_{\text{obj}}} \tag{9.14}$$

A deconvolution algorithm may be used to obtain super-resolved image of the object. A fluorescence microscope is also configured as a confocal and multi-photon confocal microscope.

QUESTIONS

9.1 A 45× microscope objective with an *NA* of 0.8 is used for imaging an object. The microscope uses visible light of 550 nm for illumination. Calculate the spatial (lateral) and axial resolution obtained with this objective.

9.2 Calculate the spatial and axial resolutions obtained with the objective (45× and $NA = 0.8$) when it is used as an oil immersion objective with oil that has a refractive index of 1.5.

9.3 A 45× microscope objective with an *NA* of 0.8 is used in confocal microscopy. The microscope uses visible light from the He-Ne laser ($\lambda = 632.8$ nm). Calculate the axial and lateral resolution.

10 Measurement of Refractive Index

10.1 INTRODUCTION

There are several methods of measuring the refractive index of a liquid and a solid substance. These can be grouped into two methods: (1) non-interferometric and (2) interferometric. Non-interferometric methods measure the absolute refractive indices of materials. These include methods that measure the angles, critical angle, and Brewster angle either using spectrometers or goniometers or some other means. The accuracy of refractive index measurement varies over a large range and depends on the method used. When the refractive index of a film is desired, ellipsometry is used. In fact, ellipsometry is used to measure the optical constants of bulk as well as thin film specimens. Interferometric methods measure either change or difference of the refractive indices and offer high accuracy of measurement. The Rayleigh and Michelson interferometers are some of the interferometers used for this measurement.

10.2 SPECTROMETER

The material whose refractive index is to be measured is made in the form of an isosceles prism: the apex angle is usually taken 60°. The two surfaces of the prisms are optically polished, and the material is assumed to be isotropic and homogeneous. The prism is illuminated with a collimated beam of light of wavelength λ and a telescope receives the refracted beam. Figure 10.1 shows the passage of a ray through a prism of angle A. A ray incident at the entrance face suffers refraction and emerges out at an angle of emergence ε. The ray deviates from its initial direction by δ, called the angle of deviation. It can be shown that, at the condition of minimum deviation,

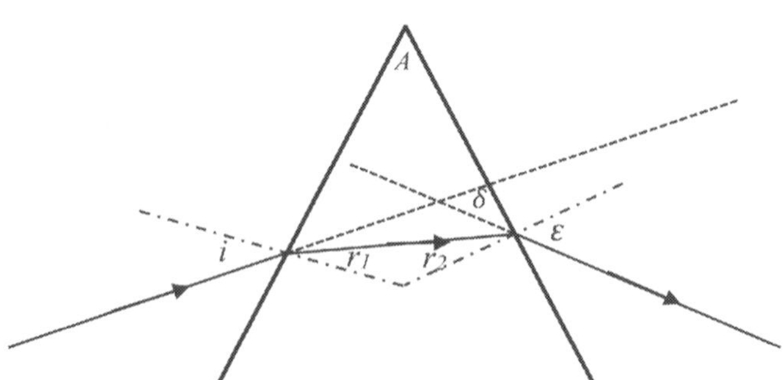

FIGURE 10.1 Passage of a ray through a prism.

DOI: 10.1201/9781003531845-10

the angle of incidence is equal to the angle of emergence and the ray inside the prism travels parallel to the base.

The refractive index μ of the material of the prism is given by

$$\mu = \frac{\sin\left[\left(A+\delta_{\min}\right)/2\right]}{\sin\left(A/2\right)} \tag{10.1}$$

The refractive index is obtained by measuring the angle of the prism A and the angle of minimum deviation $\delta_{\min}$. The angles can be measured to an accuracy better than a second of an arc. For a prism of angle $A = 60°$ and $\mu = 1.5$, this gives $d\mu/\mu = 8 \times 10^{-6}$. Better accuracy is obtainable when the angles are measured with an accuracy of a fraction of an arc of a second.

Refractive index of a liquid can be measured using a hollow prism. The prism holder is usually kept at a constant temperature. Its temperature, however, can be varied and hence the variation of refractive index with temperature can also be studied. It can be shown that the thickness of the walls or/and its refractive index have no effect on the measurement of the refractive index of the liquid provided the walls are of constant thickness and are homogenous and isotropic.

10.3 GONIOMETER

Using an autocollimator mounted on a goniometer, the refractive index of the material of the prism can also be measured. First the retro-reflection from the front face of the prism is obtained. Thereafter the retro-reflection from the back face of the prism is obtained. This measures the angle θ. The angle of the prism is separately measured. Figure 10.2 shows the schematic of the measurement process.

The refractive index is obtained from the expression

$$\mu = \frac{\sin\theta}{\sin A} \tag{10.2}$$

This method also involves the measurement of two angles, θ and A. Further, a hollow prism can be used to measure the refractive index of a liquid.

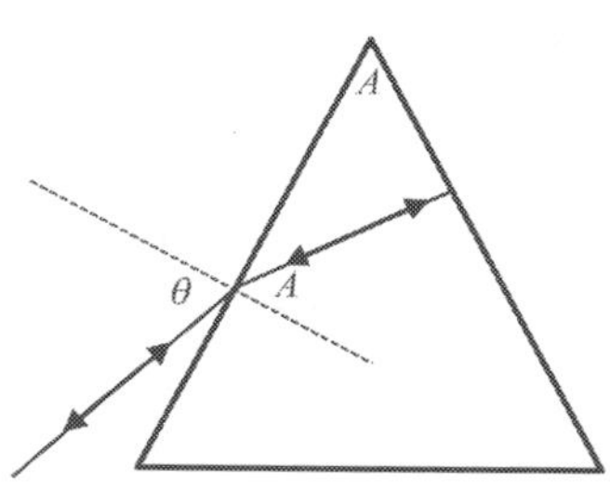

FIGURE 10.2 Measurement of the refractive index of the prism material by auto-reflection.

10.3.1 Measurement of Refractive Index of a Liquid

A hollow prism of an apex angle A, filled with a test liquid, is mounted on a goniometer. Initially the beam is retro reflected from the front surface of the prism. The retro-reflected beam passes through a pinhole and falls on the photodetector (PD) as shown in Figure 10.3. The signal from the photodetector is maximized. This is the initial setting of the prism.

The prism is now rotated to obtain a retro-reflection condition from the back surface as shown in Figure 10.3. Let this angle of rotation be θ. The refractive index μ of the liquid is obtained from

$$\mu = \frac{\sin\theta}{\sin A}$$

If we choose a right-angle hollow prism of apex angle 30°, refractive indices of liquids over a large range approaching 2 can be covered.

10.3.2 Hilger–Chance Refractometer

This refractometer is ideally suited to determine the refractive indices of liquids. Figure 10.4 shows the schematic of a unit of the refractometer that is realized by 45° prisms of refractive index μ_0 such that there is an empty prismatic volume with an apex angle of 90°. A beam of light of wavelength λ from the collimator is incident on the unit containing the liquid, and the telescope receives the emergent beam. The angle of deviation δ is measured. Initial reading corresponds to the direct see-through condition.

The refractive index of the liquid is obtained from the following expression

$$\mu = \sqrt{\mu_0^2 - \sin\delta\,\sqrt{\mu_0^2 - \sin^2\delta}} \tag{10.3}$$

The absolute refractive index can be measured to an accuracy of ~1×10^{-4} while the change in refractive index can be measured to a much higher accuracy ~2×10^{-6}. The refractive index of a solid can also be measured if it is made in the form of a right-angle prism that is placed in the unit such that there is a uniform layer of liquid around it. Obviously, when $\mu = \mu_0$, the angle of deviation δ is zero; hence, μ_0 is the value of the highest refractive index that can be measured with the refractometer.

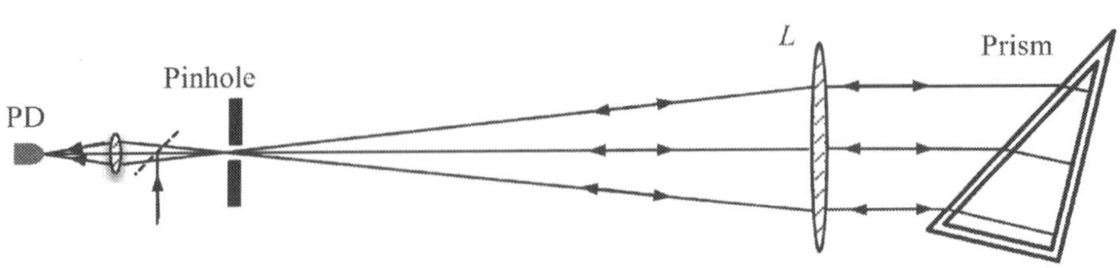

FIGURE 10.3 Measurement of refractive index of liquid using auto-reflection.

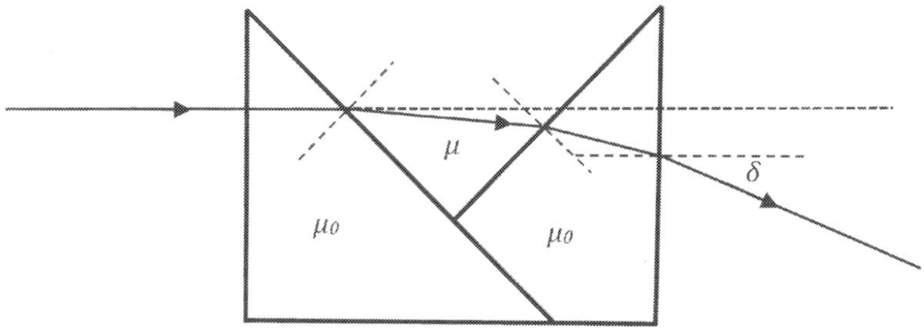

FIGURE 10.4 Ray path through a Hilger–Chance refractometer.

10.4 METHODS BASED ON THE MEASUREMENT OF CRITICAL ANGLE

Let us consider interfaces between three media of refractive indices μ_1, μ_2, and μ_3, respectively. The interfaces are parallel to each other. Let us consider a ray incident at an angle θ_3. According to Snell's law of refraction, we have

$$\mu_3 \sin\theta_3 = \mu_2 \sin\theta_2 = \mu_1 \sin\theta_1$$

where θ_1 is the angle of refraction in the medium of refractive index μ_1. Depending on the relative refractive indices, a ray may be transmitted or be totally reflected. In a particular case, the ray may travel along the boundary/interface. For such a case the angle of incidence is the critical angle. For the measurement of refractive index of a solid transparent material, a layer of liquid of suitable refractive index is used. Therefore, we consider ray propagation through three media. Figures 10.5(a–c) show three situations that lead to the critical angle. In these figures, μ_2 is the refractive index of a liquid layer and μ_3 is the refractive index of a reference material usually taken in the form of a prism. The reference material is of a very high refractive index. Refractive index μ_1 corresponds to a solid material, which is to be determined.

For the measurement of the refractive index of a solid specimen, the liquid can have a refractive index lower or higher than the refractive index μ_3 of the reference material, but it should be higher than the refractive index of the solid being measured. For the measurement of a refractive index of a liquid, the liquid layer is placed on the reference medium and the corresponding critical angle is measured as shown in Figure 10.5(c).

The refractive index of a solid material or a liquid is determined by measuring the critical angle. We will describe two refractometers that use the measurement of the critical angle. These are the Pulfrich and the Abbe refractometers. The advantages of the method based on the critical angle are that it avoids the measurement of the angle of the prism and the method can be applied even for colored liquids and solids. The disadvantage is that it measures the refractive index of the skin, which could be different than that of the bulk material.

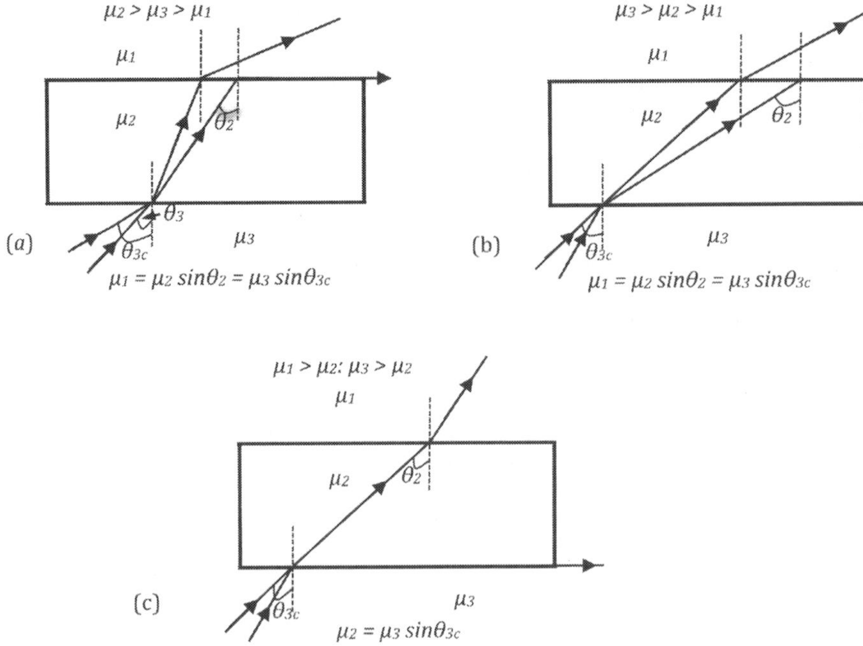

FIGURE 10.5 Critical angles for the determination of the refractive index of a solid (a and b) and liquid (c).

10.4.1 Pulfrich Refractometer

Pulfrich described a refractometer to determine the refractive index of solid materials, liquids, and materials in powdered form in 1896. It consists of a right-angle prism made of a material of high refractive index and it has two polished faces. It is placed such that one of the polished surfaces is horizontal while the other is vertical as shown in Figure 10.6. A drop of liquid whose refractive index is to be measured is placed on the top surface of the prism. The angle θ is measured on a spectrometer or a goniometer. This is the angle between the direct

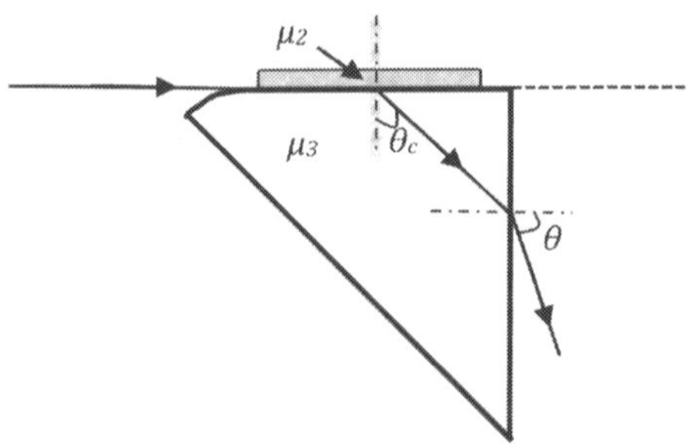

FIGURE 10.6 Ray path in the Pulfrich refractometer.

beam from the collimator and the one, which is refracted at critical angle at the top interface.

From the Figure 10.6, we obtain that $\mu_2 = \mu_3 \sin\theta_c$. Further, for the refraction at the vertical interface, we have $\mu_3 \cos\theta_c = \sin\theta$. From these two equations, we obtain

$$\mu_2 = \sqrt{\mu_3^2 - \sin^2\theta} \tag{10.4}$$

Therefore, the refractive index of the liquid is determined by the measurement of the angle θ with a telescope and the refractive index of the prism.

For measuring the refractive index of a solid, it is taken in the form of a small plate and a drop of liquid whose refractive index is higher than that of the solid material (plate) is put between the prism surface and the plate. The light is critically refracted at the solid–liquid interface. Following the theory, the refractive index μ_1 of the plate material is obtained as

$$\mu_1 = \sqrt{\mu_3^2 - \sin^2\theta} \tag{10.5}$$

When the refractive index of the material in powder form is to be measured, it is mixed in an index matching liquid, which is then placed on the upper surface of the prism and the measurement made as described in the earlier cases.

10.4.2 Abbe Refractometer

The Abbe refractometer is the most used instrument for the determination of refractive indices of liquids and solids. It consists of two prisms P_1 and P_2 of dense Flint glass (Figure 10.7). Surfaces 1, 3, and 4 are polished and surface 2 is matte. Prism P_1 is hinged at H so that it can be hinged away from P_2 or removed altogether if desired.

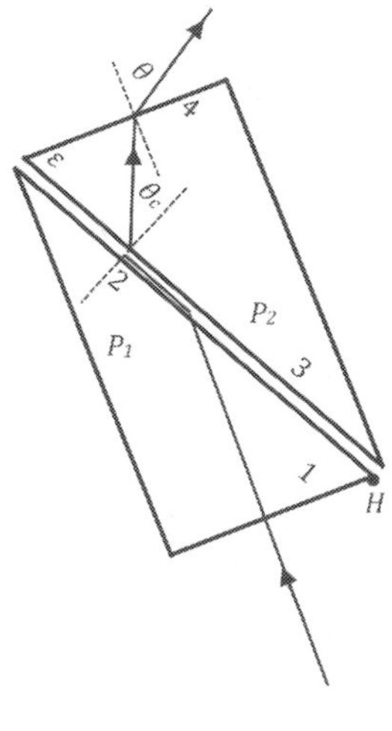

FIGURE 10.7 Ray path in prisms in an Abbe refractometer.

For determining the refractive index of a liquid, a drop of the latter is placed on surface 2, which is then closed up into contact with surface 3; the liquid is squeezed out into a thin film. The light from a suitable source, usually white, is directed toward the prism system. It strikes matt surface 2 and is scattered into the liquid film and prism P_2. Figure 10.7 shows the path for a ray that is grazing in the liquid medium and is refracted at critical angle θ_c in the prism medium. This ray exits from the prism at an angle θ.

It can be shown that refractive index μ_2 of the liquid can be expressed in terms of the measured angle θ, angle of prism ε, and its refractive index μ_3 as

$$\mu_2 = \sin\varepsilon \left(\mu_3^2 - \sin^2\theta\right)^{1/2} - \cos\varepsilon \sin\theta \tag{10.6}$$

As no ray can enter P_2 with a greater angle of refraction than that of the ray at grazing incidence, the emergent rays when collected by a telescope will all converge to points on one side of a line in the focal plane. The field will therefore by divided into dark and light portions; the edge of the bright portion corresponds to the value of θ for the critical rays.

The rotation of the telescope is registered on a graduated arc, which reads the refractive index directly. In practice, the instrument is used with white light. To compensate for the dispersion of the system, and at the same time get a rough measure of the dispersion of the substance under test, an ingenious device is adopted by introducing a reverse dispersion by means of two direct vision prisms. These are mounted one above the other in front of the objective of the telescope and are geared so that by turning a knob they may be rotated in opposite directions. They are so oriented to start with that their planes of dispersion are parallel to each other and to the principal plane of prism P_2; their dispersions are additive.

If R is the angular dispersion of one of the prisms between the C and F lines, the total dispersion in this position is $2R$. If the prisms are now rotated with equal rates in opposite directions, the resultant dispersion is still parallel to the principal plane of P_2 but is diminished. For any orientation φ, from the initial position, the resultant dispersion is $2R \cos \varphi$. When, angle φ is 90°, it is zero and it changes sign for $\varphi > 90°$ and increases to $2R$ in the reverse direction.

In using the instrument, the critical edge as first observed will probably be spread out into a short spectrum. The compensating prisms are rotated until the color disappears and a sharp achromatized edge is obtained. The cross lines in the eyepiece are set on this edge; the reading on the scale then gives the index of liquid for the sodium wavelength. The orientation of the compensating prisms gives, by reference to a set of tables supplied with the instrument, a rough value of the dispersion $(\mu_F - \mu_C)$, from C to F, of the substance under test.

For measuring the refractive index μ_1 of a solid specimen, prism P_1 is swung away, and the specimen is placed on prism P_2. Optical contact is made by putting a drop of a liquid, say α-monobromonaphthelene ($\mu_d = 1.658$). The refractive index is read off by fixing the cross lines at the edge of the bright portion in the field and reading off the scale. The value of dispersion is also obtained from the positions of compensating prisms.

10.5 MEASUREMENT OF BREWSTER ANGLE

When a beam of monochromatic light is incident at an interface between two dielectric media, or in particular at the air–glass interface at the Brewster angle, the reflected light is plane polarized. Following Fresnel relations, it can be shown that there is a finite reflectivity for transverse electric (TE) waves, whereas the transverse magnetic (TM) waves have zero reflectivity when the light is incident at the Brewster angle. Thus, the $\boldsymbol{E}$-vector in the reflected light oscillates perpendicular to the plane of incidence. The reflected and the refracted rays are perpendicular to each other. Using Snell's law, it can be shown that

$$\mu = \tan\theta_B \tag{10.7}$$

where θ_B is the angle of incidence, which is called the Brewster angle. Thus, using a linear polarizer in the reflected beam, the angle of incidence is adjusted until there is no transmitted light. The transmission axis of the polarizer is along the direction of the $\boldsymbol{E}$-vector, and the tangent of the angle of incidence gives the relative refractive index of the medium. This is a quick method and can provide refractive index value accurate to a few places to the third decimal.

It is essential that the surface of the medium is quite clean. The presence of any film would result in reflected light being elliptically polarized and hence in nonzero transmitted irradiance through a properly oriented polarizer even at the Brewster angle.

10.6 ELLIPSOMETRY

When a linearly polarized light with $\boldsymbol{E}$-vector neither in the plane of incidence nor in the plane perpendicular to the plane of incidence strikes a bare surface of a material with a complex refractive index, the reflected light is elliptically polarized: the tip of the $\boldsymbol{E}$-vector traces an ellipse. Ellipsometry is a technique used to measure the parameters of this ellipse, hence, the name. It is also used to measure thicknesses of thin films on dielectric or metallic substrates. Even multilayers can be evaluated.

Ellipsometry is used for determining the refractive index of a transparent film and to obtain the complex refractive index of the substrate. Physical and chemical absorption, oxidation, corrosion, etc., modify the state of polarization of the incident light and hence have been studied using ellipsometry. It is also used in areas like surface metrology, biology, and medicine.

Figure 10.8 shows a schematic of an ellipsometer. A narrow collimated beam of monochromatic light either from a monochromator or a laser is incident at an angle on the surface of a sample. The polarization state of the light before reflection is determined by a combination of a polarizer and a compensator. Reflected light passes through an analyzer (second polarizer) and a photodetector or an eye senses the transmitted light. The location of the compensator is arbitrary; it can also be placed between the sample and the analyzer. The compensator and both the polarizers are mounted on precision graduated circular scales (resolution of a few seconds of an arc). The polarizer azimuth reads 0° when its transmission axis lies in the plane

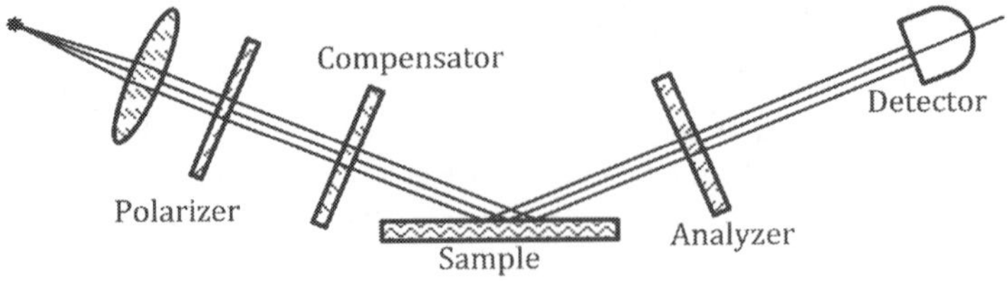

FIGURE 10.8 Schematic of an ellipsometer.

of incidence of the sample and the compensator azimuth reads 0° when its fast axis lies in the plane of incidence.

There are two methods of conducting ellipsometric measurements: (1) null ellipsometry and (2) photometric ellipsometry. In null ellipsometry the data are obtained by adjusting the polarization components for zero transmitted irradiance, whereas in photometric ellipsometry the time dependence of the transmitted irradiance resulting from a periodic rotation of one of the components (say analyzer) is studied. In both cases the general expression for the complex amplitude E_A of the electric field transmitted by the analyzer is the same. This can be obtained using Jones calculus and is given by

$$\begin{aligned} E_A = K(& r_{\|} \cos C \cos P^* \cos A - r_{\|}\ \rho_c \sin C \sin P^* \cos A \\ & + r_{\perp} \sin C \cos P^* \sin A - r_{\perp}\ \rho_c \cos C \sin P^* \sin A) \end{aligned} \tag{10.8}$$

where A, P, and C are the azimuths of the analyzer, the polarizer, and the compensator, respectively; $P^* = P - C$; $r_{\|}$, $r_{\perp}$ are the Fresnel reflection coefficients of the sample for parallel and perpendicular polarization states of the beam; and $\rho_c \left[= \rho_0\ e^{-i\delta}\right]$ is the transmittance of the compensator.

The ratio of the reflection coefficients is expressed as

$$\rho = \frac{r_{\|}}{r_{\perp}} = \left|\frac{r_{\|}}{r_{\perp}}\right| e^{i(\delta_p - \delta_s)} = \tan\psi\ e^{i\Delta} \tag{10.9}$$

Here ψ and Δ are known as the ellipsometric angles of the sample and are obtained by the ellipsometer.

10.6.1 Null Ellipsometry

In null ellipsometry the polarization components are adjusted so that the transmitted light through the analyzer is zero. Therefore, if we set the transmitted amplitude zero ($E_A = 0$) in the Equation (10.8), we are performing null ellipsometry.

Therefore, we have

$$\rho = \frac{\left(\rho_c \tan P^* + \tan C\right)}{\left(\rho_c \tan P^* \tan C - 1\right)} \tan A \tag{10.10}$$

In practice, a fixed compensator scheme is most commonly employed, and this makes the data reduction simple. In the fixed compensator scheme, the compensator azimuth is set at +45° or −45°, i.e., $C = \pm 45°$, and the polarizer and analyzer are adjusted to obtain the null condition. For further analysis we assume an ideal quarter-wave plate as a compensator, then $\rho_0 = 1$, and $\delta = 90°$ [$\rho_c = -i$]. For each setting of the compensator, two sets of polarizer and analyzer angle settings are obtained: [P_1, A_1] and [P_3, A_3] for $C = -\pi/4$, and [P_2, A_2] and [P_4, A_4] for $C = \pi/4$. From the Equations (10.9) and (10.10), we obtain

$$\Delta = 2P_1 + \frac{\pi}{2} \qquad \psi = A_1 \tag{10.11a}$$

$$\Delta = 2P_3 - \frac{\pi}{2} \qquad \psi = -A_3 \tag{10.11b}$$

$$\Delta = -2P_2 - \frac{\pi}{2} \qquad \psi = A_2 \tag{10.11c}$$

$$\Delta = -2P_4 + \frac{\pi}{2} \qquad \psi = -A_4 \tag{10.11d}$$

Therefore, the ellipsometric angles Δ and ψ are directly determined from the measured azimuths of the polarizer and the analyzer. These are to be used to determine the optical constants of the sample.

10.6.2 Photometric Ellipsometry

In this scheme, the analyzer is rotated with a constant speed while the other components are kept stationary. This results in a periodically varying detector signal, which is Fourier analyzed. The transmitted irradiance $I(t) = E_A E_A^*$ is obtained from the Equation (10.8) and for the case when $C = 0$, it is expressed as

$$I(t) = \frac{1}{2} I_0 \left(|r_{\parallel}|^2 \cos^2 P + |r_{\perp}|^2 \sin^2 P \right)\left(1 + \alpha \cos 2A + \beta \sin 2A\right) \tag{10.12}$$

where I_0 is the irradiance in the straight-through position without the sample, and A is the instantaneous analyzer azimuth; α and β are normalized Fourier coefficients that are determined experimentally from the values of the DC, cosine, and sine components of the transmitted irradiance. The coefficients α and β are related with the azimuths as

$$\alpha + i\beta = \frac{\tan^2\psi - \tan^2 P + i\, 2\tan\psi \tan P \cos(\Delta - \delta)}{\tan^2\psi + \tan^2 P} \tag{10.13}$$

The desired ellipsometric angles ψ and Δ are given by

$$\tan\psi = \sqrt{\frac{1+\alpha}{1-\alpha}} \tan P \tag{10.14a}$$

$$\cos(\Delta-\delta)=\frac{\beta}{\sqrt{1-\alpha^2-\beta^2}} \tag{10.14b}$$

Therefore, ellipsometric angles ψ and Δ are determined from the knowledge of α, β, P, and δ. The azimuth P and phase δ are known from the instrument setting while α and β are obtained by analyzing the detector signal.

10.6.3 Optical Constants of Sample

It is assumed that the sample surface is bare. In general, its refractive index μ_2 could be complex, i.e., $\mu_2=\mu-ik$, where μ is the real part and k is the imaginary part responsible for absorption. Further, assuming that the sample is surrounded by a medium of refractive index μ_1, the Fresnel reflection coefficients are expressed as

$$r_{\parallel}=\frac{-\mu_2\cos\theta_1+\mu_1\cos\theta_2}{\mu_2\cos\theta_1+\mu_1\cos\theta_2}=\frac{\tan(\theta_1-\theta_2)}{\tan(\theta_1+\theta_2)} \tag{10.15a}$$

$$r_{\perp}=\frac{\mu_1\cos\theta_1-\mu_2\cos\theta_2}{\mu_1\cos\theta_1+\mu_2\cos\theta_2}=-\frac{\sin(\theta_1-\theta_2)}{\sin(\theta_1+\theta_2)} \tag{10.15b}$$

where θ_1 and θ_2 are the angles of incidence and refraction, respectively.

Therefore

$$\tan\psi\ e^{i\Delta}=\frac{r_{\parallel}}{r_{\perp}}=\frac{-\mu_2\cos\theta_1+\mu_1\cos\theta_2}{\mu_2\cos\theta_1+\mu_1\cos\theta_2}\,\frac{\mu_1\cos\theta_1+\mu_2\cos\theta_2}{\mu_1\cos\theta_1-\mu_2\cos\theta_2} \tag{10.16}$$

This can be rearranged in the following way

$$\frac{1-\tan\psi\ e^{i\Delta}}{1+\tan\psi\ e^{i\Delta}}=\frac{\cos\theta_1\sqrt{\mu_2^2-\mu_1^2\ \sin^2\theta_1}}{\mu_1\sin^2\theta_1} \tag{10.17}$$

Squaring and then equating the real and the imaginary parts, we obtain

$$\mu^2-k^2=\mu_1^2\ \sin^2\theta_1\left[1+\tan^2\theta_1\frac{(\cos 2\psi-\sin 2\psi\sin\Delta)^2}{(1+\sin 2\psi\cos\Delta)^2}\right] \tag{10.18a}$$

and

$$2\mu k=\frac{\mu_1^2\ \sin^2\theta_1\ \tan^2\theta_1\sin 4\psi\sin\Delta}{(1+\sin 2\psi\cos\Delta)^2} \tag{10.18b}$$

Therefore, the optical constants μ and k of the bare surface can be determined if ψ and Δ are measured at an angle of incidence θ_1, and μ_1 is known.

10.6.4 Optical Constant of a Thin Film

Determination of the optical constants of a thin film deposited on a sample is of considerable importance. Let us assume that a film of thickness d and refractive index μ_1 is grown on a sample of refractive index μ_2 and the sample-film system is in an ambient medium of refractive index μ_0. The Fresnel reflection coefficients of the system when the beam of light is incident at an angle θ_0 are expressed as

$$r_{\parallel} = \frac{r_{01\parallel} + r_{12\parallel}\ e^{-2i\beta}}{1 + r_{01\parallel}\ r_{12\parallel}\ e^{-2i\beta}} \tag{10.19a}$$

$$r_{\perp} = \frac{r_{01\perp} + r_{12\perp}\ e^{-2i\beta}}{1 + r_{01\perp}\ r_{12\perp}\ e^{-2i\beta}} \tag{10.19b}$$

where the phase β introduced by the film is given by

$$\beta = \frac{2\pi}{\lambda}\ \mu_1 d \cos\theta_1 = \frac{2\pi}{\lambda}\ d\sqrt{\mu_1^2 - \mu_0^2\ \sin^2\theta_0} \tag{10.20}$$

Here r_{01} and r_{12} are the Fresnel reflection coefficients at ambient-film and film-sample interfaces, respectively, and θ_0 and θ_1 are the angles of incidence in the ambient and the film media. We now express the ratio ρ of the reflection coefficients as

$$\tan\psi\ e^{i\Delta} = \frac{r_{\parallel}}{r_{\perp}} = \rho$$

The ratio ρ is a function of several parameters. Therefore, we write this as

$$\tan\psi\ e^{i\Delta} = \rho(\mu_0,\ \mu_1, \mu_2, \theta_0, d,\ \lambda) \tag{10.21}$$

Separating it into the real and the imaginary parts and equating them we obtain

$$\psi = \tan^{-1}\left[\rho(\mu_0,\ \mu_1, \mu_2, \theta_0, d,\ \lambda)\right] \tag{10.22a}$$

$$\Delta = \arg\left[\rho(\mu_0,\ \mu_1, \mu_2, \theta_0, d,\ \lambda)\right] \tag{10.22b}$$

These equations are very complicated and can only be solved with a computer. In general, ρ is a function of nine real parameters that include real and imaginary parts of μ_0, μ_1, and μ_2 apart from θ_0, d, and λ. In special cases, simpler solutions are possible. For example, if a transparent (nonabsorbing) film is deposited on a known substrate, a solution is easily obtained.

10.7 SPECTRAL TRANSMISSION MEASUREMENT

Let us consider a thin film of thickness t and complex refractive index $\mu_2^* = \mu_2 - ik$ deposited on a transparent substrate of refractive index μ_3 as shown in the Figure 10.9(a).

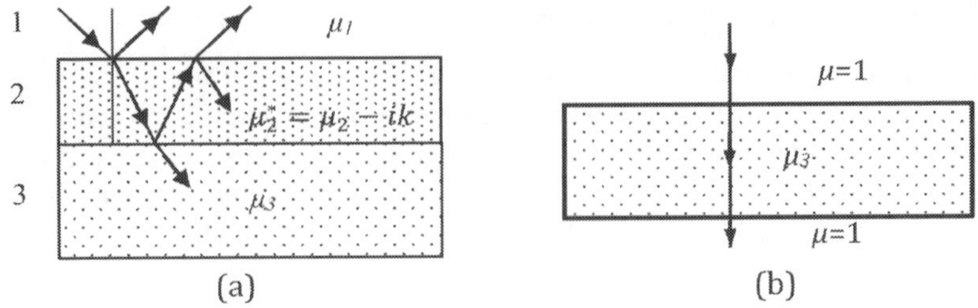

FIGURE 10.9 (a) Schematic of a thin film on a substrate. (b) Transmission through a substrate.

The transmission coefficient of the film is expressed as

$$t_f = \frac{t_{12} t_{23}\ e^{-i2\pi\mu_2^* t/\lambda}}{1 + r_{12}\ r_{23}\ e^{-i4\pi\mu_2^* t/\lambda}} \tag{10.23}$$

The transmissivity of the film in an infinite substrate is given by

$$\mathcal{T} = \frac{\mu_3}{\mu_1} \left| t_f \right|^2 \tag{10.24}$$

Assuming the film to be weakly absorbing and considering the interference effects in the substrate, the transmissivity of the system can be expressed as

$$\mathcal{T} = \frac{A\chi}{B - C\ \chi\ \cos\phi + D\ \chi^2} \tag{10.25}$$

where

$$A = 16\ \mu_1 \mu_2^2 \mu_3 \tag{10.26a}$$

$$B = \left(\mu_1 + \mu_2\right)^3 \left(\mu_2 + \mu_3^2\right) \tag{10.26b}$$

$$C = 2\left(\mu_2^2 - \mu_1^2\right)\left(\mu_2^2 - \mu_3^2\right) \tag{10.26c}$$

$$D = \left(\mu_2 - \mu_1\right)^3 \left(\mu_2 - \mu_3^2\right) \tag{10.26d}$$

$$\phi = 4\pi\mu_2 t / \lambda \tag{10.26e}$$

$$\chi = e^{-4\pi k t/\lambda} \tag{10.26f}$$

However, if we consider the substrate to be sufficiently thick so that the interference effects are neglected, the transmissivity is given by

$$\mathcal{T} = \frac{A\chi}{B' - C\ \chi\ \cos\phi + D'\ \chi^2} \tag{10.27}$$

where

$$B' = (\mu_1 + \mu_2)^2 (\mu_2 + \mu_3)^2 \tag{10.28a}$$

$$D' = (\mu_2 - \mu_1)^2 (\mu_2 - \mu_3)^2 \tag{10.28b}$$

It may be noted that only constants B and D take slightly different values under this case, while the constants A and C remain unchanged.

The transmissivity will have a maximum value when $\cos\phi = 1$, and will have a minimum value when $\cos\phi = -1$. Therefore,

$$\mathcal{T}_{\max} = \frac{A\chi}{B' - C\chi + D'\ \chi^2} \tag{10.29a}$$

$$\mathcal{T}_{\min} = \frac{A\chi}{B' + C\chi + D'\ \chi^2} \tag{10.29b}$$

Taking the inverse of Equations (10.29a) and (10.29b), and on subtracting them we obtain

$$\frac{1}{\mathcal{T}_{\min}} - \frac{1}{\mathcal{T}_{\max}} = \frac{2C}{A} \tag{10.30a}$$

It may be noted that this expression is not dependent on the interference effects of the substrate.

Substituting the expressions for C and A and simplifying, we obtain

$$\frac{1}{\mathcal{T}_{\min}} - \frac{1}{\mathcal{T}_{\max}} = \frac{2C}{A} = \frac{\mu_2^4 - \mu_2^2(\mu_1^2 + \mu_3^2) + \mu_1^2\mu_3^2}{4\ \mu_1\mu_2^2\mu_3} \tag{10.30b}$$

Writing $E = \frac{2C}{A}$: $\alpha = \mu_1^2 + \mu_3^2$: $\beta = \mu_1\mu_3$, the refractive index of the film μ_2 can be written as

$$\mu_2 = \sqrt{\frac{\alpha + 4\beta E + \sqrt{(\alpha + 4\beta E)^2 - 4\beta^2}}{2}} \tag{10.31}$$

In Equation (10.31), E is the measured quantity and $\mu_1 = 1$. In order to measure the refractive index of the film, we need the value of μ_3 and the measured transmissivities $\mathcal{T}_{\max}$ and $\mathcal{T}_{\min}$ at the same wavelength. The refractive index of the substrate can also be measured using the transmissivity measurement as would be shown later.

When we add the reciprocals of Equations (10.29a) and (10.29b), we obtain

$$\frac{1}{\mathcal{T}_{\min}} + \frac{1}{\mathcal{T}_{\max}} = 2\frac{B' + D'\ \chi^2}{A\ \chi} \tag{10.32a}$$

This can be rewritten as

$$D' \chi^2 - \frac{\mathcal{T}_{min} + \mathcal{T}_{max}}{\mathcal{T}_{min}\,\mathcal{T}_{max}} \frac{A}{2} \chi + B' = 0 \tag{10.32b}$$

Writing $F = \frac{\mathcal{T}_{min} + \mathcal{T}_{max}}{\mathcal{T}_{min}\,\mathcal{T}_{max}} \frac{A}{2}$, Equation (10.32b) is expressed as

$$D' \chi^2 - F \chi + B' = 0 \tag{10.32c}$$

The value of χ is obtained as

$$\chi = \frac{F - \sqrt{F^2 - 4B' D'}}{2 D'} \tag{10.33}$$

Therefore, both the refractive index and the absorption coefficient of a film can be obtained by measuring the transmitted irradiance as a function of wavelength.

10.7.1 Refractive Index of the Substrate

When the interference effects in the substrate are considered, the transmissivity $\mathcal{T}_s$ of the substrate is given by Equation 10.34 [Figure 10.9(b)].

$$\mathcal{T}_s = \frac{8\mu_3^2}{\left(\mu_3^2 + 1\right)^2 - \left(\mu_3^2 - 1\right)^2 \cos\phi + 4\mu_3^2} \tag{10.34}$$

The maximum transmissivity can be shown to be unity while the minimum transmissivity is given by

$$\mathcal{T}_{smin} = \frac{4\mu_3^2}{\left(\mu_3^2 + 1\right)^2} \tag{10.35}$$

This can be written in the form of a quadratic equation as

$$\mu_3^2 - 2\mu_3 \frac{1}{\sqrt{\mathcal{T}_{smin}}} + 1 = 0 \tag{10.36}$$

The solution of quadratic Equation (10.36) gives

$$\mu_3 = \frac{1}{\sqrt{\mathcal{T}_{smin}}} + \frac{1}{\sqrt{\mathcal{T}_{smin}}} \sqrt{1 - \mathcal{T}_{smin}} \tag{10.37}$$

The refractive index is, thus, obtained from the measured value of the transmissivity at a specified wavelength.

However, when we consider the substrate to be sufficiently thick so that interference effects are ignored, the transmissivity $\mathcal{T}_{sni}$ is given by

$$\mathcal{T}_{sni} = \frac{(1 - \mathcal{R})^2}{1 - \mathcal{R}^2} \tag{10.38}$$

where the reflectivity of the interface is given by

$$\mathcal{R} = \left(\frac{\mu_3 - 1}{\mu_3 + 1}\right)^2 \quad (10.39)$$

Substituting the expression of $\mathcal{R}$ in the Equation (10.39), the transmissivity $\mathcal{T}_{sni}$ is given by

$$\mathcal{T}_{sni} = \frac{2\mu_3}{1 + \mu_3^2} \quad (10.40)$$

From this the value of the refractive index is calculated as

$$\mu_3 = \frac{1}{\mathcal{T}_{snl}} + \frac{1}{\mathcal{T}_{snl}}\sqrt{1 - \mathcal{T}_{snl}^2} \quad (10.41)$$

It may be noted that these two expressions would give slightly different values of the refractive index. In practice, we consider the substrate to be thick and thus ignore interference effects.

10.8 INTERFEROMETRY

Interferometry responds to optical path difference, which is a function of the geometrical path, refractive index, and wavelength. When the wavelength and the geometrical paths are kept constant, the measurement of optical path difference can yield the difference of refractive indices. Interferometry is a highly sensitive technique to detect and measure the refractive indices. One such interferometer that has been used to measure the refractive index difference between two gases or to measure the refractive index of a gas with respect to that of the air or vacuum is the Rayleigh interferometer.

A schematic of the Rayleigh interferometer is shown in Figure 10.10, where S is a slit source, which is either illuminated by white light or monochromatic light. The lens L_1 collimates the light emanating from the slit source S. An aperture containing two slits, S_1 and S_2, is placed in the collimated beam. The slits in the aperture are aligned parallel to the slit source S. The collimated beam passing through slits S_1 and S_2 passes through two long chambers (tubes) of equal lengths. Lens L_2 combines the two beams at its focal plane. Due to the insertion of the tubes, slits S_1 and S_2 are widely separated and hence the interference pattern at the focus of lens L_2 will have very fine straight-line fringes running parallel to slit S thus requiring high magnification for viewing. A cylindrical lens L_3 accomplishes this. It offers two advantages: first, the magnification is only in one direction and hence the brightness of the fringes is not decreased much, and second, the fringes pattern extending below the tubes can be used as a fiduciary mark. A white-light source is used for locating the zeroth-order fringe. Instead of counting the fringes that get shifted due to the path difference, a pair of compensators (parallel plates), one fixed and another rotatable, is used in the interferometer. Rotation of the compensator introduces path difference,

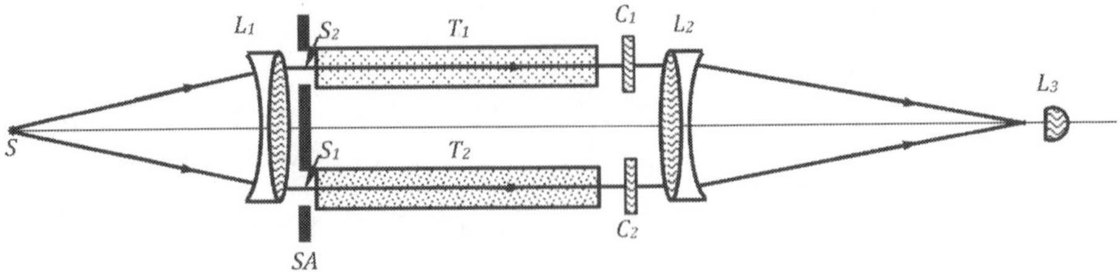

FIGURE 10.10 Schematic of a Rayleigh interferometer.

which compensates for the path difference introduced by the difference in refractive indices of the gases in the tubes. The rotation is calibrated in terms of the wavelength of the monochromatic source. Another glass plate with a sharp edge is introduced in the lower half just before lens L_2 and adjusted such that the top of the interference pattern in the lower half is made coincident with the lower end of the interference pattern in the upper half. With both tubes having air or evacuated and at the same temperature, the interference patterns in lower and upper halves coincide and appear as one.

Let the length of each tube be L. The tubes contain media of refractive indices μ_1 and μ_2, respectively. Therefore, the path difference introduced is $(\mu_1 - \mu_2)L$. This would shift the zeroth-order fringe by order m for wavelength λ. The interference equation can be written as

$$(\mu_1 - \mu_2)L = m\lambda \tag{10.42}$$

where λ is the wavelength of monochromatic light at which the refractive index difference is being measured. The order m does not have to be an integer. The objective is to find the value of m. For this purpose, white light is used to illuminate slit S, and the achromatic fringe is observed initially when both tubes are empty or filled with air. The white light is replaced by a monochromatic source and a more accurate coincidence between the lower and upper interference patterns is made. This forms the initial reading on the compensator. The tubes are filled now with gases whose refractive indices difference is to be measured, or one tube is filled with the gas and the other tube is evacuated. This would shift the zero-order fringe by order m. Using the white-light source, the zeroth-order fringe is brought into coincidence by rotating the compensator and more accurate coincidence is done with the monochromatic source. This compensator reading when subtracted from the initial reading gives the order m. The refractive index difference or the refractive index is obtained from Equation (10.42). Taking the length of the tubes as 30.0 cm, and wavelength as 546.1 nm, and assuming that the fringe order can be accurately determined to (1/20) th of the fringe width, the difference $(\mu_1 - \mu_2)$ is obtained to a value ~10^{-7}. Increasing the length of the tubes will further decrease this value. Variation of refractive index with temperature can also be carried out using this interferometer.

The Michelson interferometer also has been used to measure the refractive indices of thin sheets of transparent materials. The basic approach is the same. The insertion of the sheet in one arm of the interferometer displaces the zeroth-order fringe, which is brought again in the field of view by moving the mirror. The movement of the mirror is related to the path difference introduced by the sheet of the transparent material.

Further, spectral reflectivity measurement over a small wavelength region also can be used to measure refractive index of a thin film provided its thickness is known or measured using either multiple-beam interferometry or some other method. The details of the method are given in Chapter 14.

QUESTIONS

10.1 A refractometer with a sensing prism as shown in the Figure 10.11 is used for monitoring the refractive index of a process liquid.

The dimensions of the prism are $AB = 4$ cm, $DC = 2$ cm, $\angle DAB = \angle ABC = 60°$ and $\angle ADC = \angle DCB = 120°$. The refractive index μ_3 of the prism is $\mu_3 = 1.5473$. The refractive index μ_2 of the process liquid is 1.34. A source of quasi-monochromatic light is used for illumination. The divergence of the beam in the prism is ±1.5°. The observation plane is right to the top surface of the prism where a charge-coupled device (CCD) detector is placed. When the refractive index of the process liquid is 1.34, the middle ray (incident normally on the prism surface) is totally reflected. How much and in which direction does the demarcation line in the field of view move when the refractive index of process liquid becomes 1.35?

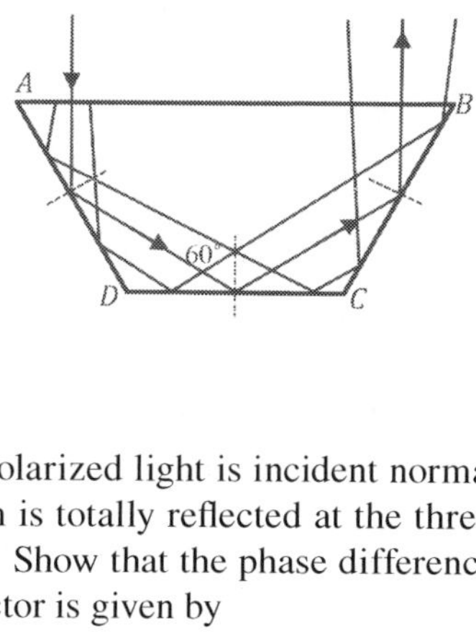

FIGURE 10.11

10.2 A beam of linearly polarized light is incident normally on the prism shown in Figure 10.11. The beam is totally reflected at the three faces and exits normally from the upper surface. Show that the phase difference between s and p components of the electric vector is given by

$$\delta_s - \delta_p = 6\tan^{-1}\left(\sqrt{\frac{1}{3} - \frac{4}{9\mu_{pr}^2}}\right) - 3\pi$$

where μ_{pr} is the refractive index of the prism material. When the bottom surface of the prism is brought in contact with a liquid, show that the phase difference $(\delta_s - \delta_p)$ is given by

$$\delta_s - \delta_p = 4\tan^{-1}\left(\sqrt{\frac{1}{3} - \frac{4}{9\mu_{pr}^2}}\right) + 2\tan^{-1}\left(\sqrt{\frac{1}{3} - \frac{4\mu^2}{9\mu_{pr}^2}}\right) - 3\pi$$

where μ is the refractive index of the liquid. Can this be used to measure the refractive index of a liquid?

10.3 Take a concave mirror of long radius of curvature and locate its center of curvature by imaging an illuminated arrow as shown in Figure 10.12. When liquid is poured inside the mirror, the image is formed at a distance d below the earlier position. Show that the refractive index μ_2 of the liquid is given by

$$\mu_2 = \frac{R}{R-d}$$

where R is the radius of curvature of the concave mirror.

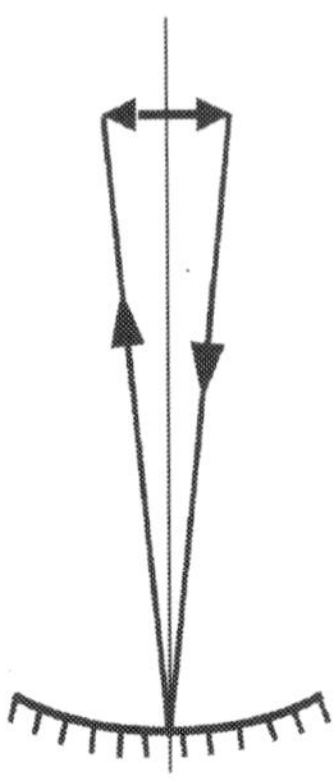

FIGURE 10.12

10.4 A black spot (point) on the stage of a traveling microscope is imaged by the microscope. A glass plate of thickness t is now placed on this black spot. The microscope is raised by d in order to focus the spot again. Show that refractive index μ_1 of the glass plate is given by

$$\mu_1 = \frac{t}{t-d}$$

10.5 Derive Equation (10.3) to measure the refractive index of a liquid with a Hilger–Chance refractometer. What is the range of refractive indices that could be measured by this refractometer? Assume that the prisms are made of dense Flint glass of refractive index 1.6705.

10.6 For measuring the refractive index of a solid sample using critical angle measurement methods, a drop of liquid is placed between the sample and the reference material. Show that it is not necessary to know the refractive index of the liquid as long as it is greater than that of the solid specimen.

10.7 Derive Equation (10.6) to measure the refractive index of a liquid with an Abbe refractometer.

10.8 An ellipsometer is used to measure the optical constants of a clean surface using null ellipsometry with a He-Ne laser. The angle of incidence at the surface is 65°, and the measured values of ψ and Δ are 29.02° and 126.35°, respectively. Calculate the optical constant of the surface of a material. For a glass substrate of refractive index 1.515, what would be the measured values of the ψ and Δ angles?

10.9 A thin sheet 10.0 μm thick of polymethyl methacrylate (PMMA) is inserted in one arm of the Michelson interferometer resulting in the shift of the white-light fringe pattern. The pattern is brought back in the field by moving the mirror 4.94 μm. If this experiment is done with the light of wavelength 546.1 nm, how many fringes are shifted by the introduction of the sheet and what is the refractive index at this wavelength?

10.10 The diameter of a certain ring in Newton's ring experiment with the light of 546.1 nm is measured as 10.0 mm. When an unknown liquid is poured into the gap between the plano-convex lens and the glass base, the diameter of this ring shrinks to 8.45 mm. What is the refractive index of the liquid?

10.11 Show that the transmissivity $\mathcal{T}$ of a thick plate of transparent material of refractive index μ placed in air when the interference effects in multiply reflected beams are ignored is given by

$$\mathcal{T} = \frac{(1-\mathcal{R})^2}{1-\mathcal{R}^2}$$

where $\mathcal{R}$ is the reflectivity of the interface given by

$$\mathcal{R} = \frac{(\mu-1)^2}{(\mu+1)^2}$$

11 Measurement of Radius of Curvature and Focal Length

11.1 INTRODUCTION

All optical components are bound by the combination of plane and curved surfaces. Due to the ease of fabrication, the curved surfaces are spherical, though nonspherical and free-form surfaces are gaining importance in some sophisticated designs. The curved surfaces provide power to the optical component. The spherical surfaces need to be produced with tight tolerances on the surface curvature. The measurement of radius of curvature and focal length of lenses and mirrors is a very important activity in the production environment. It is the range, and the accuracy and precision required that make these measurements often very complicated. In most cases, the measurement procedure is straightforward.

11.2 MEASUREMENT OF RADIUS OF CURVATURE

Radii of curvature are measured both by the mechanical and the optical means. Each of these methods can be grouped into two categories: (1) indirect method in which either the sagitta or the slope of the surface is measured and (2) direct method in which distance between the positions of the vertex and the center of curvature is measured.

11.2.1 Indirect Method: Measurement of Sagitta

Sagitta can be measured either using a mechanical spherometer or an optical spherometer. It can also be measured interferometrically such as in Newton's ring method.

11.2.1.1 Mechanical Spherometer

Usually, this is a three-leg instrument; the three pointed legs being at the vertices of an equilateral triangle. A central plunger is used to measure the sagitta. The instrument is adjusted by placing it on a plane surface, contacting the surface with the plunger, and then taking the reading. This reading acts as a reference. The instrument is then placed on a spherical surface and the reading is taken when the plunger touches the surface. The difference between these two readings is the sagitta. It can be shown that the radius of curvature R is related to the measured sagitta h by the following equation

$$R = \frac{a^2}{6h} + \frac{h}{2} \tag{11.1}$$

DOI: 10.1201/9781003531845-11

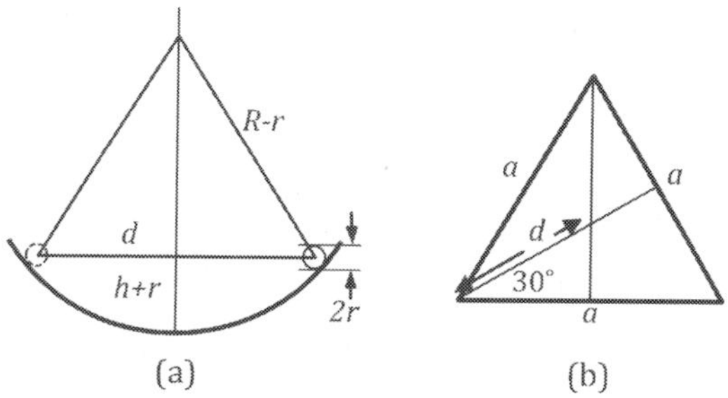

FIGURE 11.1 (a) Determination of sagitta. (b) Relation between a and d.

where a is the length of an arm of the equilateral triangle. Often the pointed legs are replaced by the legs carrying balls of radii r. In this case, a small correction is made to Equation (11.1), and the radius of curvature is now given by

$$R = \frac{a^2}{6h} + \frac{h}{2} \pm r \tag{11.2}$$

where the + sign is used for the concave surface and – is used for the convex surface. Equation (11.2) is easily obtained following Figure 11.1(a) as

$$(R - r)^2 = d^2 + [R - (h + r)]^2 \Rightarrow R = \frac{d^2}{2h} + \frac{h}{2} + r \tag{11.3}$$

But the value of d in terms of the length of the arm a of the equilateral triangle as seen in Figure 11.1(b) is obtained as

$$\frac{a}{2} = d \cos 30° = d \frac{\sqrt{3}}{2} \Rightarrow d = \frac{a}{\sqrt{3}} \tag{11.4}$$

Substituting for d in Equation (11.3), we obtain

$$R = \frac{a^2}{6h} + \frac{h}{2} + r \tag{11.5}$$

When a spherometer with a ring base is used, the radius of curvature is given by

$$R = \frac{y^2}{2h} + \frac{h}{2} \tag{11.6}$$

where y is the radius of the ring. The ring has a sharp edge, which contacts with the spherical surface and may scratch it. Instead of the ring with a sharp edge, one can have three balls of radius r mounted on it. In that case, the radius of curvature is given by

$$R = \frac{y^2}{2h} + \frac{h}{2} \pm r \tag{11.7}$$

Equation (11.7) also applies when a ring of circular section of radius r is used in place of three balls.

Assuming that y and r are accurately known, the uncertainty ΔR in R due to the uncertainty Δh in the measurement of the sagitta h can be expressed as

$$\Delta R = \frac{\Delta h}{2}\left(1 + \frac{y^2}{h^2}\right) \tag{11.8}$$

where the uncertainties have been added.

A bar spherometer is used for the evaluation of astigmatism. This spherometer is shaped like a bar with the contact points at the ends and the spindle for measurement at the center. It can measure the curvature along any diameter of the test piece. The Geneva gauge is a commercial version of the bar spherometer for optometric work. Its scale is directly calibrated in diopters assuming that the refractive index of the glass is 1.523.

Methods based on physical contact require an application of constant pressure for each measurement. It is implied that the pressure must not deform the surface. Therefore, noncontact methods are preferred for the measurement of radius of curvature of surfaces.

11.2.2 Direct Methods

11.2.2.1 Image Formation of a Distant Object

A concave mirror forms an image of an object with unit magnification when the object is placed at the center of curvature, i.e., the image is formed where the object is, as shown in Figure 11.2. Then, the distance between the object plane and the vertex of the mirror is the radius of curvature of the mirror.

11.2.2.2 Differences in Conjugate Positions

There could be a significant error in locating the vertex of the mirror; hence, the distances of the object and the image could also be in error resulting in a significantly large error in the value of the measured curvature. However, if we could locate a reference point with respect to which the distances are measured, a more accurate value of the radius of curvature can be obtained. For this purpose, the center of curvature is taken as the reference point, which is located by making an image of a point object on itself, as shown in Figure 11.3. We can now place an object at any position O_1 and obtain its image by sliding a small screen (or a detector) from

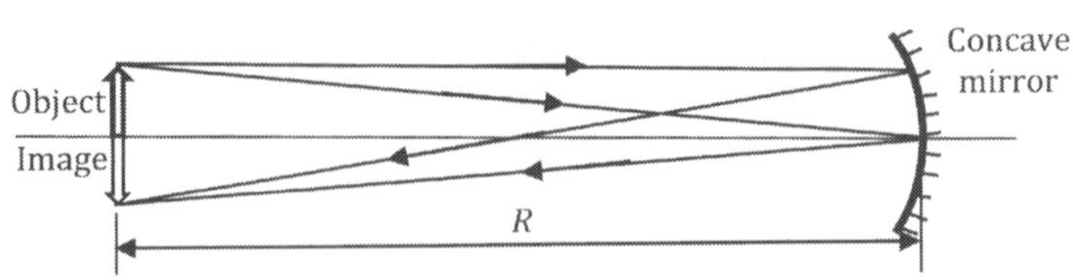

FIGURE 11.2 Measurement of radius of curvature of a concave spherical mirror.

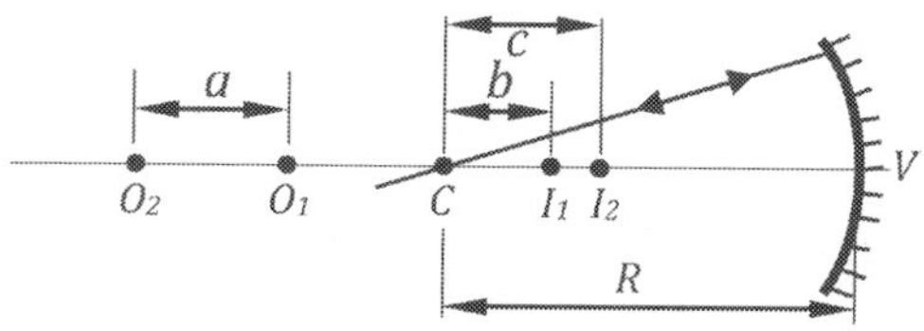

FIGURE 11.3 Imaging by a spherical concave mirror.

the plane passing through the center of curvature. Let this position be I_1. The point object is now moved to a different location O_2 and its image is obtained at I_2. Let the distances between the center of curvature C and I_1 be b and the distances between C and I_2 be c. Further, let the separation between object positions O_1 and O_2 be a. These distances can be measured accurately as these are the differences between the two positions.

Using imaging conditions, we have

$$\frac{1}{p}+\frac{1}{R-b}=\frac{2}{R} \tag{11.9a}$$

and

$$\frac{1}{p+a}+\frac{1}{R-c}=\frac{2}{R} \tag{11.9b}$$

where p is the distance of the object point at O_1 as measured from the vertex V and is unknown and R is the distance between vertex V and center of curvature C, which is to be determined.

Eliminating p in Equations (11.9a) and (11.9b) and rearranging, we obtain

$$R^2(a+b-c)-2Ra(b+c)+4abc=0 \tag{11.10}$$

The solution of the quadratic Equation (11.10) gives the value of R as

$$R=\frac{a(b+c)+\sqrt{a^2(b+c)^2-4abc(a+b-c)}}{a+b-c} \tag{11.11}$$

By inserting the numerical values of a, b, and c, the radius of curvature R is calculated. Note a, b, and c are the difference values and known to an accuracy of the least count of the optical bench on which the experiment is conducted.

11.2.2.3 Optical Spherometer

A point focus microscope along with a suitable measuring rail constitutes the optical spherometer. The microscope and the test surface are mounted on the rail. For a concave surface, two positions can be found that give the point image on retro-reflection. The separation between these positions is the radius of curvature of the surface. For making measurement on the convex surface, an additional lens is required that must

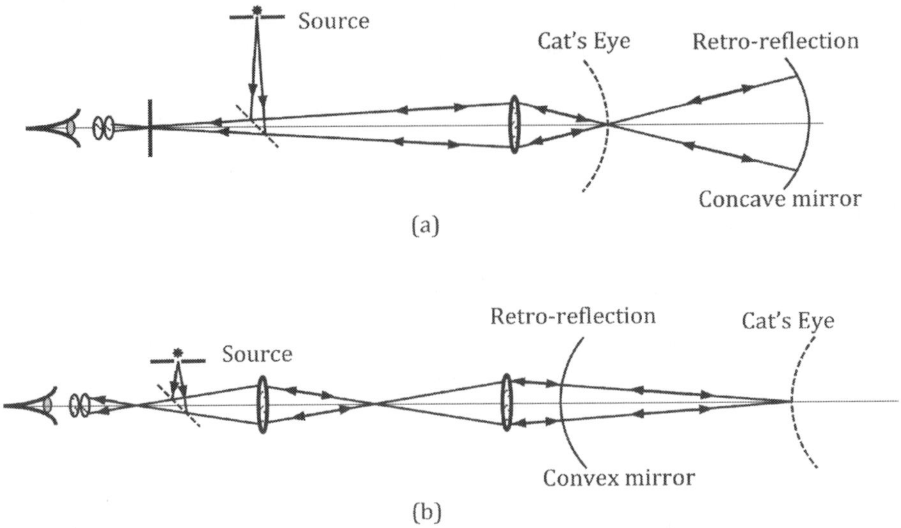

FIGURE 11.4 Determination of radius of curvature of a (a) concave and (b) convex spherical surface.

have a focal length larger than the radius of curvature of the surface being measured. Figure 11.4 shows the experimental arrangements.

For measuring radii of curvature within a moderate range (~1 m to 2 m) an autocollimator with a suitable lens can also be used to obtain cat's-eye and confocal positions. A Twyman–Green or Fizeau interferometer can also be used to determine these locations by observing null in the interference pattern. A long coherence length laser is used as a source. With careful implementation, the radii of curvature of spherical surfaces can be measured to an accuracy of a few parts in 10^5.

11.2.2.4 Measurement of Long Radius of Curvature

For the measurement of very long radii of curvature of the spherical surfaces, the methods employing measurement of sagitta by mechanical means or the direct distance measurement are not suitable: the first one due to the accuracy of measurement of the sagitta and the second due to the availability of space and optics considerations.

However, the sagitta can be measured interferometrically to a good accuracy. For measuring long radii of curvature, interference methods such Newton's ring method along with phase shifting could be used. The radius of curvature R, when Newton's ring method is used, is given by

$$\rho_m = \sqrt{m\lambda R} \text{ or } R = \frac{\rho_m^2}{m\lambda} \tag{11.12}$$

where ρ_m is the radius of the mth ring.

Talbot effect has also been utilized to measure moderate to long radii of curvature of the spherical surfaces.

11.2.2.5 Cavity Method: Measurement of Long Radius of Curvature of a Concave Surface

Gerchman and Hunter presented a method to measure a long radius of curvature in which a cavity is formed between the concave surface under test and a plane surface. A parallel (collimated) beam is incident on the concave surface, which focuses it at a point where a plane surface is placed, which retro-reflects the beam. In fact many such positions can be found by changing the separation between the plane surface and the concave surface such that the beam is focused on either of the two surfaces resulting in retro-reflection. This greatly reduces the working space required for measurement.

When a collimated beam is incident on a concave surface, it is brought to a point focus at a distance $R/2$ where the plane surface is placed. This arrangement is called the $n = 1$ *configuration*. When the plane surface is moved toward the concave surface such that the beam is focused on the concave surface, this results in the $n = 2$ *configuration*. Further shift of the plane mirror results in focusing the beam at the plane surface, which is the $n = 3$ *configuration*. The process can continue for higher-order configurations. When n is odd, the focus is on the plane surface and for when n is even it is on the concave surface. The separation between two successive such positions is used to obtain the value of the radius of curvature.

Let z_n be the separation between the concave and the plane surfaces (cavity length) for the nth *configuration*. The equations that relate the cavity length z_n and the radius of curvature R of the concave surface are derived from paraxial ray analysis. This is accomplished by repeated application of the Gaussian image equation given by

$$\frac{1}{p_m} + \frac{1}{q_m} = \frac{2}{R} \tag{11.13}$$

and the conjugate recursion formula

$$p_m = 2\, z_n - q_{m-1} \tag{11.14}$$

where $m = k, k - 1, k - 2, \ldots, 0$, and $k = (n-1)/2$ for odd n and $k = (n-2)/2$ for even n.

An appropriate initial condition is determined for each configuration depending on where the system comes to focus. These initial conditions are for n odd, $z_n = q_k$ and for n even, $z_n = q_k / 2$.

We will now apply Equations (11.13) and (11.14) to obtain the cavity length for the fifth-order configuration. Therefore, for this configuration n is 5, and k is 2. This gives m values as 2, 1, and 0. Further we have the following equations:

$$p_2 = 2z_5 - q_1 \tag{11.15a}$$

$$p_1 = 2z_5 - q_0 \tag{11.15b}$$

$$z_5 = q_2 \tag{11.15c}$$

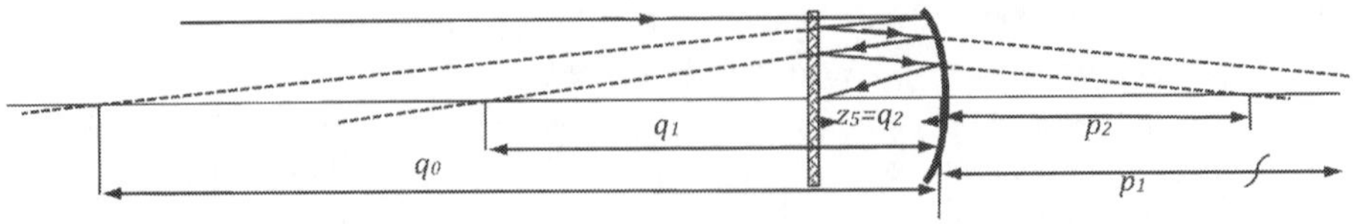

FIGURE 11.5 Ray path for five reflections by the cavity method.

Figure 11.5 shows the geometry of the fifth-order cavity with the ray paths showing positions p_m and q_m. Using the Gaussian imaging relation $\frac{1}{p_2}+\frac{1}{q_2}=\frac{2}{R}$, and substituting for q_2, we solve it for p_2 as

$$p_2 = \frac{R\, z_5}{2z_5 - R} \tag{11.16}$$

Using Equation (11.15a), we obtain

$$p_2 = \frac{R\, z_5}{2z_5 - R} = 2z_5 - q_1 \Rightarrow q_1 = \frac{4\, z_5^2 - 3R\, z_5}{2z_5 - R} \tag{11.17}$$

We now apply the Gaussian imaging condition for $m = 1$, i.e.,

$$\frac{1}{p_1}+\frac{1}{q_1}=\frac{2}{R} \Rightarrow \frac{1}{p_1}=\frac{2}{R}-\frac{2z_5 - R}{4\, z_5^2 - 3R\, z_5} \Rightarrow p_1 = \frac{4R\, z_5^2 - 3R^2\, z_5}{8\, z_5^2 - 8\, R\, z_5 + R^2} \tag{11.18}$$

Using Equation (11.15b), we have

$$p_1 = \frac{4R\, z_5^2 - 3R^2\, z_5}{8\, z_5^2 - 8\, R\, z_5 + R^2} = 2z_5 - q_0 \Rightarrow q_0 = \frac{16\, z_5^3 - 20\, R\, z_5^2 + 5\, R^2\, z_5}{8\, z_5^2 - 8\, R\, z_5 + R^2} \tag{11.19}$$

Now applying Gaussian imaging condition for $m = 0$, we have

$$\frac{1}{p_0}+\frac{1}{q_0}=\frac{2}{R} \Rightarrow \frac{1}{p_0}=\frac{2}{R}-\frac{8\, z_5^2 - 8\, R\, z_5 + R^2}{16\, z_5^3 - 20\, R\, z_5^2 + 5\, R^2\, z_5} \tag{11.20}$$

This gives

$$p_0 = \frac{16\, R\, z_5^3 - 20\, R^2\, z_5^2 + 5\, R^3\, z_5}{32\, z_5^3 - 48\, R\, z_5^2 + 18\, R^2\, z_5 - R^3} \tag{11.21}$$

Because a collimated beam is incident on the concave surface, $p_0 = \infty$. Therefore, the denominator in Equation (11.21) must be zero, i.e.,

$$32\, z_5^3 - 48\, R\, z_5^2 + 18\, R^2\, z_5 - R^3 = 0 \tag{11.22}$$

The solution of this cubic equation is $z_5 = 0.0669873R$. Essentially, one can solve for the cavity length for any value of n. The values for z_3 and z_4 are $z_3 = 0.1464466R$ and $z_4 = 0.0954915R$. Thus, the cavity length z_n is related with the radius of curvature

TABLE 11.1
Radius of Curvature from Differential Cavity Lengths

Cavity Length (z_n)		Radius vs. Differential Cavity Length
n	z_n	
1	$0.5R$	$R = 4\ (z_1 - z_2)$
2	$0.25R$	$R = 9.65685\ (z_2 - z_3)$
3	$0.1464466R$	$R = 19.62512\ (z_3 - z_4)$
4	$0.0954915R$	$R = 35.08255\ (z_4 - z_5)$
5	$0.0669873R$	$R = 57.23525\ (z_5 - z_6)$
6	$0.0495156R$	$R = 87.29584\ (z_6 - z_7)$
7	$0.0380603R$	$R = 126.47741\ (z_7 - z_8)$
8	$0.0301537R$	$R = 175.99437\ (z_8 - z_9)$
9	$0.0244717R$	

R of the concave surface through a relation $z_n = C_n\, R$, where the values of C_n for the first nine configurations are given in Table 11.1. This table also lists the relationships between the radius of curvature and the differential cavity length $(z_n - z_{n\text{-}1})$.

11.2.2.6 Measurement of Very Long Radii of Curvature

Sagitta of a concave or a convex surface can also be measured interferometrically. A flat surface is placed on the spherical surface. It makes a contact at the center when the spherical surface is convex creating a thin film of air. For the concave surface, the contact is at the edge of the flat or at the edge of the concave surface, whichever is smaller. A collimated beam illuminates this arrangement, and the interference takes place between the beams reflected from the top and the bottom of the air film. The fringes are fringes of constant thickness and hence are circular in form. The order of fringe is zero at the center for a convex spherical surface and increases outwardly. The center fringe is black due to a phase change of π on reflection at the air–glass interface. For a concave surface, the order is zero at the contact circle and increases toward the center.

Consider a convex surface placed on top of a flat surface. It can be shown that the radius r_n of the nth dark ring due to interference between waves from the plane and the convex surfaces is given by

$$r_n = \sqrt{n\ \lambda\ R} \tag{11.23}$$

where R is the radius of curvature of the convex surface and λ is the wavelength of light. This is valid when the radius of curvature is very large. This equation can be modified to yield the radius of curvature of the convex surface as

$$R = \frac{D_n^2 - D_{n-1}^2}{4\lambda} \tag{11.24}$$

where D_n and D_{n-1} are the diameters of the nth and $(n - 1)$th fringes. The diameters of rings can be measured using a traveling microscope. A graph between D_n^2 versus n will be linear and its slope is $4\lambda R$. Using Equation (11.23), it can be shown that

$$\frac{D_{n+1} - D_n}{D_{n+2} - D_{n+1}} \approx 1 + \frac{1}{2n} \tag{11.25}$$

The fringe width, for large value of n, is practically constant.

For a concave surface, the shape of the air-film is as shown in the Figure 11.6. It has a maximum air thickness t_0 at the center, which decreases toward the edge.

From the Figure 11.6, $(2\ R - t_x)t_x = x^2$, where $t_x = t_0 - t(x)$. Under the condition $R \gg t_0$, we have

$$2\left[t_0 - t(x)\right] = \frac{x^2}{R} = m'\lambda = (m_0 - m)\lambda \tag{11.26}$$

where m_0 is the fringe order corresponding to the thickness t_0, which is not necessarily an integer and is not known and m is the order at the thickness $t(x)$. From Equation (11.26), the diameters $D^2_{m_0-m}$ and $D^2_{m_0-m-1}$ corresponding to $(m_0 - m)$th and $(m_0 - m - 1)$th order fringes are expressed as

$$D^2_{m_0-m} = 4\ R(m_0 - m)\ \lambda \tag{11.27a}$$

$$D^2_{m_0-m-1} = 4\ R(m_0 - m - 1)\ \lambda \tag{11.27b}$$

On subtraction, we obtain

$$D^2_{m_0-m} - D^2_{m_0-m-1} = 4\ R\ \lambda \tag{11.28}$$

Equation (11.24) is the same formula that was obtained for the convex surface. The radius of the concave surface is obtained from the slope of $D^2_{m_0-m}$ versus $(m_0 - m)$ plot.

The method works fine when there are several circular fringes in the interference pattern; the circular shape of the fringes implies that the surface is spherical. However, when the radius of curvature is large, there are fewer fringes, but the method can still be applied. However, when the number of fringes becomes less than

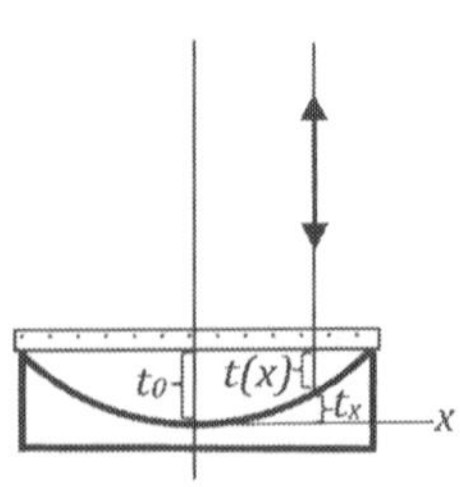

FIGURE 11.6 Calculation of film thickness between flat and concave surfaces.

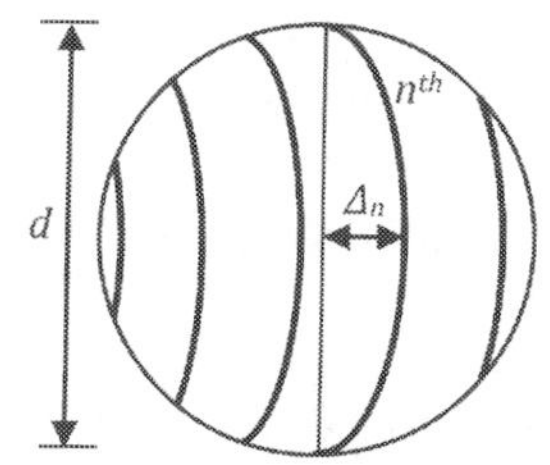

FIGURE 11.7 Interferometric measurement of large radius of curvature.

one, the method breaks down. In such a situation, the center of the fringe pattern is shifted by lightly tilting the test piece. The fringes now become arcs of circles. These arcs of circles are nearly equidistant. Let us assume that the nth fringe passes through the middle of the test surface as shown in the Figure 11.7.

Therefore

$$D_{n+1}^2 - D_n^2 = \left(D_{n+1} - D_n\right)\left(D_{n+1} + D_n\right) = 4\lambda R \tag{11.29}$$

and

$$\Delta_n \; D_n = \frac{d^2}{4} \tag{11.30}$$

It may be noted that in the above Equation (11.29) $\left(D_{n+1} - D_n\right)$ is two times the fringe width, $\bar{x}$, and $D_{n+1} \approx D_n$, and hence we have

$$\frac{\Delta_n}{\bar{x}} = \varepsilon = \frac{d^2}{4\;\lambda\;R} \Rightarrow R = \frac{d^2}{4\;\lambda\;\varepsilon} \tag{11.31}$$

The radius of curvature is obtained by measuring the sag Δ_n and fringe width $\bar{x}$. Note that the Newton's ring method can be applied to measure short to very large radii of curvature.

11.2.2.7 Radius of Curvature with a Test Plate

Test plates are made by grinding and polishing two identical circular glass plates. The process generates a spherical surface, whose radius of curvature can be measured by some other independent procedure. The process of grinding and polishing is continued until the required radius of curvature is achieved. Either of plate is called a test plate, which is used to check the radius of curvature of the component being fabricated in the production shop. A convex test plate is used to check a concave surface and vice versa.

If the radius of curvature of the test surface is different than that of the test plate, the circular fringes are observed. One can also determine whether the radius of curvature of the test surface is smaller or larger than that of the test plate. Let the radius of curvature of the test plate be R, and that of the test surface be $R + \Delta R$. Further, let the diameter of the test plate be d. It can be shown that the gap perpendicular to one of the surfaces at a distance r_n from the point of contact is given by

$$\Delta_n = \Delta R(1 - \cos\theta) \tag{11.32}$$

where the angle θ is given by $\sin\theta = \frac{r_n}{R}$, where r_n is the radius of the nth circular fringe. Fringe of nth order will be formed when $2\ \Delta_n = n\ \lambda$. Substituting for Δ_n under the condition $\frac{\Delta_n^2}{\Delta R^2} \ll \frac{\Delta_n}{\Delta R}$ and with a little manipulation, we obtain

$$r_n^2 = n\lambda \frac{R^2}{\Delta R} \Rightarrow D_n^2 = 4n\lambda \frac{R^2}{\Delta R} \tag{11.33}$$

where D_n^2 is the diameter of the nth fringe. In practice, instead of several circular fringes, one would like to have the radius of curvature of test surface as close to that of the test plate as possible and hence there may be less than one fringe. It is under this condition that one is required to determine the departure from the expected radius of curvature. Therefore, the center of the fringe pattern is shifted by slightly tilting the test surface. The fringes now are the arcs of circles, and their sag is used to determine the value of ΔR. As before like Equations (11.29) and (11.30), we set up the following two equations:

$$D_{n+1}^2 - D_n^2 = \left(D_{n+1} - D_n\right)\left(D_{n+1} + D_n\right) = 4\lambda \frac{R^2}{\Delta R} \tag{11.34}$$

and

$$\Delta_n\ D_n = \frac{d^2}{4} \tag{11.35}$$

Using these two equations, we obtain

$$\frac{\Delta_n}{\overline{x}} = \varepsilon = \frac{d^2\ \Delta R}{4\ \lambda\ R^2} \Rightarrow \Delta R = \frac{4\ \lambda\ \varepsilon\ R^2}{d^2} \tag{11.36}$$

This equation gives the departure of the radius of curvature of the test surface from the expected value R.

11.2.2.8 Newton's Method

When the distances of an object and its image are measured from the foci, we obtain a formula, which is termed as Newton's lens equation or simply Newton's formula. Mathematically

$$z\ z' = f\ f' \tag{11.37}$$

where z and z' are the extra focal object and image distances, respectively, and measured from the front and back focal points. Equation (11.37) has been very cleverly used to measure the long radius of curvature. Figure 11.8 shows the schematic of the measurement principle. In Figure 11.8, a point object O is imaged at O', and a

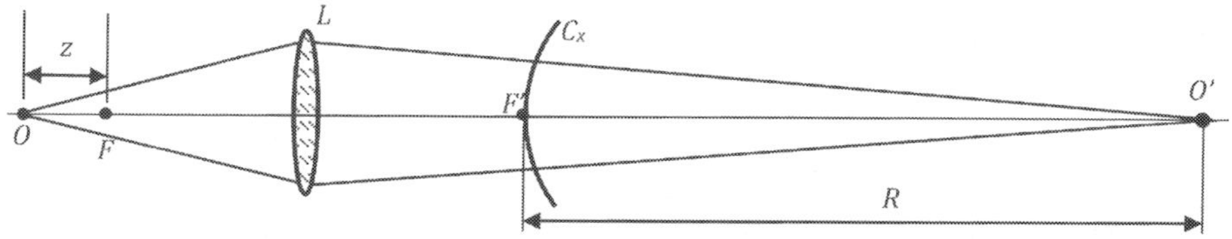

FIGURE 11.8 Schematic to measure the radius of curvature using Newton's lens equation.

convex surface is inserted such that its vertex coincides with the back focus point F'. In this case, $z' = R$ and the focal lengths f and f' are equal. Thus $z = f^2/R$ is the working formula. If z could be measured, then the radius of curvature is obtained from the single measurement.

The basic question is how to place the surface exactly at the back focus and measure z accurately. To locate the points O, F, and F', a Fizeau interferometer is utilized, and the distance z is measured on a precision linear scale. The procedure is explained in the next paragraph with the help of Figure 11.9.

A laser beam is expanded and then collimated using a lens, L_c. Plate P is a partially coated plate, which provides a reference beam. The procedure to locate point F', F, and O, in that order, is as follows.

A lens L is placed in the collimated beam, which focuses the beam to a diffraction focus. The convex surface is now placed in the convergent beam. When properly placed, the beam is retro-reflected (cat's-eye position) and a null is obtained between the interference of the reference and the retro-reflected beams.

Now the lens L_1 is placed in the beam and a plane mirror is inserted between L and C_x without disturbing the positions of lens L and convex surface C_x. Lens L_1 is translated until the beam exiting from lens L is collimated. This position is obtained when the beam reflected from the plane mirror, on interference, gives the null. In this case, the back focal point of lens L_1 and front focal point of lens L coincide. Thus, the position F is found and the location of L_1 is noted. The plane mirror is removed and lens L_1 is translated away until the null is obtained. In this case, the rays from lens L strike the surface C_x normally and are retro-reflected (confocal position). This position of lens L_1 is noted and the difference between these two positions is the distance z that is required to determine R. The relative error in the measurement of radius of curvature is given by

$$\frac{\Delta R}{R} = \sqrt{4\left(\frac{\Delta f}{f}\right)^2 + \left(\frac{\Delta z}{z}\right)^2} = \sqrt{4\left(\frac{\Delta f}{f}\right)^2 + \left(\frac{R}{f^2}\right)^2 (\Delta z)^2} \tag{11.38}$$

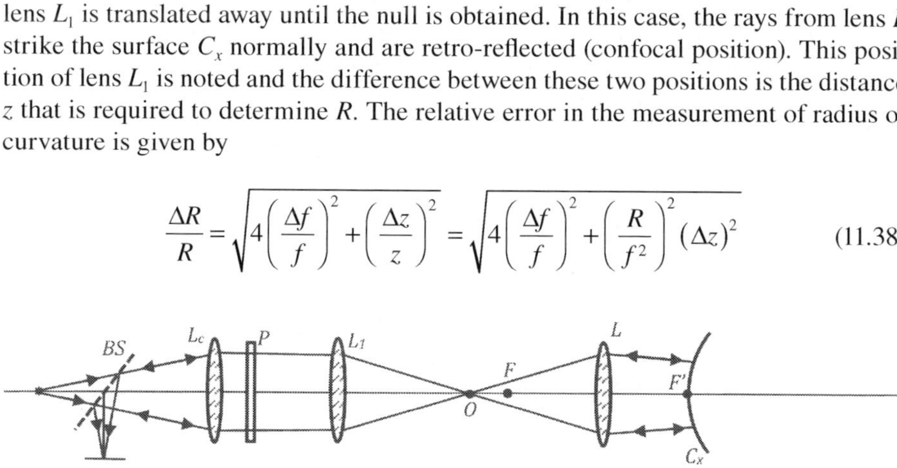

FIGURE 11.9 Experimental arrangement to locate points O and F.

where Δf is the accuracy in the measurement of focal length of lens L and Δz is the accuracy in the measurement of distance z.

11.3 SCANNING PROFILOMETRY

The surface profile can be obtained by measuring the slope along a diameter of a symmetric surface. Let the surface be defined by $z = f(x)$ along the diameter and let $f(0) = 0$. The slope is measured at locations x_k, and the consecutive positions are separated by Δx. Thus $f(x_{k+1}) = f(x_k) + \frac{\Delta x}{2}\left[f'(x_k) + f'(x_{k+1})\right]$. Thus, integrating the slope values can generate the profile of the surface. The radius of curvature is obtained from

$$R = \frac{f''(x)}{\left\{1 + \left[f'(x)\right]^2\right\}^{3/2}} \tag{11.39}$$

The slope can be measured by an autocollimator in conjunction with a pentaprism, which scans the surface. Higher accuracy is obtained when an interferometric arrangement, say a phase measuring interferometer, is used.

11.4 RADIUS OF CURVATURE MEASUREMENT BY TALBOT INTERFEROMETRY

It has been shown that a periodic structure repeats itself when illuminated by a coherent beam. We can make use of this phenomenon to determine the radius of curvature of a spherical surface. Figure 11.10 shows the schematic of the experimental setup where only the confocal part is shown.

The grating G_1 of pitch p is illuminated by a diverging wave of radius of curvature r. The Talbot planes are formed at distances z_s where $z_s = \frac{2N\,p^2\,r}{\lambda\,r - 2N\,p^2}$, where $N = 1, 2, 3, \ldots$ for different Talbot planes. The spacing between the successive Talbot planes increases with the order N. The pitch of the grating also increases as if it is geometrically projected, i.e., $p' = p\,\frac{r + z_s}{r}$, where p' is the pitch at a $(r + z_s)$ plane. If grating G_2 is placed at the first Talbot plane, a moiré pattern due to pitch mismatch is formed. The pitch of the moiré fringe pattern is $p_m = \frac{\lambda}{2}\frac{r}{p}$. The radius of curvature of the wave at the plane of the grating G_1 is thus obtained. This could be related to the radius of curvature of the mirror surface.

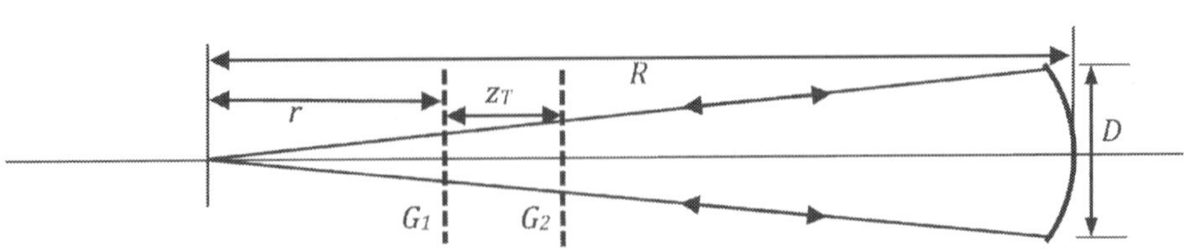

FIGURE 11.10 Measurement of the radius of curvature by Talbot phenomenon.

Assuming that the illuminated sizes of grating G_1 and G_2 are y and y_1, then

$$p' = p\,\frac{y_1}{y} = p\,\frac{y+\alpha\, z_T}{y} = p\left(1+\frac{\alpha\, z_T}{y}\right) \tag{11.40}$$

where $\alpha = D/R$. The pitch of the moiré fringe is $p_m = \frac{p\,p'}{p'-p} = \frac{p'\,y}{\alpha\, z_T}$. If there are n moiré fringes in the pattern formed, then

$$n = \frac{y}{p_m} = \frac{\alpha\, z_T}{p'} = \frac{D}{R}\,\frac{z_T}{p'} \Rightarrow R = \frac{D}{n}\,\frac{z_T}{p'} = \frac{D\, z_T}{n\, p}\left(1-\frac{n\, p}{y}\right) \tag{11.41}$$

The radius of curvature can be obtained by measuring D, z_T, and noting the number of moiré fringes formed. Often the Talbot phenomenon is used for setting purposes.

11.5 MEASUREMENT OF FOCAL LENGTH

11.5.1 Focal Length of a Thin Lens

For a single thin lens, the effective focal lengths f and f' are defined as the distances from the lens vertex to the front focal point and the rear focal point, respectively. However, for a thick lens or a lens combination, the effective focal lengths f and f' are the distances between the front focal point and the front principal point, and the rear principal point and the rear focal point. The back focal length is the distance between the rear vertex to the rear focal point and the front focal length is the distance between the front focal point and the front vertex of the lens. It is easier to measure the back and the front focal lengths.

11.5.1.1 Focal Length by Imaging

The simplest though slightly inaccurate method to determine the focal length of a lens is to image a distant object. Because the object is far away, the distance between the lens and the image is the focal length. One could use sun as an object and make its image. Alternately, an incandescent or a fluorescent lamp far away in the room could be taken as the luminous object. Using this method, focal length can be assigned to within a few millimeters.

A simple arrangement to measure the focal length of a thin positive lens is to use a mesh (gauge) illuminated by a diffuse light source. The mesh is placed flush on a screen that is located on the optical bench. The test lens and a plane mirror behind it are also placed on the bench and their heights are properly adjusted. The lens is moved toward or away from the mesh until its sharp image is formed on the screen. The light retro-reflected by the plane mirror forms this image. The mirror may be tipped a little so that the image is formed alongside of the mesh. The distance between the screen and the vertex of the lens is the focal length of the lens.

11.5.1.2 $y'/\tan\theta'$ Method

Figure 11.11 shows an off-axis collimated beam incident on the lens. The image is formed at its focal plane.

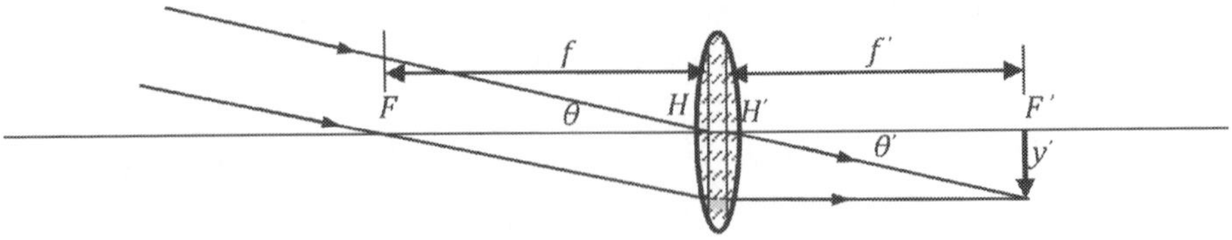

FIGURE 11.11 Ray path to measure focal length of a lens.

From the Figure 11.11, $y' = -f\ \tan\theta = f'\ \tan\theta'$. When the lens is in air, its nodal planes are coincident with the principal planes and hence its focal lengths f and f' are equal, and the angle θ' in image space is equal to the angle of incidence θ. In practice, the $y' / \tan\theta'$ method is implemented by placing a reticle at the focal plane of a collimator. The graduations on reticle and the focal length of the collimator are accurately known, implying that the angular size of the object is accurately known. The test lens is placed coaxially with the collimator; the reticle appears at infinity to the test lens and a microscope measures its image formed at the focal plane of the lens. If the measurement is done directly using a translation stage on which the microscope is mounted, the focal length is obtained as

$$f = \frac{y'}{\tan\theta} = \frac{y'}{y_o} f_c \tag{11.42}$$

where y_0 is the size of the reticle whose image is y' and f_c is the focal length of the collimator. If the graticule in the eyepiece of the microscope is used for measuring y', then the measured value of y' is to be divided by the magnification m of the microscope.

11.5.1.3 Magnification Method

Using the lens imaging formula, the lateral magnification M when the media on both sides of the lens are the same can be expressed as

$$M = -\frac{q}{p} = 1 - \frac{q}{f} \tag{11.43}$$

where p and q are the object and the image distances as measured from the vertex of a thin lens. Further, if the distance between the object plane and the image plane is more than $4f$, an image can be formed for several positions of the lens. Therefore, keeping the object and image planes fixed, the test lens is translated to form an image. Let the distance between the vertex of the lens and the image plane be q_i and the corresponding magnification of the image be M_i. The slope of a plot between q_i and M_i gives the inverse of the focal length.

Alternately, the lens can be kept fixed, and the object position is varied and the magnification measured at the corresponding image position. Let q_1 and q_2 be the image positions at which the measured magnifications are M_1 and M_2, respectively. Then

$$M_1 = 1 - \frac{q_1}{f} : M_2 = 1 - \frac{q_2}{f} \Rightarrow (M_2 - M_1) = \frac{|q_2 - q_1|}{f} \tag{11.44}$$

The separation $(q_2 - q_1)$ can be measured accurately. The method does not require the knowledge of the locations of the principal planes.

A variant of this method involves the measurement of the displacement of the object. Let M_1 and M_2 be the magnifications for the object positions p_1 and p_2, respectively. Mathematically

$$1+\frac{1}{M_1}=\frac{p_1}{f} \text{ and } 1+\frac{1}{M_2}=\frac{p_2}{f}$$

This gives

$$f=(p_2-p_1)\frac{M_1M_2}{M_1-M_2} \tag{11.45}$$

The method is due to Abbe.

There is another interesting method credited to Bessel. Here the object and the image planes are fixed. It is ascertained that the separation L between the two planes is equal to or greater than four times the focal length of the lens to be measured. This gives two positions of the lens where the image of the object is in focus as shown in Figure 11.12. The distance between these positions is taken as d. Then the focal length of the lens is given by

$$f=\frac{L^2-d^2}{4L} \tag{11.46}$$

This method requires measurement of two distances only.

11.5.1.4 Focal Length of a Negative/Diverging Lens

The focal length of a diverging lens cannot be determined directly. There are other indirect methods to determine the focal length of a diverging lens. One of the methods, known as the virtual object method, requires a positive lens of shorter focal length than that of the diverging lens. Figure 11.13 shows the schematic of image formation by a combination of a positive lens and a negative lens. It may be noted that the lenses do not have to be in contact.

Positive lens L makes an image of a source at O at the image position O'. When the test lens L_T is inserted in the light path and properly positioned, the image shifts to O″. The focal length of the negative lens is calculated from the relation

$$f=\frac{p'q'}{p'-q'} \tag{11.47}$$

where the distances p' and q' are shown in Figure 11.13.

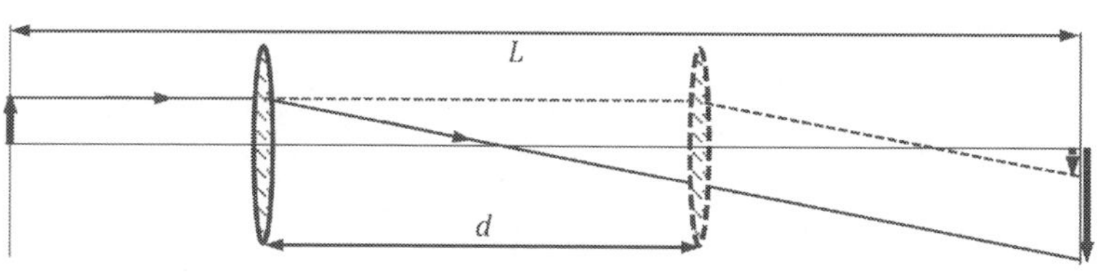

FIGURE 11.12 Bessel method to determine focal length of a positive lens.

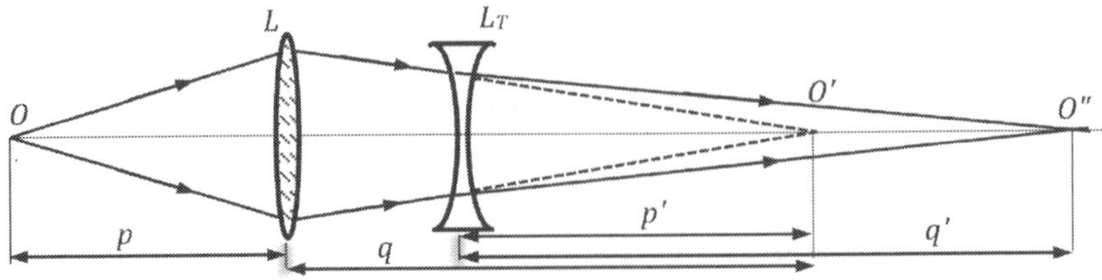

FIGURE 11.13 Determination of the focal length of a diverging lens.

11.5.1.5 Nodal Slide Method

The nodal slide method is based on the fact that the nodal planes and the principal planes are coincident when the media on both sides of the lens are the same and the rotation of the lens about the rear nodal point does not shift the image.

In practice, an object, say a fine mesh, is illuminated by a light source and imaged by a test lens, which is placed on a nodal slide. The lens is placed on a translation stage, which is mounted on a rotatable mount. The axis of rotation can be made to pass through any portion of the lens by moving the translation stage. This arrangement is called a nodal slide. The lens is so positioned that the axis of rotation passes through the nodal point. In such a situation, rotation of the lens does not shift the image. Indeed, this criterion is used to locate the nodal plane. Because the nodal plane is coincident with the principal plane, the distance between the image plane and the axis of rotation is the focal length.

11.5.1.6 Focal Length Measurement from the Difference between Conjugate Positions

Several methods described here assume the lens to be a thin lens, which is a mathematical idealization. A lens is bound by two surfaces and has a finite central thickness. The focal length is the distance between the respective principal plane and the focal plane. The nodal slide method locates the nodal plane. There is a method that uses the difference between the object positions and the image positions. However, to obtain these difference data, we need to find a reference with respect to which the measurements are made. Fortunately, the second focal point can be found by placing the lens in a collimated beam and this point is taken as a reference with respect to which the measurements are made for different object positions. Figure 11.14 shows the various locations of the point object.

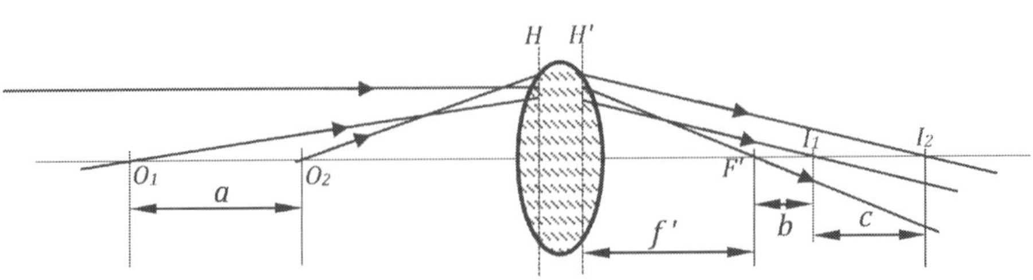

FIGURE 11.14 Imaging by a lens.

When the point object is at infinity, its image is formed at the second focal point and the distance f' is the distance between the principal plane H' and the focal plane F'. The image of the point object O_1 is formed at I_1 and that of O_2 is formed at I_2. To receive the images, the screen/detector is to be moved by the distances b and c, which are measured on the optical bench. Similarly, the distance between O_1 and O_2 is measured on the optical bench, which is equal to a. The focal length f' is calculated from the measured values a, b, and c. The formula is obtained by using the imaging conditions, that is

$$\frac{1}{p}+\frac{1}{f'+b}=\frac{1}{f'} \tag{11.48a}$$

and

$$\frac{1}{p-a}+\frac{1}{f'+c}=\frac{1}{f'} \tag{11.48b}$$

where p is the object distance for the point O_1 as measured from the principal plane H. Eliminating p from Equations (11.48a) and (11.48b) and rearranging, we obtain

$$f'=\sqrt{\frac{abc}{(c-b)}} \tag{11.49}$$

Thus, the focal length is obtained by using the measured difference values and hence the focal length is determined with higher accuracy.

11.6 MOIRÉ DEFLECTOMETRY

Consider a situation where the lens is illuminated by a collimated beam, which is brought to focus by the test lens. In the convergent beam are placed two identical Ronchi gratings of pitch p. Let their separation be Δ as shown in the Figure 11.15. They are placed such that their rulings are parallel.

Grating G_1 is projected on grating G_2 and hence its pitch p' becomes smaller. If the illuminated size of grating G_2 is L and that of G_1 is $L + y$, then the pitch p' is related to pitch p as

$$p'=\frac{L}{L+y}p \Rightarrow \frac{p}{p'}=1+\frac{y}{L} \tag{11.50}$$

The moiré fringes are formed due to pitch mismatch. The pitch d of the moiré fringes is $d=\frac{p\,p'}{p-p'}=\frac{p\,L}{y}$. Further, from Figure 11.15 we have $\frac{a}{f}=\frac{y}{\Delta}$. Substituting for y, we obtain

$$f=\frac{a\,\Delta}{p\,N} \tag{11.51}$$

where N (=L/d) is the number of fringes observed over grating G_2 of size L. Therefore, the focal length is determined by noting the number of moiré fringes observed with identical Ronchi gratings.

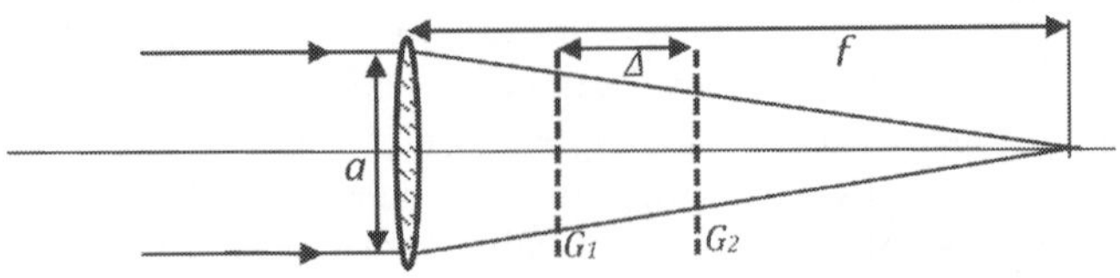

FIGURE 11.15 Moiré formation for focal length measurement.

QUESTIONS

11.1 A three-leg mechanical spherometer, instead of forming an equilateral triangle, makes a triangle of sides a_1, a_2, and a_3. The plunger is equidistant from the vertices of the prism. When used to measure the radius of curvature, show that the radius of curvature R is given by

$$R = \frac{(a_1 + a_2 + a_3)^2}{54h} + \frac{h}{2}$$

11.2 Show that the cavity length for the third-order configuration is obtained from the solution of the quadratic equation

$$8\ z_3^2 - 8\ R\ z_3 + R^2 = 0$$

The solution of this equation gives $z_3 = 0.1464466R$.

11.3 Show that the cavity length for the fourth-order configuration is obtained from the solution of the quadratic equation

$$16\ z_4^2 - 12\ R\ z_4 + R^2 = 0$$

The solution of this equation gives $z_4 = 0.0954915R$.

11.4 A collimated beam illuminates a thin plano-convex lens that has a long radius of curvature. The light reflected from the front plane surface and the back spherical surface interfere to produce circular fringes, which may be termed as modified Newton's rings. The reflected beam from the spherical surface is focused at a distance f_0. Show that the refractive index of the material of the lens is given by

$$\mu = \frac{R}{2f_0}$$

where R is the radius of curvature of the convex surface as measured from modified Newton's rings.

11.5 Enumerate the steps required to determine the radius of curvature of a concave surface using Newton's formula and show the schematic of the experimental setup.

11.6 For determining the focal length using the Talbot phenomenon, a Ronchi grating G_1 of period a with grating lines running parallel to the y-axis is kept in contact with the test lens as shown in Figure 11.16. A similar grating G_2 is placed at the nth Talbot plane making a small angle θ with the y-axis. Show that the moiré fringe pattern will make an angle ϕ_n with the x-axis, where $\tan\phi_n = \frac{\frac{f}{f-z_n}-\cos\theta}{\sin\theta}$. Here f is the focal length of the test lens and z_n is the nth Talbot plane. Further, show that focal length f is obtained from the inclination of the moiré fringes as $f = \frac{1}{\cos\theta+\tan\phi_n\sin\theta-1}\frac{n\,a^2}{\lambda}$.

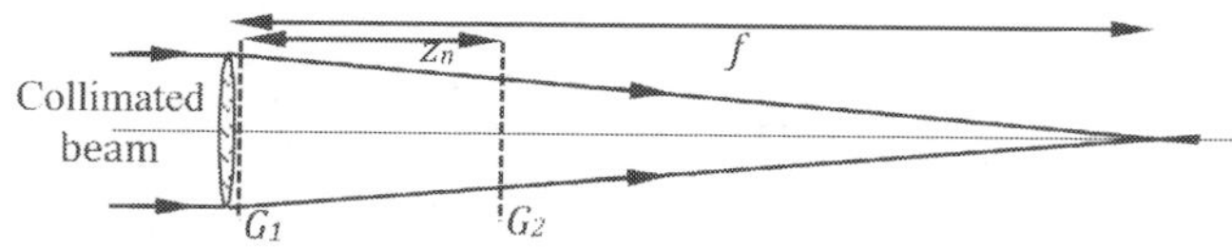

FIGURE 11.16

11.7 Show that

$$\frac{D_{n+1}-D_n}{D_{n+2}-D_{n+1}} \approx 1+\frac{1}{2n}$$

and

$$\frac{D_{n+1}^2-D_n^2}{D_{n+2}^2-D_{n+1}^2} = 1$$

where D_n is the diameter of the nth ring in Newton's ring experiment.

11.8 A Geneva gauge calibrated for a refractive index of 1.523 is used to measure the curvature of a plano-convex lens having a refractive index of 1.5. The distance between the two endpins on the Geneva gauge is 2 cm. The Gauge measures a power of 2 diopters. How many microns of sag is the Geneva gauge measuring?

11.9 The size of an image of an object of linear size 12 mm placed 12.0 cm in front of the lens is measured to be 8 mm. When the location of the object is changed to 17.0 cm in front of the lens, the image size becomes 6 mm. Calculate the focal length of the lens.

11.10 The back focal length and the focal length of a 150 mm diameter lens and nominal focal length of 600 mm is to be measured to an accuracy of ± 3 μm. Can this be done? Explain.

11.11 The distance between the object plane and the image plane along with the distance of two object positions along with the measured magnifications are given. Derive a formula for the focal length of a thick lens under test. What additional information is required to obtain both focal lengths of the lens and its principal planes?

11.12 A collimator is used to measure the focal length of a given lens. The collimator has been characterized by the ratio of focal length of the objective to the line's size of the reticle, which is 1000. The image of reticle as formed by the test lens is 2.0 ± 0.01 mm. What is the focal length of the test lens? Draw a sketch of the experimental arrangement.

11.13 Newton's rings are observed between a plano-convex lens and a plano-concave lens as shown in Figure 11.17. The radius of curvature of the convex surface is 50.0 cm. If the 25th dark ring observed in reflected light of wavelength 564.1 nm has a diameter of 15.0 mm, what is the radius of curvature of the concave surface?

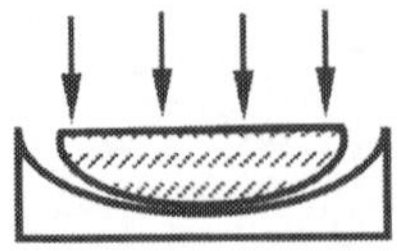

FIGURE 11.17

12 Optical Testing

There is a synergy among design, production, and measurement. You cannot produce optics any better than you can test it. The test methods in the workshop environment have to be rugged and should meet the tolerances specified in the design. This chapter introduces the basic principles of optical testing of surfaces and components.

12.1 TESTING OF A FLAT SURFACE

Many optical components have at least one flat surface; several of them have many. These components are part of an optical system or an instrument and thus directly influence its performance. It is therefore necessary to measure them, i.e., how flat are they? For this purpose, they are to be compared with a reference surface, whose surface flatness is known.

12.1.1 Liquid Surface as a Reference

A liquid is kept in a reasonably large-size open container. It is assumed that at equilibrium the surface of the liquid will be exactly spherical, having the same radius of curvature as that of the Earth's at 6371 km. Therefore, liquid in a container of a 50 cm diameter would also have the same radius of curvature as the Earth, ignoring the edge effects; hence it can be said to be flat. Assuming a 40-cm-diameter liquid surface not affected by edges, its sagitta would be 3.1 nm. Taking a wavelength of 500 nm for measurement, the surface could be said to have a flatness of $\lambda/161$, which is much better than would be required in a workshop environment. Therefore, the surface of the liquid could be taken as a reference flat against which other surfaces made in the workshop could be compared. The liquid in question, however, has to satisfy certain requirements, such as have a high viscosity, be nonhygroscopic, and have a very low vapor pressure. One liquid that suitably matches these properties is silicone oil, DC705, which is employed in evaporation plants.

The optical flat to be measured against this liquid reference surface is placed on a mount, usually a three-point support mount, and an interference pattern is observed using a Fizeau interferometer with phase-shifting capabilities.

The main drawback of this method is the instability problems associated with the liquid itself. Any disturbance of the oil, resulting from making a mechanical adjustment or the vibration transmitted to its surface, etc., would take hours and sometimes days to dissipate. This problem is lessened when one uses amalgamated mercury for the liquid, but it has other associated problems.

DOI: 10.1201/9781003531845-12

12.1.2 Calibration by Three-Flat Method

Three flats (A, B, and C) of nominal flatness and of the same circular shape are compared in pairs using a Fizeau-type phase-shifting interferometer. Let the departure from flatness for these three flats be $w_A(x,y)$, $w_B(x,y)$, and $w_C(x,y)$, respectively. Figure 12.1 shows the procedure by which these three flats are paired for measurement.

When flats A and B are paired for measurement, the Fizeau interferometer measures the wave front function $w(x,y)$, which is given by $w(x,y) = 2\left[w_B(-x,y) + w_A(x,y)\right]$. The interference takes place between the beams reflected from the lower surface of flat B and the top surface of flat A. The factor of 2 appears due to the interfering beam passing twice. We will ignore the factor 2 in the rest of the discussion, but it should be included when the final result of the surface errors is displayed. For these three pairs of flats, the measured wave front errors are, therefore, related to the surface errors of the flats through the following equations:

$$w_1(x,y) = w_B(-x,y) + w_A(x,y) \tag{12.1a}$$

$$w_2(x,y) = w_C(-x,y) + w_A(x,y) \tag{12.1b}$$

$$w_3(x,y) = w_C(-x,y) + w_B(x,y) \tag{12.1c}$$

It may be noticed that these three equations contain four unknowns: $w_A(x,y)$, $w_B(x,y)$, $w_B(-x,y)$, and $w_C(-x,y)$ and hence no point-by-point solution can be obtained for all of the surfaces. The problem arises due to the inherent nature of the test. When one surface is compared with the other, a left-to-right inversion takes place, which introduces a new set of unknowns.

Along the axis of inversion ($x = 0$), Equations (12.1a–c) reduce to

$$w_1(0,y) = w_B(0,y) + w_A(0,y) \tag{12.2a}$$

$$w_2(0,y) = w_C(0,y) + w_A(0,y) \tag{12.2b}$$

$$w_3(0,y) = w_C(0,y) + w_B(0,y) \tag{12.2c}$$

There are now only three unknowns and hence the solution is found as

$$w_A(0,y) = \left[w_1(0,y) - w_3(0,y) + w_2(0,y)\right]/2 \tag{12.3a}$$

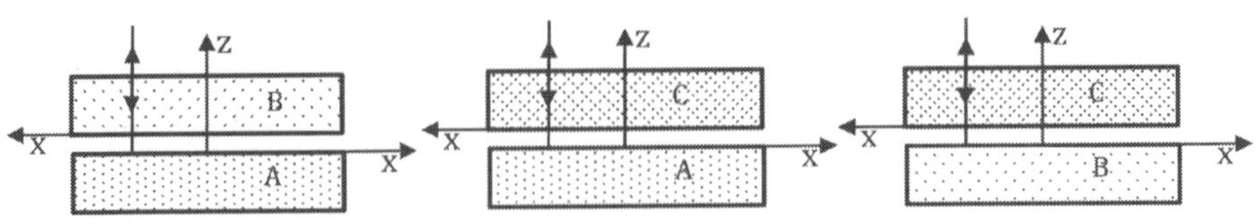

FIGURE 12.1 Three-flat method for measuring flatness.

$$w_B(0,y) = [w_1(0,y) - w_2(0,y) + w_3(0,y)]/2 \tag{12.3b}$$

$$w_C(0,y) = [w_2(0,y) - w_1(0,y) + w_3(0,y)]/2 \tag{12.3c}$$

This yields the surface errors along a diameter. This is the classical three-flat test. However, if the surface errors over the whole surface are to be determined, additional data is required, which is generated by rotating one of the plates in a pair in its own plane. (See Bernard S. Fritz, Absolute calibration of an optical flat, *Opt. Eng.*, 23(4), 379–383, 1984, for details.)

Having realized an optical flat, it is now used to test flat surfaces of components in the optical workshop or in the laboratory. Earlier, the Twyman–Green (TG) interferometer was the workhorse in the production facility. Presently, the Fizeau phase-measuring interferometer is routinely used. However, the basic principle of testing remains the same and hence we will first discuss optical testing by the TG interferometer shown in Figure 12.2. A beam from a quasi-monochromatic source is

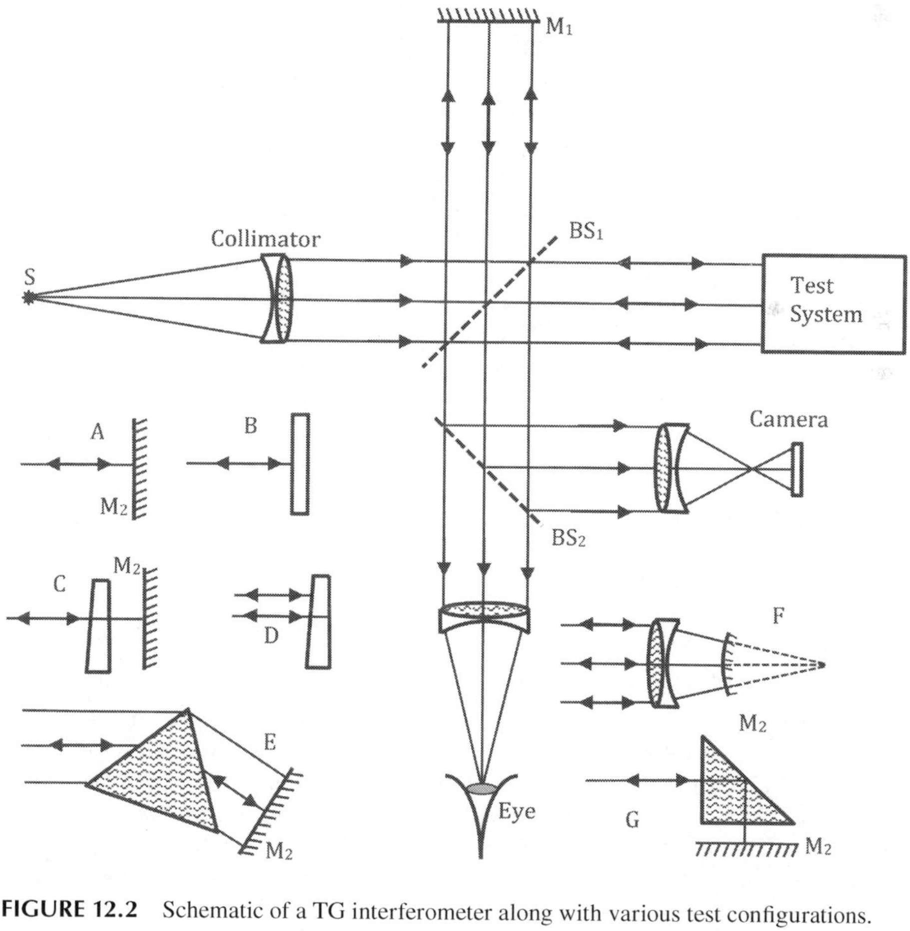

FIGURE 12.2 Schematic of a TG interferometer along with various test configurations.

collimated and then split into two beams. One beam is reflected back by mirror M_1, which acts as a flat reference, while the other beam passes through the test system, is retro-reflected, and passes again through the test system. These two beams are combined by the beam splitter BS_1. The fringes can either be observed or recorded by the camera. Mirror M_1 can be mounted on a lead zirconate titanate (PZT) translator for phase shifting, and the fringe pattern can be recorded using a charge-coupled device (CCD) and processed.

Some of the test systems are also shown in the Figure 12.2 and are labeled *A* to *G*. In *A* mirror M_2 is tested for flatness, whereas in *B* a flat is tested for flatness. A beam splitter in the form of a plane parallel plate can be tested in two ways. In the first arrangement, it is inserted in the test arm, which has a mirror M_2. Mirror M_2 is a flat. The beam splitter and the mirror combination constitutes the test system. When properly aligned, the fringes arise due to the phase difference ϕ_1 introduced by the beam splitter, where $\phi_1 = 2(2\pi / \lambda)(\mu - 1)t$. Here μ and t are the refractive index and thickness of the beam splitter. If the refractive index is taken constant throughout, the fringes arise due to the thickness variation; one or both the surfaces may depart from flatness. Therefore, the surfaces first should be checked for flatness and this test should be carried out to check the homogeneity of the material.

In the second arrangement, the beam from mirror M_1 is blocked and the interference is observed between the beams reflected from the front and the back surfaces of the beam splitter. These two beams have a phase difference ϕ_2, where $\phi_2 = 2(2\pi / \lambda)\ \mu t$. This arrangement is more sensitive than the earlier one. Again, the fringes arise due to the variation of both μ and t. If the flatness has been checked, then this method gives information about the inhomogeneity of the material with higher sensitivity.

The angle of a wedge can also be measured the same way. Assuming that the wedge is bound by the flat surfaces and the material is homogenous, the interference fringes will be straight lines. If the wedge edge is along the *x*-direction and the wedge angle is α, the fringes run parallel to the *x*-direction and the fringe widths in configurations *C* and *D* are, respectively, given by the equations

$$\bar{y}_1 = \frac{\lambda}{2(\mu - 1)\alpha} \tag{12.4a}$$

$$\bar{y}_2 = \frac{\lambda}{2\ \mu\ \alpha} \tag{12.4b}$$

The wedge angle α is determined by measuring the fringe widths.

The prisms are used both for deviation and dispersion of the beam. In both applications, an incident collimated beam should exit as collimated. Nonflat surfaces and inhomogeneity of the medium can distort the beam. Therefore, the prisms are tested to evaluate the distortions and corrected by figuring the surfaces. Configurations *E* and *G* are used to test a 60° and a 45° prism. The 60° prism is used in the minimum deviation position, whereas the 45° prism is used in a position where it either deviates the beam by 90° or retro-reflects it.

Configuration *F* is used to test a lens for its on-axis aberration. A spherical mirror with cat's-eye configuration retro-reflects the beam. In order for the rays to travel the same zone in their return journey, the radius of curvature of the convex mirror must be properly chosen; the mirror should be located as close to the lens as possible.

12.2 TESTING OF SPHERICAL SURFACES

Because it is easier to realize either a flat surface or a spherical surface, the basic philosophy of testing a surface is such that either it is compared with a flat surface or a spherical surface of nearly the same radius of curvature. Further, it may be noted that as an aberrated wave propagates it changes in shape, and it is particularly so when the wave is aspherical, therefore, the aberration of the wave front is tested at the pupil plane.

Consider the test system shown in configuration *F* of Figure 12.2. This configuration is meant to evaluate the on-axis aberrations of a lens and the convex mirror used is assumed to be good. Instead, if the lens is diffraction limited so that it outputs a spherical beam, then this configuration could be used to evaluate the convex surface/mirror. If the surface is spherical, one should obtain straight-line fringes in a properly aligned interferometer. There are, however, restrictions on the radius of curvature of the surface. Concave mirror/surface of a long radius of curvature could not be tested in this configuration.

12.2.1 Scatter-Plate Interferometer

The usefulness of this interferometer arises from the fact that it can be used to test large optics with very high accuracy using white light. This is an excellent example of a common path interferometer. The interference takes place between direct-scattered and scattered-direct beams. This essentially requires two identical scatter plates suitably oriented.

Figure 12.3 shows a scatter-plate interferometer due to Burch using two identical scatter plates S_1 and S_2, respectively. The scatter plates are placed at the plane passing through the center of curvature of the mirror under test. Plate S_1 is illuminated by a convergent beam, which is focused on mirror *M*. A direct beam reflected from mirror *M* passes through plate S_2. Scattered beam from plate S_1 fills the whole surface of the mirror and on reflection passes through plate S_2. There are in fact four beams that pass through plate S_2: (1) scattered by S_1 and on reflection from *M* passes direct through S_2

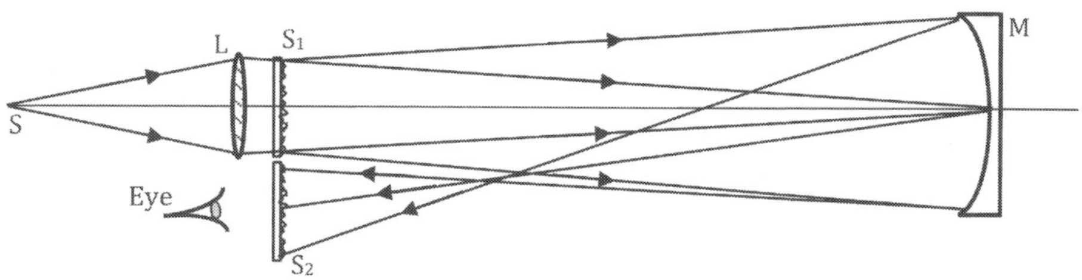

FIGURE 12.3 Schematic of a scatter-plate interferometer.

(scattered-direct), (2) direct through S_1 and on reflection passes scattered by S_2 (direct-scattered), (3) direct through S_1 and on reflection passes direct through S_2 (direct-direct), and (4) Scattered by S_1 and on reflection passes scattered through S_2 (scattered-scattered). Of these scattered-direct is the test beam and direct-scattered is the reference beam. These two beams interfere to produce fringes of equal path difference.

Plate S_2 is an inverted replica of plate S_1. It is therefore possible to combine the functions of both plates into one. This composite scatter plate can be made either photographically or by a computer. To make it photographically, a ground glass plate is illuminated by a laser beam and first exposure is made by placing the photographic plate at a suitable distance. The plate is rotated by 180° in its own plane and another record is made. On development, one has the composite scatter plate. Care, however, should be taken while recording with respect to speckle size, and the rotated position should be within 1 arcsec at about 180°.

Figure 12.4 shows the experimental arrangement with a single scatter plate. Scatter plate S_1 is kept at the center of curvature of test mirror M. Lens L_2 images the scatter plate on to the observation plane where the interferogram is recorded. The scatter-plate interferometer is a common path interferometer; hence, it is insensitive to vibrations and refractive index variations caused by temperature. As it is a common path interferometer, it is very difficult to introduce the phase-shifting method. However, there are two possibilities: a tiny mirror may be mounted on a PZT on which the reference beam falls when an annular mirror/surface is to be tested, or the reference and the object beams could be of perpendicular polarization states and the polarization-based phase shifting is introduced. Both these methods have been used in practice. Further, it may be noted that a lateral displacement of the scatter plate introduces tilt fringe while the longitudinal displacement introduces defocus.

12.2.2 Point Diffraction Interferometer

This is another common path interferometer, which is used to test large radii of curvature mirrors or surfaces using a white light or a low coherence source. This is the simplest interferometer and is very compact.

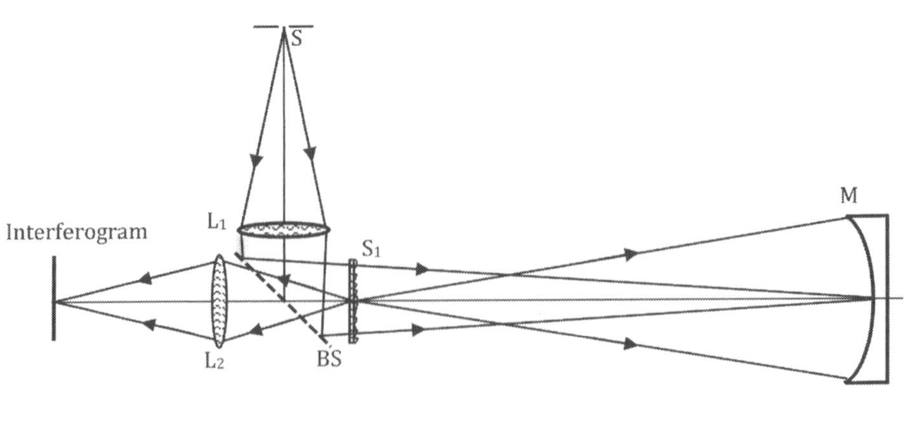

FIGURE 12.4 Schematic of a scatter-plate interferometer with a single scatter plate.

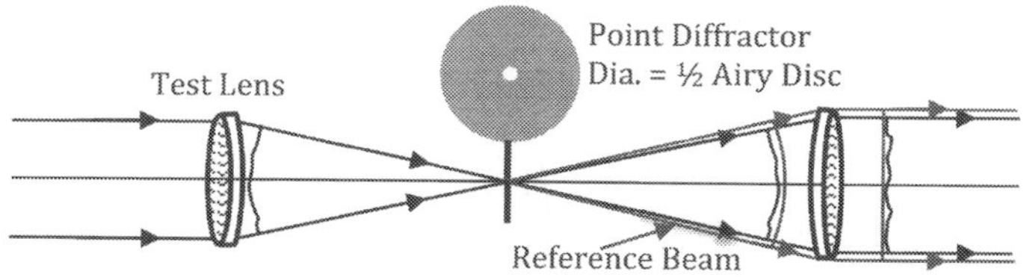

FIGURE 12.5 Schematic of a point diffraction interferometer.

Linnik first described it nearly 85 years ago. It consists of a coated transparent sheet/thin plate with an average transmission of a couple of percent. It may have either a hole or a fully transparent aperture of roughly half of the Airy disc pertaining to the optics under test. An aberrated wave front converges on to this sheet. The aperture in the sheet diffracts a tiny portion of the wave front and thus generates a synthetic spherical reference wave as shown in Figure 12.5. The aberrated wave is transmitted through the sheet and thus emerges as aberrated spherical wave. The transmittance of the sheet is such that both these waves have roughly equal amplitudes thereby produce fringes of good contrast. Lateral shift of the point diffractor produces tilt fringes while the axial shift results in fringes due to defocus. The major disadvantage of this compact interferometer is that the amount of light in the reference beam depends on the position of the pinhole; lateral shift of the pinhole results in rapid decrease of contrast of the fringes.

A point diffraction interferometer can also be incorporated into a microscope and interference microscopy can be performed with a point diffraction interferometer. As this is a common path interferometer, the reference and the object waves cannot be isolated and hence direct phase-shifting methods cannot be employed. Two indirect methods have been employed for phase shifting: one uses gratings, whereas the other is based on polarization. Several variants of this interferometer have been developed and used.

12.2.3 Laser Unequal Path Interferometer (LUPI)

Until the advent of the laser, large optics could be tested either with a scatter plate or a point diffraction interferometer. The laser provided a long coherence length source. Because the coherence length of a laser can be very long, long radii surfaces could be tested provided the path difference between the object beam and the reference beam is well within the coherence length. This results in two arms of the interferometer being quite unequal and hence the name laser unequal path interferometer (LUPI). Therefore, a suitable setup of the TG interferometer for testing a long radius concave surface is shown in Figure 12.6.

A beam from the laser is collimated and then split by the beam splitter *BS*. One beam travels to the reference mirror M_1 while the other beam passes through the diverger, which produces a good spherical wave. This wave illuminates the test surface, which retro-reflects the beam. This beam is aberrated due to the aberrations of the surface. Passage through the diverger results in an aberrated plane wave. The reflected collimated beam from mirror M_1 is now combined with the aberrated

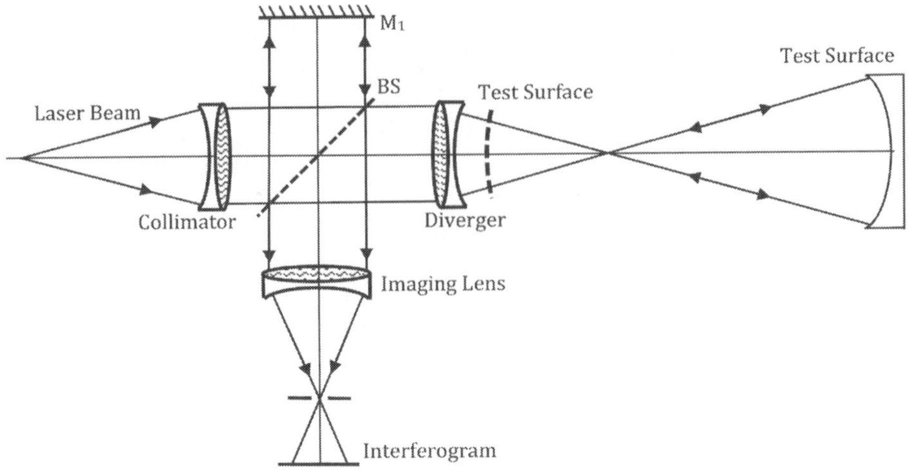

FIGURE 12.6 Laser unequal path TG interferometer for testing a concave mirror.

collimated beam by the beam splitter. These two beams interfere. The resulting interferogram is analyzed to characterize the surface. This procedure uses the well-known confocal position for testing a concave surface. The surface figure of convex surface/mirror, as shown by a dotted line, can also be evaluated by placing it in the cat's-eye position.

A good feature of LUPI is that the two arms of the interferometer can be unequal as long as the coherence requirement is met. The influence of vibrations and temperature variation could be serious as the two beams travel orthogonal paths, and for long radius optics the interferometer and test optics are widely separated.

When tabletop computer power became available, reference mirror M_1 was placed on a PZT so that three or more phase-shifted interferograms could be captured on CCD and processed to give the phase profile or surface errors.

12.2.4 Fizeau Interferometer

The Fizeau interferometer can be assembled in a number of ways. Figure 12.7 shows a configuration to test a concave surface where the reference beam is generated by a spherical surface: reflection from the convex surface generates the reference. Instead of a spherical reference surface, one can employ a flat reference surface to be suitably placed. If a high reflectivity surface (mirror) is to be tested, an attenuator can be placed in thc path of the test beam so that high-contrast fringes are obtained.

Alternately, one can use a partially coated spherical reference surface thereby resorting to multiple-beam interferometry. This results in sharper fringes, but the information between the fringes is not available unless one scans the surface. Further, a tilt between the reference beam and the test beam may lead to walk-off and produce some unsymmetrical features.

The Fizeau interferometer, like the TG interferometer, can be used to test a variety of optical components: it is as versatile as the TG interferometer but comparatively

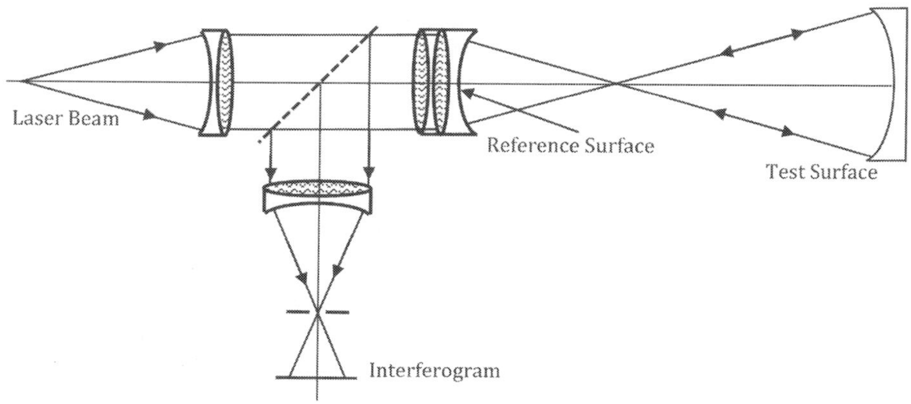

FIGURE 12.7 Schematic of a Fizeau interferometer to test a concave surface.

less expensive and less sensitive to vibrations. On the other hand, it is less flexible and less light efficient than the TG interferometer.

12.2.5 Shack Cube Interferometer

The heart of this interferometer is a beam-splitting cube as shown in Figure 12.8. On one side of the cube is a cemented a plano-convex lens whose spherical surface is of high quality; the radius of curvature of the spherical surface is such that its center of curvature is either on the opposite face or very near to it. On the perpendicular face is placed a spatial filter, *SF*, to which the laser beam is focused. The laser beam, after passage through the spatial filter, is reflected to the spherical surface, and part of the beam is reflected by the spherical surface, which focuses on the rear face of the cube. The part transmitted through the surface falls over the test surface. If its center of curvature coincides with that of the spherical reference surface, the beam retraces the path. The cube with the cemented plano-convex lens may be considered a thick lens, which images the test surface at I_m. The fringes formed at this plane are seen by the eyepiece *E*.

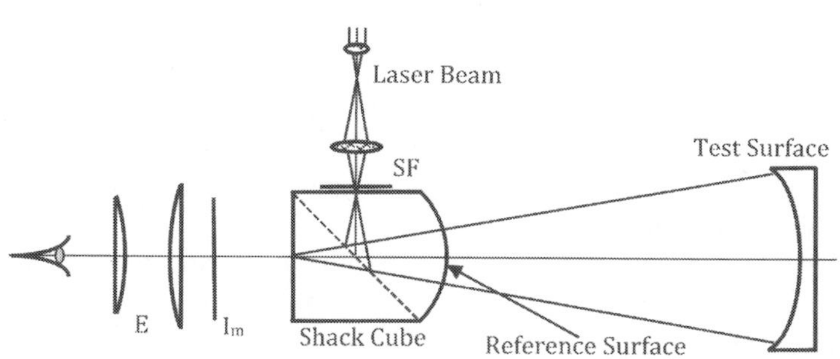

FIGURE 12.8 Shack cube for testing a concave mirror.

The Shack cube interferometer is quite compact but is less flexible than the TG interferometer or the Fizeau interferometer. Because beams share the same path inside the cube, the interferometer can be of average commercial quality and hence is less expensive. It has only one high-quality surface. However, to test a lens or a convex surface, auxiliary optics is required.

12.3 TESTING OF ASPHERICAL SURFACES

Aspherical surfaces are used in optical instruments as their usage results in the reduction of a number of components and hence lesser weight, better image quality, and lower cost. A conic may be considered a special case of an aspherical surface. A conic is a surface of revolution defined, in the Cartesian coordinate system with origin at the vertex, by the equation

$$x^2 + y^2 - 2Rz + (k+1)z^2 = 0 \tag{12.5}$$

where z is the axis of revolution and R is the radius of curvature at the vertex. The constant k ($k \neq 0$) is the conic constant: $k = -1$ for a paraboloid surface, $k > -1$ for an ellipsoid surface, and $k < -1$ for a hyperboloid surface.

The sag of the conic surface is given by

$$z = \frac{(x^2 + y^2)/R}{1 + \sqrt{\left\{1 - (k+1)\left[(x^2 + y^2)/R^2\right]\right\}}} \tag{12.6}$$

When we consider an aspherical surface, its sag can be expressed as

$$z = \frac{(x^2 + y^2)/R}{1 + \sqrt{\left\{1 - (k+1)\left[(x^2 + y^2)/R^2\right]\right\}}} + A_4\left(x^2 + y^2\right)^2 + A_6\left(x^2 + y^2\right)^3 + \cdots\ldots \tag{12.7}$$

where A's are aspheric coefficients. It is thus seen that an aspheric surface departs considerably from the reference spherical surface of radius of curvature R. This results in the crowding of fringes, sometimes too many to be resolved. Obviously, the testing of aspherical surfaces poses problems. These problems, however, can be tackled using one of the following techniques:

Use of null optics: Null optics can be either conventional or computer generated. Even a hologram of the tested wave front could be used as null optics.

Use of shearing interferometry: This gives slope fringes and hence the number of fringes gets considerably reduced.

Use of long wavelength interferometry: This automatically reduces the number of fringes. But the wavelength may be in a region where the detectors are difficult to find. Alternately, the synthetic wavelength could be generated using two wavelengths. Depending on the choice of wavelengths, a very wide variation of synthetic wavelength is possible.

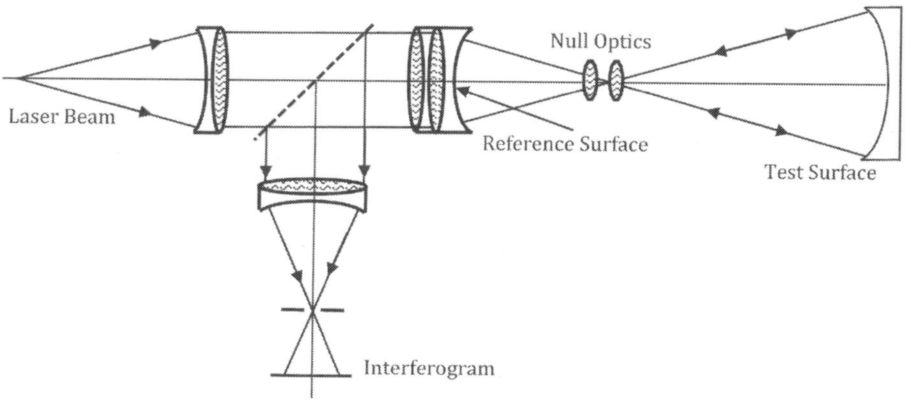

FIGURE 12.9 Schematic of a Fizeau interferometer to test an aspherical surface.

Sub-Nyquist interferometry (SNI): This is applied to testing aspherics with a simple assumption that the surface under test is smooth and continuous and has continuous derivatives. This assumption allows the analysis to interpret fringes that occur at frequencies well above the Nyquist frequency. Surfaces with departures of several hundreds of waves from a reference sphere have been measured using sub-Nyquist interferometry.

Figure 12.9 shows a Fizeau interferometer in which an aspherical surface is to be tested. The reference surface is spherical. Null optics is introduced in the path such that the wave front from the test surface is converted to a spherical one, which is then compared with the spherical reference. The interferogram is captured at the image of the pupil.

12.3.1 Null Test with a Computer-Generated Hologram

Making a conventional null optics is a very expensive and time-consuming process. Given the design data of the test surface/optics, it is far easier to make a computer-generated hologram (CGH). The CGH is generated usually with a plane reference beam and the test beam at the pupil plane. When reconstructed with the test beam, it generates several beams and one of them is a plane beam. The plane beam is isolated by spatial filtering. However, if the test surface departs from the designed one, the reconstruction creates a distorted plane beam, which is examined with a plane reference beam.

Figure 12.10 shows the schematic arrangement of a TG interferometer. The DC transmittance of CGH will allow the plane reference beam through while the test beam is diffracted into various beams; one of them is a plane beam, which is filtered by the spatial filter. The two plane beams interfere to produce an interference pattern. If the test beam departs from the true surface, the diffracted plane beam will be distorted. The fringes then depict the departure from the true form.

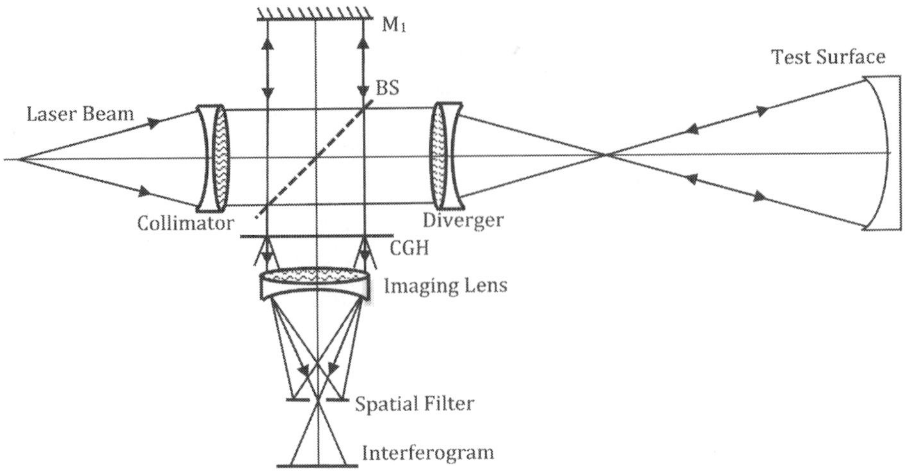

FIGURE 12.10 Schematic of a TG interferometer to test an aspherical surface with a CGH.

Because the CGH is mechanically produced, there are errors due to the printer, distortion, etc., in the reduction process. Alternately, we may record a hologram of the test beam at the pupil plane with a plane reference beam. After processing, the hologram is placed at a location occupied by the CGH as in Figure 12.10. It will now generate several beams when reconstructed with the test beam; one of them will be the plane beam. The process is identical to that in a CGH.

12.4 OBLIQUE INCIDENCE INTERFEROMETER

There are some practical applications in which the surfaces depart from flatness by several tens of wavelengths. The technique is, therefore, to be desensitized so that a reasonable number of fringes is observed. Use of a longer wavelength is one such desensitizing technique. Another technique is to use oblique incidence so that the height $h(x,y)$ is seen as $h(x,y)\cos\theta$, where θ is the angle of incidence. Interferometric evaluation of rough surfaces can also be done with an oblique incidence interferometer.

Figure 12.11(a) shows a schematic of the interferometer. A pair of identical linear gratings is used for the dual function of beam splitting and beam combining. Grating G_1 generates the diffracted beams: the first-order diffracted (−1 order) beam is incident on the test surface at an angle θ. A grating can be designed that would give only direct and first-order beams. Grating G_2 diffracts both the direct beam and oblique beam from the substrate. The direct beam through grating G_2 and first-order diffracted (+1 order) beam are used for observing interference.

The path difference between the interfering beams, apart from a constant path, is expressed as

$$\Delta = 2h(x,y)\cos\theta \tag{12.8}$$

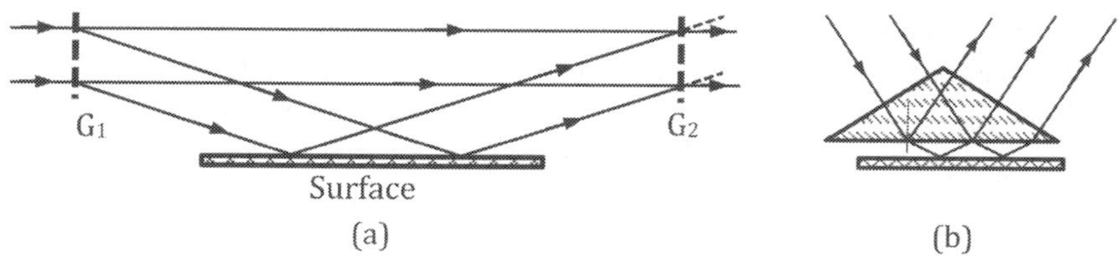

FIGURE 12.11 Schematic of an oblique incidence interferometer with (a) gratings and (b) a prism.

where $h(x,y)$ represents the height at a point (x,y) on the surface. The successive fringes differ in height variation of $\bar{h}$, where

$$\bar{h} = \frac{\lambda}{2\cos\theta} \tag{12.9}$$

Instead of height variation of the half-wavelength in the Fizeau interferometer, the oblique incidence interferometer has been desensitized by $\cos\theta$. Alternately, it is equivalent to using a wavelength of $\lambda/\cos\theta$.

The interferometer can be made very compact using a prism as shown in Figure 12.11(b). The base surface of the prism provides the reference beam. Also, the perspective error is smaller compared with the grating interferometer.

12.5 SHEAR INTERFEROMETRY

Shear interferometry does not require a reference beam but compares a test beam with its sheared replica. There are many types of shears, but the most important and most often used is lateral shear. Here the test wave front with wave front deformation $W(x,y)$ is compared with its laterally displaced replica $W(x+S, y)$ as shown in Figure 12.12, and the interference takes place between the overlapped beams. S is the shear.

The optical path difference between the two wave fronts is given by

$$OPD = W(x+S,y) - W(x,y) = \frac{\partial W}{\partial x} S + \cdots \tag{12.10}$$

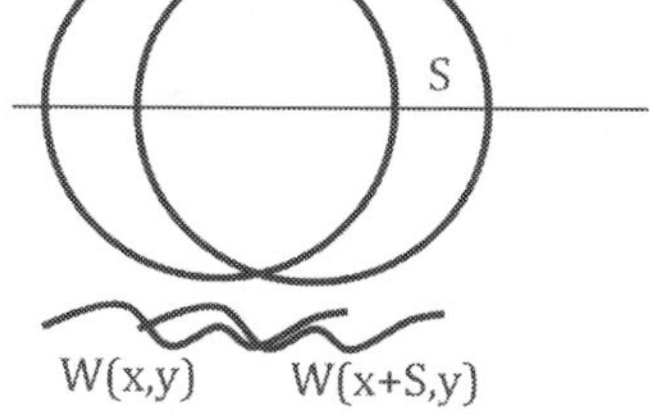

FIGURE 12.12 Sheared pupils.

The term $\frac{\partial W}{\partial x}$ represents the gradient or the slope of the wave front deformation. The *OPD* depends on the derivative of the wave front deformation $W(x,y)$ or slope rather than on the wave front deformation. The fringe formation is governed by the equation

$$\frac{\partial W}{\partial x} S = m\lambda \tag{12.11}$$

This gradient $\frac{\partial W}{\partial x}$ is proportional to the component of transverse aberration TA_x in the direction of shear. In fact it is equal to $\frac{TA_x}{r}$ $\left(\frac{\partial W}{\partial x} = \frac{TA_x}{r}\right)$, where r is the radius of curvature of the reference wave front, or the distance from the vertex of the wave front to the plane where transverse aberration is measured, if this plane is not at the center of curvature of the wave front. Therefore, the lateral shear interferometer measures the transverse aberration TA_x in the direction of shear.

Equation (12.11) can be rewritten as

$$\frac{\partial W}{\partial x} S = \frac{TA_x}{r} S = m\lambda \tag{12.12}$$

This greatly reduces the number of fringes in the interferogram. One of the simplest shear interferometers is Murty's plane parallel plate interferometer where the shear is introduced by the thickness of the plane parallel plate as shown in Figure 12.13. Shear S is given by

$$S = \frac{t \sin 2\theta_i}{\sqrt{\mu^2 - \sin^2\theta_i}} \tag{12.13}$$

where θ_i is the angle of incidence, and t and μ are the thickness and the refractive index of the plane parallel plate. Usually, the plate is used at an angle of incidence of 45°. The shear then is given by $S = \sqrt{2}\, t \left(2\mu^2 - 1\right)^{-1/2}$.

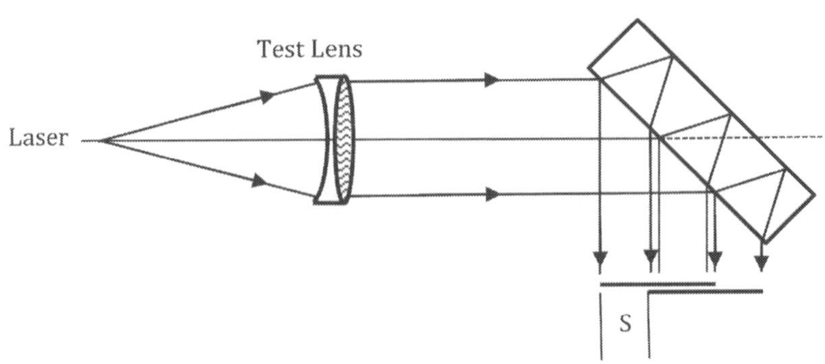

FIGURE 12.13 Schematic of Murty's plane parallel plate shear interferometer.

12.6 LONG WAVELENGTH INTERFEROMETRY

The use of long wavelength in interferometry reduces the number of fringes. A carbon dioxide laser (λ = 10.6 μm) can be used as a source because it has excellent coherence properties. Zinc selenide and germanium could be used to make transmission optics like a reference sphere, and pyroelectric vidicon can be used as a detector. Techniques used with conventional interferometry are applicable at this wavelength as well. The reduction in fringes would be in the ratio of wavelengths, i.e., $\lambda_{CO_2} / \lambda_{visible}$, which turns out to be approximately 16.7 when a He-Ne laser wavelength of 632.8 nm is taken for $\lambda_{visible}$.

12.6.1 Dual-Wavelength Interferometry

It is also possible to generate a long wavelength using two wavelengths. If the wavelengths of two waves used for testing are λ_1 and λ_2, then the synthetic wavelength λ_s is given by $\lambda_s = \frac{\lambda_1\lambda_2}{|\lambda_1-\lambda_2|} = \frac{\bar{\lambda}^2}{\Delta\lambda}$. Therefore, a very large synthetic wavelength can be generated if these two wavelengths are very close to each other. In principle, each wavelength produces its own interference pattern when the reference and the test wave fronts interfere. The moiré between these interference patterns is the fringe pattern that is generated by the synthetic wavelength.

A variant of this, called two-wavelength holography by some researchers, uses Gabor in-line geometry. It uses a collimated reference beam and applies to both transmission and opaque objects. As an example, consider a lens as an object on which a collimated beam is incident. A hologram is recorded at a plane between the focal plane and the lens by adding a collimated beam axially. First exposure is done with light of wavelength λ_1 and the second exposure is done with light of wavelength λ_2. This double-exposure hologram when reconstructed with light of either wavelength will generate fringes that will be due to the aberrations of the lens. The fringes are due to the synthetic wavelength $\lambda_1\lambda_2/|\lambda_1 - \lambda_2|$.

Assuming the amplitudes of both waves are equal, the resultant amplitude is expressed as

$$A(x,y) = a_0 e^{\frac{2\pi i}{\lambda_1}W(x,y)} + a_0 e^{i\delta_0} \tag{12.14}$$

where $W(x,y)$ is the path function representing axial aberrations of a lens and δ_0 is a constant phase of the collimated reference wave. The irradiance recorded can be expressed as

$$I_1(x,y) = 2I_0\left[1+\cos\left(\frac{2\pi}{\lambda_1}W(x,y)+\delta_0\right)\right] \tag{12.15a}$$

When a second recording using the light of wavelength λ_2 is carried out, the irradiance distribution can be expressed as

$$I_2(x,y) = 2I_0\left[1+\cos\left(\frac{2\pi}{\lambda_2}W(x,y)+\delta_0\right)\right] \tag{12.15b}$$

The resultant irradiance distribution is

$$I(x,y) = I_1(x,y) + I_2(x,y) = 2I_0\left[2 + \cos\left(\frac{2\pi}{\lambda_1}W(x,y) + \delta_0\right) + \cos\left(\frac{2\pi}{\lambda_2}W(x,y) + \delta_0\right)\right]$$

$$I(x,y) = 4I_0\left\{1 + \cos\left[\pi\left(\frac{1}{\lambda_1} + \frac{1}{\lambda_2}\right)W(x,y) + \delta_0\right]\cos\left[\pi\left(\frac{1}{\lambda_1} - \frac{1}{\lambda_2}\right)W(x,y)\right]\right\}$$

$$I(x,y) = 4I_0\left\{1 + \cos\left[\pi\left(\frac{|\lambda_1 + \lambda_2|}{\lambda_1\lambda_2}\right)W(x,y) + \delta_0\right]\cos\left[\pi\left(\frac{|\lambda_1 - \lambda_2|}{\lambda_1\lambda_2}\right)W(x,y)\right]\right\} \quad (12.16)$$

The term $\cos\left[\pi\left(\frac{|\lambda_1-\lambda_2|}{\lambda_1\lambda_2}\right)W(x,y)\right]$ generates fringes that are due to the aberration function $W(x,y)$.

12.7 METROLOGY TECHNIQUES FOR ASPHERIC AND FREE-FORM OPTICS

The free-form surfaces differ from conventional surfaces due to loss of rotational symmetry. The complexity of the surface for the manufacturing increases due to the deviation from the best-fit sphere. The manufacturing of free-form optical surfaces with better accuracy depends on suitable metrology feedback. However, no suitable standard and traceable method of measurement of complex surfaces has yet been established. The available metrology techniques are broadly placed into three categories: single point probe-based techniques, noninterferometric (slope- or curvature-based) methods, and interferometry-based techniques. Each metrology technique has its inherent advantages and limitations, which are briefly described in the following.

Point probe-based measurement techniques are used for the coarse measurement of complex shapes and provide the global shape measurement of the surface by using a touch probe. These techniques are categorized mainly into the following types based on their resolution.

A coordinate measurement machine (CMM) is one of the most popular metrology tools for the measurement of free-form optics. It is a touch probe-based method that typically has a ruby sphere attached to the scanning head with a constant force ~ 0.3 N and an accuracy of approximately 1 μm. The measurement can be done either by discrete probing or by scanning continuously as the stylus tip scans across the surface. The advantage of the CMM is that it can measure the surface relative to the datum fiducial features of the surface part. The main drawback of the system is its low resolution and large measurement time to capture high-density data.

The mechanical profilers are tactile-based probes made of ruby or diamond with a cone angle of 60° or 90° and a tip radius in the range of 1 μm to 10 μm. This stylus makes physical contact with the surface under test to measure the statistical properties of smooth optical surfaces. The advantages are surface independence, and high resolution, i.e., the stylus tip can be as small as 20 nm, and vertical resolution is typically sub-nanometer. The spherical and aspherical components can be measured with high-resolution data up to the order of 10 nm rms (i.e., stylus radius dependent). The measurement method is limited by the surface slope and susceptible to slip error and lateral force error for off-axis measurement of high departure free-form surfaces. This method has another disadvantage: it can damage the surface under test and have difficulty aligning the free-form test surface during the measurement.

The noncontact optical probe techniques are noninterferometric point-based sensors used for precision metrology. MarForm MFU200 from Mahr GmbH, Germany, and LUPHOScan from Taylor Hobson, UK, are such point-based sensors where the optical probe can be placed normally to the surface during measurement using two linear stages and one rotary stage. The sensors are developed to measure free-form shapes, but the measurement uncertainties are larger for the measurement of high-slope surfaces. Recently, a noncontact point cloud metrology technique based on swept source optical coherence tomography has been developed for the measurement of free-form optics. It is coupled with an interferometric system with multiple scanning stages. However, the available point-based sensors require multiple measurements (points or line scans) to cover the entire surface, which requires longer measurement or computational time.

The noncontact optical profilers use an optical beam to probe the surface by using the principle of phase-shifting interferometry. The optical profiler makes use of a high-resolution microscope objective to record the height variation of the surface. It measures a small area of the test surface; however, a full surface profile of spherical, aspherical, and free-form surfaces is possible by measuring multiple small apertures that are stitched together to get a complete three-dimensional (3D) profile. Extracting the complete topography of a surface using these methods becomes very time-consuming and hence is not suitable for surfaces with high slopes.

The slope measurement techniques are very useful for the metrology of asymmetrical optical surfaces such as conics, off-axis geometries, and free-form surfaces. These surfaces have a variation in slope over their geometry. Noninterferometric slope measurements are suitable methods for surface figure metrology. Some of the important techniques that are being used for free-form metrology are briefly described in the following.

Phase-measuring deflectometry (PMD), also known as the fringe reflection measurement technique, is used to measure the local slope of the optical surface under test by using phase information. PMD is a noncontact, less sensitive to external vibrations and temperature instabilities, low cost, and robust technique for the measurement of free-form optics. The local slopes can be calculated from the phase shift of the deviated fringes, and then the wave front shape is obtained by integrating the surface slopes. This technique is primarily applied for testing the local curvature of the free-forms of automobiles optics and progressive eyeglasses.

The Shack–Hartmann sensor (SHS) is one of the most well-established and powerful approaches to wave front sensing for large wave fronts and has been reported recently for free-form surface measurement. The micro lens array in front of the CCD camera is arranged to capture the focus spots of the small samples of the wave fronts. The local slopes of the wave front can be measured as a lateral shift of the focus position and are integrated to reconstruct the wave front. The small sub-apertures are measured by scanning the SHS either in raster or spiral mode; the misalignment inaccuracies are minimized and stitched together to get the full free-form wave front. The sensor has the capability to be integrated into the machine axis for in situ measurement during the polishing process. The advantages of the SHS-based measurement include that it is less sensitive to vibration, compact, wavelength independent, easy to operate, and provides a 3D profile map from the slope data. Even with all these advantages, this technique has some limitations, like low resolution, alignment inaccuracies, and a long measurement time.

The interferometric measurement of aspherical and free-form surfaces is difficult due to the nonavailability of the null corrector required for their test. These null correctors are test surface specific, and their development is time-consuming and costly. Therefore, non-null interferometric methods are suitable for the measurement of such surfaces. A brief description of these methods is given in the following.

The sub-aperture stitching technique was primarily developed to overcome the aperture size limitations of conventional interferometry. As the size of the optics increases, stitching interferometry becomes more promising compared with the expense and fabrication time of large aperture interferometers and transmission spheres (TSs). In addition, the technique overcomes the metrology challenges involved in high steep slopes and high departure surfaces in non–null-based interferometric configurations. In the recent past, research work has been reported for the measurement of free-form surfaces using noninterferometric stitching techniques using SHS.

The well-established interferometry-based high-precision metrology of conventional optics to get the full surface information, i.e., global and local, are widely used. It measures the relative height between the test surface and the reference surface of a standard TS. It is a very useful testing method where the surface under test is compared with a standard reference surface. Interferometric techniques have an additional advantage in that they measure large areas with high lateral resolution, gathering up to hundreds of thousands of data points per measurement. However, for the metrology of aspheric and free-form surfaces, there is a need for specific reference or auxiliary optics.

In free-form surface profiles the measurement of high slopes with conventional interferometric configurations yields errors such as retracing error, high fringe data density, complicated interferogram, and noise/pixilated interferogram. These issues can be solved by incorporating customized auxiliary null corrector optics or stitching techniques. For testing free-form surfaces with interferometric methods, a refractive or diffractive null corrector, or a combination of both, is incorporated into the standard interferometric configurations.

The null correctors are specially designed groups of optical elements that are placed in the interferometer cavity as per the test plan to compensate for the deviation of the free-form profile with respect to the best-fit spherical or plane wave front. It generates the shape of the wave front from the incoming free-form wave front as the one that matches the surface under test to get the null fringes. The wave front is retraced through the common path after being reflected from the test surface. If the surface under test is perfect, then a null interferogram is obtained; otherwise, the fringe pattern is analyzed for wave front deviation. The null corrector design is test wave front specific, making it expensive to fabricate. The null corrector is classified as conventional, reflective, or refractive optics and holographic null corrector (CGH or diffractive optical null). A diffractive optical element or a CGH provides the compensating wave front. These CGH have been used in interferometric testing of aspheric surfaces.

Long wavelength interferometry is of interest for several reasons. First, it makes it possible to test infrared (IR) transmitting materials as it uses an IR source (wavelength = 3.8 μm). Second, as the deviation from the best-fit reference sphere (in units of the probing wavelength) is reduced, long wavelength interferometry is useful in obtaining interferograms of deep surfaces and even ground surfaces to measure higher slopes than a He-Ne-based interferometer. The most common source used is a laser diode with a wavelength of 3.8 μm and another option is a CO_2 interferometer working at 10.6 μm.

Self-referencing interferometry is a very important class of interferometry in which the wave front under study is slightly displaced by a small amount and the interference fringes are obtained between the original reference beam and the shifted reference beam. This is the well-known shear interferometry, which responds to the derivative of the path function. There are various types of shearing, viz., lateral shearing, radial shearing, and rotational shearing. When a convergent beam is incident normally on a transmission-type diffraction grating at a point on the grating plane, a diffraction pattern is generated. A double-frequency grating was devised by Wyant to implement a small amount of shear. This is accomplished by using a grating with two distinct frequencies recorded on the same substrate. Shear in the tangential and sagittal directions is generated simultaneously by inserting two gratings. If the double-frequency gratings are written holographically on the photographic plate, it is known as a holo-shear lens. The holo-shear lens has been utilized to measure the aberrations of the spherical lenses and test the off-axis parabola. The in situ measurement scheme using a lateral shear interferometer is developed and demonstrated for the shape measurement of parabolic mirrors. A multifrequency spatial-carrier and phase-shifting fringe-projection system based on two-wavelength lateral shearing interferometry is used for the measurement of the 3D shape of a discontinuous object.

12.7.1 Tilted Wave Interferometry

The tilted wave interferometer (TWI) was invented by a research team from the Institute for Applied Optics at the University of Stuttgart, Germany. It is a flexible tool for the high precision and fast measurement of aspheres and free-forms. It combines the high accuracy and traceability of the interferometric principle with a high dynamic range of up to 10° gradient deviation from the spherical form, without the need for compensation optics and moving parts. Furthermore, like all full-field interferometric methods, it has a high lateral resolution. An outstanding feature is its short measurement time, which is achieved by a highly parallelized data acquisition. In principle, it is a single-shot technique that allows the robust measurement of the entire surface in few seconds with a height resolution better than 1 nm and a lateral resolution of several microns.

Figure 12.14 shows a schematic of a tilted wave front interferometer; the tilted wave fronts are realized with the help of a matrix of point sources, which consist of a specially designed micro lens array (*PSA*) that is followed by a movable pinhole array (*MA*). The spherical wave front from each point source, after passing the beam splitter (BS_1), is collimated by lens L_2 resulting in a set of plane wave fronts with different well-defined amounts of tilt. The *tilted wave fronts* are transformed to spherical wave fronts by the objective lens (*OBJ*) to compensate for the basic spherical form of the test surface. After reflection at the test surface, the wave fronts propagate back to the beam splitter (BS_2), where they are reflected to the camera arm of the interferometer. Here the wave fronts coming from different parts of the test surface interfere with the reference wave, which is propagating along the reference arm. The interference pattern is recorded at the chip of the camera. An aperture (*AP*) at the Fourier plane blocks all the light that would generate fringes with a density that violates the Nyquist criterion. In the case of a strong non-null test, this limits the lateral extension of an interferogram (sub-interferogram or patch) that is produced by one single wave front from the *PSA* on the chip.

There are multiple patches generated by the different point sources across the test surface. To measure the whole test surface, the patches must cover the whole clear aperture (CA) of the test surface. In practice, an overlap of the sub-interferograms (patches) from different point sources occurs at least for some of the patches in practice. In order to overcome this, every second row and every second column of the array of point sources is blocked by a movable mask array (*MA*), leading

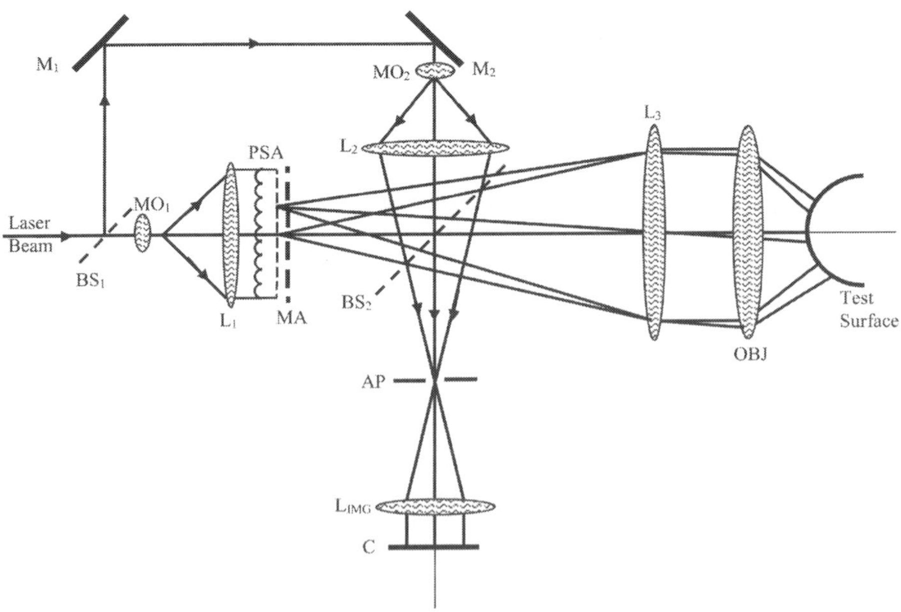

FIGURE 12.14 A schematic of titled wave interferometer. BS_1 and BS_2, beam splitters; M_1 and M_2, mirrors; MO_1 and MO_2, mirrors; L_1, L_2, and L_3, lenses; *PSA*, point source array; *MA*, movable aperture array; *AP*, aperture; L_{IMG}, imaging lens; *C*, camera; *OBJ*, objective.

to measurement data without overlap/cross talk, but with the requirement that the *MA* needs to be shifted subsequently into four different positions. The interferometer uses a very powerful calibration procedure to provide the requisite accuracy of measurement.

QUESTIONS

12.1 The wedge angle of a wedge plate of refractive index 1.5153 is to be measured with a TG interferometer equipped with mercury lamp ($\lambda = 546.1$ nm). The measured fringe width is 5.3 mm. What is the wedge-angle? What would be the fringe width if this wedge plate is tested with the Fizeau arrangement?

12.2 Show that the sag of a conic surface defined by $x^2 + y^2 - 2Rz + (k+1)z^2 = 0$ is given by

$$z = \frac{(x^2 + y^2)/R}{1 + \sqrt{\{1 - (k+1)[(x^2 + y^2)/R^2]\}}}$$

12.3 For the plane parallel plate interferometer find the angle of incidence for which the shear is maximum. Assume the refractive index and thickness of the plate as μ and t, respectively.

12.4 Figure 12.15 gives an interferogram obtained from a Fizeau interferometer at a wavelength of 632.8 nm. Obtain the profile of the optical path difference across the center of the interferogram with and without the tilt. Is it possible to determine the sign of error?

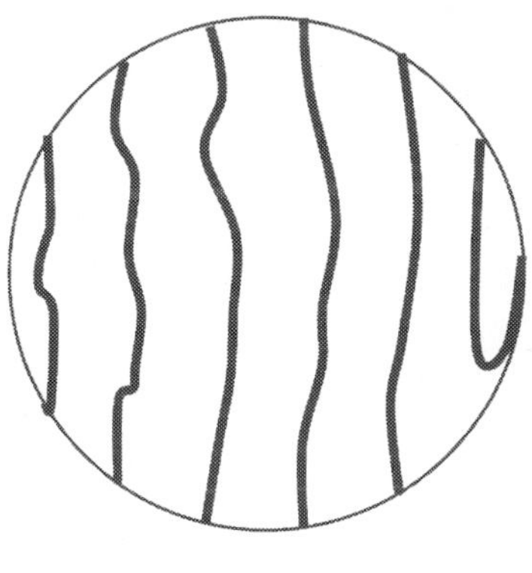

FIGURE 12.15

12.5 A TG interferometer is used to test a mirror at an angle of incidence of 45° with radiation from a He-Ne laser (632.8 nm) as shown in Figure 12.16(a). The recorded interferogram is given in Figure 12.16(b). What is the peak/valley error in micrometers of the mirror surface? When the reference mirror is translated toward the beam splitter, the fringes move to the right in the interferogram. Is the center of mirror a valley or hill?

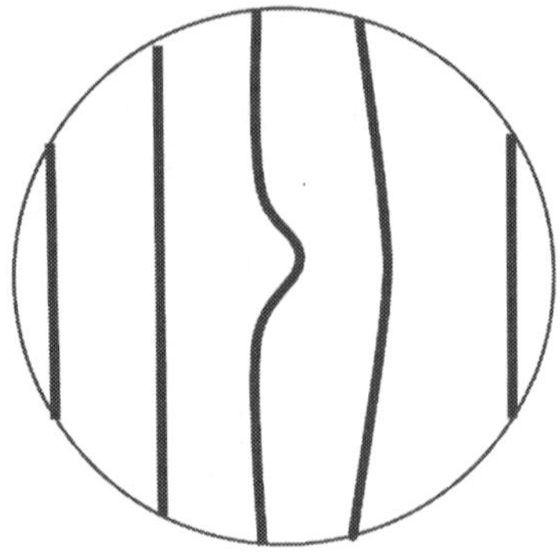

FIGURE 12.16

12.6 An interferogram as shown in Figure 12.17 of a sample is obtained with a two-beam interferometer using a He-Ne laser (632.8 nm).

a. If the sample is a flat mirror, what is the departure from flatness if (1) it is tested at normal incidence and (2) at an angle of incidence of 60° in a TG interferometer? Sketch the interferometer setup. (3) When the sample is tested in a Mach–Zehnder interferometer by using it as one of its mirrors. Sketch the interferometer setup.
b. If the sample is a plane parallel plate of refractive index 1.515, what is the thickness variation when it is tested in (1) a TG interferometer and (2) in a Mach–Zehnder interferometer.

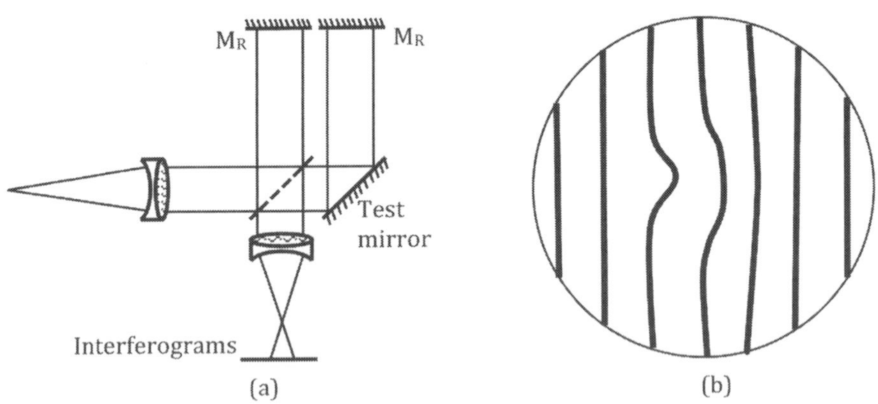

FIGURE 12.17

12.7 Figure 12.18 gives three interferograms obtained by testing a nearly spherical mirror in a Fizeau interferometer using a He-Ne laser (632.8 nm).

a. Identify the aberration present.
b. What is peak to valley surface height error in units of micrometers? State any assumptions that you may like to make.

c. How was the interferometer adjusted in going from interferogram to the left to the interferogram to the right?
d. Describe the motion of the fringes in the middle of the interferogram as you move the mirror closer to the reference surface. State any assumptions that you made.

FIGURE 12.18

12.8 Five interferograms given in Figure 12.19 are obtained when testing a mirror in a TG interferometer. Calculate the number of waves of astigmatism, focus, x-tilt, and y-tilt for each of these interferograms.

a. Dark fringe at top and bottom
b. Dark fringe at left and right
c. Dark fringe at top and bottom
d. Dark fringe at top and bottom
e. Dark fringe at top and bottom and bright fringe at left and right

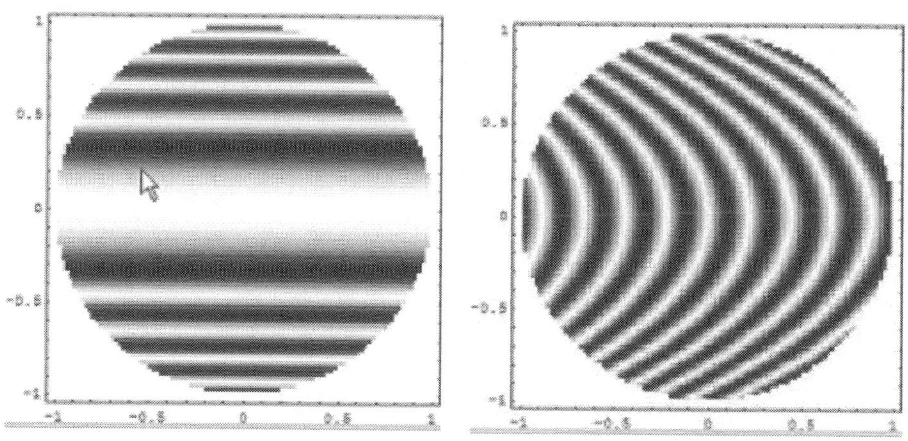

(a) (b)

FIGURE 12.19 *(Continued)*

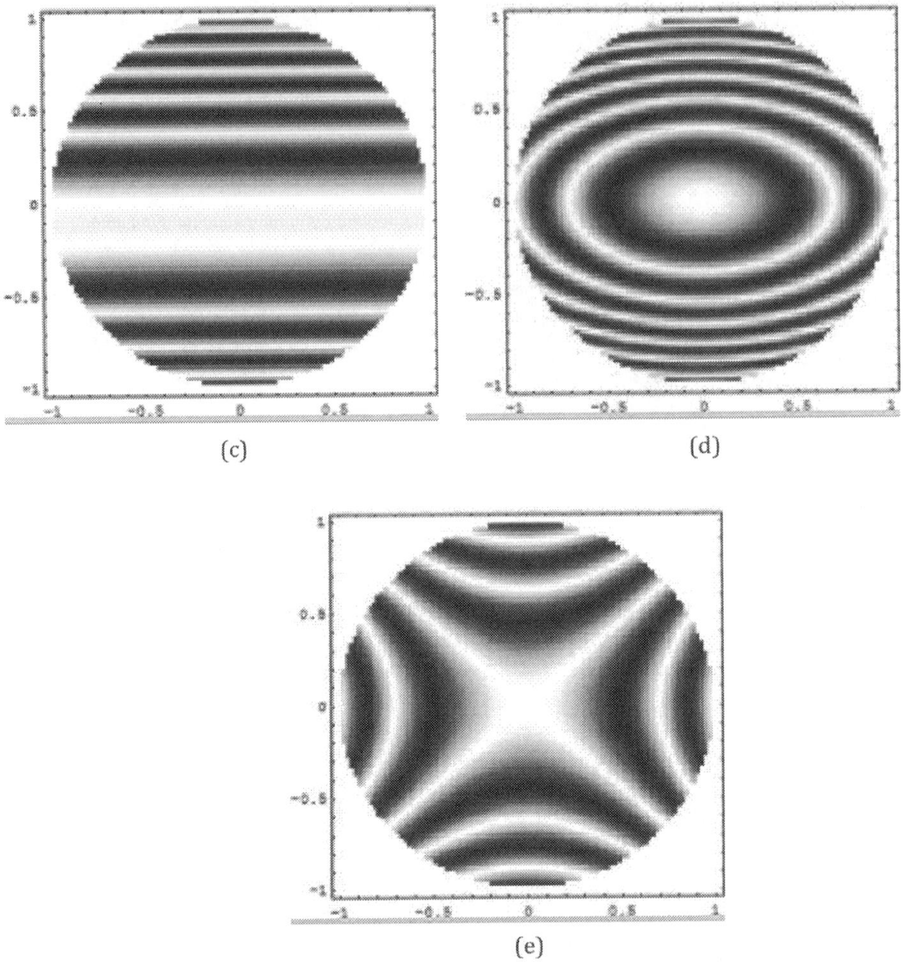

FIGURE 12.19

13 Angle Measurement

13.1 DEFINITION OF AN ANGLE

The circumference of a circle makes an angle of 360° or 2π radians at its center. If the circumference is divided into 360 equal parts, each segment will make an angle of 1°. The angle is the ratio of circumference to the radius of the circle. Each degree is divided into 60 minutes of an arc and each minute is divided into 60 seconds of an arc. Therefore, the angle measurement is linked to the length and hence is traceable to the standard meter through various kinds of techniques and instrumentation. Angle gauges with different tolerances are used to transfer secondary length standards to the laboratories and industry. We will discuss optical methods to measure angles. These include autocollimator, goniometer, and interferometric methods.

13.2 AUTOCOLLIMATOR

Figure 13.1 is a schematic of an autocollimator, which essentially is a telescope.

Filtered radiation from a lamp illuminates a graticule (a cross wire), which is placed at the focal plane of a good objective lens. The beam exiting from the lens (objective) is thus collimated and falls on a reflecting object. If the object is perpendicular to the optical axis of the autocollimator, the beam retraces its path and falls on a reticle properly centered, which is also located at the focal plane of the objective. An observer sees the image of the reticle. Alternately, an array detector could be used to sense the position of the image. When the object is not perpendicular to the optical axis but tilted by an angle α, the reflected beam deviates by 2α and the image of the graticule shifts by Δy ($=2f\alpha$) at the reticle. The reticle is graduated in terms of angles and hence the autocollimator gives direct angle readings. The range of the instrument is limited but it has very high accuracy (~0.1 arcsec or ~0.1"). It is, therefore, used either for differential measurement or for setting purposes.

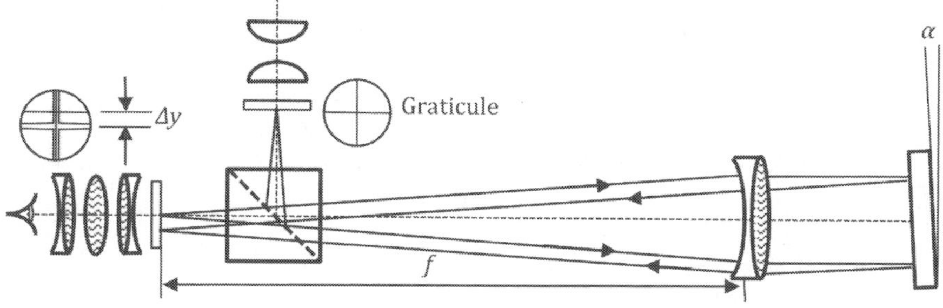

FIGURE 13.1 Schematic of a collimator.

DOI: 10.1201/9781003531845-13

13.2.1 Measurement of Angle of a Glass Wedge

Measurement of an angle of a glass wedge requires a pair of autocollimators that are symmetrically placed on the worktable as shown in Figure 13.2. The glass wedge of refractive index μ is placed on the table. Let the angle of incidence be θ_i. The beam is reflected from the upper surface of the wedge at angle θ_i while the beam reflected from the bottom surface emerges out with a slightly different angle β (= $\theta_i + \Delta\theta$) due to the wedge angle α. It can be shown using Snell's law that

$$\sin\beta = \mu \sin(\theta_r + 2\alpha) \tag{13.1}$$

where θ_r is the angle of refraction in the wedge. When α is small so that $\cos 2\alpha = 1$ and $\sin 2\alpha = 2\alpha$, we obtain

$$\alpha = \frac{\cos\theta_i}{\sqrt{\mu^2 - \sin^2\theta_i}} \frac{\Delta y}{2f} \tag{13.2}$$

Here $\frac{\Delta y}{2f}$ is directly read as an angle and angle α calculated using Equation (13.2) or Δy may be measured and the angle α calculated when f is given.

The measurement of the angle of a glass wedge can also be done with a single autocollimator. In this case, the autocollimator is aligned perpendicular to the upper surface, i.e., $\theta_i = 0$, and hence $\alpha = \frac{1}{\mu}\frac{\Delta y}{2f}$. In fact, in both the cases, the wedge plate is to be rotated in its plane such that there is deflection only along one direction. In other words, the wedge edge is perpendicular to the plane of incidence. Further, it is noted that the knowledge of the refractive index of the material of the wedge is required in order to determine the wedge angle. This may not be needed if the wedge is placed on a steel platen and the autocollimator is aligned so that the reflection from the platen nicely centers on the reticle. Now if the reflections from the platen and upper surface of the wedge are simultaneously viewed, the reflection from the wedge's upper surface will result in the shift of the cross wire; this shift gives the angle of the wedge. It may again be noted that if there is no shift of the cross wire ($\Delta y = 0$), then the two surfaces are parallel.

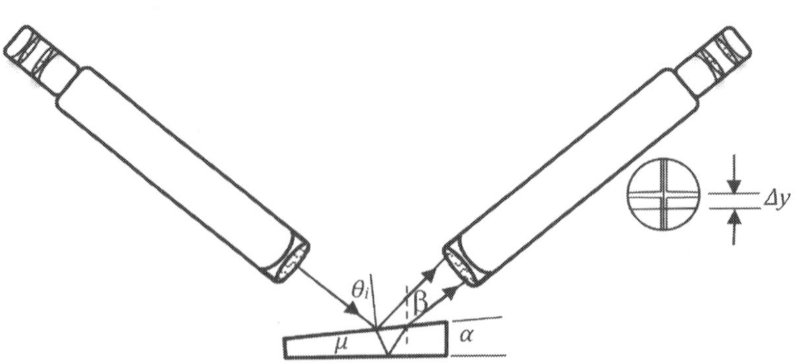

FIGURE 13.2 Experimental arrangement to measure angle of a glass wedge with autocollimators.

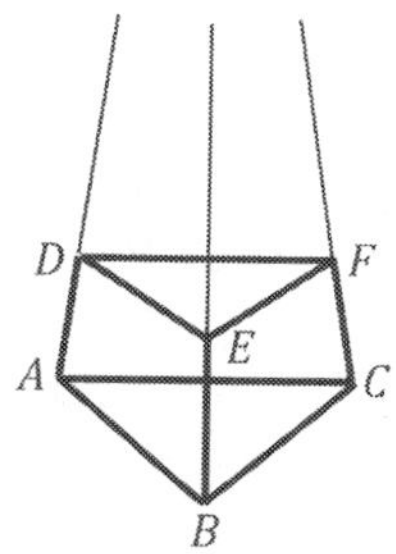

FIGURE 13.3 Representation of pyramidal error.

13.2.2 Angle of a Prism

Before we discuss the measurement of an angle of a prism, let us first discuss the pyramidal error. We consider a 60° prism, which is often used in the laboratory as shown in Figure 13.3. It has five faces; *ABC* and *DEF* are the bottom and top surfaces, which are generally ground. The other three surfaces should be perpendicular to these faces. If they are not, the prism has a pyramidal error; the image of the cross wire in the autocollimator will shift to one of the quadrants. We can measure this pyramidal error; however, we will assume the prisms are free from pyramidal error.

The angle of the prism is measured by comparing it with an angle standard. Both are placed side by side and both reflect the beam from the autocollimator. The shift between the two cross wire images will give the difference in angles.

13.2.3 Measurement of Error in 90° of a Right-Angle Prism

Figure 13.4 shows a right-angle prism with the right angle having an error of α. Either of the two 45° angles may have an error of $-\alpha$ or it may be distributed on both the angles. The ray path in the right-angle prism is shown in Figure 13.4(a). The beam from the autocollimator enters the front surface and is totally reflected at the

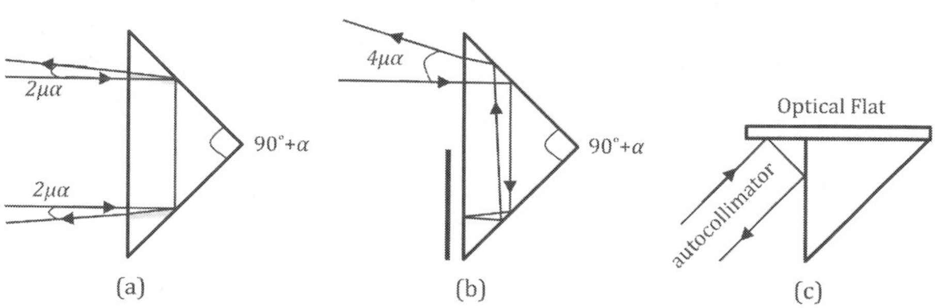

FIGURE 13.4 Measuring the error in the right angle of a right-angle prism (a) by retro-reflection, (b) by reflecting with a mirror, (c) using an optical flat.

two surfaces and emerges out angularly shifted by $2\mu\alpha$. The two images formed are angularly shifted by $4\mu\alpha$. If the shift measured in the reticle is Δy, then

$$\frac{\Delta y}{f} = 4\mu\alpha \Rightarrow \alpha = \frac{\Delta y}{4\mu f} \tag{13.3}$$

Alternately, we may block half of the diagonal face and get light reflected internally from this surface as shown in Figure 13.4(b). Further, when the right-angle prism is placed on an optical flat, we may also autocollimate and find the departure from right angle as shown in Figure 13.4(c).

13.2.4 Measurement of Error in 45° Angle of a Right-Angle Prism

Figure 13.5 shows the ray path in a right-angle prism in which all the angles depart from their correct values. Beam from the autocollimator is totally reflected from the diagonal, partially reflected from the lower face, and then retraces its path back toward the autocollimator. On exiting from the prism, the beam is angularly deflected by $2\mu(\alpha + 2\beta)$, which results in the displacement Δy of the cross wire at the reticle placed on the focal plane. Therefore

$$\frac{\Delta y}{f} = 2\mu(\alpha + 2\beta) \Rightarrow \beta = \frac{\Delta y}{4\mu f} - \frac{\alpha}{2} \tag{13.4}$$

13.2.5 Testing of a Pentaprism

A pentaprism (optical square) is a solid five-sided prism with angles of 90°, 112.5°, 112.5°, 112.5°, and 112.5° as shown in Figure 13.6. However, when the two side arms are extended, they enclose an angle of 45° as shown in Figure 13.6. The incident ray enters normally to the entrance face and exits perpendicularly through the exit face of an ideal pentaprism; the deviation is exactly 90°. Further, if the incident ray makes an angle ε with the normal to the entrance aperture, it still exits deviated by 90° provided the angle ε is small ($\sin \varepsilon \sim \varepsilon$). Thus, an ideal pentaprism is an optical square and quite tolerant to misalignment. If there is any error in the angles of the pentaprism, the deviation of the beam is not equal to 90°; the actual deviation depends on the angles and the refractive index of the prism.

Let us first consider an ideal pentaprism, which deviates the beam by 90°. The pentaprism is placed on a turntable. Light beam from an autocollimator is incident

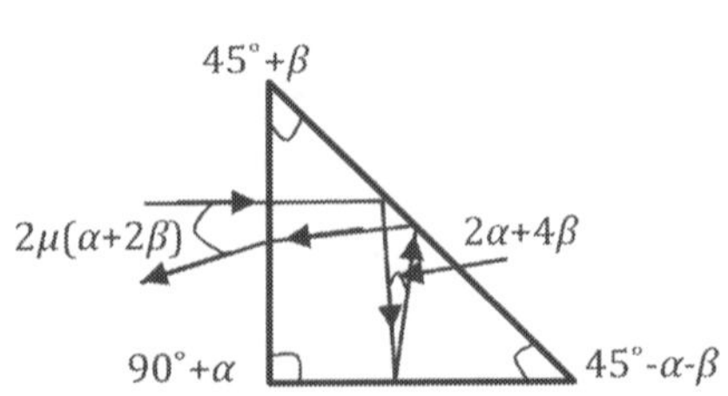

FIGURE 13.5 Measuring error in 45° angle of a right-angle prism.

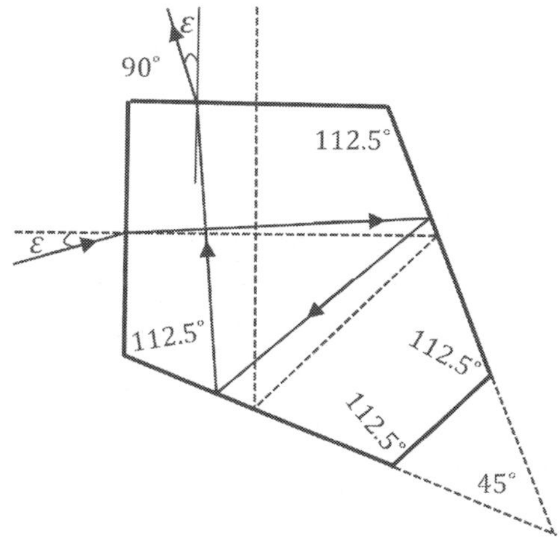

FIGURE 13.6 Pentaprism and its angles.

on its face, which is adjusted such that the reflected beam makes the image of the graticule on itself. By this procedure the beam is retro-reflected from the front face. A plane mirror is placed perpendicular to the exiting beam from the pentaprism and is adjusted such that the image of the graticule again falls on itself. In this situation, the plane of the mirror is parallel to the optical axis of the autocollimator. Now the other face of the pentaprism is made to retro-reflect the beam from the autocollimator and the second mirror is aligned such that the beam from it is also retro-reflected. The two mirrors are now parallel to each other.

Let us now examine the situation when there is an error in the pentaprism. The method of measurement of the deviation from 90° is schematically illustrated in Figure 13.7(a and b). A beam of light from an autocollimator is incident on the

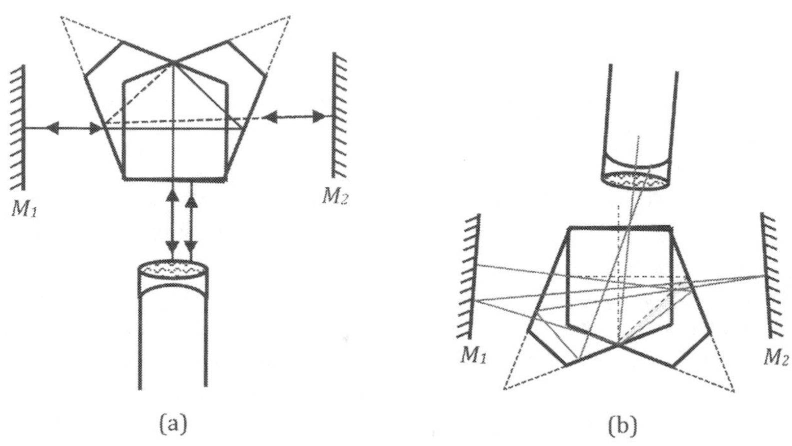

FIGURE 13.7 (a and b) Testing of a pentaprism.

pentaprism, which is adjusted such that the image of the graticule formed due to light reflected from the front face is centered on the reticle; the front face is now normal to the autocollimator axis. The beam from the pentaprism is incident on the mirror M_1, which is adjusted to retro-reflect the beam so that the image of the graticule is centered on the reticle. If the pentaprism is perfect, the beam is deviated by 90° and the plane of the mirror M_1 is parallel to the autocollimator axis or the perpendicular to the mirror surface is at a right angle to the optical axis of the autocollimator. If the pentaprism has an error and deviates the beam by $90° + \theta$, the plane of the mirror makes an angle θ with the autocollimator axis. Similarly, another mirror, M_2, is adjusted for retro-reflection on the other side. The two mirrors, M_1 and M_2, will be parallel if the pentaprism is perfect. However, if the pentaprism angle suffers with an error, then the emergent ray makes an angle of θ with its ideal counterpart. The mirror is to be oriented at an angle θ with its ideal position in order to retro-reflect the ray. These two mirrors now enclose an angle of 2θ as shown in Figure 13.7(a). In short, the mirrors M_1 and M_2 are parallel if the angle of pentaprism is 90° and enclose an angle of 2θ when the angle of pentaprism departs by θ from 90°. Without disturbing the mirrors, the autocollimator is placed in the opposite space as shown in Figure 13.7(b), and the pentaprism is repositioned such that the beam retro-reflected from mirror M_1 makes an image on the graticule. Next the pentaprism is placed such that the image reflected from mirror M_2 is seen. This would usually happen if the deviation θ is not large. The angular shift of this image is noted. This angular shift will be 4θ. The ray paths have been shown in Figure 13.7(b).

It may be noted that due to the finite measurement range of the autocollimator, it cannot be used for measuring angles greater than a certain value. In fact, the method should be used when the departures from the ideal performance are very small. Further, if the sequence of angular deviations is not properly followed, the beam may be deviated in the other direction in the last measurement step, thus escaping the autocollimator.

13.3 GONIOMETER

13.3.1 Absolute Measurement of Angle

For the measurement of large angles, either a goniometer or a spectrometer is used. The heart of the instrument is a circular scale either written/engraved on a glass or a stainless steel material. This may have say 36,000 divisions, with each degree divided into 100 divisions. The scale is read photo-electronically, and accuracy in the second of an arc or smaller is achievable. An autocollimator is used for setting purposes. The axis of the turntable and that of the scale are coincident.

Let us consider a prism whose apex angle is to be measured. The prism is placed on the turntable and rotated until the retro-reflected image from one of the faces of the prism is centered on the reticle of the autocollimator. The angular position is noted. The turntable is rotated again until the retro-reflected image from the other face of the prism is received and centered on the reticle. The angular position is again noted. These two readings are used to calculate the angle of the prism. Another method to measure the 60° angle of a prism has been described in Chapter 10. This

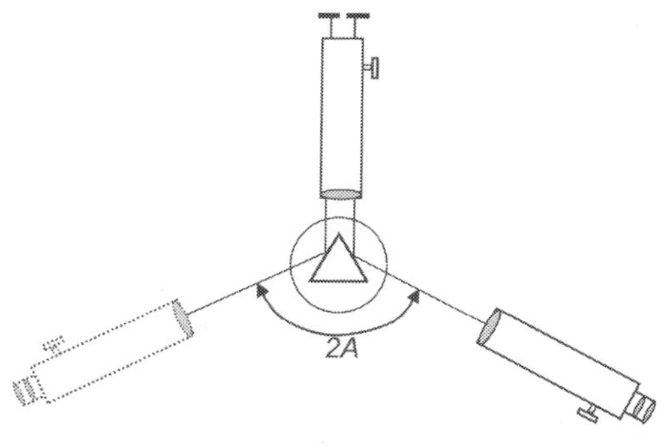

FIGURE 13.8 Laboratory method used to measure the angle of a prism.

method requires polished surfaces, which are normal to the base; there should not be any pyramidal error.

There is yet another method to measure the 60° angle of a prism that is routinely used in the laboratory (Figure 13.8). The prism is properly positioned on the turntable of a spectrometer. The collimator is adjusted such that the light from the collimator falls on both surfaces of the prism that enclose the 60° angle. Light reflected from one of the surfaces is received by the telescope, which is adjusted so that the image of the slit of the collimator falls on the cross wire. The angular position of the telescope is noted. The telescope is moved to receive the light reflected from the other surface and it is adjusted so that the image of the slit falls on the cross wire. The angular position of the telescope is noted. The angular difference between these two positions of the telescope is twice the angle of the prism.

Angle encoders are available that could be used for angle measurement with proper instrumentation.

13.4 INTERFEROMETRY

13.4.1 Angle of a Wedge Plate

A wedge is a relatively thin plate in which the two surfaces enclose a small angle; the angle could vary from a few seconds of an arc to almost a degree. It is used as a beam splitter and also as a shear plate for collimation testing. It is important to know the angle of the wedge plate. It can be determined two ways, either using a Fizeau interferometer or a Twyman–Green (TG) interferometer. When a Fizeau interferometer is used, the plate is placed almost normal to the incident beam; the reflected beam from the front and back surfaces will enclose an angle $2\mu\alpha$. Thus, two plane waves with enclosed angle of $2\mu\alpha$ interfere producing a straight-line fringe pattern with spacing/fringe width $\bar{x} = \frac{\lambda}{2\mu\alpha}$. By measuring the fringe width, we obtain the angle α of the wedge as $\alpha = \frac{\lambda}{2\mu\,\bar{x}}$. We can also look at this in another way. Assume that the wedge edge is along the y-direction and one of the surfaces of the wedge is perpendicular to the z-direction. The thickness t of the wedge at any x-coordinate is $\alpha\,x$. The path difference between the rays reflected from the front and back surfaces is $2\mu t = 2\mu\alpha\, x$. Therefore, the condition for fringe formation is

$$2\mu\alpha\ x = m\lambda \Rightarrow \alpha = \frac{\lambda}{2\mu\bar{x}}$$

where $\bar{x}$ is the fringe width. The fringes are perpendicular to the x-direction.

When the wedge plate is placed in a TG interferometer, it introduces a path difference of $2(\mu - 1)t$, where $t = \alpha x$. Again, a straight-line fringe pattern is formed. The fringe spacing is $\bar{x} = \frac{\lambda}{2(\mu-1)\alpha}$. Hence, the angle of wedge is obtained as $\alpha = \frac{\lambda}{2(\mu-1)\bar{x}}$. The method based on TG interferometer is more sensitive than that based on Fizeau interferometer.

13.4.2 Angle between the Surfaces of an Opaque Plate or a Long Cylinder/Bar

Figure 13.9 shows an interferometric arrangement in which G is an opaque plate/bar with end faces polished. When the surfaces are parallel to each other and the interferometer is properly aligned, the retro-reflected beams on superposition yield a uniform field. However, if the end surfaces are not parallel but enclose a small angle α, the reflected plane beams also enclose an angle α resulting in a straight-line fringe pattern. The angle α between the surfaces is related to the fringe width as $\alpha = \frac{\lambda}{2\bar{x}}$, where $\bar{x}$ is the fringe width. This configuration is insensitive to the misalignment of the test object as can be seen from Figure 13.9. Here G' is the test object, which is misaligned, but the reflected beams from the two surfaces of the plate/bar travel in the same direction giving uniform illumination when the surfaces are parallel to each other. The arrangement is sensitive only to the nonparallelism of the surfaces.

13.4.3 Interferometric Testing of Prisms

13.4.3.1 Testing of a Right-Angle Prism

Prisms can be tested using either a Fizeau interferometer or a TG interferometer. Figure 13.10(a) shows a schematic of a Fizeau interferometer to test either a

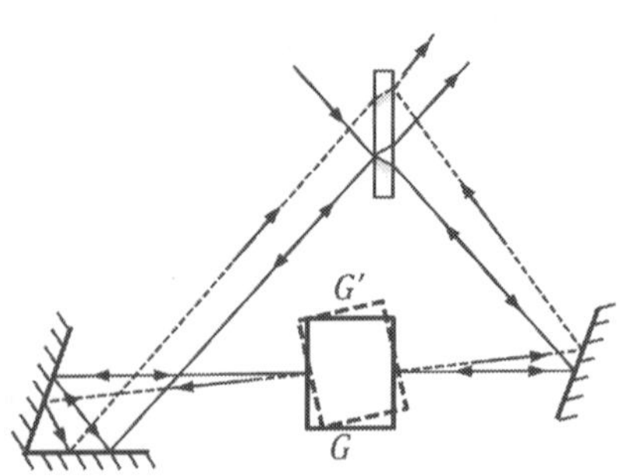

FIGURE 13.9 Schematic to measure the angle between the surfaces of an end block.

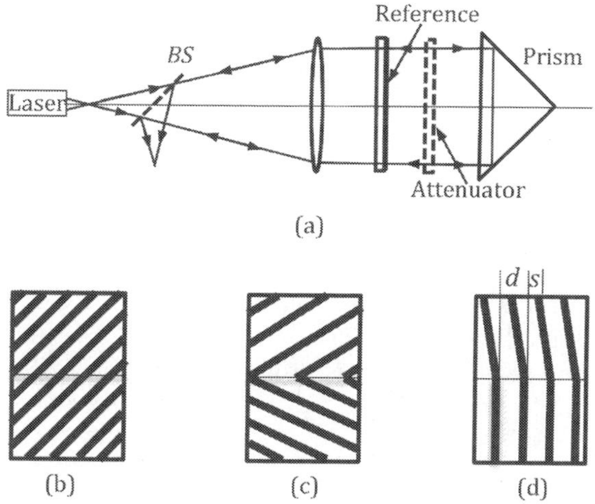

FIGURE 13.10 Testing of a 90° prism. (a) Fizeau interferometer. (b) Interferogram of a good prism. (c) Interferogram of a prism with an angle error. (d) Calculation of the error.

right-angle prism or a cube corner. Other prisms can also be tested, when these were properly aligned.

An attenuator is inserted to match the amplitude of the wave from the test object to that from the reference surface. For a 90° prism with no angular errors, a fringe pattern as shown in Figure 13.10(b) is obtained: the orientation and number of fringes can be changed by proper adjustment of the reference mirror. However, when there is an angle error, a fringe pattern as shown in Figure 13.10(c) may be obtained. This fringe pattern is aligned as shown and the fringe width d and fringe deviation s are measured. Taking L as the width of the prism, the angle error is calculated as

$$\alpha = \frac{s}{d}\,\frac{\lambda}{4\mu L}$$

When a cube corner is tested using a Fizeau interferometer, an interferogram as shown in Figure 13.11(a) is obtained when it is free from the angular errors. A corner cube with an angular error gives an interferogram as shown in Figure 13.11(b).

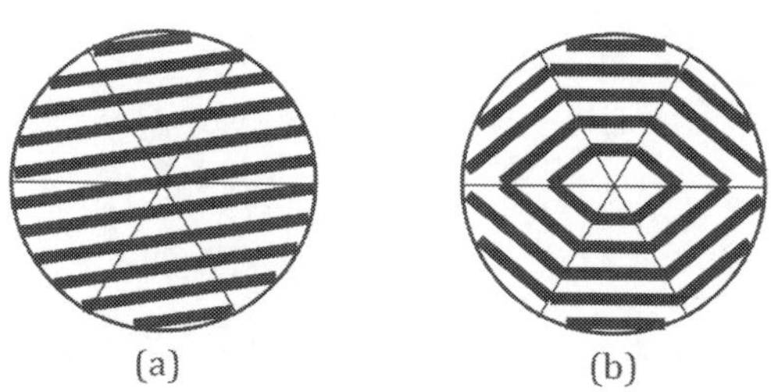

FIGURE 13.11 Interferogram of (a) a good cube corner and (b) of a cube corner with angle error (drawn).

QUESTIONS

13.1 Derive Equation (13.2) from Equation (13.1).

13.2 Show that the error β in a 45° angle of a right-angle prism when using autocollimator is given by

$$\beta = \frac{\Delta y}{4\mu f} - \frac{\alpha}{2}$$

where α is the error in the 90° angle, μ is the refractive index of the prism, and Δy is the shift measured at the focal plane of a collimator of focal length f.

13.3 The angles of a pentaprism are in error and their values are given by $90° + \delta$, $112.5° + \alpha$, $112.5° + \beta$, $112.5° + \gamma$, and $112.5° - \delta - \alpha - \beta - \gamma$. Show that a ray incident normally at one face exits making an angle of $\mu(2\beta + 2\gamma - \delta)$ with the normal to the exit face.

13.4 A plane parallel plate interferometer is used to measure the coefficient of linear thermal expansion of a metal bar of nominal length L_0. A tiny mirror is attached at the end of the bar and a lens of focal length f focuses the beam on the mirror. On retro-reflection, a uniform field is obtained in the interference pattern. When the temperature of the bar is changed by ΔT, an n number of fringes is observed in the field of view of diameter D. Assuming operation at wavelength λ and the shear S, show that the expansion coefficient α is given by

$$\alpha = \frac{1}{2}\frac{\lambda}{S}\frac{n}{D}\frac{f^2}{L_0\,\Delta T}$$

13.5 Given a pentaprism (Figure 13.12) with the measured angles as shown. Calculate the deviation of the beam from 90°.

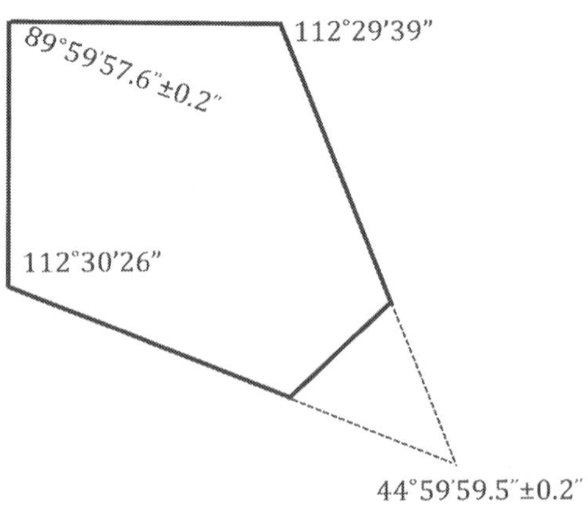

FIGURE 13.12

13.6 Can a right-angle prism with one 45° angle measured as 45°8″ and other as 44°59′52″ meet the deviation tolerance of 2″? Why?

13.7 A 100 mm diameter glass window of refractive index 1.515 is tested in a TG interferometer using a He-Ne laser (632.8 nm). The interferogram shown in Figure 13.13 is obtained. Before the window was placed in the interferometer, it was adjusted for a single interference fringe (uniform field).

a. What is the wedge angle of the window?
b. If the interferometer had 11 vertical fringes in the field of view (extreme bright fringes just at the edge of field of view) before the window was placed in the interferogram, how would the interferogram now look?

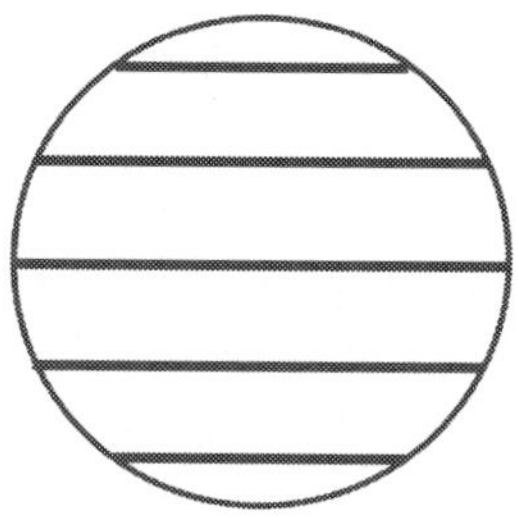

FIGURE 13.13

13.8 A 90° prism made of glass of refractive index 1.515 is tested in retro-reflection in a TG interferometer using a He-Ne laser (632.8 nm) as the source. With unit magnification between the prism and the interferogram, the straight parallel fringes in the interferogram have a spacing of 1 mm and 3 mm.

a. What is the error in 90° angle? State the assumptions that you make.
b. How could you determine if the 90° angle is too large or too small?

13.9 How could you use Haidinger fringes for testing a right-angle prism for departure from a 90° angle?

14 Thickness Measurement

Measurement of thickness of a thin film, while being deposited or after being deposited, and of surface undulation is required in many applications. A number of techniques to cover a large range of thickness have been developed. These techniques are based on the triangulation principle, interference, polarization, etc. Some of these techniques are described in this chapter.

14.1 TRIANGULATION-BASED PROBE

A triangulation-based probe makes use of laser radiation as it can be focused to a tight spot on the surface of an object that is diffuse. A well-corrected lens images the spot on a linear array detector, which is aligned perpendicular to the optical axis of the imaging lens. Figure 14.1 shows the schematic of the instrument.

If the object is axially translated by d, the image of the spot moves by $md \sin\theta$ on the detector, where m is the magnification and θ is the angle between the surface normal and the optical axis of the imaging system. In order to use this probe to measure thickness or step height, it should be calibrated. The system, however, suffers from defocus error, which will be small if the angle θ is small, but it cannot be made arbitrarily small. Therefore, the detector array is inclined at an angle ϕ with the optical axis rather being perpendicular to it as shown in the Figure 14.2. In this case the displacement, s, of the spot on the detector is related to the axial shift, d, of the object surface as

$$s = md\frac{\sin\theta}{\sin\phi} \tag{14.1}$$

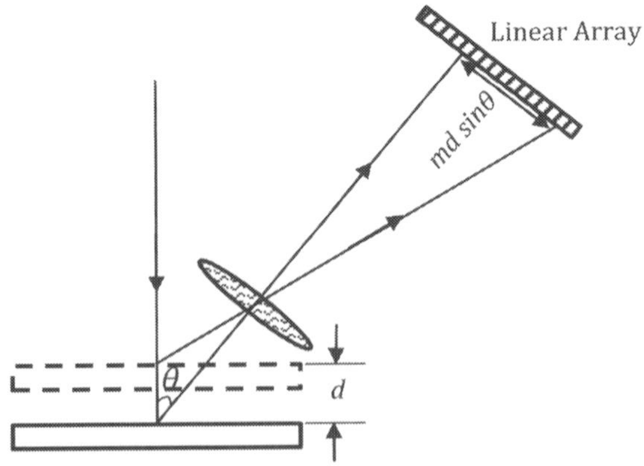

FIGURE 14.1 Schematic of an experimental setup to measure displacement d or step height d.

DOI: 10.1201/9781003531845-14

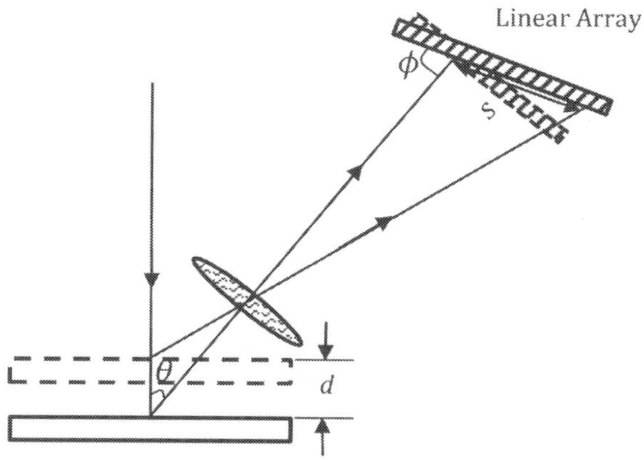

FIGURE 14.2 Schematic of another setup to measure step height d or displacement d.

Again, this is not an exact formula and hence the probe has to be calibrated. However, the correct plane of focus can be obtained using the Scheimpflug principle.

14.2 SPECTRAL REFLECTOMETRY

Both the refractive index and the thickness of a thin film deposited on a substrate can be determined from reflectivity measurement. Further, the growth of the film can also be determined by measuring the reflectivity at one wavelength. Figure 14.3 is a schematic of a thin film deposited on a substrate. It is assumed that the film is transparent. The reflectivity R of the structure is given by

$$R = \left| \frac{r_{12} + r_{23}\ e^{-i\delta}}{1 + r_{12}\ r_{23}\ e^{-i\delta}} \right|^2 = \frac{r_{12}^2 + r_{23}^2 + 2r_{12}r_{23}\ \cos\delta}{1 + r_{12}^2 r_{23}^2 + 2r_{12}r_{23}\ \cos\delta} \tag{14.2}$$

where r_{12} and r_{23} are the Fresnel reflection coefficients at the boundary 1–2 and 2–3, respectively, and δ is the phase change between the two consecutive rays in the reflected beam. The phase difference δ is given by

$$\delta = \frac{2\pi}{\lambda}\ 2\mu_2 t\ \cos\theta_r \tag{14.3}$$

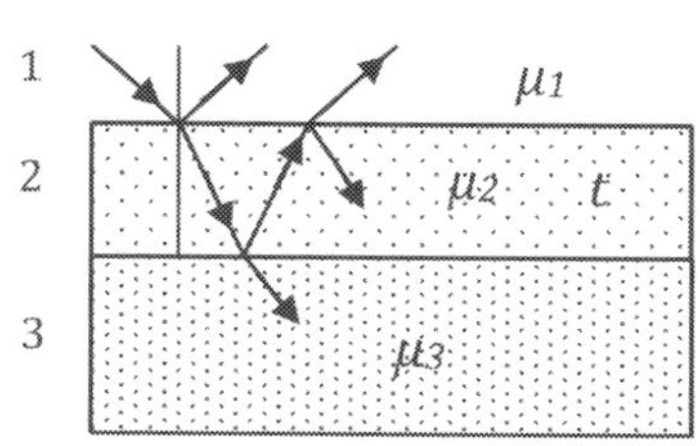

FIGURE 14.3 Thin film structure.

where μ_2 and t are the refractive index and the thickness of the thin film, and θ_r is the angle of refraction in the film.

To measure the film thickness, the film-substrate structure is illuminated normally, and the reflectivity is measured. In this case, $\delta = \frac{4\pi}{\lambda}\mu_2 t$. When the illumination is by a monochromatic beam of wavelength λ, the reflectivity will oscillate if the film is growing. Otherwise, a fixed value of reflectivity is obtained. Assuming $\mu_3 > \mu_2 > \mu_1$, the reflectivity will be maximum whenever $\mu_2 t = m\frac{\lambda}{2}$. If the refractive index is known, the value of thickness can be obtained.

For a film of unknown thickness, the reflectivity is measured as a function of wavelength. Reflectivity again oscillates; it is maximum when the condition $2\mu_2 t = m\lambda$ is satisfied, where m is an integer.

Assuming that the refractive index is constant over a small wavelength interval, and two consecutive maxima are observed at wavelength λ_1 and λ_2, then

$$2\mu_2 t = m\lambda_1 = (m+1)\lambda_2 \tag{14.4a}$$

This gives

$$2\mu_2 t = \frac{\lambda_1\lambda_2}{\lambda_1 - \lambda_2} \tag{14.4b}$$

Therefore, only the optical thickness can be found by this method. If the refractive index is known, then the actual/geometrical thickness can be calculated. However, when the reflectivity measurements are made at oblique incidence, θ_i, then

$$2\mu_2 t \cos\theta_r = m\lambda_1 = (m+1)\lambda_2 \tag{14.5}$$

Here λ_1 and λ_2 are the wavelengths of adjacent peaks in the spectral reflectivity curve. From Equation (14.5), the thickness t is obtained as

$$t = \frac{1}{2\sqrt{\mu_2^2 - \sin^2\theta_i}}\frac{\lambda_1\lambda_2}{\lambda_1 - \lambda_2} \tag{14.6}$$

Alternately, the reflectivity R can be written as

$$R = A + B\cos\left(\frac{4\pi}{\lambda}\mu_2 t\right) \tag{14.7}$$

and the refractive index could be modeled using the Sellmeier equation

$$\mu_2^2(\lambda) - A = \sum_k \frac{G_k\,\lambda^2}{\lambda^2 - \lambda_k^2} \tag{14.8}$$

The Sellmeier's coefficients A, G_k, and λ_k can be found in any good optics book. By varying μ_2 and t, the reflectivity is computed and matched with the experimental data. A correct match then gives the value of μ_2 and t.

14.3 ELLIPSOMETRY

Ellipsometry is used to determine the optical constants and the thickness of the film on a substrate. It first measures the optical constants of the bare substrate. Thereafter, the ellipsometric parameters are obtained with the film on the substrate. If the film is transparent, both refractive index and thickness can be obtained from the ellipsometric measurements as described in Chapter 10. However, when the film is absorbing, the experimental data and the theoretical values obtained by varying μ_2, k_2, and d are compared and when the correct fit is realized, these are taken as the values corresponding to the film.

14.4 INTERFEROMETRY

14.4.1 Fringes of Equal Chromatic Order (FECO)

Figure 14.4(a) shows a schematic of the experimental setup for fringes of equal chromatic order (FECO) interferometry. The sample illuminated with a collimated beam of white light is imaged on the entrance slit of a spectrograph. The FECO fringes are observed at infinity. The FECO interferometer is a multiple-beam interferometer; thus, the film (sample) whose thickness is to be determined is deposited on a glass plate and coated with a highly reflective material.

The irradiance distribution in the transmitted light is given by

$$I(\delta) = \frac{I_0}{1 + F\,\sin^2(\delta/2)} \tag{14.9}$$

where $\delta = \frac{2\pi}{\lambda}\,2t + 2\phi$, ϕ is the phase change on reflection from each surface for normal illumination, and $F = \frac{4R}{(1-R)^2}$ with R being the reflectivity of the surfaces.

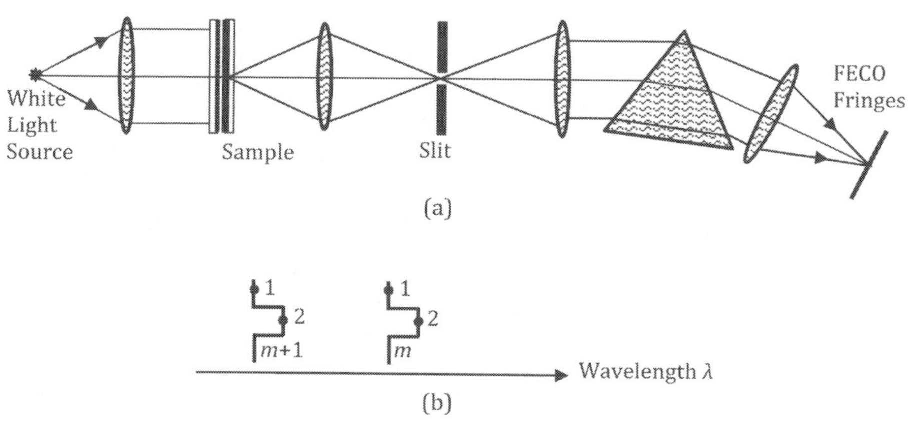

FIGURE 14.4 (a) Experimental arrangement to measure thickness with FECO fringes. (b) Procedure to calculate thickness.

The phase change ϕ is assumed to be independent of wavelength over a short wavelength range. A bright fringe of order m is formed when $\delta = 2m\pi$. Thus

$$2\frac{t}{\lambda} + \frac{\phi}{\pi} = m \tag{14.10}$$

Therefore, for a given fringe, (t/λ) is constant. The fringe order decreases with increasing wavelength.

The procedure to determine the thickness from the FECO fringes is described by using Figure 14.4(b).

From Equation (14.10) and Figure 14.4(b), we write

$$t_1 = \left(m - \frac{\phi}{\pi}\right)\frac{\lambda_{1,m}}{2} \tag{14.11a}$$

and

$$t_2 = \left(m - \frac{\phi}{\pi}\right)\frac{\lambda_{2,m}}{2} \tag{14.11b}$$

From Equations (14.11a) and (14.11b), we obtain

$$t_2 - t_1 = \left(m - \frac{\phi}{\pi}\right)\frac{\lambda_{2,m} - \lambda_{1,m}}{2} \tag{14.12}$$

Here $(t_2 - t_1) = t$ is the thickness of the film.

For point 1 on fringes of order m and $m + 1$, we have

$$t_1 = \left(m - \frac{\phi}{\pi}\right)\frac{\lambda_{1,m}}{2} = \left(m + 1 - \frac{\phi}{\pi}\right)\frac{\lambda_{1,m+1}}{2} \tag{14.13}$$

This gives

$$\left(m - \frac{\phi}{\pi}\right) = \frac{\lambda_{1,m+1}}{\lambda_{1,m} - \lambda_{1,m+1}} \tag{14.14}$$

Substituting for $\left(m - \frac{\phi}{\pi}\right)$ in Equation (14.12), the thickness, t, of the film is given by

$$t = t_2 - t_1 = \left(\frac{\lambda_{1,m+1}}{\lambda_{1,m} - \lambda_{1,m+1}}\right)\frac{\lambda_{2,m} - \lambda_{1,m}}{2} \tag{14.15}$$

The thickness is obtained in terms of wavelengths so the wavelengths must be precisely measured.

As $(t_2 - t_1)$ is proportional to $\lambda_{2,m} - \lambda_{1,m}$, the profile of the cross section of an unknown surface is obtained by plotting a single fringe on a scale proportional to the wavelength. There are no ambiguities if the region is a hill or a valley. Further, if the film is several wavelengths thick, there is no ambiguity, which arises in assigning the fringe order when monochromatic light is used. This technique, however, suffers from the disadvantages of obtaining data along a line and the requirement of high-reflectivity surfaces to obtain multiple-beam fringes.

14.4.2 Fizeau Fringes

Fizeau fringes are used to measure the thickness of both transparent/dielectric and absorbing films. Multiple-beam interferometry is used to obtain sharper fringes and hence better accuracy of thickness measurement. Figure 14.5(a) shows a schematic of the experimental arrangement. A monochromatic source of light is used to illuminate the arrangement with a collimated beam. Straight-line fringes running parallel to the wedge edge are formed. Due to the thickness of the film, the interference pattern gets shifted as shown in Figure 14.5(b). Let the shift be x and the fringe width be X. The film thickness, t, is obtained from

$$t = \frac{x}{X}\frac{\lambda}{2} \tag{14.16}$$

For the films of thicknesses less than $\lambda/2$, there is no ambiguity. However, for thicker films, it may not be easy to assign the fringe order. However, white-light interferometry can be used to estimate the fringe order from the shift of the achromatic fringe. The thickness is then measured using monochromatic light and Equation (14.16).

Two-beam interferometry with phase shifting can also be used to determine film thickness. The film does not have to be coated, and neither does the reference surface. For the measurement of thickness of a metallic film it is advisable to use a reference surface of high reflectivity.

14.4.3 Michelson Interferometer

Thickness of a thin sheet can also be determined using a Michelson or any other two-beam interferometer provided the refractive index of the sheet is known. Alternately,

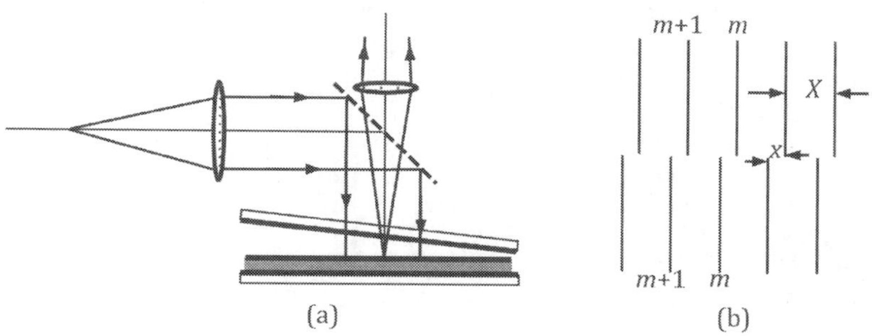

FIGURE 14.5 (a) Schematic of experimental setup for observing Fizeau Fringes. (b) Calculation of thickness.

the refractive index can be determined if the thickness is known. The Michelson interferometer is set up for white-light straight-line fringes. When a sheet of refractive index, μ, and thickness, t, is introduced in one of the two paths, it results in a path difference of $2(\mu-1)t$. The white-light fringe pattern is thus displaced. Translation of the mirror of the interferometer brings the fringe pattern back in the field of view. If the mirror is translated by L, then $L=(\mu-1)t$. The distance L can be calibrated using a monochromatic source with a known wavelength.

14.4.4 Haidinger Fringes

The fringes formed at infinity when an extended source illuminates a thick transparent plate of constant thickness are known as Haidinger fringes, which are fringes of constant inclination. However, a laser source can also be used for this purpose. Figure 14.6 shows a schematic of an experimental setup to use Haidinger fringes for the determination of plate thickness. The laser beam is focused by a microscope objective on a tiny hole in a screen placed at the focal plane. The diverging beam illuminates the plates under test. The light is reflected from the back and the front surfaces of the plate. The reflected light interferes to produce circular fringes. Alternately, one could consider the waves coming from two virtual sources arising due to reflections from the front and the back surfaces of the plate. These sources are separated by $2t/\mu$, where t and μ are the thickness and the refractive index of the plate.

Formation of fringes at the screen is governed by

$$\frac{2t}{\mu}-\frac{t}{\mu}\frac{D_m^2}{16p^2}=m\lambda \Rightarrow \frac{t}{\mu}\frac{D_m^2}{16p^2}=(m_0-m)\lambda=m'\lambda \tag{14.17}$$

where p is the distance between the screen and the front face of the plate, D_m is the diameter of the m'th fringe, and m_0 is the order at the center of fringe pattern. It may be noted that the order of fringe keeps on decreasing as one moves away from the center. If a plot is made between D_m^2 and m', the slope is $16\ p^2\lambda\frac{\mu}{t}$. Because p, λ, and μ are known, the thickness t can be calculated from the measured slope. This method is a quick way to determine whether the thickness is constant, as the fringes will then be centered at the pinhole. If the plate has a wedge of angle α, the fringe pattern shifts toward the increasing thickness of the wedge. It can be seen that the reflections of the beam from the front and the back surfaces of the plate generate virtual sources, which are shifted both axially and laterally as shown in the Figure 14.7.

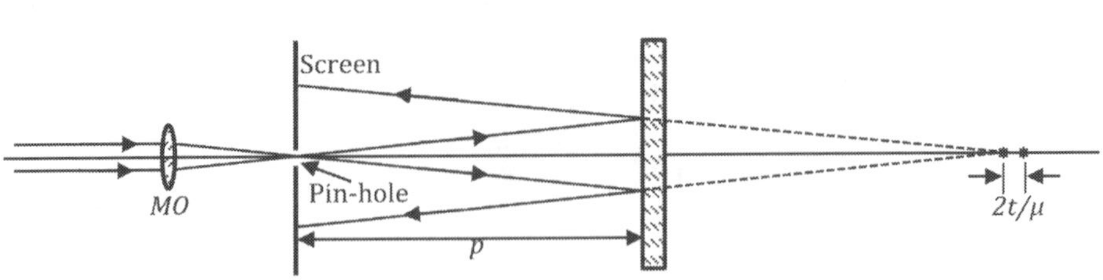

FIGURE 14.6 Schematic of an experimental setup to observe Haidinger fringes.

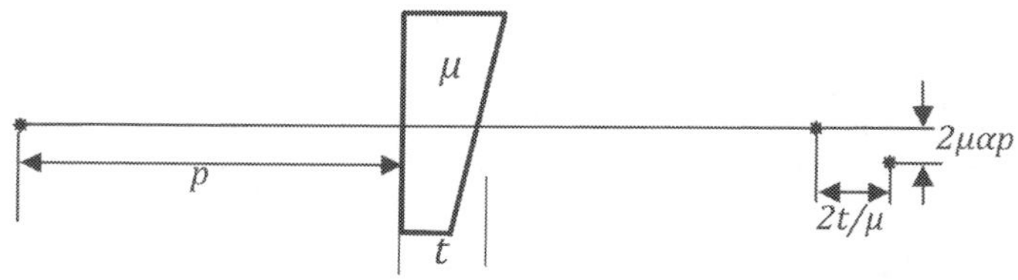

FIGURE 14.7 Generation of virtual sources by a wedge plate to explain the shift of Haidinger fringes.

If the distance between the pinhole and the center of the nearly circular fringe pattern is d, then the angle of the wedge α can be calculated from an approximate formula as

$$\alpha = \frac{t\ d}{2\mu^2\ p^2} \tag{14.18}$$

The center of the fringe pattern shifts toward the thicker edge of the plate. The method is quite sensitive; however, it requires that the surfaces of the plate be polished.

Constancy of the thickness of the silicon wafers is also checked using a wavelength of 1550 nm and an infrared (IR)-sensitive detector.

14.5 LOW COHERENCE INTERFEROMETRY (LCI)

Low coherence interferometry (LCI) is another way to measure the thickness of a film. An interference pattern is observed when the path difference between the two arms of a two-beam interferometer is zero or nearly zero. Therefore, the interfaces of a film can be located with this technique. Figure 14.8(a) shows a Michelson interferometer that is adapted for thickness measurement. The light source, S, is either a while light source or a light-emitting diode (LED). The mirror, M, is on a precision stage to measure the displacement.

When the beam is focused on the lower surface of the film or on the substrate, an interference pattern is observed when the path difference between the two arms

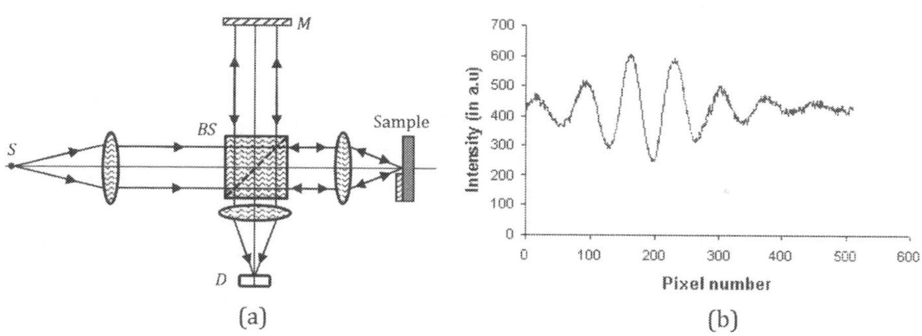

FIGURE 14.8 (a) Michelson interferometer. (b) Trace of an interference pattern.

is zero. Figure 14.8(b) shows the trace of the interference pattern, and it is easier to locate the center of the interference pattern. The beam is now focused on the top surface of the film. In order to bringe the interference pattern in the field, the mirror M is translated by L. The film thickness is μL or L depending on whether the film is transparent or metallic.

Axial resolution R_{axial} of the low coherence interferometer for a source with a Gaussian spectral distribution is given by

$$R_{\text{axial}} = \frac{l_c}{2} = \frac{2\ln(2)}{\pi}\frac{c}{\Delta\nu} = \frac{2\ln(2)}{\pi}\frac{\bar{\lambda}^2}{\Delta\lambda} \approx 0.44\frac{\bar{\lambda}^2}{\Delta\lambda} \quad (14.19)$$

where l_c is the coherence length, R_{axial} is the full width at half maximum of the autocorrelation function, $\Delta\lambda$ is the full width at half maximum of source spectral distribution, and $\bar{\lambda}$ is the central wavelength.

There are several other types of interferometers that could be used for LCI. In the Linnik interferometer, two matched microscope objectives are used. While one objective focuses the laser beam on the sample, the other focuses the laser beam on a reference mirror. An elegant method is to use a Mirau objective, which in itself is an equal path interferometer, and hence, white-light fringes are easily obtained. The objective can be focused on the upper and the lower interfaces and the displacement of the objective, when multiplied by the refractive index of the film, gives its optical thickness.

14.6 CONFOCAL MICROSCOPY

There are several applications in semiconductor and polymer industries where thickness measurement of films ranging from several micrometers to several tens of micrometers is to be carried out. Confocal microscopy, due to its excellent depth discrimination characteristics, is quite well suited to this application. Figure 14.9 shows a schematic of a confocal microscope.

Laser light from source S is collimated and the microscope objective MO focuses the light at a point in the sample. The image of the point focus is made on the pinhole P, and the detector D detects the light passing through the pinhole. The lateral and axial resolutions in confocal microscopy are given by $\left(0.61\ \lambda/\sqrt{2}\ NA\right)$ and $\left(1.4\ \lambda/NA^2\right)$, respectively.

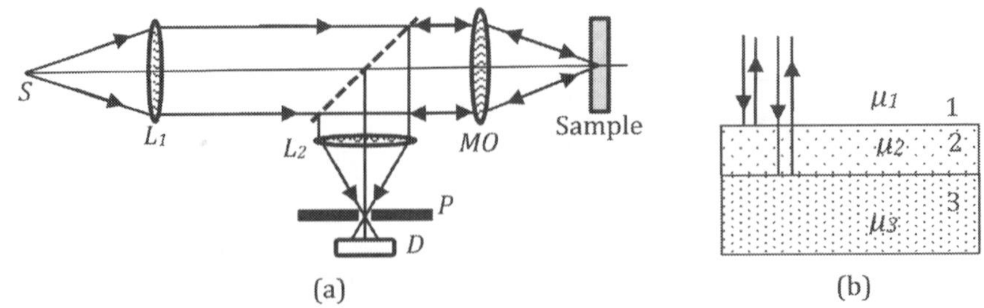

FIGURE 14.9 (a) Schematic of a confocal microscope. (b) Reflectivity calculations.

For measuring the thickness of the film, the microscope is focused on the top surface of the film. It is then lowered and focused on the lower surface of the film. The optical thickness of the film is obtained by multiplying the distance as measured by the confocal microscope with the refractive index of the film. Obviously, the film must be transparent. If the film is metallic or absorbing, then the distance measured by the confocal microscope by focusing the objective on the upper surface of the film and then on the surface of the substrate gives the thickness of the film.

Assuming the film to be transparent, the irradiance of the light reflected from the upper surface is given by

$$I_{\text{upper}} = I_0 \left(\frac{\mu_2 - \mu_1}{\mu_2 + \mu_1} \right)^2 \tag{14.20}$$

where I_0 is the irradiance of the incident light. This happens because the multiply reflected light does not contribute to the signal due to its depth discrimination property. In order to find the light reflected by the lower surface of the film, we should note that transmission occurs twice through the 1–2 interface. Thus, the irradiance of light reflected from the lower surface of the film is given by

$$I_{\text{lower}} = I_0 \left[\left(\frac{2\mu_1}{\mu_2 + \mu_1} \right) \left(\frac{2\mu_2}{\mu_2 + \mu_1} \right) \left(\frac{\mu_3 - \mu_2}{\mu_3 + \mu_2} \right) \right]^2 \tag{14.21}$$

If the refractive indices of the first and third media are known, the refractive index of the film can be obtained from the ratio of the reflected irradiances from lower and upper interfaces.

14.7 LIGHT SECTION MICROSCOPY

This method is used to examine the topography of the surface and to measure the thickness of both the transparent and the metallic films. An adjustable narrow slit is imaged on the sample using a microscope objective at an angle of 45° with the normal to the surface and is sectioned by the top and bottom surfaces of the film as seen in Figure 14.10(a). The projected slit image is seen by another microscope inclined at an angle of 45° with the normal to the surface; the slit image appears as shown in Figure 14.10(b).

Because the film is transparent and imaging of the slit takes place due to reflection from both the top and the bottom surfaces of the film, the bottom surface is apparently lifted due to refraction. The microscope measures the distance AB' by making a measurement of $A_gB'_g$ at the graticule. The distance AB', assuming the angle of incidence as θ_i, is given by

$$AB' = \frac{t}{\sqrt{\mu^2 - \sin^2\theta_i}} \tag{14.22a}$$

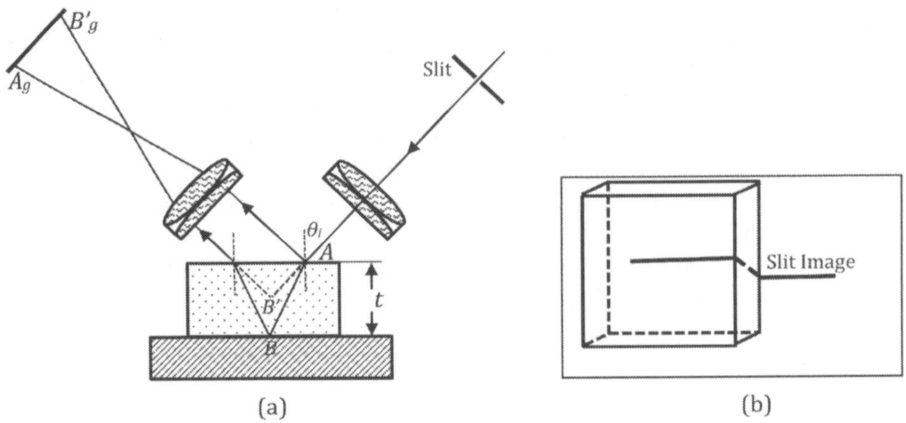

FIGURE 14.10 (a) Schematic of a light section microscope. (b) Projected slit image.

The thickness t of the film is obtained as

$$t = AB'\sqrt{\mu^2 - \sin^2\theta_i} \tag{14.22b}$$

When θ_i is 45°, it reduces to $t = AB'\sqrt{\mu^2 - 0.5}$. For a metallic film, the slit image is sectioned by the upper surface of the film and the surface of the substrate. Its thickness is then given by $t = AB'\cos\theta_i$, where AB' is the distance measured by the microscope.

QUESTIONS

14.1 An oil film of index 1.47 rests on a pool of water of refractive index 1.33. A beam of light strikes the film at angle of 60°. The light of wavelength at 570 nm is strongly reflected by the film. What is the thickness of the film?

14.2 A broadband source emitting a Gaussian beam in the frequency range of 640 nm to 690 nm is used in LCI. What is the axial resolution of the interferometer? To what accuracy can the optical thickness of a film be measured with this interferometer?

14.3 A wedge plate of glass of refractive index 1.515 and nominal thickness of 5 mm is examined using Haidinger fringes. It is observed that the center of the nearly circular fringes is shifted vertically upward by 15.0 mm with respect to the illuminating pinhole. The distance between the wedge plate and the screen is 80.0 cm. What is the angle of the wedge? Which is the thicker side of the wedge?

14.4 Multiple-beam interferometry is used to measure the thickness of a thin film whose thickness is less than half the wavelength of light of wavelength 546.1 nm. The fringe width is measured to be 12.34 mm and the fringe pattern due to the thickness of the film is shifted by 10.46 mm. What is the thickness of the film? If the space between the film and the substrate is filled with a liquid of refractive index 1.47, what are the new fringe width and shift of the fringe pattern?

14.5 Obtain the irradiance distribution in the fringe pattern in transmission formed due to multiple reflections in a wedge of refractive index μ, wedge angle α, and nominal thickness t. Assume the angle of incidence as θ_i.

14.6 A Haidinger fringe pattern is observed from a plate of thickness t and refractive index μ on a screen that is at a distance p from the front face of the plate. The illumination beam diverges from a pinhole in the center of the screen. Show that the path difference between the two interfering beams is given by

$$\frac{2t}{\mu} - \frac{t}{\mu}\frac{\left(x^2 + y^2\right)}{4p^2}$$

where x and y are the Cartesian coordinates of a point on the screen with the origin at the pinhole.

15 Measurement of Velocity

15.1 INTRODUCTION

Velocity is defined as the rate of change of the position vector. It thus requires the measurement of positions at two time intervals; this yields the average velocity. If the time interval is exceedingly small, we obtain instantaneous velocity. Measurement of velocity is required in several areas, notable of these are fluid mechanics and aerodynamics. The measurement of blood flow in biomedicine is required, and velocity measurement is indispensable for process control and optimization of many industrial processes.

It is known that the frequency of light reflected from a moving object is Doppler-shifted, and the Doppler shift is directly proportional to the velocity. The application of this fact for measuring velocity became possible only after the advent of the laser. Several other velocity-measuring techniques were also developed later. Almost all the optical techniques of velocity measurement require a transparent or nearly transparent fluid, which is seeded with particles. For their operation they rely on the detection of scattered light from these seeded particles. These techniques fall in the domain of laser anemometry. These techniques really do not measure the velocity of the fluid instead they measure the velocity of the particles (scatterers). Therefore, the scatterers are assumed to follow the fluid flow faithfully. The particle density of scatterers should not be less than 10^{10} particles/m^3. The particle size for measuring the gas flow ranges from 1 μm to 5 μm and for measuring the liquid flow, say of water, from 2 to 10 μm.

Laser anemometry offers several advantages over hot-wire anemometry: noncontact measurement avoiding any interference to the flow, excellent spatial resolution, fast response and hence fluctuating velocities can be measured, no transfer function as the output voltage is linearly related to velocity, large measurement range, and can be used to measure both the liquid and the gas flows. Another advantage is that it can be used both in forward and backward scattering directions. Though the irradiance of light scattered in the backward direction is several orders of magnitude smaller than that in the forward direction, in several situations where through ports are not possible, backward scattering geometry is used for measurement.

15.2 SCATTERING FROM A MOVING PARTICLE-DOPPLER SHIFT

Consider a particle moving with a velocity $\bar{V}$, illuminated by a wave of angular frequency ω_i propagating in the direction given by the propagation vector $\bar{k}_i$ as shown in Figure 15.1.

Let us observe the scattered wave with the propagation direction $\bar{k}_s$. The angular frequency ω_s of the scattered wave is given by

$$\omega_s = \omega_i + \mu\, \bar{V} \cdot \left(\bar{k}_s - \bar{k}_i\right) \tag{15.1}$$

DOI: 10.1201/9781003531845-15

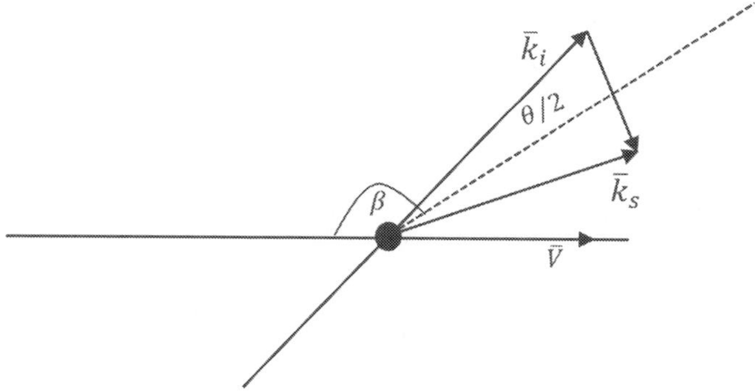

FIGURE 15.1 Geometry to calculate the Doppler shift.

where μ is the refractive index of the medium in which the scatterer is moving. It is seen that the frequency of the scattered wave is different than that of the incident wave, i.e., the scattered wave is Doppler-shifted. The Doppler shift, $\Delta\nu_D$, can be expressed as

$$\Delta v_D = \frac{\omega_s - \omega_i}{2\pi} = \frac{2}{\lambda}\,\mu\,V\,\sin\frac{\theta}{2}\,\sin\beta \tag{15.2}$$

where θ is the angle between the propagation vectors of the incident and the scattered waves and β is the angle between the direction of scatterer's motion and the bisector of propagation vectors as shown in the Figure 15.1. When $\beta = \pi/2$, the Doppler shift/frequency, Δv_D, is given by

$$\Delta v_D = \frac{2}{\lambda}\,\mu\,V\,\sin\frac{\theta}{2} \tag{15.3}$$

The velocity V of the scatterer and hence of the flow is obtained from this relation as

$$V = \frac{\lambda}{2\,\mu\,\sin(\theta/2)}\,\Delta v_D \tag{15.4}$$

The velocity of fluid flow is directly proportional to the Doppler frequency Δv_D. In practice, the observable Doppler shifts lie between 10^6 Hz and 10^8 Hz, whereas the frequency of the laser light is in the range of 10^{14} Hz. Thus, the Doppler shift is a very small fraction of the frequency of the incident wave and hence direct measurement of the Doppler shift introduces large uncertainties in its measurement and consequently in the determination of velocity. Thus, the Doppler frequency is measured by heterodyning: the scattered light is mixed with the direct light. There are a number of methods based on heterodyning, which are explained in the following sections.

15.3 METHODS OF VELOCITY MEASUREMENT

15.3.1 Reference Beam Mode

The light scattered by the moving scatterer is mixed with that of the unscattered light on a photodetector. The output of the detector is a Doppler signal. Figure 15.2 shows the schematic of the experimental setup. A beam from a laser is divided into two beams: the illumination beam and the reference beam. The illumination beam, which is stronger than the reference beam, is focused on a point of interest in the flow field.

The reference beam need not pass through the flow field but must be collinear with the scattered beam. A mask selects the direction. These two beams then mix at the photodetector, say a photomultiplier tube (PMT). The signal from the PMT is processed to obtain the velocity.

Let the reference beam be expressed by

$$E_r(r,t) = E_{r0} e^{i(\omega_i t - \bar{k}_i \cdot \bar{r})} \tag{15.5a}$$

Similarly, the scattered field is expressed as

$$E_s(r,t) = E_{s0} e^{i(\omega_s t - \bar{k}_s \cdot \bar{r})} \tag{15.5b}$$

where $\omega_s = \omega_i + \mu\, \bar{V} \cdot (\bar{k}_s - \bar{k}_i) = \omega_i + \mu\, V\, \frac{2\pi}{\lambda} \sin\theta$ for the geometry given in Figure 15.2. Assuming that the laser beam has sufficiently large coherence length so that the influence of any path difference between the two beams can be ignored, the photocurrent $i(t)$ from the PMT will be proportional to

$$i(t) \propto \left| E_r(r,t) + E_s(r,t) \right|^2 \tag{15.6}$$

While working out this expression, it may be noted that the scattered beam and the reference beam have the same propagation vector. Taking the constant of proportionality as $\mathcal{B}$, we have

$$i(t) = \mathcal{B}\left[E_{r0}^2 + E_{s0}^2 + 2E_{r0}\, E_{s0} \cos(2\pi\, \Delta\nu_D t) \right] \tag{15.7}$$

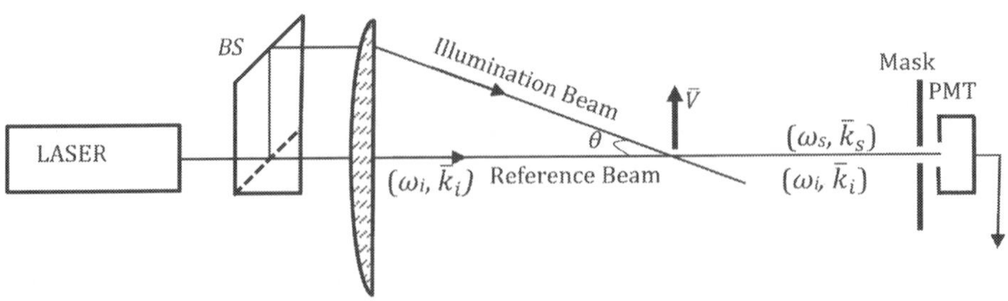

FIGURE 15.2 Reference beam mode to measure flow velocity.

It has been mentioned that the illumination beam is stronger than the reference beam due to the scattered light being very weak. Therefore, $E_{r0}^2 \gg E_{s0}^2$ and hence the expression for the photocurrent can be written as

$$i(t) = \mathcal{B}\left[E_{r0}^2 + 2E_{r0}\ E_{s0}\cos\left(2\pi\ \Delta\nu_D t\right)\right] \tag{15.8}$$

This expression gives the photocurrent due to a single scatterer moving through an illuminating beam of an infinite size. In practice, the illumination beam is focused to a small size, say 0.1 mm in the flow field; hence, the scatterer remains in the beam for a finite time. Further, the irradiance distribution in the beam is Gaussian. In order to account for these, we introduce a function $\mathcal{P}(t)$, called the blocking function such that $\mathcal{P}(t) \neq 0$ when the scatterer is in the illumination beam. In practice, there are N scatterers distributed in the sample volume, and the photocurrent will then depend on the relative phases of all the scattered waves. Processing of the photocurrent to get velocity information will be discussed in a later section.

15.3.2 Fringe Mode

As the name suggests a fringe pattern is created in the sample volume. In order to create a fringe pattern of maximum contrast, the two interfering beams should be of equal irradiance. Therefore, the incident beam is split into two beams of equal irradiance. The beams are superposed in the region of interest. As the laser beam has a Gaussian profile, the beam waist of both the beams should coincide in the region of interest. Figure 15.3 shows the schematic of the experimental arrangement.

The amplitudes of the waves in the region of interest can be written as

$$E_{r1}(r,t) = E_{r01}e^{i\left(\omega_i t - \bar{k}_{i1}\cdot\bar{r}\right)} = E_{r01}e^{i\left\{\omega_i t - \frac{2\pi}{\lambda_m}\left[z\ \cos(\theta/2) - x\ \sin(\theta/2)\right]\right\}} \tag{15.9a}$$

$$E_{r2}(r,t) = E_{r01}e^{i\left(\omega_i t - \bar{k}_{i2}\cdot\bar{r}\right)} = E_{r01}e^{i\left\{\omega_i t - \frac{2\pi}{\lambda_m}\left[z\ \cos(\theta/2) + x\ \sin(\theta/2)\right]\right\}} \tag{15.9b}$$

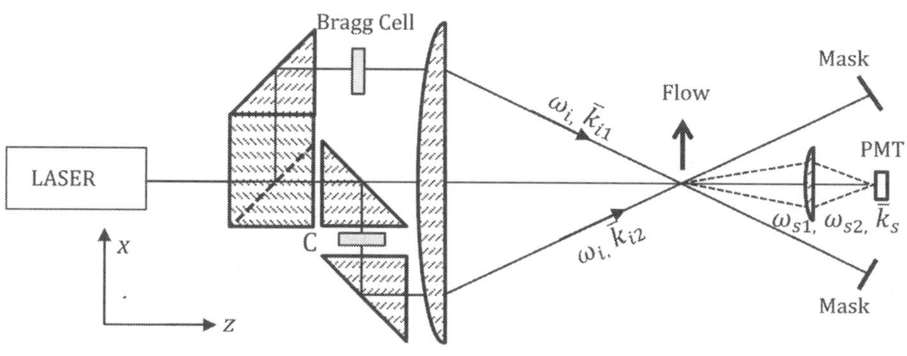

FIGURE 15.3 Schematic of the experimental arrangement to measure velocity using the fringe mode.

where λ_m is the wavelength in the medium in which scatterers are moving and θ is the angle between the two beams. Further, the amplitudes of both the beams are taken equal.

These beams will interfere and produce a stationary interference pattern. The irradiance distribution in the interference pattern is given by

$$I = \left|E_{r1}(r,t) + E_{r2}(r,t)\right|^2 = 2E_{r01}^2\left\{1 + \cos\frac{2\pi}{\lambda_m}\left[2\ x\sin(\theta/2)\right]\right\} \tag{15.10}$$

The irradiance maxima are formed when

$$\frac{2\pi}{\lambda_m}\left[2\ x\sin(\theta/2)\right] = 2m\pi \tag{15.11}$$

The fringes run perpendicular to x-axis and their spacing is

$$\bar{x} = \lambda_m / 2\ \sin(\theta/2) = \lambda / 2\mu\ \sin(\theta/2) \tag{15.12}$$

Essentially, the interference planes are formed in the sample volume. A scatterer moving with velocity V will scatter maximum light whenever it is at the bright interference plane and less or no light when it is at the dark plane. Therefore, the scattered light will fluctuate as the scatterer traverses the interference planes. The scattered light is collected and received by a PMT. Therefore, the current will vary with a frequency $\Delta\nu_D$, which is given by

$$\Delta\nu_D = \frac{V_x}{\bar{x}} = \frac{2\ \mu\ V_x\ \sin(\theta/2)}{\lambda} \tag{15.13}$$

This is the Doppler shift that one would expect from this geometry. We can arrive at this relation, Equation (15.13), by considering that the scattered light by a moving scatterer is Doppler-shifted. Because the scattered light of propagation vector $\bar{k}_s$ is collected, the frequencies of the scattered waves can be expressed as

$$\omega_{s1} = \omega_i + \mu\ \bar{V}\cdot\left(\bar{k}_s - \bar{k}_{i1}\right) \tag{15.14a}$$

$$\omega_{s2} = \omega_i + \mu\ \bar{V}\cdot\left(\bar{k}_s - \bar{k}_{i2}\right) \tag{15.14b}$$

where ω_{s1} and ω_{s2} are the angular frequencies of scattered light when the scatterer is illuminated by the light waves with the propagation vectors $\bar{k}_{i1}$, and $\bar{k}_{i2}$, respectively. The Doppler frequency is given by

$$\Delta\nu_D = \frac{\omega_{s1} - \omega_{s2}}{2\pi} = \frac{1}{2\pi}\mu\ \bar{V}\cdot\left(\bar{k}_{i2} - \bar{k}_{i1}\right) = \frac{2\ \mu\ V_x\ \sin(\theta/2)}{\lambda} \tag{15.15}$$

This is the same equation that we derived by considering scattering from the fringe planes.

The drawback of the laser Doppler anemometer (LDA) is that it does not distinguish between the velocities V_x and $-V_x$: it has directional ambiguity. The directional ambiguity is removed by using a Bragg cell that introduces a fixed frequency ν_B in one of the laser beams and consequently the Doppler frequency observed is given by

$$\Delta\nu_D = \nu_B + \frac{2\,\mu\,V_x\,\sin(\theta/2)}{\lambda} \tag{15.16}$$

Therefore, the negative velocities down to

$$V_x > -\frac{\nu_B\lambda}{2\,\mu\,\sin(\theta/2)} \tag{15.17}$$

can be measured. In order to compensate for the additional path introduced by the Bragg cell, a compensating plate, *C*, or another Bragg cell is introduced in the second beam.

Figure 15.4 shows the interference fringes in the superposition region. Because the Bragg cell introduces an additional frequency of ν_B between the beams, the fringe planes are no longer stationary but move with a velocity $\bar{V}_B$, where $\bar{V}_B = \nu_B\lambda/2\mu\,\sin(\theta/2)$. Introduction of the Bragg frequency removes the ambiguity in direction detection of the flow velocity.

These two beams are superposed at the focus in the flow field; the region of superposition is called the sample volume in which the velocity is measured. Assuming the beam waist of diameter *d* in the region of superposition, then the sample volume can be approximately defined in terms of its size d_x, d_y, and d_z along *x*-, *y*- and

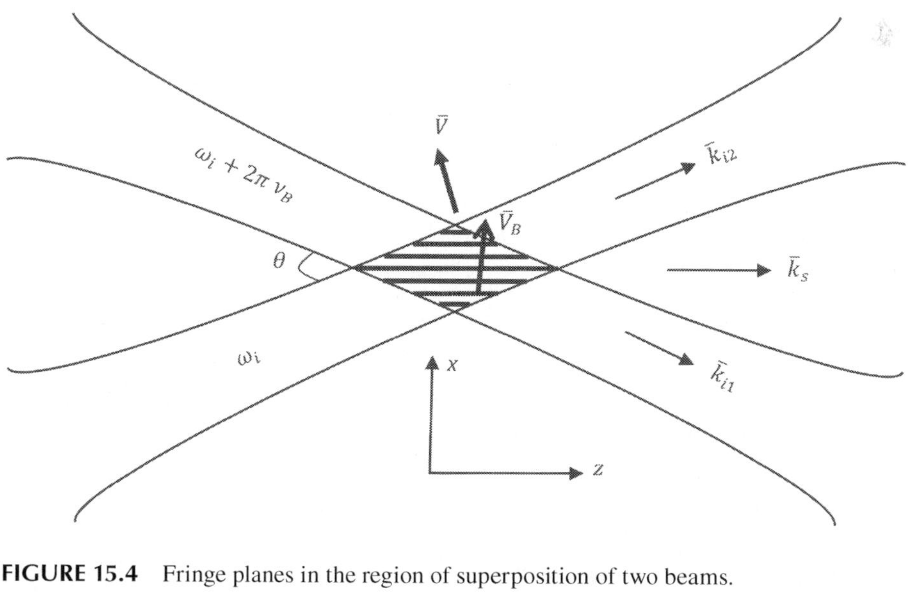

FIGURE 15.4 Fringe planes in the region of superposition of two beams.

z-directions where $d_z = d/\sin(\theta/2)$, $d_x = d/\cos(\theta/2)$, and $d_y = d$. The number of fringe planes, N, is obtained by dividing d_x by the fringe width, i.e.,

$$N = \frac{d_x}{\bar{x}} = \frac{d}{\cos(\theta/2)}\frac{2\mu\sin(\theta/2)}{\lambda} = \frac{2\mu d}{\lambda}\tan(\theta/2) = \frac{8\mu f}{\pi D}\tan(\theta/2) \quad (15.18)$$

where D is the laser beam diameter and f is the focal length of the lens that brings the two beams to superposition at the focus.

15.4 SCATTER LIGHT BEAMS ANEMOMETRY

Instead of using two beams to intersect at the region of interest in the flow field, one can use a single laser beam that is focused in the region of interest and the scattered light is picked up in two different directions, preferably symmetric to the incident beam. These scattered beams then are directed to proceed in the same direction as shown in Figure 15.5.

The angular frequencies ω_{s1} and ω_{s2} of the light scattered in the directions $\bar{k}_{s1}$ and $\bar{k}_{s2}$ can be expressed, respectively, as

$$\omega_{s1} = \omega_i + \mu\,\bar{V}\cdot\left(\bar{k}_{s1} - \bar{k}_i\right) \quad (15.19a)$$

$$\omega_{s2} = \omega_i + \mu\,\bar{V}\cdot\left(\bar{k}_{s2} - \bar{k}_i\right) \quad (15.19b)$$

Therefore, the Doppler frequency $\Delta\nu_D$ as seen by the detector on which both the scattered beams are superposed is

$$\Delta\nu_D = \frac{\omega_{s1} - \omega_{s2}}{2\pi} = \frac{1}{2\pi}\mu\,\bar{V}\cdot\left(\bar{k}_{s1} - \bar{k}_{s2}\right) = \frac{2\,\mu\,V_x\,\sin(\theta/2)}{\lambda} \quad (15.20)$$

where θ is the angle between the scattered beams, which are assumed to be symmetric with respect to the incident beam. It may be noted that the two expressions in Equations (15.15) and (15.20) are identical. However, this geometry gives the freedom to choose any value of θ albeit with weak irradiance of scattered light while the angle is limited by the *F#* of the lens in the former case. The Bragg cell can also be introduced in the scattered beam, but caution should be exercised.

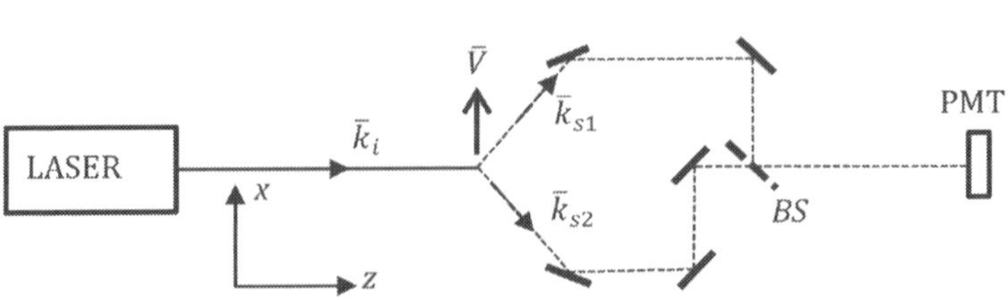

FIGURE 15.5 Laser anemometry with scattered beams.

15.5 MULTICHANNEL LDA SYSTEMS

In order to measure the components of the velocity, the same sample volume should present the possibility to sense each component separately. This is accomplished either by the use of different wavelengths of laser light or using different polarization states. We describe an LDA system that measures two components of the velocity. It may be recalled that when the laser beams lie in the y–z plane, their superposition creates fringe planes that lie perpendicular to the y-axis. Similarly, when the laser beams lie in the x–z plane, the fringe planes will lie normal to the x-axis. Using lasers outputting different wavelengths, the scattered signal from this set of planes can be isolated. Therefore, a two-velocity component LDA consists of two laser sources, optics to create two beams from each laser that are orthogonal to each other, and an achromatic lens to bring these to a common focus where two orthogonal fringe planes are created in the sample volume. Two photodetectors with appropriate filters receive the signal from each laser. The photo-signal is then processed to obtain velocity components.

For obtaining all the three components of the velocity vector, a two-component LDA and a one-component LDA, properly oriented, are used simultaneously.

15.6 SIGNAL PROCESSING

If the diameter of the scatterer is much smaller than the spacing between the fringe planes in the sample volume, it can be safely assumed that the amplitude and phase of the beam are constant over the diameter of the scatterer. Further, the scatterer identifies clearly the bright and dark regions in the sample volume. With a single scatterer moving through the flow field, a typical signal as shown in the Figure 15.6 is received. This is called Doppler burst as seen in Figure 15.6(a). It has a DC component arising due to the nonuniform irradiance distribution in the beam: this is called the Doppler pedestal. It is removed by high-pass filtering that results in a Doppler pulse as seen in Figure 15.6(b). The Doppler signal consists of a sequence of randomly occurring nonoverlapping pulses of random amplitudes. These pulses appear

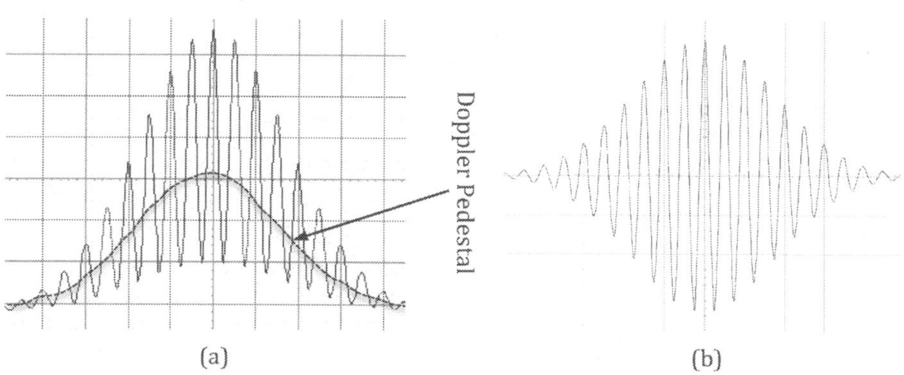

FIGURE 15.6 (a) Doppler burst. (b) Doppler pulse.

whenever a scatterer crosses the sample volume. This train of pulses or the Doppler signal is then presented to the processor.

When there is more than one scatterer in the sample volume, the Doppler signal is the sum of the Doppler bursts from each scatterer. Because the scatterers are distributed randomly in the sample volume, the Doppler bursts are added with random phases. Therefore, the envelope and the phase of the Doppler signal will vary randomly: the quality of signal deteriorates. The LDA will then give an estimate of the velocity as weighted average of the velocities as seen by the scatterers in the sample volume.

However, when the size of the scatterer is larger than the fringe spacing, the scatterer integrates the fringe pattern resulting in almost no modulation. Further, the irradiance and the phase of the light wave is not constant over the particle.

There are a number of signal processing methods employed with an LDA: spectrum analyzer, tracker, and counter. The choice of a particular method is dictated by the signal-to-noise ratio of the input signal, the type of information required, accuracy of measurement, etc. It is often difficult to decisively recommend a particular processing method, but for highly noisy Doppler signals, it is advised to use frequency domain processing techniques. Currently, fast Fourier transform (FFT) of the digitized Doppler signal can be performed at the rate of 100 kHz and higher. The peak of the Fourier spectrum gives the Doppler frequency.

15.7 PARTICLE IMAGE VELOCIMETRY (PIV)

The particle image velocimetry (PIV) is a technique to determine the velocity of a fluid by imaging the tracer particles (scatterers) at two instants of time. The average velocity is computed as

$$\bar{V}_{av} = \frac{x_2 - x_1}{t_2 - t_1} = \frac{\Delta x}{\Delta t} \qquad (15.21)$$

where x_1 and x_2 are the positions of a tracer particle at time t_1 and t_2, respectively. The PIV is a simple, nonintrusive and full-field optical technique, but its experimental realization is involved due to several considerations. The tracer particles should faithfully follow the fluid flow and should scatter sufficient light for imaging. Therefore, the choice of the tracer particles is to be judiciously made. A thin sheet of very bright light illuminates the fluid and the image is photographed. A short time later, the fluid is illuminated again with the same sheet of light and another photograph is made. This double-exposure photograph, after development, can be Fourier-filtered with a narrow beam to find the displacement of particles. The velocity at each interrogation point is thus computed but the direction ambiguity remains. During the 1990s, the digital cameras and later the digital interline transfer cameras were introduced in PIV. This gave birth to digital particle image velocimetry (DPIV) though it did not match the resolution of the film. It, however, provided the potential to perform PIV in real time. Further, DPIV removed the direction ambiguity, which was inherent in the film-based PIV.

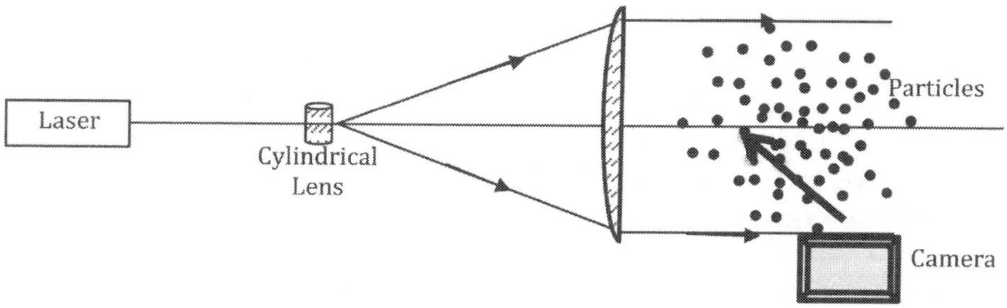

FIGURE 15.7 Schematic of a PIV setup.

Figure 15.7 shows a schematic of a PIV setup. Laser beam from a double-pulse laser is expanded in a sheet by using a cylindrical lens of a very short focal length and then collimated using a large size lens. The pulse duration is usually around 10 ns and the pulse separation can be varied over a large range. This helps in tuning the sensitivity of flow measurement. Measurement accuracy and dynamical range increase with increasing time difference between the two pulses. The motion of the tracer particles is frozen due to the short pulse duration. The image as captured by the interline charge-coupled device (CCD) camera is transferred to the interline register very fast (~1 μs) and then the second image due to the second pulse is captured.

The preferred method to calculate the displacement of the scatterer in between two pulses is to perform cross-correlation between the two images. The cross-correlation has a single peak providing magnitude and direction of the velocity. For this purpose, each image is divided into small regions. Let a small region of interest in the first image be called the test pattern and a slightly larger region but centered around the same location be called a reference pattern. The target pattern is shifted pixel by pixel to each location on the reference pattern, and the values of two patterns are multiplied pixel by pixel to yield the normalized cross-correlation coefficient $c_{m,n}$ as

$$c_{m,n}(x,y) = \frac{\sum_m \sum_n g_{m,n} \, f_{m+x,n+y}}{\sqrt{\sum_m \sum_n g_{m,n}^2 \, \sum_m \sum_n f_{m,n}^2}} \tag{15.22}$$

where $g_{m,n}$, $f_{m,n}$, and $c_{m,n}$ are the values in the test pattern, reference pattern, and correlation coefficient at pixel m,n, respectively, and x, y are the pixel off-set. When the cross-correlation has been performed, the location of the peak is found: its position with respect to the center of the reference pattern gives the magnitude of the displacement. This is the direct numerical correlation method requiring N^4 operations to compute cross-correlation where the test pattern has $N \times N$ pixels. The FFT method is often used to calculate the cross-correlation. This requires N^2 operations but is restricted to the test pattern of $N \times N$ where $N = 2^n$, n being an integer. Usually, for the computation of cross-correlation, a test pattern is taken as 32×32 pixels. The PIV community accepts the process of computing cross-correlation by using FFT algorithms.

15.8 MEASUREMENT OF VERY HIGH VELOCITY

For measuring velocity of projectiles and to record its time history, and for the study of impact phenomena, the Doppler-shifted light is fed to a Michelson interferometer with unequal arm lengths: the sensitivity of the interferometer is proportional to the path difference between the two arms. With a slight modification, this interferometer can be used to record the time history of diffuse objects using the velocity interferometer system for any reflector (VISAR). A schematic of the interferometer is shown in Figure 15.8.

The frequency of the light retro-reflected from a target moving with velocity $\bar{V}$ in air is given by

$$\nu_D(t) = \nu_i + \nu_i \frac{2V(t)}{c} \tag{15.23}$$

where ν_i is the frequency of the light incident on a moving target.

Because the light of frequency $\nu_D(t)$ is fed to the Michelson interferometer that has a path difference Δ, the condition for constructive interference can be written as

$$\Delta = m\ \lambda \tag{15.24}$$

where λ is the wavelength of light entering the interferometer. Because Δ is constant and the frequency, consequently the wavelength of the incident light, is changing, the fringe order should change. Thus

$$dm\ \lambda + m\ d\lambda = 0 \Rightarrow |dm| = \frac{\Delta}{\lambda^2}|d\lambda| = \frac{\Delta}{c}\ |d\nu_D| = \frac{\Delta}{c}\ \frac{2\nu_i\ |dV|}{c} = \frac{2\ \Delta}{\lambda\ c}\ |dV| \tag{15.25}$$

In other words, the change in velocity results in the change of fringe order, i.e.,

$$\left|\frac{dV}{dm}\right| = \frac{\lambda\ c}{2\ \Delta} \tag{15.26}$$

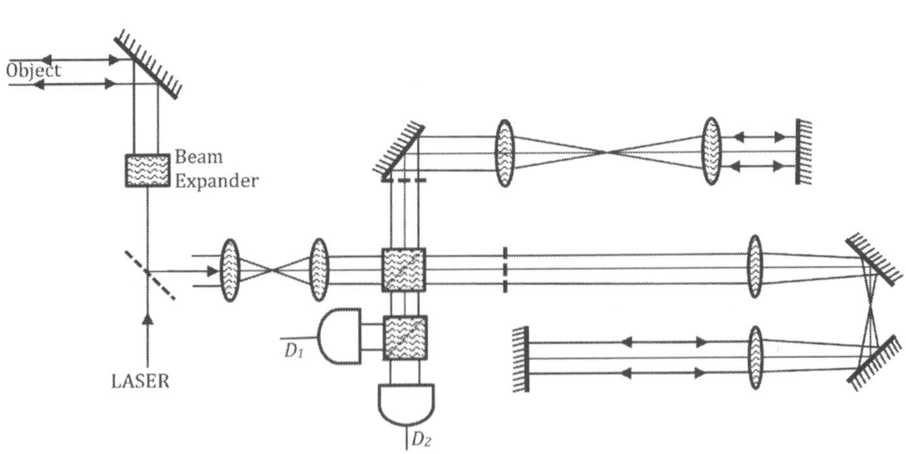

FIGURE 15.8 Schematic of a VISAR.

Due to the change in the fringe order, the irradiance in the interference pattern would vary as

$$I(t) = 2I_0\{1+\cos[2\pi m(t)]\} \tag{15.27}$$

The velocity $V(t)$ would change by $\lambda c/2\Delta$ for a change of one order in the interference fringe pattern. Therefore, the record of irradiance variation with time is also a record of the velocity history of the projectile or the propagation of the shock wave.

As the interferometer is used with any object, diffuse or specularly reflecting, a $4f$ arrangement is used so that the pupils in both the arms of the interferometer are superposed resulting in high-contrast fringes. There is no moving part in the interferometer. One defines the sensitivity of the interferometer as the velocity change that changes the fringe count by one. It is thus given by $\lambda c/2\Delta$. Higher sensitivity requires longer path difference. Therefore, a laser source with a very long coherence length is needed.

Laser beam delivery to the object and collection of the reflected beam can also be done through fiber optics. The interferometer senses only the longitudinal (axial) component of the velocity. In impact studies it is desirable to get information about the transverse component of the velocity as well. This information is obtained by imprinting a grating on the specimen (object). Symmetrically diffracted beams are collected and fed to the interferometer. The signal obtained on impact will carry information about the change in the velocity of the in-plane component that lies in the plane of the diffracted beams.

QUESTIONS

15.1 A laser Doppler system uses two beams from an Ar-ion laser operating at 514 nm. The beams illuminate the test section at angles of ±5° to the viewing direction. A scatterer moves across the test section in the plane of the beams with a velocity component of 1.0 m/s at a right angle to the viewing direction. What is the frequency of the beat signal? If the beams enclose an angle of 15°, what would be the frequency of the beat signal?

15.2 A laser light of 532 nm reflected from a projectile in the direction of its motion is fed into a Michelson interferometer with an optical path difference of 1 m. The velocity of the projectile from its initial value of 1000 m/s is brought to zero linearly over a period of 1 ms. How many fringes are observed in the interferometer? Show the variation of output irradiance with time. What is the sensitivity of the interferometer?

15.3 A two-color laser Doppler system using 488 nm and 514 nm radiation from an Ar-ion laser is used to measure the particle velocity. The system uses an achromatic lens of focal length 800 mm and an aperture of 120 mm. The beam separations at the lens are 100 mm. If a scatterer moves with a velocity of 1.2 m/s at a right angle with the viewing direction but making an angle of 30° with the horizontal plane (plane of beams of 488 nm and 514nm), what is the frequency of the beat signals obtained from each color?

15.4 A pair of parallel pencil beams of wavelength λ is incident on a moving target as shown in Figure 15.9. The scattered light is picked up by the photodetector, which is placed symmetrically with respect to the incident beams. How does the frequency of the photodetector signal depend on the velocity of the target? How is the direction discrimination ability built into the system?

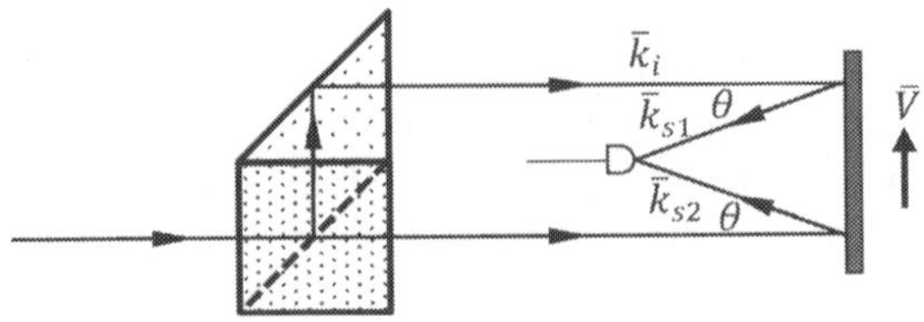

FIGURE 15.9

15.5 It is desired to design a sensor to measure blood velocity in vivo. Figure 15.10 shows a sketch of the experimental arrangement. The beam from a laser is coupled to a multimode fiber by a SELFOC lens. The fiber illuminates a small region in the flow field. The flow velocity makes an angle of θ with the direction of illumination. The scattered beam is picked up by the fiber. The beam reflected by the end face of the fiber is mixed with the scattered beam at the PMT. Show that the Doppler frequency is given by

$$\nu_D = 2\mu V \cos\theta / \lambda$$

where V is the velocity of flow of fluid of refractive index μ and λ is the wavelength of laser.

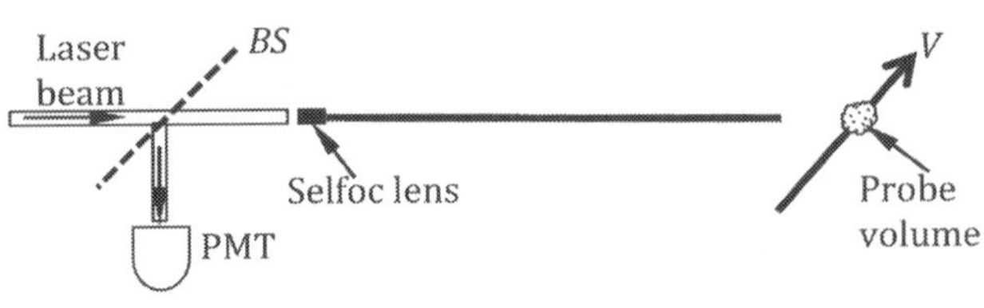

FIGURE 15.10

16 Pressure Measurement

Pressure is defined as force per unit area. Its SI unit is N/m^2 (also called Pascal [Pa]). Therefore, the force and the pressure measurements are synonymous. Optical methods of pressure measurement include the deformation measurement of elastic members like diaphragm, cantilever, etc.; photoelastic method; acousto-optic effect; pressure-sensitive paint (PSP), and pressure-dependent thickness change.

16.1 PRESSURE-SENSITIVE PAINT (PSP)

A typical PSP consists of a luminescent molecule (luminophore) in an oxygen-permeable binder to physically adhere the luminophore to the surface of the test specimen. Pressure measurement with PSP depends on the sensitivity of the specific luminophore to the presence of the oxygen molecules or on the partial pressure of oxygen. This can be understood with the help of the Jablonski diagram given in Figure 16.1.

All the luminescent molecules at the room temperature are in the ground state (S_0). When the luminophore is illuminated with light, it absorbs a photon, raising it to vibrational manifold above the first singlet state (S_1). It then relaxes from S_1

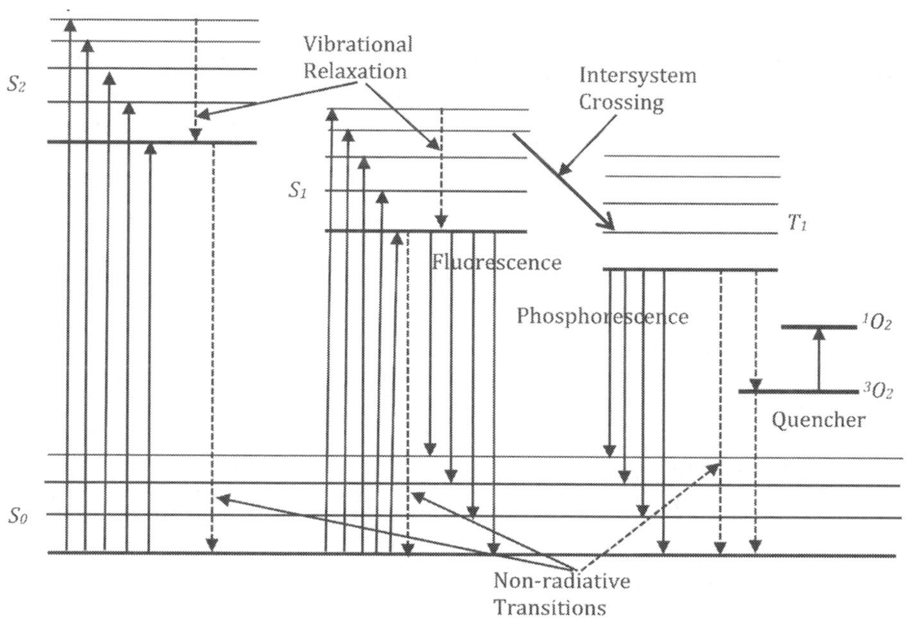

FIGURE 16.1 Jablonski diagram.

DOI: 10.1201/9781003531845-16

to vibrational levels above S_0 emitting a photon of a longer wavelength. Not all the absorbed photons result in the emission of photons.

On absorption of photons from the excitation light, the luminescent molecules are excited to the higher singlet states with their vibration manifolds. Due to vibrational relaxation, these molecules lose some energy and relax to the lowest energy level in the excited singlet states. The relaxation from S_1 to S_0 states gives rise to fluorescence. Due to intersystem crossing, some molecules relax to the excited triplet state T_1. The relaxation from T_1 to S_0 gives rise to phosphorescence.

Some molecules relax to the ground state with nonradiative transition. When oxygen interacts with the molecule such that the transition to the ground state is nonradiative, the process is called oxygen quenching.

The photophysical process in the presence of thermal and oxygen quenching can be described by considering luminescence efficiency Φ, defined as

$$\Phi = \frac{\text{rate of luminescence emission}}{\text{rate of excitation}} = \frac{I_L}{I_a} \tag{16.1}$$

where I_L is the luminescent irradiance and I_a is the irradiance absorbed. As there are several processes involved, this can be expressed as

$$\frac{I_L}{I_a} = \frac{k_L}{k_L + k_D + k_Q\left[O_2\right]} = k_L\ \tau \tag{16.2}$$

where k_L, k_D, and k_Q are the rate constants for luminescence emission, deactivation due to thermal quenching, and nonradiative transition due to oxygen quenching, and $\left[O_2\right]$ is the local oxygen concentration. The lifetime τ of the excited molecule is given by $\tau = 1/\left\{k_L + k_D + k_Q\left[O_2\right]\right\}$. The luminescent irradiance I_L will be maximum when the oxygen concentration is zero $\{\left[O_2\right] = 0\}$ and the corresponding lifetime will be τ_0. Thus,

$$\frac{I_{max}}{I_a} = \frac{k_L}{k_L + k_D} = k_L\ \tau_0 \tag{16.3}$$

Dividing Equation (16.3) with Equation (16.2), we obtain

$$\frac{I_{max}}{I_L} = 1 + \frac{k_Q\left[O_2\right]}{k_L + k_D} = 1 + k_Q\ \tau_0\left[O_2\right] = 1 + K_Q\left[O_2\right] \tag{16.4}$$

The local concentration of oxygen $\left[O_2\right]$ is proportional to the partial pressure of oxygen $P_{[O_2]}$ and hence the above Equation (16.4) can be re-written as

$$\frac{I_{max}}{I_L} = 1 + K\ P_{[O_2]} \tag{16.5}$$

where K is the quenching constant with respect to the partial pressure $P_{[O_2]}$. In practice, it is not possible to realize the condition of zero concentration of oxygen.

Instead, the luminescent irradiance is measured at some reference value where the oxygen concentration or the partial pressure of oxygen is known. Following the same steps as earlier, we obtain the following equation:

$$\frac{I_{ref}}{I_{L(p,T)}} = A(T) + B(T)\frac{p}{p_{\text{ref}}} \tag{16.6}$$

where $A(T) = \frac{k_L + K_D(T)}{k_L + K_D(T_{\text{ref}}) + k_Q(p_{\text{ref}},\, T_{\text{ref}})}$ and $B(T) = \frac{K_Q(p_{\text{ref}},\, T)}{k_L + K_D(T_{\text{ref}}) + k_Q(p_{\text{ref}},\, T_{\text{ref}})}$. It may be seen that $\left[A(T) + B(T)\right]_{T=T_{\text{ref}}} = 1$. Equation (16.6) is known as the Stern–Volmer equation, which is frequently used in PSP measurements. I_{ref} is the luminescent irradiance measured at $p = p_{\text{ref}}$ and $T = T_{\text{ref}}$.

Metalloporphyrin, ruthenium complexes, and pyrene derivatives are commonly used as luminophores for PSP measurements. Metalloporphyrins have very high quantum yield for phosphorescence and hence are very sensitive to oxygen. Platinum tetra (pentafluorophenyl) porphine (PtTFPP) is used widely for PSP measurements. It has a primary absorption peak at 392 nm, weak absorption peaks at 506 nm and 538 nm, and emits at 662 nm. In ruthenium complexes, $[Ru(dpp)_3]^{2+}$ (tris(4,7-diphenyl-1, 10-phenanthroline) ruthenium (II) dichloride) is also frequently used for PSP measurements. It has a broad absorption band with a peak at 500 nm and has an emission peak at 646 nm. One of the pyrene derivatives that is used for PSP measurements is pyrene-1-butyric acid (PBA). It has an absorption spectrum with twin peaks at 328 nm and 344 nm. The emission spectrum has a global maximum at 464 nm.

There could be two different schemes for the measurement of partial pressure of oxygen. In one scheme, the luminescent irradiance with respect to a reference value is to be measured, whereas in the other scheme the decay time is to be measured. The experimental setup would consist of an excitation source, optics to expand the beam to illuminate the PSP-coated object, a charge-coupled device (CCD) camera equipped with a suitable filter, and a processing system. The PSP sample is first calibrated, and calibration curves are then generated, which are used to arrive at the pressure based on the value of luminescent irradiance. In the experimental scheme in which decay time is to be measured, the PSP-coated object is scanned with an unexpanded narrow beam and a photodetector with a suitable filter in front is used. The decay time at different locations on the object is measured and related to the oxygen pressure.

16.2 MEASUREMENT OF PRESSURE WITH PHOTOELASTIC MATERIAL

When stress is applied to some isotropic materials such as glass or plastics, they become anisotropic. This condition is temporary and disappears when the stress is removed. The stress produces physical deformation, which completely alters the isotropic state of the material. Let us consider a plate of thickness d of an isotropic material. This is subjected to uniaxial stress. The stress-optic law takes a very simple form

$$(\mu_1 - \mu_2) = C\sigma \tag{16.7}$$

where μ_1 and μ_2 are the principal refractive indices, C is the stress-optic constant specific to the material and σ is the applied stress. The plate behaves like a uniaxial crystal.

Now let us consider a beam of linearly polarized light of wavelength λ incident normally on a plate of thickness d of the photoelastic material. The beam is propagating along the z-direction. Within the plate two linearly polarized beams, one vibrating in the x–z plane and other in the y–z plane, are supported. These two beams acquire a phase difference in traversing the plate and emerge at the exit surface with a phase difference δ, which is given by

$$\delta = \frac{2\pi}{\lambda}|\mu_1 - \mu_2|d = \frac{2\pi}{\lambda}C\sigma d \tag{16.8}$$

The phase change δ depends linearly on the difference of the principal refractive indices, on the thickness of the plate, and inversely on the wavelength of light used. This phase change introduced by the plate of a known thickness is measured at a particular wavelength. Therefore, the stress and consequently the pressure is measured.

Let us consider an experimental setup as shown in the Figure 16.2.

The transmission direction of polarizer P makes an angle of 45° with the x-axis. The photoelastic material (*PEM*) is loaded along the y-direction. The slow axis of the quarter-wave plate makes an angle of 45° with the x-axis. The polarizing beam splitter (*PB*) allows the field along the slow axis of the quarter-wave plate. The irradiances detected by the detectors D_1 and D_2 are those corresponding to the parallel and perpendicular components of the electric field.

Let us assume that the field transmitted by the polarizer is described by

$$E(z;t) = E_0\cos(\omega\, t - kz) \tag{16.9}$$

The field can be written in its components along the x- and y-axes as

$$E_x(z;t) = \frac{E_0}{\sqrt{2}}\cos(\omega\, t - kz) \tag{16.10a}$$

$$E_y(z;t) = \frac{E_0}{\sqrt{2}}\cos(\omega\, t - kz) \tag{16.10b}$$

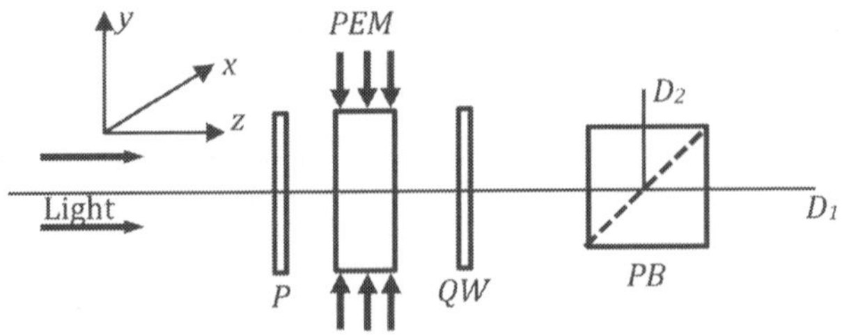

FIGURE 16.2 Schematic of experimental setup for pressure measurement using photoelastic material (*PEM*).

These field components are incident on the photoelastic plate of thickness *d*. The field components transmitted by the *PEM* of thickness *d* are

$$E_{xM}(z;t)=\frac{E_0}{\sqrt{2}}\cos(\omega\, t-kz-\psi_2) \tag{16.11a}$$

$$E_{yM}(z;t)=\frac{E_0}{\sqrt{2}}\cos(\omega\, t-kz-\psi_2-\delta) \tag{16.11b}$$

where $\psi_2=\frac{2\pi}{\lambda}(\mu_2-1)d$ and $\delta=\frac{2\pi}{\lambda}(\mu_1-\mu_2)d$.

These components are resolved along the slow and the fast axes of the quarter-wave plate. The field components just before the quarter-wave plate are

$$E_s(z;t)=\frac{E_0}{2}\cos(\omega\, t-kz-\psi_2)+\frac{E_0}{2}\cos(\omega\, t-kz-\psi_2-\delta) \tag{16.12a}$$

$$E_f(z;t)=-\frac{E_0}{2}\cos(\omega\, t-kz-\psi_2)+\frac{E_0}{2}\cos(\omega\, t-kz-\psi_2-\delta) \tag{16.12b}$$

The field components after the quarter-wave plate are

$$E_{se}(z;t)=\frac{E_0}{2}\cos\left(\omega\, t-kz-\psi_2-\psi_1-\frac{\pi}{2}\right)+\frac{E_0}{2}\cos\left(\omega\, t-kz-\psi_2-\psi_1-\delta-\frac{\pi}{2}\right) \tag{16.13a}$$

$$E_{fe}(z;t)=-\frac{E_0}{2}\cos(\omega\, t-kz-\psi_2-\psi_1)+\frac{E_0}{2}\cos(\omega\, t-kz-\psi_2-\psi_1-\delta) \tag{16.13b}$$

where $\psi_1=\frac{2\pi}{\lambda}(\mu_e-1)d'$ and $\frac{\pi}{2}=\frac{2\pi}{\lambda}(\mu_0-\mu_e)d'$; μ_0, μ_e are the refractive indices along slow and fast axes; and d' is the thickness of the quarter-wave plate.

Writing $\tau=\omega\, t-kz-\psi_2-\psi_1$, the field components exiting from the quarter-wave plate are rewritten as

$$E_{se}(z;t)=\frac{E_0}{2}\sin\tau+\frac{E_0}{2}\sin(\tau-\delta)=E_0\sin\left(\tau-\frac{\delta}{2}\right)\cos\frac{\delta}{2} \tag{16.14a}$$

$$E_{fe}(z;t)=-\frac{E_0}{2}\cos\tau+\frac{E_0}{2}\cos(\tau-\delta)=E_0\sin\left(\tau-\frac{\delta}{2}\right)\sin\frac{\delta}{2} \tag{16.14b}$$

The analyzer with its transmission direction parallel to that of the polarizer will transmit $E_{se}(z;t)$, and with axis perpendicular will transmit $E_{fe}(z;t)$. Therefore the irradiance detected by D_1 is proportional to $\cos^2\frac{\delta}{2}$ and the irradiance detected by detector D_2 is proportional to $\sin^2\frac{\delta}{2}$, i.e., $I_1=I_0\cos^2\frac{\delta}{2}$ and $I_2=I_0\sin^2\frac{\delta}{2}$. When a pressure p is applied to the material, it induces stress, which in turn introduces the phase difference δ. The phase difference δ is related to the pressure p as

$$\delta=\frac{2\pi}{\lambda}CpA\, d\Rightarrow p=\frac{\lambda}{2\pi}\frac{1}{CAd}\delta \tag{16.15}$$

where A is the area of the *PEM* plate. By measuring δ and with the knowledge of the geometry of the plate and its stress-optic coefficient, the pressure can be measured. For measurement of δ, we can follow either of the following approaches:

A: The signals from the detectors D_1 and D_2 are subtracted, which then gives

$$\Delta I(\delta) = I_1 - I_2 = I_0 \cos\delta \tag{16.16}$$

This value can be divided by the sum $(I_1 + I_2)$ of I_1 and I_2. This would reduce the influence of fluctuations in the signals arising due to environmental factors.

B: The signal I_2 from detector D_2 is divided by the signal I_1 from detector D_1 resulting in

$$\frac{I_1}{I_2} = \tan^2\frac{\delta}{2} \tag{16.17}$$

The response is nonlinear.

16.3 RUBY PRESSURE STANDARD

Fluorescence of R_1 and R_2 lines of ruby (Cr^{3+} doped Al_2O_3) has been used to measure high hydrostatic pressures. Pressure can be measured very accurately up to pressures of 10 MPa and with moderate accuracy up to 100 GPa. Ruby is excited with a laser typically in the blue or green region of the spectrum and the fluorescence spectrum is observed. Ruby or ruby grains are placed along with hydrostatic pressure medium in a diamond anvil cell, which is subjected to pressure. It is observed that the wavelength of the R_1 line of ruby shifts with an application of hydrostatic pressure P as

$$P = \frac{A}{B}\left[\left(\frac{\lambda}{\lambda_0}\right)^B - 1\right] \text{GPa} \tag{16.18}$$

where λ is the measured wavelength at pressure P (in GPa) of the ruby R_1 line, wavelength $\lambda_0 = 694.24$ nm is the zero pressure value at 298 K temperature, and $A = 1904$ and $B = 5$. This formula fits the experimentally obtained pressure data very well but is quite ad hoc, as it is not based on an expected variation of energy levels of ruby with the pressure. The equation fits the data well because the shift of the ruby R_1 line is almost linear with the pressure, with a small nonlinear correction.

The experimental data also fit well to the following quadratic equation:

$$P = a\left[\left(\frac{\lambda - \lambda_0}{\lambda}\right) + b\left(\frac{\lambda - \lambda_0}{\lambda}\right)^2\right] GPa \tag{16.19}$$

with $a = 1798 \pm 8.4$ and $b = 8.57 \pm 0.15$.

In some respects, the ultimate diamond anvil apparatus for the measurement of hydrostatic pressure using fluorescence of ruby is that of Barnett, Block, and Piermarini. They have described an optical system for rapid routine pressure measurement, which utilizes a pressure shift in the sharp R line fluorescence of ruby. A metal gasket (simply a relatively large square of Inconel about 0.13 mm thick in which a hole of about 0.20 mm in diameter has been drilled) is centered between the diamond anvils, and a 4:1 mix of methanol and ethanol is placed in the hole. A speck of ruby is added as well as a speck of whatever else is desired, and the anvils are closed on the gasket. Whatever is inside is trapped, and further advance of the anvils subjects the contents to hydrostatic pressure all the way up to 100,000 atm. The pressure is determined by the fluorescence of the speck of ruby!

16.4 FABRY–PEROT ETALON AS PRESSURE SENSOR

The Fabry–Perot (FP) etalon consists of two glass or quartz plates separated by a fixed distance and having the inner surfaces coated for high reflectivity. A collimated beam bounces back and forth resulting in multiple-beam interference pattern both in reflection and transmission. The condition for the formation of bright fringe or high transmission is given by

$$2\mu d + \frac{\phi_1}{2\pi} + \frac{\phi_2}{2\pi} = m\lambda \tag{16.20}$$

where μ is the refractive index of the medium between the plates, d is the distance between the plates, ϕ_1 and ϕ_2 are the phase changes at reflection at the two high reflecting surfaces, m is the fringe order, and λ is the wavelength of radiation incident on the etalon. We may consider ϕ_1 and ϕ_2 to be equal and can also set them equal to zero without any loss: these being constant simply shift the positions of maxima of the interference pattern.

The irradiance distribution in transmission is given by

$$I(\delta) = I_0 \frac{1}{1 + F\,\sin^2(\delta/2)} \tag{16.21}$$

where $\delta\left(=\frac{2\pi}{\lambda}2\mu d\right)$ is the phase difference and $F = 4R/(1-R)^2$ with R being the reflectivity of each surface. The transmitted irradiance varies with δ being maximum equal to I_0 when $\delta = 2m\pi$ and minimum equal to $I_0/(1+F)$ when $\delta = (2m+1)\frac{\pi}{2}$. The phase difference δ can be changed either by changing μ or d, or both. Therefore, two types of pressure sensors can be conceived: in one, the distance d changes with an application of the pressure and in the other, the refractive index of the medium inside the etalon changes.

16.4.1 FP Etalon with Flexible Mirrors

Let us assume that the mirrors of the etalon are thin so that they can be flexed by an application of the pressure. Let us further assume that the mirrors are of circular geometry and rigidly clamped at the edges. Let a pressure p be applied to one of the mirrors, which flexes it. The central deflection of the mirror z_0 can be expressed as

$$z_0 = \frac{3(1-\nu^2)p}{16E\,h^3}a^4 \tag{16.22}$$

where a and h are the radius and the thickness of the mirror, E and ν are the Young's modulus and the Poisson's ratio of the material of the mirror, and p is the applied pressure. This expression is valid if the deflection is less than 30% of the thickness. It is, thus, seen that the central deflection is proportional to the applied pressure. If a narrow collimated laser beam is used, then for a small deflection, the central region of the mirror may be considered nearly flat. Therefore, the phase difference δ between the beams can now be expressed as

$$\delta = \frac{2\pi}{\lambda}2\mu(d - z_0) \tag{16.23}$$

This will cause the transmitted irradiance to change, which may be related to the applied pressure. Though the relation between z_0 and δ is linear, the relation between z_0 and the transmitted irradiance is highly nonlinear. Therefore, the operation point should be carefully chosen so that this relationship is nearly linear.

16.4.2 Change of Refractive Index

The density of all the materials, solid, liquid, and gases shows dependence on the applied pressure. We consider a simple case of air, assuming it a single component gas, which fills the space between the mirrors of the etalon.

The mean polarizability α of molecules in a medium can be expressed as

$$\alpha = \frac{3}{4\pi N}\frac{\mu^2 - 1}{\mu^2 + 2} \tag{16.24}$$

where N is the number of molecules per unit volume and μ is the refractive index of the medium. This relation is known as the Lorentz–Lorenz formula. Instead of the mean polarizability, we can also use the total polarizability A, which is the polarizability of one mole of a substance. The total polarizability, also called the molar or atomic polarizability is defined as

$$A = \frac{4\pi}{3}N_{AG}\alpha \tag{16.25}$$

where N_{AG} is the Avogadro number, i.e., the number of molecules in a mole of a substance.

Therefore, the refractive index of the medium is now related to the molar polarizability through the following relation

$$\frac{\mu^2 - 1}{\mu^2 + 2} = A\frac{N}{N_{AG}} \tag{16.26}$$

We now use the ideal gas law, $PV = nRT$, where P is the absolute pressure of the gas, V its volume, n is the number of moles, R is the universal gas constant, and T is the absolute temperature. Therefore Equation (16.26) can be rewritten as

$$\frac{\mu^2 - 1}{\mu^2 + 2} = A\frac{\rho}{M} = A\frac{P}{RT} \tag{16.27}$$

where ρ is the density and M is the molar mass of the gas. For the gases, the refractive index is nearly unity and therefore the Equation (16.27) is simplified to

$$\mu \approx 1 + \frac{3}{2}A\frac{P}{RT} \tag{16.28}$$

It may be noted that A is practically constant over a moderate pressure range. Under such a situation, the refractive index of the gas is linearly related to the applied pressure. Therefore, if the pressure of the gas inside the FP etalon varies, the transmitted irradiance would vary and hence it can be adapted for the measurement of pressure. Again the operating point should be judiciously chosen.

This principle is applied in the design of an optical microphone. The medium between the mirrors of the FP etalon is air. Its pressure varies due to the acoustical waves. However, in this case the process is not isothermal as has been assumed in the foregoing analysis. If we consider the process to be adiabatic, then $P^{1-\gamma}T^{\gamma} = \text{constant}$, where γ is the adiabatic constant.

When adiabatic process is considered, then a change in pressure dP can be related to the change in the refractive index $d\mu$ through the relation

$$d\mu = \frac{3}{2}\frac{A}{\gamma\, RT}dP \tag{16.29}$$

This change in the refractive index modulates the transmitted irradiance of the FP etalon. Again, the operating point should be carefully chosen.

QUESTIONS

16.1 It is observed that the wavelength of the R_1 line of ruby shifts with an application of hydrostatic pressure P as

$$P = \frac{A}{B}\left[\left(\frac{\lambda}{\lambda_0}\right)^B - 1\right] \text{ GPa}$$

where λ is the measured wavelength at pressure P (in GPa) of the ruby R_1 line, wavelength $\lambda_0 = 694.24$ nm is the zero pressure value at 298 K temperature, and $A = 1904$ and $B = 5$. How much is the shift of fluorescence in line R_1 when the pressure is 100,000 atm? Instead of this expression, if we use the following expression

$$P = a\left[\left(\frac{\lambda - \lambda_0}{\lambda}\right) + b\left(\frac{\lambda - \lambda_0}{\lambda}\right)^2\right] \text{ GPa}$$

with $a = 1798 \pm 8.4$ and $b = 8.57 \pm 0.15$, what pressure would we get for the shift calculated earlier?

16.2 A polariscope is assembled with various components arranged as

$$P_{\frac{\pi}{2}}\ Q_{\frac{\pi}{4}}\ M_{\alpha,\delta}\ Q_{\frac{\pi}{2}}\ A_{\frac{\pi}{4}}$$

where P, Q, M, and A stand for the polarizer, the quarter-wave plate, the model, and the analyzer, respectively, and the subscripts give their orientations. Show that the irradiance transmitted is given by

$$I(\delta) = A(1+\cos\delta)$$

where δ is the phase difference introduced by the model and A is a constant.

16.3 An FP etalon consists of two thin mirrors of BK7 glass of a diameter of 25 mm and coated with a reflectivity of 80%. Assuming the thickness of the mirrors to be 2 mm, what pressure applied to a mirror causes the transmitted irradiance to fall to 50% of its peak value? Phase change due to reflection at the mirror surfaces may be taken as zero. Take $E = 85 \times 10^3 \text{N/mm}^2$, $\nu = 0.202$, density $\rho = 2.43$ g/cc^3, and the wavelength of operation $\lambda = 632.8$ nm.

17 Temperature Measurement

17.1 INTRODUCTION

Temperature is one of the seven base/fundamental quantities in the SI system and its unit is Kelvin (K). Many physical properties of materials are influenced by the temperature, hence, its measurement is crucial. Further, many physical processes are also influenced by the temperature. In fact, temperature is one of the process variables. There are several scales that are used to express temperature of a body. Of these, the thermodynamic scale or the Kelvin scale has absolute zero as its null. Kelvin is defined as the fraction 1/273.16 of the thermodynamic temperature of the triple point of water. Other scales to express temperature are the Celsius scale and the Fahrenheit scale. In the Centigrade scale the triple point of water is expressed as 0.01 degree Celsius (0.01°C). To convert Kelvin to degrees Celsius, following relation is used

$$t_{\text{Celsius}} = T_{\text{Kelvin}} - 273.15$$

because the thermodynamic temperature of the triple point of water is 0.01°C. The following relation is used when converting Fahrenheit (°F) to Celsius (°C)

$$t_{\text{Celsius}} = \frac{5}{9}\left(t_{\text{Fahrenheit}} - 32\right)$$

It may be noted that the Fahrenheit scale and the Celsius scale intersect at −40, i.e., −40°F = −40°C.

It has been mentioned earlier that many physical properties of materials are influenced by the temperature; hence, there are a large number of methods to measure the temperature. In this chapter, we discuss only those methods that utilize light.

17.2 BLACKBODY RADIATOR

Planck's law describes the spectral radiant emittance $M(\lambda,T)$ of a perfect blackbody radiator as a function of its temperature and the wavelength of the emitted radiation as

$$M(\lambda,T) = \frac{C_1}{\lambda^5\left(e^{C_2/\lambda T} - 1\right)} \ (\text{W/m}^2\ \mu\text{m}) \tag{17.1}$$

DOI: 10.1201/9781003531845-17

where λ is in micrometers and the constants $C_1 = 2\pi hc^2 = 3.74177\times10^8\left(\text{W } \mu\text{m}^4/\text{m}^2\right)$ and $C_2 = hc/k = 1.43878\times10^4\left(\mu\text{m K}\right)$. The spectral radiance $E(\lambda,T)$ of the blackbody is given by

$$E(\lambda,T)=\frac{M(\lambda,T)}{\pi}=\frac{C_1}{\pi\lambda^5\left(e^{C_2/\lambda T}-1\right)}\left(\text{W}/\text{m}^2\ \mu\text{m sr}\right) \tag{17.2}$$

This is also called the spectral emissive power of a blackbody when the wavelength is taken in meters. The values of the constants then are $C_1 = 3.74177\times10^{-16}\left(\text{W m}^2\right)$ and $C_2 = 1.43878\times10^{-2}\,(\text{m K})$.

By integrating $M(\lambda, T)$ over all the wavelengths, we obtain the Stefan–Boltzmann law

$$M(T)=\int_0^\infty M(\lambda,T)\,d\lambda=\sigma\,T^4 \tag{17.3}$$

where $\sigma=\left(2\pi^5\,k^4\right)/\left(15\,c^2h^3\right)=5.672\times10^{-8}\ \text{W}/\text{m}^2\ \text{K}^4$. $M(T)$ is also known as radiant emittance. It represents the power emitted per unit area by the blackbody.

In practice, real bodies, also called gray bodies, are not blackbodies and emit less radiation than predicted by Planck's law by a factor of ε, the emissivity of the surface. The spectral radiant emittance of a gray body radiator is given by

$$M_g(\lambda,T)=\varepsilon(\lambda)\frac{C_1}{\lambda^5\left(e^{C_2/\lambda T}-1\right)}=\varepsilon(\lambda)M(\lambda,T) \tag{17.4}$$

where emissivity $\varepsilon(\lambda)$ is a function of wavelength.

The Stefan–Boltzmann law for gray bodies with emissivity ε takes the form

$$M(T)=\varepsilon\,\sigma\,T^4 \tag{17.5}$$

There are many optical methods to measure temperature of bodies, some of them are described in the following sections.

17.3 METHODS OF TEMPERATURE MEASUREMENT

17.3.1 Pyrometers

A pyrometer is a radiation thermometer to measure the temperature of distant objects. The radiation emitted by the body is used to measure temperature. Several kinds of radiation pyrometers exist. Some of them include the following:

Total radiation pyrometers
Photoelectric pyrometers
Dual-wavelength or ratio pyrometers
Multi-wavelength pyrometers

17.3.1.1 Total Radiation Pyrometers

Figure 17.1 shows a schematic of a total radiation pyrometer. The radiation from the body/target is focused on the detector by a lens. The material of the lens is such that it has a transmission over a broad wavelength range. The detector is a thermal detector (bolometer/thermocouple/thermopile). The radiation reaching the detector raises its temperature. The pyrometer is used to measure temperature in the wavelength band of 0.2 μm to 14 μm.

Because the ambient temperature T_{amb} around the body cannot be ignored, the radiant emittance $M_g(T)$ from the body is expressed as

$$M_g(T) = \varepsilon\,\sigma\left(T^4 - T_{\text{amb}}^4\right) \tag{17.6}$$

Not all the radiation emitted by the body reaches the detector. Further, there is a heat exchange with the housing of the detector. The pyrometer is calibrated with the blackbody radiation. The sensitivity of the pyrometer is obtained from

$$\frac{\Delta T}{T} = \frac{1}{4}\frac{\Delta\varepsilon}{\varepsilon} \tag{17.7}$$

17.3.1.2 Photoelectric or Spectral Band Pyrometers

The pyrometer uses a photoelectric detector and operates in the wavelength band in which the signal is generated by photons incident on the detector. The spectral band pyrometer measures the radiance over a relatively narrow wavelength band within the spectral range of 0.5 μm to 25 μm. The choice of wavelength mostly depends on the temperature range.

The lens (L_e) collects the radiation from the surface of the body and focuses on the detector (D) (Figure 17.2). Before reaching the detector, the radiation passes through a narrowband filter F to select the wavelength of interest. Two apertures AA and AB define the target area or field of view, and the acceptance angle of the pyrometer. The output of the detector is a complex function of the dimensions of the apertures, the transmission of the various optical components, and the detector responsivity.

The pyrometer measures the spectral radiance L_m given by

$$L_m = \varepsilon(\lambda)\,L(\lambda, T_s) - L(\lambda, T_d) \tag{17.8}$$

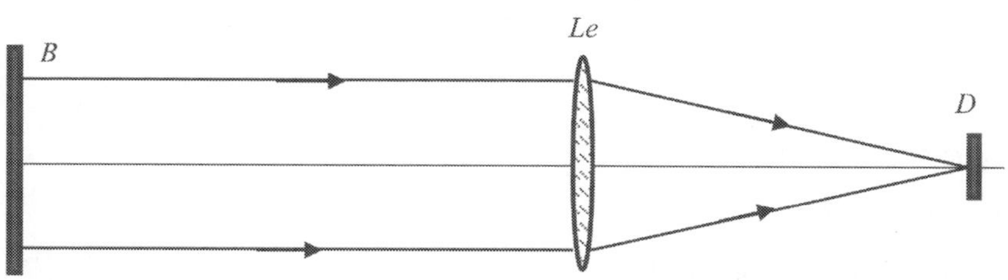

FIGURE 17.1 Schematic of a radiation pyrometer. B, body/target; Le, lens; D, detector.

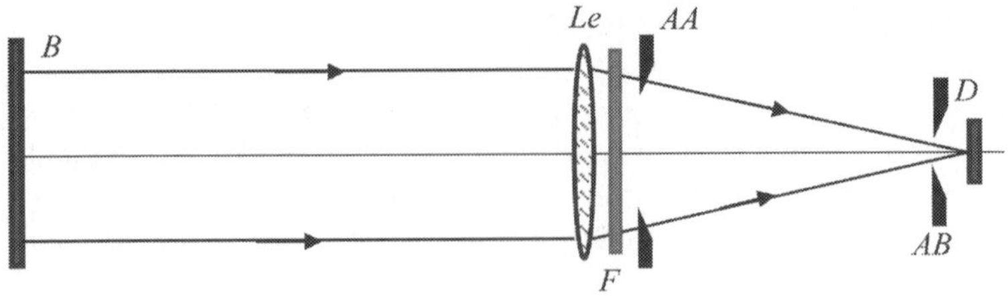

FIGURE 17.2 Schematic of a spectral band pyrometer. *B*, body/target; *Le*, lens; *F*, filter; *AA*, angle-defining aperture; *AB*, target-defining aperture; *D*, detector.

where $\varepsilon(\lambda)$ is the emissivity of the target surface, T_s is the surface temperature of the target, and T_d is the temperature of the detector. The emissivity of the detector is taken as unity. For the measurement of temperatures ≥150°C, the radiance term due to the detector makes a negligibly small contribution and can be dropped. However, for lower temperatures it should be taken in account.

Alternately the concept of an effective wavelength λ_{eff} representation of the bandwidth can be used. For a narrowband pyrometer, the effective wavelength can be approximated by

$$\lambda_{\text{eff}} = A + \frac{B}{T} \tag{17.9}$$

where A and B are constants, and the effective wavelength λ_{eff} is in micrometers. When $\lambda T \ll C_2$, the output of the detector $V(T)$ is obtained using Wien's law as

$$V(T) = C\, e^{-\frac{C_2}{AT+B}} \tag{17.10}$$

where C is another constant.

17.3.1.3 Dual-Wavelength Pyrometers

Dual-wavelength pyrometers, also called ratio pyrometers, compare the spectral radiance from the target/body in two wavelengths using photoelectric detectors. The ratio R of the spectral radiances from the body at the two wavelengths λ_1 and λ_2 is given by

$$R = \frac{E(\lambda_1,T)}{E(\lambda_2,T)} = \frac{\varepsilon(\lambda_1)}{\varepsilon(\lambda_2)}\left(\frac{\lambda_2}{\lambda_1}\right)^5 \frac{\left(e^{\frac{C_2}{\lambda_2 T}} - 1\right)}{\left(e^{\frac{C_2}{\lambda_1 T}} - 1\right)} \tag{17.11}$$

Alternately

$$\ln\left[R\frac{\varepsilon(\lambda_2)}{\varepsilon(\lambda_1)}\left(\frac{\lambda_1}{\lambda_2}\right)^5\right] = \ln\left[\frac{\left(e^{\frac{C_2}{\lambda_2 T}} - 1\right)}{\left(e^{\frac{C_2}{\lambda_1 T}} - 1\right)}\right] \tag{17.12}$$

If the temperature of the body is low such that $\lambda_1 T \ll C_2$ and $\lambda_2 T \ll C_2$, then

$$\ln\left[R\frac{\varepsilon(\lambda_1)}{\varepsilon(\lambda_2)}\left(\frac{\lambda_1}{\lambda_2}\right)^5\right] = C_2\frac{\lambda_1 - \lambda_2}{\lambda_1\lambda_2}\frac{1}{T} \tag{17.13}$$

From this the temperature of the body is obtained as

$$T = C_2\frac{\lambda_1 - \lambda_2}{\lambda_1\lambda_2}\frac{1}{\ln\left[R\frac{\varepsilon(\lambda_1)}{\varepsilon(\lambda_2)}\left(\frac{\lambda_1}{\lambda_2}\right)^5\right]} \tag{17.14}$$

In order to get the temperature of the body, its spectral emissivities at the wavelengths are to be measured. If the body is a metal, its spectral emissivity can be calculated using the formula

$$\varepsilon(\lambda) = K\sqrt{\frac{\rho}{\lambda}} \tag{17.15}$$

where K is a constant and has a value $K = 0.365\ \Omega^{-1/2}$, ρ is the resistivity in ohm-centimeters, and λ is the wavelength in centimeters. If the body is nonconducting, its emissivity is calculated using the formula

$$\varepsilon(\lambda) = \frac{4\mu(\lambda)}{\left[1+\mu(\lambda)\right]^2} \tag{17.16}$$

where $\mu(\lambda)$ is the refractive index of the body at the wavelength λ. If we assume that the emissivity of the body is the same at the two wavelengths, i.e., $\varepsilon(\lambda_1) = \varepsilon(\lambda_2)$,

$$T = C_2\frac{\lambda_1 - \lambda_2}{\lambda_1\lambda_2}\frac{1}{\ln\left[R\left(\frac{\lambda_1}{\lambda_2}\right)^5\right]} \tag{17.17}$$

This can be recast as

$$T = B\frac{1}{\ln\left[\frac{R}{A}\right]} \quad \text{where } B = C_2\frac{\lambda_1 - \lambda_2}{\lambda_1\lambda_2} \text{ and } A = \left(\frac{\lambda_2}{\lambda_1}\right)^5$$

Alternately the ratio R can be expressed as

$$R = Ae^{\frac{B}{T}} \tag{17.18}$$

It should be noted that dual-wavelength pyrometers are less sensitive than the spectral band pyrometers.

17.3.2 Refractive Index-Based Methods

The refractive index of a transparent medium changes with temperature. This dependance of refractive index on temperature is the basis of the methods that fall under this category. Temperature distribution in the medium results in its refractive index variation; a homogeneous medium becomes inhomogeneous. The ray propagation in an inhomogeneous medium is described by the eikonal equation

$$\frac{d}{ds}\left(\mu\frac{d\mathbf{r}}{ds}\right)=\nabla\mu$$

where ds is the arc length defined as $ds^2 = dx^2 + dy^2 + dz^2$, $\mathbf{r}$ is the position vector $\mathbf{r} = x\hat{i} + y\hat{j} + z\hat{k}$, and $\mu(x, y, z)$ is the refractive index at a point in the medium. Let us consider that the inhomogeneous medium is confined between $z = 0$ and $z = L$ (test section) and refractive index variation is in the y–z plane. A ray is propagating along the z-direction in the inhomogeneous medium. As it propagates, the arc length changes to dz and each displacement dz produces a small deflection of the ray. The angle of deflection α is given by $\alpha = \frac{1}{\mu}\int_0^L \frac{\partial\mu}{\partial y}dz$ as shown in Figure 17.3.

The schlieren method, shadowgraphy, interferometry, speckle interferometry, and speckle shear and hologram interferometry (HI) are the methods that can be listed under this broad category. The schlieren method responds to the derivative of the refractive index, shadowgraphy responds to the second derivative of the refractive index, and the remaining methods respond to the refractive index variations. It may also be mentioned that the shadowgraphy is the least expensive and simple method but requires quite a bit of computation.

We shall apply these methods to measure the temperature distribution in flames or a similar combustion process. In many practical situations, the density ρ of the gas can be determined using ideal gas equation, i.e., $p = \rho RT / M$, where R is the gas constant, T its absolute temperature, and M is the molecular weight. Further the density of the gas is related to its refractive index through the Gladstone–Dale relation $\mu - 1 = \mathbb{K}\rho$, where $\mathbb{K}$ is the Gladstone–Dale constant. If pressure is considered constant during the combustion process, we have

$$\frac{\mu-1}{\mu_0-1}=\frac{\rho}{\rho_0}=\frac{p}{p_0}\frac{T_0}{T}$$

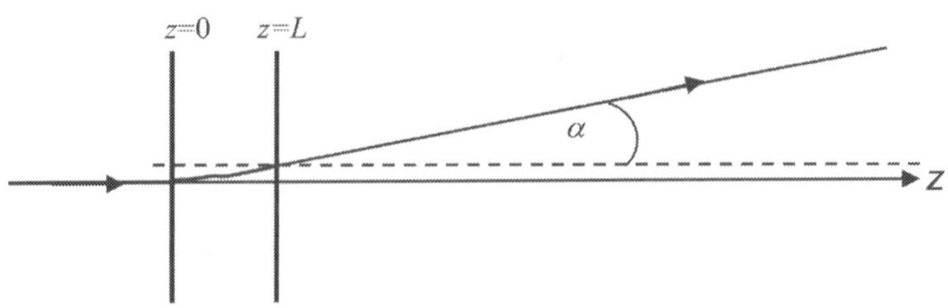

FIGURE 17.3 Ray propagation in an inhomogeneous medium.

where p_0 is the pressure (Pa) and T_0 is the temperature (K) at some reference point. This gives

$$T = \frac{\mu_0 - 1}{\mu - 1}\frac{p}{p_0}T_0$$

On differentiating with respect to y,

$$\frac{\partial T}{\partial y} = -\frac{\mu_0 - 1}{(\mu - 1)^2}\frac{p}{p_0}T_0\frac{\partial \mu}{\partial y} = -\frac{\mu_0 - 1}{(\mu - 1)^2}\frac{\mu - 1}{\mu_0 - 1}T\frac{\partial \mu}{\partial y} = -\frac{1}{\mu - 1}T\frac{\partial \mu}{\partial y}$$

Alternately

$$\frac{\partial \mu}{\partial y} = -(\mu - 1)\frac{1}{T}\frac{\partial T}{\partial y}$$

Differentiating again with respect to y yields

$$\frac{\partial^2 \mu}{\partial y^2} = \mathbb{K}\left[\frac{2\rho}{T^2}\left(\frac{\partial T}{\partial y}\right)^2 - \frac{\rho}{T}\frac{\partial^2 T}{\partial y^2}\right]$$

17.3.2.1 Schlieren Method

There are a large number of experimental arrangements; some use lenses and others use mirrors. The source could be an extended source or a point source. A knife is used at the image plane of the source. We will, however, discuss an experimental arrangement that uses two good quality positive lenses, and the test section is in the collimated beam as shown in Figure 17.4. A knife edge K is placed at the focal plane of lens L_2. When the knife edge is properly located, the image at the screen Sc will be dark when there is no flame or flow in the test section. A minute downward displacement of the knife edge will make the image of the test section bright. Having properly positioned the knife edge (dark image), and with combustion or flow on, the inhomogeneities due to refractive index variation will deflect the incident ray. One such ray has been shown in the Figure 17.4. This ray is not blocked by the knife edge

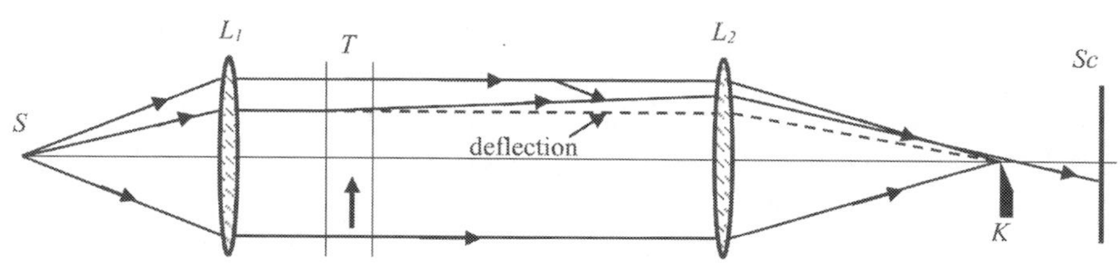

FIGURE 17.4 Schematic of the schlieren experimental setup. S, source; L_1 and L_2, lenses; T, test section; K, knife edge; Sc, screen (image plane to test section).

and that region of the test section appears bright in the image. The angle of deflection α is given by

$$\alpha = \frac{1}{\mu}\int_0^L \frac{\partial \mu}{\partial y}\,dz$$

For 2D schlieren of extent L along the optical axis, this becomes

$$\alpha = \frac{L}{\mu}\frac{\partial \mu}{\partial y}$$

Because the knowledge of the angle of deflection allows the refractive index to be evaluated, the temperature can also be evaluated.

It is mentioned that the temperature range for schlieren measurements is approximately 0°C to 2000°C, with a sensitivity of the order of 0.1°C and an accuracy of 10% of the range.

17.3.2.2 Shadowgraphy

The experimental setup to determine temperature distribution in a flame is extremely simple. A collimated beam passes through the test section as shown in Figure 17.5. Depending on the temperature distribution that causes the refractive index inhomogeneities, different rays in the collimated beam undergo different deflections, thereby creating variation in the irradiance at the screen. The linear displacement of the rays in the beam is difficult to measure as it is usually very small. Instead, the contrast in the shadowgraphs is measured. The relationship between the irradiance I at the test section and the irradiance I_{sc} at the screen is

$$\frac{I_{sc}}{I} = \frac{\Delta y}{\Delta y_{sc}}$$

The contrast C at the screen is defined as

$$C = \frac{I_{sc} - I}{I_{sc}} = \frac{\Delta I}{I_{sc}} = 1 - \frac{\Delta y_{sc}}{\Delta y} = 1 - \frac{\Delta y + d\alpha\, z_{sc}}{\Delta y} = -z_{sc}\frac{d\alpha}{\Delta y} \cong -z_{sc}\frac{\partial \alpha}{\partial y} = -\frac{z_{sc}}{\mu}\int \frac{\partial^2 \mu}{\partial y^2}\,dz$$

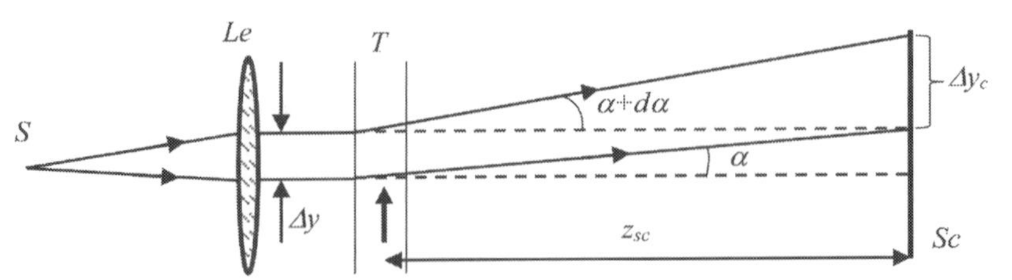

FIGURE 17.5 Schematic of the shadowgraphy experimental setup. *S*, source; *L*, lens; *T*, test section; *Sc*, screen.

If the pressure in the test section is assumed constant and the gas law holds, then the contrast in the shadowgraph can be related to the temperature variations as

$$\frac{\Delta I}{I_{sc}} = -\frac{z_{sc}}{\mu}\mathbb{K}\int\left[\frac{2\rho}{T^2}\left(\frac{\partial T}{\partial y}\right)^2 - \frac{\rho}{T}\frac{\partial^2 T}{\partial y^2}\right]dz$$

Shadowgraphy is less sensitive compared with the schlieren method because $\frac{\partial^2 \mu}{\partial y^2}$ is much smaller than $\frac{\partial \mu}{\partial y}$. On the other hand, $\frac{\partial^2 \mu}{\partial y^2}$ can be much larger than $\frac{\partial \mu}{\partial y}$ in gas flows involving shock waves or turbulence both of which around supersonic projectiles. For this reason, the shock waves stand out as stark lines in the shadowgraph.

17.3.2.3 Interferometry

Interferometry measures the phase change that the light experiences when it passes through the test section. This phase change is converted into irradiance variations, which we call an interferogram. There are a variety of interferometers. We will use the Mach–Zehnder interferometer because the light beam passes through the test section only once as shown in the Figure 17.6. Monochromatic light from a point source S is collimated by the lens L_e. This collimated beam is split into two beams by the beam splitter BS_1. One of these beams passes through the test section, while the other beam propagates in space. These two beams are folded by mirrors M_1 and M_2 and recombined by the beam splitter BS_2. Interference takes place between the two beams both in reflection and transmission and is observed on screens Sc_1 and Sc_2. The interference patterns observed on the two screens are complimentary; any one of them can be used for evaluation.

The interferometer can be used either in a fringe free mode or a straight line fringe mode. The fringe field is created by introducing a very small angle between the two interfering beams. Assuming a 2D thermal field in the test section in which

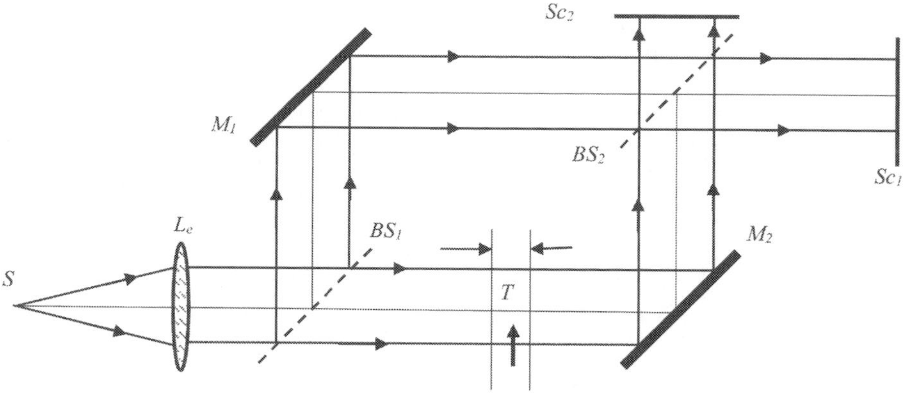

FIGURE 17.6 Schematic of a Mach–Zehnder interferometer. S, source; BS_1 and BS_2, beam splitters; M_1 and M_2, mirrors; T, test section; Sc_1 and Sc_2, screens.

refractive index variation is perpendicular to the propagation direction of the beam, the path difference Δ introduced is given by

$$\Delta = (\mu - \mu_0) L = \varepsilon \lambda_0$$

where μ_0 is the ambient refractive index outside the test section and also in the other arm of the interferometer, λ_0 is the wavelength of monochromatic light, and ε is the fringe order. The fringe order ε gives the fringe displacement in terms of the fringe width and hence its value contains integer as well as fraction. This path difference results in the displacement of the fringe; refractive index variation due to temperature variation causes the distortion in the fringes. The fringe order ε is measured. Using the Gladstone–Dale equation along with the ideal gas equation, the temperature T is given by

$$T = \left(\frac{\lambda_0 R}{\mathbb{K} PLM} \varepsilon + \frac{1}{T_0} \right)^{-1}$$

where R is the gas constant, P is the absolute pressure (Pa) in the test section, M is the molecular weight, T_0 is the ambient temperature (K), and L is the length (m).

17.3.2.4 Hologram Interferometry (HI)

Figure 17.7 is a schematic to record a hologram. Both single-exposure and double-exposure HI can be performed for temperature measurement. Confining to the double-exposure HI, the first exposure is made when there is no source of temperature, say a flame, and the second exposure is made with the temperature source.

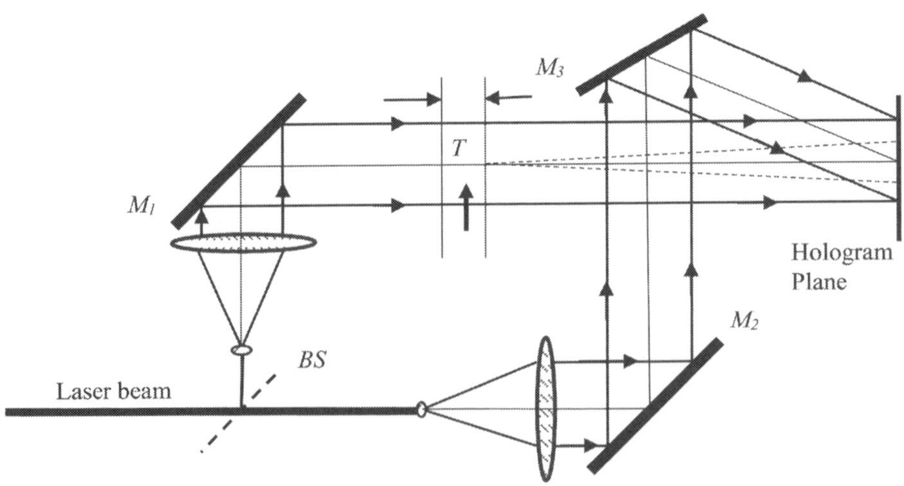

FIGURE 17.7 Schematic to record a hologram. BS, beam splitter; M_1, M_2, and M_3, mirrors; T, test section.

The irradiance of the reference beam is much higher than the irradiance of the object beam to have a linear recording. On reconstruction, an interference pattern is observed from which temperature distribution is constructed.

Unlike interferometry, HI can be performed with poor quality optical elements in the path of light. This is because HI records the changes in the object wave and imperfections of the optical components automatically cancel out. Parallel illumination of the test section can be used if only 2D distribution of the temperature of the flame is to be studied. However, when 3D distribution of temperature is desired, diffuse illumination of the test section (object) is preferred. Diffuse illumination is also chosen if the size of the object is larger than the size of the holographic plate. Diffuse illumination is achieved by placing a diffuser before the object in the light path. Further, if highly transient processes are to be studied, a pulse laser is used for the hologram recording.

Digital HI is currently used to measure temperature distribution in flames. The hologram is recorded on a charge-coupled device (CCD) and the two exposures are independently handled.

17.3.2.5 Electronic/Digital Speckle Pattern Interferometry (ESPI/DSPI)

Figure 17.8 shows a configuration of electronic/digital speckle pattern interferometry (ESPI/DSPI) that is used to measure the temperature of flames. A reference beam is added axially to the object beam. The object (flame) is illuminated with diffuse light, and it is imaged on the *CCD* plane. The speckle size is controlled by an aperture *LA* in front of the imaging lens.

First exposure is made when there is no object present in the test section, and the second exposure is made with the flame. These exposures are independently handled. The first exposure is subtracted from the second exposure pixel by pixel and then squaring results in an interferogram (the correlation fringes). Phase distribution across the image plane is obtained by phase shifting.

Let us consider an axisymmetric phase object, such as a micro-flame, in which the *y*-direction is the axis of symmetry. The phase difference $\Delta\phi$ at a point is the

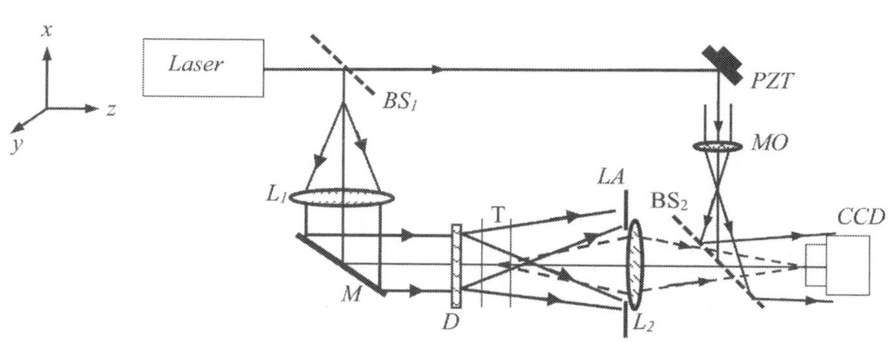

FIGURE 17.8 Schematic for electronic speckle pattern interferometry. BS_1, BS_2, beam splitters; L_1, L_2, lenses; *M*, mirror, *T*, test section; *LA*, lens aperture; *MO*, microscope objective; *CCD*, charged-coupled device.

difference of phases at that point between the second and first exposures. The refractive index difference $\Delta\mu$ is obtained using the inverse Abel transform as

$$\Delta\mu(r) = \frac{-\lambda}{2\pi^2} \int_r^{R_0} \frac{d(\Delta\phi)/dx}{(r^2 - x^2)^{1/2}} dx$$

where R_0 is the maximum radial distance, $r = (x^2 + z^2)^{1/2}$ is the radial coordinate in the cylindrical polar coordinate system, (x,y,z) are the coordinates of a point in a Cartesian coordinate system, z is the direction of light propagation, and y is along the axis of symmetry.

Using the refractive index difference, the temperature distribution in the microflame is calculated using the Lorentz–Lorenz formula as

$$T = T_0 \left[\left(\frac{\mu - \mu_0}{\mu_0} \right) \left(\frac{3PA + 2RT_0}{3PA} \right) + 1 \right]^{-1}$$

$$T = T_0 \left[\left(\frac{\mu - \mu_0}{\mu_0} \right) \left(1 + \frac{2RT_0}{3PA} \right) + 1 \right]^{-1}$$

where T_0 is the reference temperature, μ_0 is the refractive index of air at the reference temperature, P is the atmospheric pressure, R is the gas constant, and A is the molar refractivity of air. It is seen from the equation that reference temperature T_0 is the temperature at a location where $\mu = \mu_0$ and hence it is the ambient temperature.

There are several other optical techniques that have been used to measure the temperature of flames. Shakher and Nirala described these techniques in detail in their review article (see Bibliography). All of the techniques are based on the measurement of the refractive index.

QUESTIONS

17.1 Using the formula

$$M(\lambda,T) = \frac{C_1}{\lambda^5 \left(e^{C_2/\lambda T} - 1\right)} \left(\text{W}/\text{m}^2\ \mu\text{m}\right)$$

calculate the spectral radiant emittance of a blackbody at a temperature of 1000 (K) and at a wavelength of 1.0 μm, 1.2 μm, and 4.0 μm. Given $C_1 = 3.74177 \times 10^8 \left(\text{W}\ \mu\text{m}^4/\text{m}^2\right)$ and $C_2 = 1.43878 \times 10^4 \left(\mu\text{m K}\right)$

17.2 The emissive power $M(T)$ of a blackbody is given by

$$M(T) = \int_0^\infty M(\lambda,T)\, d\lambda = \sigma T^4$$

Show that the constant σ is given by $\sigma = (2\pi^5 k^4)/(15 c^2 h^3)$.

17.3 If the emissivity of a hot body is overestimated by 10%, show that the temperature of the hot body using a total radiation pyrometer will be underestimated by 2.5%.

17.4 A total radiation pyrometer measures the temperature of a hot body as 700°C assuming its surface emissivity as 0.8. However, the correct surface emissivity of the body is 0.86. What is the corrected temperature of the body?

17.5 A dual-wavelength pyrometer measures the spectral radiance of a hot body at two wavelengths of 1.0 μm and 1.2 μm and their ratio is 0.22. Obtain the temperature of the body assuming (1) it a blackbody and (2) it is a body whose ratio of emissivities is 0.9. Take $C_2 = 1.43878 \times 10^4\ (\mu\text{m K})$.

17.6 Spectral radiance of a blackbody at its most intense radiation of 2.0 μm is $1.0 \times 10^5\ (\text{W/m}^2\ \text{sr}\ \mu\text{m})$. Calculate its spectral radiance when its temperature is raised so that the most intense radiation is at 4.0 μm.

17.7 In an experiment using HI to measure the temperature at a point inside the flame, the measured change in the refractive index is -2.5×10^{-4}. The ambient temperature is 27°C and the refractive index of the air at this temperature is 1.00029. Calculate the temperature in the flame at this point. The molar refractivity of air is taken as 4.36×10^{-6} m^3/mol.

18 Fiber-Optic and MEMs-Based Measurements

18.1 INTRODUCTION

During the last couple of decades, fiber-optic devices have been developed to sense and measure displacement, strain, flow, pressure, rotation, electrical and magnetic fields, and a host of other variables. They offer the advantage of small sizes and are highly cost-effective; hence, have found applications in a large number of areas. Though the basic principle of operation of many of the fiber-optic devices is similar to those used in conventional measurement devices, the sensing portion could be drastically different.

Fiber-optic devices are grouped under two categories: extrinsic or hybrid and intrinsic or all fiber. In the former, the measurand modifies the properties outside the fiber and the fiber is used only as the carrier of information, whereas the latter also performs the sensing part. In many applications there is no measurement involved, and the fiber-optic device is just a sensor and forms a part of a control system. We will only discuss those devices that are used for measurement.

The measurand may change the irradiance, amplitude, phase, frequency, and polarization of the light wave. Therefore, there are a large number of measuring devices that use one of these many characteristics of the light wave. We will first discuss those that are based on irradiance modulation.

18.2 IRRADIANCE MODULATION

18.2.1 Displacement Measurement: Lateral Shift between the Fibers

Let us consider two step-index fibers of identical core diameters $2a$. Displacement of one fiber relative to other results in a loss of coupling as shown in Figure 18.1(a): the coupling efficiency decreases. The coupling efficiency η is given by

$$\eta = \frac{2}{\pi}\left\{\cos^{-1}\left(\frac{d}{2a}\right)-\left(\frac{d}{2a}\right)\left[1-\left(\frac{d}{2a}\right)^2\right]^{1/2}\right\} \tag{18.1}$$

where d is the lateral displacement. Because the fibers are in contact, i.e., there is no axial misalignment, the power coupled will depend on the common area A of the two fibers; (1/4)th part of the common area is shown hatched in the Figure 18.1(b).

DOI: 10.1201/9781003531845-18

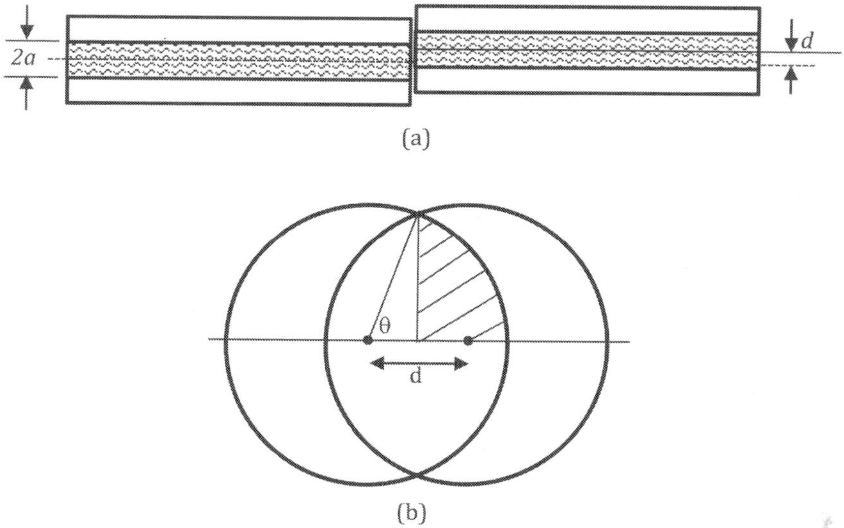

FIGURE 18.1 (a) Lateral misalignment. (b) Calculation of overlap area.

The common area A is four times the area of the sector minus that of a triangle of base $d/2$ and radius a. This is therefore written as

$$A = 4\left\{\frac{\theta}{2\pi}\,\pi a^2 - \frac{1}{2}\frac{d}{2}\left[a^2 - \left(\frac{d}{2}\right)^2\right]^{1/2}\right\} \tag{18.2}$$

where $\cos\theta = \frac{d}{2a}$.

Coupling efficiency is obtained by dividing the common area by the area of the core, which on little manipulation yields the expression given in Equation (18.1).

The variation of η as a function of $(d/2a)$ is shown in Figure 18.2. It can be seen that the coupling efficiency will fall from unity to 0.94 (94%) when one of the fibers is displaced by $d/2a = 0.05$, which corresponds to 2.5 μm when a multimode fiber of 50 μm is used. Therefore, this provides a very good sensitivity for the displacement measurement. It may be noted that the sensitivity will improve with the decrease in core diameter but at the expense of linear range, which can be inferred from the plot of η with $(d/2a)$.

Instead of step-index fibers, if the graded index fibers with parabolic refractive index profile are used, then the coupling efficiency η is given by

$$\eta = \frac{2}{\pi}\left\{\cos^{-1}\left(\frac{d}{2a}\right) - \frac{1}{3}\left(\frac{d}{2a}\right)\left[1-\left(\frac{d}{2a}\right)^2\right]^{1/2}\left[5-2\left(\frac{d}{2a}\right)^2\right]\right\} \tag{18.3a}$$

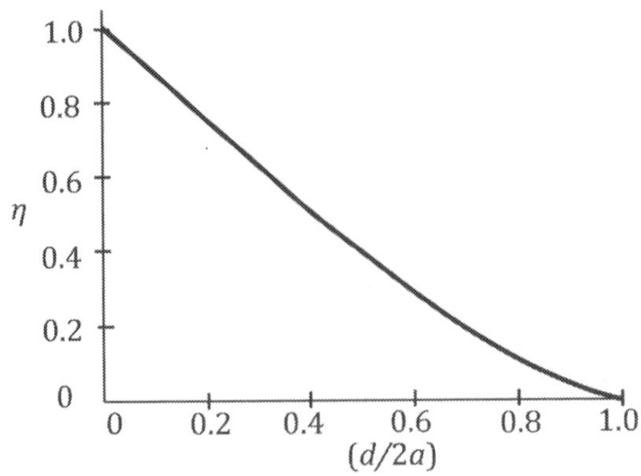

FIGURE 18.2 Variation of efficiency η with $(d/2a)$.

When $d/2a$ is small, Equation (18.3a) can be approximated to

$$\eta \approx 1 - \frac{16}{3\pi}\frac{d}{2a} \tag{18.3b}$$

Equation (18.3b) is accurate to 1% for $\frac{d}{2a} < 0.1$.

One can think of several other applications of this simple arrangement. A fiber-optic microphone is one such application wherein one of the fibers is coupled with the diaphragm of the microphone. Still another application is to measure the pressure with high sensitivity. In fact, one can think of arrangements to measure and control all process variables.

18.2.2 Displacement Sensor: Beam Attenuation

Displacement can be sensed using irradiance modulation when the displacement results in irradiance attenuation. A simple setup, as shown in Figure 18.3, consists of two fibers with collimating lenses so that the beam to be intercepted is larger in size. An opaque screen moving in the beam will cut down the light flux passing through and hence the light output can be calibrated in terms of the displacement of the opaque screen.

Assuming the beam of a circular cross section of radius R and an opaque screen of rectangular shape bigger than the beam diameter, the area A intercepted when the screen enters the beam by d is given by

$$A = R^2\left\{\cos^{-1}\left(1-\frac{d}{R}\right) - 2\left(1-\frac{d}{R}\right)\left[\frac{d}{2R}\left(1-\frac{d}{2R}\right)\right]^{1/2}\right\} \tag{18.4}$$

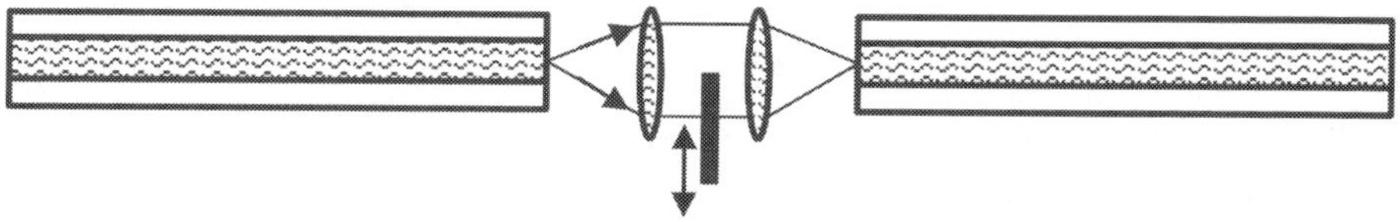

FIGURE 18.3 Displacement sensor using a plate to block the light.

Irradiance of light received by the second fiber will be proportional to the unintercepted area of the beam. Normalizing it with the area of the beam, the efficiency can be expressed as

$$\eta = 1 - \frac{1}{\pi}\left\{\cos^{-1}\left(1-\frac{d}{R}\right) - 2\left(1-\frac{d}{R}\right)\left[\frac{d}{2R}\left(1-\frac{d}{2R}\right)\right]^{1/2}\right\} \tag{18.5}$$

The efficiency $\eta = 1$ when $d = 0$, $\eta = 0.5$ when $d = R$, and $\eta = 0$ when $d = 2R$. In this arrangement the irradiance decreases from its maximum value as the screen intercepts the beam. This therefore can be used to measure displacements with the largest measurable displacement equal to the beam diameter.

The sensitivity of the arrangement can be increased considerably albeit at the expense of range provided two Ronchi grating screens are used: one grating is fixed while the other is translated as shown in Figure 18.4.

It can be seen that half the irradiance of the beam is blocked, and the irradiance varies between 0 and $I_0/2$ when the grating moves by one period. For the measurement of larger displacement, the periodicity of the signal should be taken into consideration. Displacements smaller than the pitch of the grating can be easily measured.

18.2.3 Proximity Probe

Proximity probes are extremely sensitive for the measurement of displacement and are used in a variety of sensors like pressure transducer, etc. It may have an input fiber and an output fiber, or an input fiber and two output fibers or vice versa, or

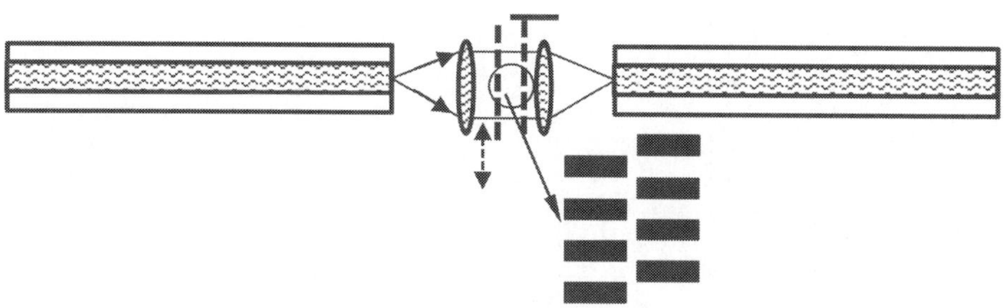

FIGURE 18.4 Displacement measurement with Ronchi gratings.

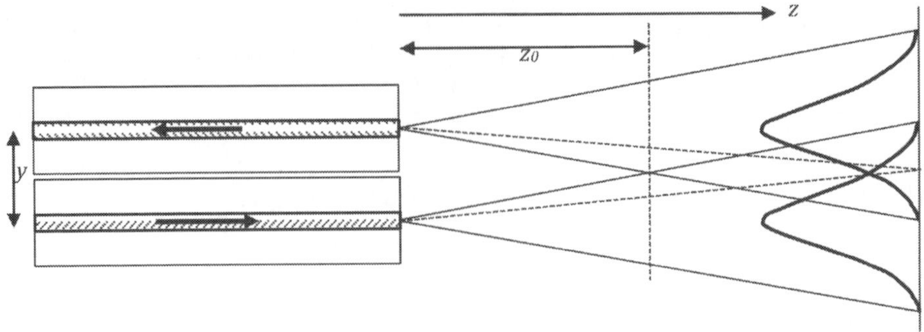

FIGURE 18.5 Schematic of a proximity sensor.

multiple input and multiple output fibers. We shall, however, discuss its principle of operation considering only two identical fibers, one input fiber and one output fiber with their ends in the same plane. These two fibers are laterally separated. Figure 18.5 shows the schematic of the proximity probe that has only two fibers.

Optical energy from the transmitting (input) fiber is confined in its numerical aperture and generally has a bell-shaped distribution; this may be called the attenuation curve. Because the two fibers are identical, the attenuation curve of the output fiber is identical to that of the input fiber. The attenuation curves for both fibers are shown in Figure 18.5. The energy coupled to the output fiber will be proportional to the square of the attenuation curve. Up to the distance z_0 as measured from the exit tip of the fiber, there is no coupling. As the reflecting surface moves further, intense parts of the attenuation/response curve of input fiber feed the energy to the output fiber; hence, the output increases, though there is an inverse square decrease. The output reaches a maximum value at some distance z and thereafter decreases mainly due to the inverse-square decrease. A typical output behavior is shown in Figure 18.6. The linear portion of the curve as indicated by points A and B can be used for the displacement measurement.

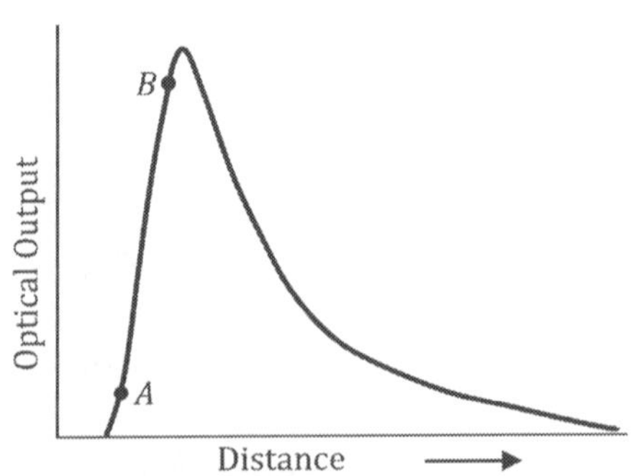

FIGURE 18.6 Typical output versus distance of a proximity sensor.

18.2.4 Microbend Displacement or Pressure Sensor

The bends in fibers lead to radiation losses. This loss is generally represented by an absorption coefficient α_B, which is expressed as

$$\alpha_B = K\ e^{-\frac{R}{R_c}} \tag{18.6}$$

where K is a constant, R is the radius of curvature of the bend, and R_c is another constant, which depends on the radius a of the fiber and its numerical aperture (NA) through the relation $R_c = \frac{a}{NA^2}$. A bend with a radius equal to the fiber radius R_c will give rise to appreciable loss, but at such a small bend radius, the fiber is likely to break. On the other hand, appreciable losses may occur if there is a continuous succession of small deformations or microbends. If Λ is the period of deformations, then the modes with propagation constants β_1 and β_2 in a step-index fiber will strongly couple together provided that

$$\frac{1}{\Lambda} = \frac{|\beta_1 - \beta_2|}{2\pi} \tag{18.7a}$$

If the period is of the order of 1 to 2 mm, the higher order modes couple strongly and are radiated out resulting in the drop of the light output of the fiber. However, when the graded refractive index fiber with quadratic variation of refractive index is used, all modes couple together provided that

$$\Lambda = \frac{2\pi a}{\sqrt{2\Delta}} \tag{18.7b}$$

where $\Delta = \frac{\mu_{\text{core}} - \mu_{\text{clad}}}{\mu_{\text{clad}}}$.

Figure 18.7 shows a fiber microbend device wherein an application of force or any other external agent that displaces the corrugations will result in a change in the output of the fiber. The device can be calibrated and used for measurement. It can also be used to detect presence of an external agent.

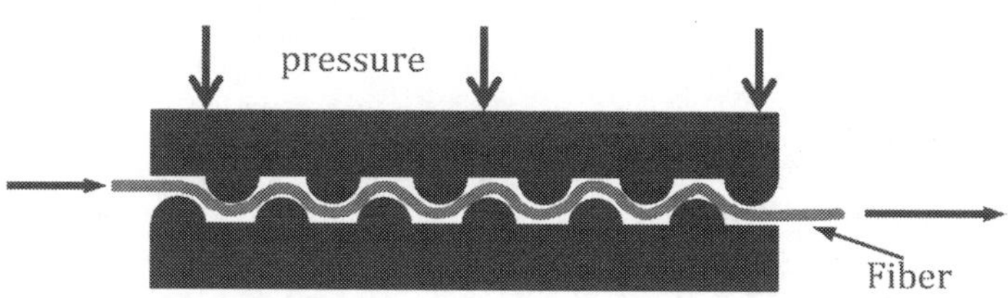

FIGURE 18.7 A fiber microbend device.

18.2.5 Measurement of the Refractive Index of Liquids: Fiber-Optic Refractometer

Consider a multimode fiber of core radius a_1, which has been tapered to radius a_2 over a small length L, which is the sensing length. The tapered end of radius a_2 is butt-coupled with another fiber of core radius a_2. Essentially, we have two fibers, 1 and 2, of core radii a_1 and a_2 with a taper of length L in between as shown in Figure 18.8. When a Lambertian source is used to inject power into input fiber 1, then the power coupled to fiber 2 through the taper will vary linearly with the dielectric constant, ε (= μ^2) of the medium in which the taper is immersed. Power coupled to the fiber is given by

$$P = P_0 \frac{\mu_1^2 - \mu^2}{\mu_1^2 - \mu_2^2}\left(\frac{a_2}{a_1}\right)^2 \tag{18.8}$$

where μ_1 and μ_2 are the refractive indices of the core and the cladding of both the fibers.

It may be seen that the power coupled to fiber 2 decreases linearly with μ^2. When the refractive index of the surrounding medium is equal to that of the core, the power coupled becomes zero and hence the largest value of the refractive index that can be measured with this method is equal to the refractive index of the core of the fiber. On the other hand, all the power launched to fiber 1 will be coupled to fiber 2 when

$$\mu^2 = \mu_1^2 - \left(\frac{a_1}{a_2}\right)^2 (\mu_1^2 - \mu_2^2) \tag{18.9}$$

This, therefore, gives the lowest value of refractive index that can be measured with this arrangement. By varying the taper one can change this lowest value. In any case, the range over which the refractive index can be measured lies within the refractive indices of the core and the cladding. Nevertheless, it is an attractive method to measure refractive indices of variety of liquids and bio-fluids.

18.3 PHASE MODULATION

18.3.1 Interferometric Sensors

Like the conventional interferometers used for the measurements of variables that alter the phase of the light wave, interferometers designed with single-mode fibers

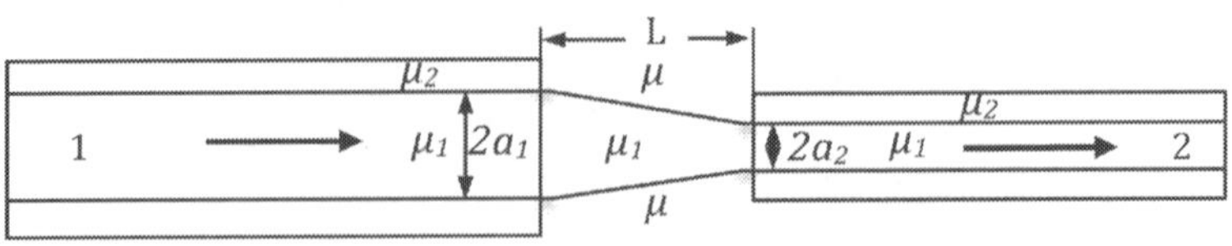

FIGURE 18.8 Schematic of a fiber-optic refractometer.

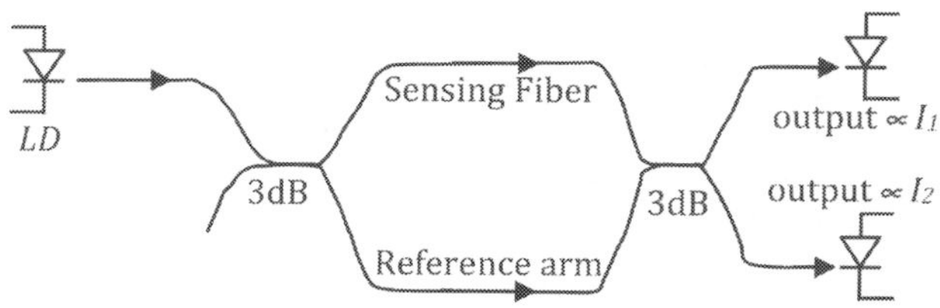

FIGURE 18.9 Schematic of a fiber Mach–Zehnder interferometer.

can also be used to measure such variables. Common interferometric configurations like Michelson, Mach–Zehnder (M-Z), Fabry–Perot, etc., have been realized and used for measurement. They all use single-mode fiber and have a sensing arm and a reference arm.

Figure 18.9 shows an arrangement of an M-Z interferometer. Output from the semiconductor laser (*LD*) is coupled to a single-mode fiber. A 3-dB coupler divides the amplitude of the light wave into a sensing fiber and a reference fiber. After traversing some length of the fiber, these light wave amplitudes are recombined by another 3-dB coupler and then the light irradiance is detected by the photodiodes giving outputs I_1 and I_2, respectively.

Let the amplitude of the wave through the reference fiber be

$$E_r = E_{r0} e^{i(\omega t + \phi_r)} \tag{18.10a}$$

Similarly, let the amplitude of the wave through the sensing fiber be

$$E_s = E_{s0} e^{i(\omega t + \phi_s + \Delta\phi)} \tag{18.10b}$$

where ϕ_r and ϕ_s are the phases of the waves in the reference and the sensing fibers, respectively, and $\Delta\phi$ is the phase introduced in the wave through the sensing fiber by the external agent. The output will be proportional to irradiance I_1 in the interference pattern where I_1 is given by

$$\begin{aligned} I_1 &= \left|E_r + E_s\right|^2 = E_{r0}^2 + E_{s0}^2 + 2E_r E_s \cos\left(\phi_s + \Delta\phi - \phi_r\right) \\ &= 2E_0^2\left[1 + \cos\left(\phi_s + \Delta\phi - \phi_r\right)\right] \end{aligned} \tag{18.11}$$

where we have assumed the amplitudes of both the waves to be equal, i.e., $E_{s0} = E_{r0} = E_0$.

The response of the interferometer is periodic, and the sensitivity is variable; it has the maximum sensitivity at the quadrature point. It may be noted that the operation of the interferometer can always be done around the quadrature point by changing the phase of the reference wave. This is achieved by winding the reference fiber on a lead zirconate titanate (PZT) cylinder to which a voltage is applied resulting in elongation of the fiber.

The irradiance I_2 is complimentary of that of I_1, i.e., it is phase shifted by π; hence, the output difference from the two photodetectors will result in twice the sensitivity.

Let us see how the external agent influences the phase of the wave through the sensing fiber. Assuming it to be a step-index fiber of length L and core refractive index μ_1, the phase ϕ is given by

$$\phi = \frac{2\pi}{\lambda}\mu_1 L \tag{18.12}$$

where λ is the wavelength from a semiconductor laser. The phase of the wave can be influenced either by a change in the refractive index of the core of the fiber or the length of the fiber or both. Therefore, the phase change is expressed as

$$\Delta\phi = \frac{2\pi}{\lambda}\frac{d\mu_1}{dX}\Delta X\ L + \frac{2\pi}{\lambda}\mu_1\frac{dL}{dX}\Delta X \tag{18.13}$$

where X is the quantity to be measured.

It is known that force, hydrostatic pressure, and temperature can be measured with fiber-optic interferometers. In fact, the interferometer can be used to measure any quantity that effects the phase of the wave. For example, the magnetic field can be measured through magnetostriction effect.

It may be noted that an interferometer can be operated with a resolution in phase of the order of 10^{-6} rad. But the limiting factors are shot noise in the detector and fluctuations in the irradiance and the frequency of the laser output.

18.3.1.1 Temperature Measurement

A small length L of a fiber can be used to measure the temperature. Temperature changes the dimensions (both the length and the diameter) of the fiber as well as its refractive index. The phase of the light wave as it traverses the length L of the fiber is given by

$$\phi = \beta\ L = \frac{2\pi}{\lambda}\mu_1 L \tag{18.14}$$

where β is the propagation constant. The phase change due to temperature change ΔT can be expressed as

$$\begin{aligned}\Delta\phi &= \frac{2\pi}{\lambda}\frac{d\mu_1}{dT}\Delta T\ L + \frac{d\beta}{da}\alpha\ a\ \Delta T\ L + \frac{2\pi}{\lambda}\mu_1\alpha\ \Delta T\ L \\ &= \frac{2\pi}{\lambda}\Delta T\ L\left[\left(\mu_1 + \frac{\lambda}{2\pi}a\frac{d\beta}{da}\right)\alpha + \frac{d\mu_1}{dT}\right]\end{aligned} \tag{18.15}$$

Here $\frac{d\beta}{da}$ is the rate of change of the propagation constant with the radius a of the core, and α is the linear expansion coefficient. If a silica fiber is used as the sensing element, we can obtain the phase change per unit length and per degree centigrade temperature by taking $\frac{1}{\mu_e}\frac{d\mu_e}{dT} = 8.6 \times 10^{-6}\ /°C$, $\mu_e = 1.491$, and $\alpha \approx 0.55 \times 10^{-6}/°C$ at 1300 nm. Here μ_e is the effective refractive index of silica and is used in place of μ_1.

Ignoring the contribution of rate of change of propagation constant with the radius, we obtain the phase change nearly of 66 rad/m °C.

The sensing fiber lies in one arm of an M-Z interferometer. The fringes are counted electronically. In order to determine the direction of fringe movement, and hence whether the temperature is above or below the set point, two detectors in quadrature are used. Use of four detectors compensates for source irradiance variations. Due to extremely high sensitivity, temperature stability of the interferometer is desired.

18.3.1.2 Fiber-Optic Pressure Sensor

Application of hydrostatic pressure to a fiber changes the phase of the wave propagating inside the fiber through the photoelastic effect. Considering a sensing length L of a fiber with a core diameter of $2a$ and refractive index μ_1, it can be shown that the phase change $\Delta\phi$ when the pressure changes by ΔP can be expressed as

$$\Delta\phi = \frac{\pi}{\lambda} L\ \Delta P\left[\frac{\lambda}{2\pi} a\frac{d\beta}{da} - \mu_1^3\left(P_{11}+P_{12}\right)\right]\frac{\left(1-\nu-\nu^2\right)}{E} \tag{18.16}$$

where P_{11} and P_{12} are the photoelastic constants, E is the Young's modulus, and ν is the Poisson's ratio of the fiber core material. Silica fiber coated with a medium that transforms the pressure field into longitudinal strain is used as a sensing fiber. The sensing fiber lies in one arm of the M-Z interferometer. Due to its very high sensitivity to pressure changes, the M-Z interferometer is used to detect extremely weak acoustical signals; it is used as a hydrophone. It may be seen that a fiber 10 m long will detect an acoustical field at the threshold of hearing. Further, it has a good response over the frequency range from DC to 20 MHz, although at frequencies higher than 100 kHz the efficiency of conversion of pressure field into longitudinal strain decreases; hence the use of coated silica fiber as a sensor is restricted to frequencies less than 100 kHz. In the range of 100 kHz to 20 MHz the bare silica fiber may be used as a sensor.

A fiber-optic pressure sensor, apart from its very high sensitivity, offers the advantage of geometrical flexibility; this advantage is available to any fiber-optic sensor. An omnidirectional acoustic sensor can be realized by winding the fiber in a coil whose diameter is much less than the acoustical wavelength. Two such coils can be used as pressure gradient sensor; one coil is placed in the reference arm and the other in the sensor arm of the interferometer. Similarly, a highly directional acoustic sensor can be realized by winding the fiber on a thin long cylinder whose diameter is much smaller than the acoustic wavelength and its length several times the length of the acoustic wavelength.

18.3.1.3 Fiber-Optic Strain Sensor

The interferometric strain sensor measures the phase change caused by the longitudinal strain in a fiber. Let a mono-mode fiber of length L be subjected to a longitudinal strain ε. This causes a phase change $\Delta\phi$ of the wave propagating in the fiber, which is given by

$$\Delta\phi = \Delta\left(\beta\ L\right) = \Delta\beta\ L + \beta\ \Delta L \tag{18.17}$$

The first term contains the phase change due to the change in the propagation constant β, which can arise from the strain-optic effect in which the strain changes the refractive index of the core of the fiber and the waveguide mode dispersion effect due to the change in the core diameter (2a) produced by the longitudinal strain. Thus

$$\Delta\beta\ L = L\frac{d\beta}{d\mu_1}\Delta\mu_1 + L\frac{d\beta}{da}\Delta a \tag{18.18}$$

Usually, $\frac{d\beta}{d\mu_1} = \frac{\beta}{\mu_1}$, and $\Delta\mu_1$ for a homogenous isotropic medium is given by

$$\Delta\mu_1 = -\frac{1}{2}\mu_1^3\left[\varepsilon(1-\nu)P_{12} - \nu\ \varepsilon\ P_{11}\right] \tag{18.19}$$

where P_{11} and P_{12} are the photoelastic coefficients. Further, $\Delta a = -\nu\ \varepsilon\ a$.

The second term in Equation (18.17) represents the phase change due to change in length, ΔL (= ε L). Substituting Equation (18.19) in Equation (18.18) and then in Equation (18.17), we obtain

$$\Delta\phi = -\frac{1}{2}\mu_1^3\left[\varepsilon(1-\nu)P_{12} - \nu\ \varepsilon\ P_{12}\right]L\frac{\beta}{\mu_1} - L\frac{d\beta}{da}\nu\ \varepsilon\ a + \beta\ \varepsilon\ L \tag{18.20}$$

Alternately

$$\frac{\Delta\phi}{\varepsilon\ L} = \beta\left[1 - \frac{1}{2}\mu_1^2\left[(1-\nu)P_{12} - \nu\ P_{11}\right] - \frac{\nu}{\beta}\ a\frac{d\beta}{da}\right] \tag{18.21}$$

The sensing fiber is placed in one arm of the M-Z interferometer and the phase change is measured by counting the number of fringes produced in the interference pattern. The sensitivity of the measurement of strain can be doubled if the fibers in the sensing and the reference arms are arranged such that the sensing fiber experiences tensile strain and the reference fiber experiences compressive strain or vice versa.

18.3.1.4 Fiber-Optic Accelerometers

Fiber-optic accelerometers utilize the property that the acceleration-induced stress produces a strain in the fiber, which in turn introduces a phase change in the wave propagating in a mono-mode fiber. In one of the configurations, the accelerometer is in the form of a simple harmonic oscillator consisting of a mass suspended between two fibers or from a single fiber. When the accelerometer is accelerated along the length of the fiber, a strain is developed in the supporting fiber. The force acting on the mass m is ma_g, where a_g is the acceleration. Using Hooke's law, we obtain the strain in the fiber, which is given by

$$\varepsilon = \frac{\Delta L}{L} = \frac{m\ a_g}{E\ A} \tag{18.22}$$

where A is the area of the fiber of diameter d, and E is Young's modulus of the material of the fiber. The change in length causes a phase change in the fiber, which is given by

$$\Delta\phi = \frac{2\pi}{\lambda}\,\mu_1\,\Delta L = \frac{2\pi}{\lambda}\,\mu_1\,\varepsilon\,L = \frac{8}{\lambda}\,\mu_1\,\frac{m\,a_g}{E\,d^2} \tag{18.23}$$

The phase change can be measured using the M-Z interferometer.

18.3.1.5 Fiber-Optic Gyroscope or Rotation Rate Sensors

Gyroscopes, popularly known as gyros, are based on a principle known as the Sagnac effect. The Sagnac effect can be understood by considering an ideal circular interferometer of radius R in which the light enters at point A where it is divided into two counter-propagating beams as shown in Figure 18.10(a). If the interferometer is not rotating, the counter-propagating beams arrive at point A after time t ($=2\pi R/c$). However, if the interferometer is rotating about an axis through its center and perpendicular to its plane clockwise with an angular velocity Ω, the beam traveling clockwise will take longer than the beam traveling anticlockwise. The times taken by the beams traveling clockwise (t_+) and counterclockwise (t_-) can be expressed as

$$t_+ = \frac{2\pi R + R\Omega t_+}{c} \tag{18.24a}$$

$$t_- = \frac{2\pi R - R\Omega t_-}{c} \tag{18.24b}$$

The time difference between the beams at the beam splitter where they combine is

$$\Delta t = t_+ - t_- \approx \frac{4\pi R^2\Omega}{c^2} \tag{18.25}$$

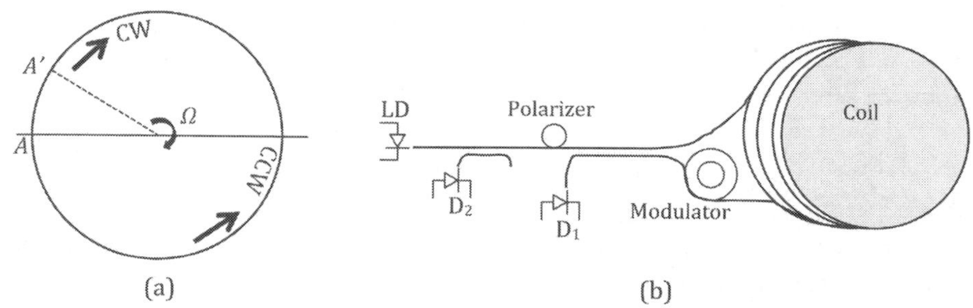

FIGURE 18.10 (a) Understanding the Sagnac effect. (b) Schematic of a fiber-optic gyroscope.

where we have used the condition $c^2 \gg R^2\Omega^2$. This corresponds to a difference in round-trip optical path difference $\Delta L = c\ \Delta t = \frac{4\pi R^2\Omega}{c} = \frac{4\ A\Omega}{c}$, where A is the area of the interferometer loop. This can be written also in terms of the phase difference $\Delta\phi$ between the two beams as

$$\Delta\phi = \frac{2\pi}{\lambda}\,\Delta L = \frac{8\pi A\Omega}{\lambda c} \tag{18.26a}$$

The phase difference is proportional to the area of the loop and for a practical device the sensitivity is low. To increase the sensitivity, a low-loss long fiber (say, 1 km length) wound in N loops is used in optical gyros as in Figure 18.10(b), and various schemes of noise reduction are employed. The phase difference due to rotation then can be expressed as

$$\Delta\phi = \frac{8\pi A\Omega\ N}{\lambda c} \tag{18.26b}$$

It may be noted that the expression Equation (18.26b) is independent of the medium in which the light beam is traveling. The simple analysis presented here does not show this fact. Using relativistic considerations, it can be shown that the Equation (18.26b) is valid in any medium. With a typical gyroscope, which has a 200 m long fiber wound on a 0.1 m diameter coil, the measurement of earth angular rotation rate (Ω_e = 15°/hr = 73 μrad/s) requires detection of phase difference of $\Delta\phi = 36\ \mu$rad. This corresponds to an optical path difference of the order of 10^{-12}m. We now discuss the way with which such small phase change could be measured.

The counter-propagating beams on interference produce a signal, which is given by

$$I = I_0\left(1+\cos\Delta\phi\right) \tag{18.27}$$

Due to the cosine dependence of the signal, it is highly nonlinear near $\Delta\phi \approx 0$ with sign ambiguity and has poor sensitivity for low rotation rates. This can be overcome by phase modulation. For this purpose, a phase modulator is constructed by winding and cementing a few turns of the fiber on a short piezo-ceramic (PZT) tube. Application of modulation voltage results in optical path-length variations due to the electro-optic effect. As a result, counterclockwise and clockwise propagating beams will experience phase delays of $\phi(t)$ and $\phi(t + \tau)$ where $\tau = L/v$; L is the length of the fiber and v is the velocity of light in the fiber. The net phase difference can now be expressed as

$$\phi_{CCW} - \phi_{CW} = \phi_{sag} + \phi(t) - \phi(t+\tau) \tag{18.28a}$$

where $\phi_{sag} = |\Delta\phi|$ is the Sagnac phase shift. The phase shift can also be written, by shifting the reference point, as

$$\phi_{CCW} - \phi_{CW} = \phi_{sag} + \phi(t-\tau/2) - \phi(t+\tau/2) \tag{18.28b}$$

By applying a phase modulation at an angular frequency ω_m and amplitude ϕ_m such that $\phi(t) = \phi_m \cos \omega_m t$, the phase difference in Equation (18.28b) can be expressed as

$$\phi_{CCW} - \phi_{CW} = \phi_{sag} + 2\phi_m \sin(\omega_m \tau / 2) \sin \omega_m t = \phi_{sag} + \phi_{m0} \sin \omega_m t \quad (18.29)$$

where $\phi_{m0} = 2\phi_m \sin(\omega_m \tau / 2)$ is the modulation amplitude, which can be maximized by selecting the modulation frequency such that $f_m = \omega_m / 2\pi = 1/2\tau$.

The output signal now can be expressed as

$$\begin{aligned} \frac{I}{I_0} &= \left[1 + \cos(\phi_{sag} + \phi_{m0} \sin \omega_m t)\right] \\ &= 1 + \cos\phi_{sag} \cos(\phi_{m0} \sin \omega_m t) - \sin\phi_{sag} \sin(\phi_{m0} \sin \omega_m t) \end{aligned} \quad (18.30)$$

Using Bessel's theorem, this can be expressed as

$$\begin{aligned} \frac{I}{I_0} = 1 &+ \left[J_0(\phi_{m0}) + 2\sum_{k=1}^{\infty} J_{2k}(\phi_{m0}) \cos 2k\omega_m t\right] \cos\phi_{sag} \\ &- \left[2\sum_{k=1}^{\infty} J_{2k-1}(\phi_{m0}) \cos(2k-1)\omega_m t\right] \sin\phi_{sag} \end{aligned} \quad (18.31)$$

The output contains all the harmonics of the modulating signal in addition to the DC term. The amplitudes of all even harmonics are multiplied by $\cos\phi_{sag}$, whereas those of odd harmonics are multiplied by $\sin\phi_{sag}$. By choosing $\phi_{m0} = 1.8$, the J_1 function is maximized and the Sagnac phase shift with sign ambiguity removed is recovered by a lock-in amplifier by measuring the amplitude of the first harmonic of the modulating signal.

Finally, the fiber-optic gyroscopes have several unique features such as low cost, less maintenance, quick warm-up, wide dynamic range, large bandwidth, reliability, etc. The fiber-optic gyroscopes suffer from noise arising due to backscattering, optical Kerr effect, polarization fluctuations, Faraday effect, uneven thermal fluctuations etc. Counter measures have been incorporated to reduce the noise. Low coherence source efficiently and effectively reduces the noise contributions due to backscattering and the Kerr effect. Polarization fluctuations are controlled by the use of a polarizer and polarization-maintaining (PM) fibers. Because PM fiber is very expensive, some novel configurations have been used that need only a few meters' length of PM fiber. Noise contribution due to the Faraday effect is also reduced by the use of the polarizer and PM fiber. Thermal fluctuations are suppressed by special windings.

18.3.1.6 Fiber-Optic Fabry–Perot Interferometer

A fiber-optic Fabry–Perot interferometer, in the form of a tip sensor, is used in a number of applications such as measurement of pressure, temperature, etc. There are a number of designs of this sensor; three of which are shown in the Figure 18.11.

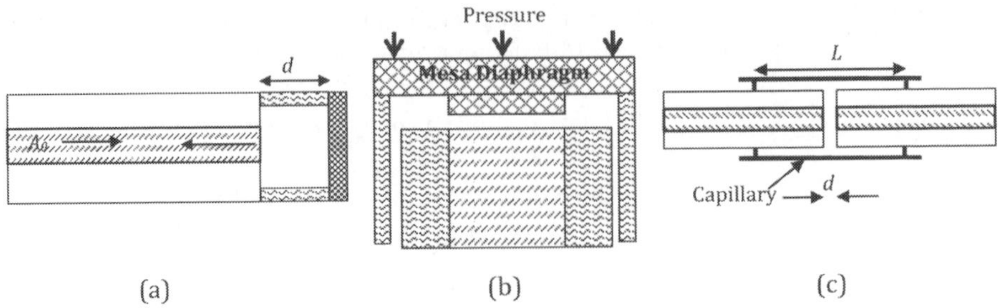

FIGURE 18.11 Fiber-optic Fabry–Perot interferometers: (a) membrane type, (b) mesa diaphragm type, and (c) capillary tip type.

This is a low-finesse interferometer in which a beam is generated by reflection from the end face of the fiber and the second beam is generated by reflection from the surface of the membrane or the face of another fiber. The reflectivities of these two surfaces are shown by R_1 and R_2, respectively. Therefore, the amplitudes of these two beams can be expressed as $A_1 = A_0\sqrt{R_1}$ and $A_2 = A_0\left(1-R_1\right)\sqrt{\alpha(z)R_2}\; e^{-\frac{4\pi i d}{\lambda}}$, where $\alpha(z)$ is the attenuation due to expansion of the beam and finally a collection of the small portion of the beam by the fiber, and d is the separation. Using a ray model, it can be shown that the attenuation $\alpha(z)$ can be expressed as

$$\alpha(z) = \frac{a^2}{\left[a + 2d\tan\left(\sin^{-1}NA\right)\right]^2} \tag{18.32}$$

where a is the radius of the fiber of numerical aperture NA.

The irradiance distribution in the interference pattern created by the superposition of these two beams is given by

$$\mathrm{I} = I_0'\left[1 + V\cos\left(4\pi d / \lambda\right)\right] \tag{18.33}$$

where the irradiance $I_0' = I_0\left[R_1 + \left(1-R_1\right)^2 R_2\ \alpha(z)\right]$ with I_0 being the irradiance of the incident beam, and the visibility V is $V = \frac{2\sqrt{R_1\ R_2\ \alpha(z)}\ (1-R_1)}{R_1\ +(1-R_1)^2 R_2\ \alpha(z)}$. The visibility can be maximized by choosing a proper value of R_2. It can be shown that the visibility will be maximum when $R_2 = \frac{R_1}{(1-R_1)^2\ \alpha(z)}$. This suggests that the reflectivity of the second surface should be high to offset the influence of divergence of the beam from the fiber. The detector signal will be proportional to the irradiance and would vary as the separation d changes.

18.4 PRESSURE SENSOR: MEMBRANE TYPE

The separation d would change by an application of pressure at the membrane as shown in Figure 18.11(a). It can be shown that the central deflection z_0 of the membrane assuming it to be circular and rigidly clamped at the periphery is given by

$$z_0 = \frac{3(1-\nu^2)p}{16E\ h^3}R^4 \tag{18.34}$$

where R and h are the radius and the thickness of the membrane, E and ν are the Young's modulus and the Poisson's ratio of the material of the membrane, and p is the applied pressure. This expression is valid if the deflection is less than 30% of the film thickness; the film thickness should also be less than 20% of the membrane radius. For a given tip sensor, the pressure range is determined by the thickness and the radius of the membrane, the Young's modulus, and the Poisson's ratio of its material. For the linear range between the central deflection and the pressure, the pressure should be less than the maximum value p_{max} given by

$$p_{max} = \frac{8E\ h^4}{5(1-\nu^2)R^4} \tag{18.35}$$

It is seen that the central deflection is directly proportional to the pressure. This means that the application of pressure will change the gap length d. Unfortunately, the gap length d does not change uniformly over the illuminated region leading to further attenuation of the reflected beam. However, this problem could be solved when a mesa diaphragm is used. It can be shown that the deflection is almost uniform over the radius of the mesa. Further, this diaphragm can be easily fabricated by etching the silicon. Design of a tip with a mesa diaphragm is shown in Figure 18.11(b).

18.4.1 Pressure Sensor: Capillary Tip

This particular sensor uses two identical fibers that are placed inside a capillary that is bonded to the fibers effectively having a gage length of L, though the gap length is only d as shown in Figure 18.11(c). When the hydrostatic pressure is applied, the change in L is given by

$$\Delta L = \frac{L}{E}\frac{R_0^2}{R_0^2 - R_i^2}(1-2\nu)p \tag{18.36}$$

where L is the distance between two points where the fibers are fused with the capillary, R_0 and R_i are the outer and inner radii of the capillary tube, E and ν are the Young's modulus and Poisson's ratio of capillary material, and p is the applied pressure. Again, it should be noted that the gap length d changes as the pressure applied to the tube changes.

18.5 BRAGG GRATING SENSORS

A fiber Bragg grating consists of a periodic refractive index variation in the core of a germanium-doped silica fiber. The refractive index variation is created by the interference of two waves from an excimer laser. The grating can be written on a small segment (~1 mm) of the fiber and can also be written at different locations for distributed sensing. The grating reflects the wavelength λ_B, which satisfies the Bragg condition

$$\lambda_B = 2\ \mu_e \Lambda \tag{18.37}$$

where Λ is the grating period and μ_e is the effective refractive index in the core. When a beam of light from a broadband source is coupled to the fiber, the wavelength satisfying the Bragg condition is reflected and the rest of the spectrum is transmitted. The bandwidth of the reflected signal depends on various parameters such as number of grating planes, refractive index modulation, etc., and is typically in the range of 0.05 nm to 0.3 nm.

The Bragg grating sensor is usually used to monitor strain and temperature. The shift in the Bragg wavelength due to the strain and the temperature can be expressed as

$$\Delta\lambda_B = 2\left(\Lambda\ \frac{\partial \mu_e}{\partial l} + \mu_e\ \frac{\partial \Lambda}{\partial l}\right)\Delta l + 2\left(\Lambda\ \frac{\partial \mu_e}{\partial T} + \mu_e\ \frac{\partial \Lambda}{\partial T}\right)\Delta T \tag{18.38}$$

where Δl and ΔT are the length variation and temperature variation, respectively. Both strain and temperature simultaneously change the Bragg wavelength, and the shift is linearly related with the strain and the temperature change. This expression can be recast in terms of the applied strain ε, the strain-optic coefficients $P_{i,j}$, the Poisson ratio ν, and the coefficient of linear expansion α as

$$\Delta\lambda_B = 2\ \mu_e\ \Lambda\left(\left\{1 - \frac{\mu_e^2}{2}\left[P_{12} - \nu\left(P_{11} + P_{12}\right)\right]\right\}\varepsilon + \left[\alpha + \frac{1}{\mu_e}\frac{d\mu_e}{dT}\right]\Delta T\right) \tag{18.39}$$

Alternately

$$\frac{\Delta\lambda_B}{\lambda_B} = \left\{1 - \frac{\mu_e^2}{2}\left[P_{12} - \nu\left(P_{11} + P_{12}\right)\right]\right\}\varepsilon + \left(\alpha + \frac{1}{\mu_e}\frac{d\mu_e}{dT}\right)\Delta T \tag{18.40}$$

For a silica fiber, $P_{11} \approx 0.113$, $P_{12} \approx 0.252$, $\nu \approx 0.16$, and $\mu_e = 1.491$ ($\lambda = 1300$ nm). The value of effective strain coefficient $P_e = \frac{\mu_e^2}{2}\left[P_{12} - \nu\left(P_{11} + P_{12}\right)\right]$ is 0.215. Further, the value of $\frac{1}{\mu_e}\frac{d\mu_e}{dT}$ for silica fiber is $8.6 \times 10^{-6}/°\text{C}$ and $\alpha \approx 0.55 \times 10^{-6}/°\text{C}$. Substituting these values, we obtain

$$\frac{\Delta\lambda_B}{\lambda_B} = 0.795 \times \frac{10^{-6}}{\mu\varepsilon} + 9.15 \times \frac{10^{-6}}{°\text{C}}$$

For operation at 1.3 μm, this gives $\frac{\Delta\lambda_B}{\Delta\varepsilon} = 1.03\ \text{pm}/\mu\varepsilon$ and $\frac{\Delta\lambda_B}{\Delta T} = 11.9\ \text{pm}/°\text{C}$. These values correspond to the theoretical strain and the temperature sensitivities of the fiber Bragg grating sensor. If two gratings with different sensitivities are used side by side simultaneously, it may be possible to measure both the strain and the temperature change. However, if temperature resolution of 0.1°C is desired, then it requires the measurement of a Bragg shift of 0.001 nm, which may be difficult to realize. On the other hand, 1000 $\mu\varepsilon$ produces a Bragg shift of 1.03 nm, which is easily measured.

18.6 POLARIZATION MAINTAINING SINGLE-MODE FIBERS

A circularly symmetric optical fiber can support two orthogonally polarized modes, which propagate with identical velocities. However, circular symmetry is an idealization and hence an actual fiber may depart from it resulting in an inherent (intrinsic) birefringence in the fiber. Further, the bending or the squeezing of the fiber may lead to extrinsic birefringence. The two orthogonally polarized modes now have a slightly different refractive indices and hence slightly different propagation constants. The modes with slightly different propagation constants can couple together by fiber perturbations of relatively long periodicities. A fiber with large birefringence, known as *Hi-Bi fiber*, will maintain the polarization state over a long length of light propagation, say a kilometer or longer. Such fibers either have an elliptical core, or an elliptical cladding, or stress forming regions in the form of a bow tie or panda.

In such a fiber there are two preferred directions along which linear polarization propagates with very little coupling to orthogonal polarization. These are the slow and fast axes and will be along the axes of the ellipse in an elliptical core fiber. Now if a linearly polarized beam is launched in the fiber making an angle of 45° with the slow or fast axis, both polarizations are equally excited in the fiber. These propagate with velocities c/μ_x and c/μ_y and hence an increasing phase difference will develop as the modes propagate down the fiber.

We can express the amplitudes of these two modes as

$$E_x(z;t) = E_0\ e^{i(\omega t + \delta_x)} \tag{18.41a}$$

$$E_y(z;t) = E_0\ e^{i(\omega t + \delta_y)} \tag{18.41b}$$

If this type of fiber is to be used for sensing, then the phase difference between these two modes must be influenced by the external perturbation, which is being measured. Let this phase difference be $\Delta\phi_P$. Unfortunately, these two modes cannot interfere because they are in orthogonal states of polarization. However, these two modes can be made to interfere using a polarizer oriented with its direction of polarization making an angle of 45° either with the fast or slow axes. The output from the detector is then proportional to

$$i \propto \left(1 + \cos\Delta\phi_p\right) \tag{18.42}$$

where $\Delta\phi_p = \delta_x - \delta_y = \frac{2\pi}{\lambda_0}\left(\mu_x - \mu_y\right)L = \frac{2\pi}{\lambda_0} B_r L$, with L as the sensing length and B_r the birefringence.

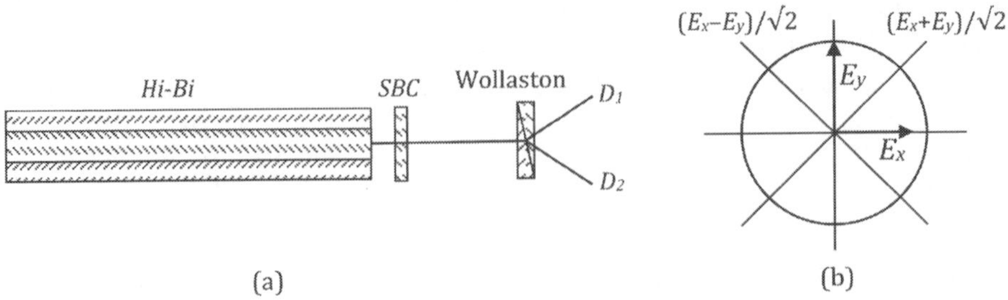

FIGURE 18.12 (a) Sensing with Hi-Bi fiber. (b) Field components after Wollaston prism.

However, this solution is susceptible to source fluctuations and has direction ambiguity. A better solution is to use a combination of a Soleil–Babinet compensator (*SBC*) and a Wollaston prism as shown in the Figure 18.12(a). The Soleil–Babinet compensator is used to introduce a phase difference of $\pi/2$ between the slow and the fast modes. The Wollaston prism splits these modes into two beams each, which are superposed. These two beams have amplitudes $\left(E_x + E_y\right)/\sqrt{2}$ and $\left(E_x - E_y\right)/\sqrt{2}$ as shown in Figure 18.12(b). These beams are incident on detectors D_1 and D_2; their outputs are given by

$$i_1 \propto \left(1 + \sin\Delta\phi_p\right) \tag{18.43a}$$

$$i_2 \propto \left(1 - \sin\Delta\phi_p\right) \tag{18.43b}$$

The outputs from the detectors are conveniently processed as

$$\frac{i_1 - i_2}{i_1 + i_2} = \sin\Delta\phi_p \tag{18.44}$$

The processed signal is proportional to $\Delta\phi_p$ for small values of $\Delta\phi_p$. Further, the sign ambiguity has been removed. External perturbation influences the phase $\Delta\phi_p$ and hence can be measured. There are large numbers of physical phenomena that change the state of polarization of the light and introduce birefringence including optical activity, Faraday rotation, magnetostriction, electrogyration, electro-optic effect, Kerr effect, Photoelastic effect, etc. A variety of physical parameters can be detected and measured using these phenomena.

Let X be the external perturbation that needs to be measured. Expressing $\Delta\phi_p$ explicitly, the effect of external perturbation on the phase can be written as

$$\frac{d\Delta\phi_p}{dX} = \frac{2\pi}{\lambda_0}\left(B_r \frac{dL}{dX} + L\frac{dB_r}{dX}\right) \tag{18.45}$$

It is obvious now that the external perturbation influences both the length of the sensing element and the birefringence; their relative contributions to the phase change could be markedly different.

18.6.1 Current Measurement: Faraday Rotation

When a beam of linearly polarized light passes through a medium placed in the magnetic field, its plane of polarization rotates: the rotation is proportional to the component of magnetic field parallel to the direction of propagation and also the length of the medium in the magnetic field. Further, the sense of rotation is independent of the direction of propagation.

Rotation of plane of polarization can be explained by resolving the incident linearly polarized beam into right-handed and left-handed circularly polarized beams, which have different refractive indices and hence propagate with different velocities. On emergence when these circularly polarized beams combine, the plane of polarization is rotated. Therefore, the application of magnetic field creates birefringence ($\mu_r - \mu_1$), where μ_r and μ_1 are the refractive indices for right-handed and left-handed circularly polarized beams. The angle of rotation θ therefore is given by

$$\theta = \frac{2\pi}{\lambda_0}\left|\mu_r - \mu_l\right| L \tag{18.46a}$$

However, generally the rotation θ is expressed as

$$\theta = VBL \tag{18.46b}$$

where V is the Verdet constant, and B is the magnetic field. The value of V for silica is 4 rad m^{-1} T^{-1}. This is also known as Becquerel's formula.

A current carrying conductor produces a magnetic field, the magnitude of which is given by

$$B(r) = \frac{\mu_0 \mu_{\text{rel}}\ I}{2\pi r} \tag{18.47}$$

where I is the current flowing in the conductor, r is the point from the center of the conductor where the field B is measured, $\mu_o = 4\pi$ x 10^{-7}H m^{-1}, and μ_{rel} is the relative permeability of the medium. The direction of the magnetic field is along the circumference of the circle of radius r and its sense depends on the direction of the current. Therefore, if a circular loop of fiber is placed around the current carrying conductor, the direction of the magnetic field will be along the light propagation in the fiber. Therefore, the rotation of plane of polarization is given by

$$\theta = V\frac{\mu_0 \mu_{\text{rel}}\ I}{2\pi r} 2\pi r = V\mu_0 \mu_{\text{rel}}\ I \tag{18.48}$$

The rotation is linearly related to the current in the conductor and is independent of the radius of the loop. In fact, the rotation is independent of both the size and the shape of the loop. This result is very useful as it shows that the current measuring device will be insensitive to vibrations. It is easily seen that the sensitivity of the device is low and hence several turns of the fiber are used. If we consider n loops around the conductor, the angle of rotation scales up directly with n. In order to use the fiber-optic current sensor, it has to be assembled around the current carrying conductor. The conductor is to be dissembled and then inserted in the fiber-optic coil before installation. However, if one end of the fiber forming the coil is coated, the reflected beam will undergo further rotation enhancing its sensitivity by two. At the same time, it is easily wound around the conductor.

In practice, a polarizer with its polarization direction inclined at 45° either with the slow or fast axes in the fiber provides a linearly polarized beam, which excites both modes equally. Due to the Faraday effect, these modes propagate with different velocities and on emergence from the magnetic field combine to yield a linearly polarized beam with its polarization rotated by θ. An analyzer, which is crossed with the polarizer, is placed at the end of the fiber. The irradiance transmitted by the analyzer, using the Malus law, is given by

$$I(\theta) = I_0 \cos^2\left(\frac{3\pi}{4} + \theta\right) = \frac{I_0}{2}(1 + \sin 2\theta) \tag{18.49}$$

The output of the detector will be proportional to $I(\theta)$. Because the angle of rotation of the plane of polarization due to the magnetic field is small, the output of the detector will be proportional to the angle of rotation θ. The measurement of θ will yield the value of current flowing in the conductor.

In order that the output of the detector is not influenced by the incident light irradiance fluctuations, a slightly different scheme is utilized. A Wollaston prism is placed after the fiber such that its principal plane is oriented at 45° with respect to the linear polarization orientation in the fiber with no current in the conductor. The Wollaston prism produces two orthogonally polarized beams of equal irradiance. However, when the plane of polarization is rotated by an angle θ, the irradiances of the two beams incident on two separate detectors are given by

$$I_1(\theta) = I_0 \cos^2\left(\frac{\pi}{4} + \theta\right)$$

$$I_2(\theta) = I_0 \sin^2\left(\frac{\pi}{4} + \theta\right)$$

Let the outputs of the detectors be i_1 and i_2, which are proportional to the light irradiances I_1 and I_2, respectively. The output of the detectors is processed as to yield the signal S_I through

$$S_I = \frac{i_1 - i_2}{i_1 + i_2} = \sin 2\theta \approx 2\theta$$

The measurement of θ gives the value of current flowing in the conductor. In practice, the fiber may not be an ideal fiber as assumed here but may have both linear and circular birefringence. This introduces a correction factor.

18.7 FIBER-OPTIC BIOSENSORS

Fiber-optic biosensors are based on the transmission of light along silica glass fiber, or plastic optical fiber to the site of analysis. Optical fiber biosensors can be used in combination with different types of spectroscopic technique, e.g., absorption, fluorescence, phosphorescence, surface plasmon resonance (SPR), etc.

Biosensors are attractive because nonspecialist personnel can easily use them, and they allow accurate determination with either no or minimal sample treatment. Therefore, fiber-optic biosensors may be especially useful in routine tests, patient home care, surgery, and intensive care, as well as in emergency situations.

The fiber-optic sensors either employ a single-fiber configuration where light travels to and from the sensing tip in one fiber, or a double-fiber configuration, where separate fibers are used for illumination and detection. The single-fiber sensor is most compact and least expensive, but additional instrumentation may be required to separate the return signal from the illumination light.

The design of intravascular catheters requires special considerations related to the sterlity and biocompatibility of the sensor: intravascular fiber-optic sensors must be sterilizable and their material nonthrombogenic and resistent to platelet and protein deposition. These catheters are made of materials covalently bound with heparin or antiplatelet agents.

Fiber-optic sensors use either plain optical fibers as a remote device for detecting changes in the spectral properties of the tissue and the blood or optical fibers tightly coupled to various indicator-mediated transducers. Only a limited number of biochemical analytes have an intrinsic optical absorption that can be measured with sufficient selectivity directly by spectroscopic methods. Other species, particularly hydrogen, oxygen, carbon dioxide, and glucose, which are of primary interest in diagnostic applications, are not susceptible to direct photometry. Therefore, indicator-mediated sensors have been developed using specific reagents that are properly immobalized on the surface of a fiber-optic sensor.

We will classify the fiber-optic sensors into two categories, direct and indirect. In direct sensors, the measurement relies on direct illumination of a sample through the end face of an optical fiber. Light emanating from the fiber end is scattered or fluoresced back into the fiber: measurement of the returning light indicates the absorption or fluorescence of the sample at the fiber tip. In indirect sensors, a miniature transducer is attached at the distal end of one or two fibers. It was suggested that such an optical transducer should be called an optode, which is similar to the term electrode used in electricity. The light sent through the fiber actuates the transducer, which in turn interacts with the sample. The output light from the transducer is analyzed by an optical processing system. If the optode is physical in nature, the parameters that can be measured are physical such as the pressure or the temperature. If the optode is chemical, the parameters are chemical such as pH or glucose content in the blood.

18.7.1 Direct Fiber-Optic Sensors

The following two methods are used:

a. *Photometry (colorimetry):* Monochromatic light is directed onto the tissue or the blood and the absorption or the reflection at this wavelength is measured. If this is performed at several wavelengths, the spectral reflectance or absorbance of the tissue/sample is obtained. Such spectra are "fingerprints" of specific compounds in the tissue or in the blood.
b. *Fluorometric method:* By measuring the excitation or emission spectrum of the tissue/sample, important information is gathered.

18.7.1.1 Direct Physical Sensors

18.7.1.1.1 Reflectance Measurement

Reflectivity of a sample interface at normal incidence is given by

$$R = \frac{(\mu - 1)^2}{(\mu + 1)^2} \tag{18.50}$$

where μ is the relative refractive index of the tissue/sample. If it is absorbing, μ is complex: its imaginary part is reponsible for the absorption. Reflectivity measurements are carried out at 700 nm because this wavelength penetrates deeply (millimeters) into the skin. It is observed that the reflectivity is strongly affected by the microcirculation of blood. It is also observed that the reflectivity of skin increases linearly initially when it is stretched.

18.7.1.1.2 Absorption Measurement

The absorption is governed by the Beer's Law, which states

$$I(z) = I_0 \ e^{-\mu_a z} \tag{18.51}$$

where $I(z)$ is the irradiance of the light beam when it has penetrated a distance z in the sample, I_0 is the irradiance at the surface, and μ_a is the absorption coefficient in cm^{-1}. The distance z is measured in centimeters. The absorbance or optical density, D_0, is defined as

$$D_0 = -\log_{10} \frac{I(z)}{I_0} \tag{18.52}$$

Therefore, the absorption coefficient μ_a can be expressed as

$$\mu_a = \frac{D_0}{z} \ln 10. \tag{18.53}$$

Often the optical density is monitored.

18.7.1.1.3 Temperature Measurement

The gray body irradiance, or the emissive power, measured as joules per unit time per unit area or watts per unit area, is given by

$$I = \varepsilon \, \sigma \, T^4 \tag{18.54}$$

where ε is the emissivity of the gray body and σ is the Stefan–Boltzmann constant ($\sigma = 5.67 \times 10^{-8}$ J s^{-1} m^{-2} K^{-4}). Emissivity of biological tissue is generally one and hence the emissive power is dependent only on the temperature. At room temperature, the emission is mainly in the far-infrared (IR) region of the spectrum. As the temperature rises, the emission shifts toward near-IR and finally to the visible. If the IR emission is measured, the temperature of the tissue can be determined. IR emission from a warmer part of the body is transmitted to a detector via an IR-transmitting fiber. This method is called IR fiber radiometry.

The fiber-optic technique has been suggested for hyperthermia and hypothermia. In the treatment of cancer using enhanced temperature (hyperthermia) by radiation from a laser, it is desired to measure the temperature to prevent thermal damage. IR fiber radiometry is the ideal method to do this. Temperature accuracy of 0.1°C has been reported.

18.7.1.2 Direct Chemical Sensors

18.7.1.2.1 Oximetry

Oxygen in blood is carried by the hemoglobin (Hb) in red cells. Blood in arteries should be more than 95% saturated, whereas in veins it's about 75% saturated. Measurement of oxygen saturation is useful in cardiology, in monitoring patients during anesthesia, and in the intensive care units.

Samples of blood taken from the right side of the heart that show an unusually high saturation ratio may indicate a congenital abnormality of the heart. A blood sample showing low saturation may be due to reduced capacity of blood to carry oxygen, low cardiac output, or reduced ability of the cardiac pulmonary system. Absorption spectrum of oxyhemoglobin (HbO_2) and reduced Hb is well documented.

The oxygen sensor contains an emitter (a light-emitting diode [LED]), which emits radiation at two different wavelengths (630 nm and 960 nm) depending on the polarity of the applied voltage, and a detector, which measures the irradiance incident on it. In practice, the emitter and the detector are placed on either side of the bodily appendage, say a finger or an earlobe, and the absorption of the light emitted at each wavelength by the intervening tissue is measured. The oxygen saturation of blood is computed using the relation

$$\text{oxygen saturation} = A - B\left\{\frac{D_0(\lambda_1)}{D_0(\lambda_2)}\right\} \tag{18.55}$$

where A and B are coefficients, which are functions of the specific absorbities of Hb and HbO_2, and D_0 is the corresponding absorbance of the blood measured at two wavelengths, λ_1 and λ_2.

In a pulse oximeter, the light source is driven by an alternating source, which turns on one wavelength while the other is off. The pulse oximeter is looking for the maximum and minimum irradiance at each wavelength to discriminate steady-state flow (venous) from the pulsating flow (arterial). The maximum (I_{max}) and minimum (I_{min}) values correspond to the diastolic and systolic arterial pressure, respectively, arithmetically filtering out the steady-state venous attenuation, as this will act as a background for both the diastolic and the systolic measurements. Pulsed oxygen saturation as measured by the pulse oximeter is expressed as

$$\mathrm{SpO_2} = f\left\{\frac{\ln\left(I_{min}/I_{max}\right)_{630}}{\ln\left(I_{min}/I_{max}\right)_{960}}\right\} \tag{18.56}$$

where f is a calibration factor that takes into account bone, pigmentation, thicknesses of the various layers of tissue, and potentially nail polish for a finger pulse oximeter.

In case the absorption coefficients for HbO_2 at 960 nm and for the reduced Hb at 630 nm are μ_{a,HbO_2_960} and μ_{a,Hb_630}, respectively, the oxygen saturation can be written as

$$\mathrm{SaO_2} = \frac{\mu_{a,HbO_2\_960} \times \mathrm{SpO_2} - \mu_{a,Hb\_630}}{\left(\mu_{a,Hb\_960} - \mu_{a,HbO_2\_960}\right) \times \mathrm{SpO_2} - \left(\mu_{a,Hb630} - \mu_{a,HbO_2\_630}\right)} \times 100\% \tag{18.57}$$

The oxygen concentration is found by multiplying the HbO_2 saturation with the total Hb concentration, which can be derived from the hematocrit. The hematocrit is the volume percentage of red blood cells in whole blood and is expressed as a percentage. For example, a hematocrit of 40% means that there is 40 mL of erythrocytes in 100 mL of blood.

The oxygen concentration is found using the following equation

$$\left[\mathrm{O_2}\right] = 1.39\ [\mathrm{Hb}]\ \frac{\mathrm{SaO_2}}{100} \tag{18.58}$$

The constant 1.39 (mL/g) is called the Hüfner's number. This represents the volume of oxygen in milliliters that can be collected per gram of Hb.

18.7.2 Indirect Fiber-Optic Sensors

18.7.2.1 Indirect Physical Sensors

18.7.2.1.1 Pressure Measurement

For pressure measurement, a mechanical optode is attached to the distal tip of the fiber. The mechanical optode consists of a membrane on a hollow tube. At normal pressure, the membrane is flat and hence couples a certain amount of the input light to the fiber. When the outside pressure is greater than the normal pressure P_o ($P > P_o$), the membrane takes a convex shape, and the input light coupled to the fiber decreases. On the other hand, when the outside pressure is lower than the normal pressure ($P < P_o$), the membrane takes a concave shape, and couples more light to the fiber. The output of the fiber is calibrated in terms of pressure.

In another arrangement, instead of a single-fiber bundle, a bifurcated fiber bundle is used. It is observed that the signal varies linearly in the pressure range of 100 mm of Hg to 300 mm of Hg and exponentially between 300 mm of Hg and 3000 mm of Hg for a typical fiber-optic sensor.

18.7.2.1.2 Temperature Measurement

It is well established that the controlled heating of tumor tissue to certain temperature levels (hyperthermia) leads to tumor regression. In a typical treatment, the temperature of tumor tissue is elevated to 42.5°C to 43.5°C for a period of 20 minutes to 60 minutes. Accurate measurement of temperature, along with a heat-generating source, is needed for the success of the procedure. There are several ways to provide localized heating, which include ultrasound waves and radiofrequency (RF) or microwave fields. The Nd:YAG laser has also been used for localized heating. Fiber-optic sensors based on optodes are employed for temperature measurement, in which some property changes as a function of temperature. Two different types of optodes are used for this purpose.

Liquid crystal (LC)-based optode: LCs show dramatic change in color when the temperature changes. The LCs are attached to the distal tip of the fiber and the reflected light is monitored, which depends on the temperature. In one fiber-optic sensor, red light (670 nm) from an LED is transmitted through a fiber bundle onto a layer of cholesterol compound LC. The reflected light irradiance is dependent on temperature. This sensor has a resolution of 0.1°C and a response time of about 4 seconds.

Luminescent probe: Luminescence of some materials depends strongly on temperature in some temperature range. Therefore, the temperature can be measured by measuring luminescence emission irradiance. Lanthanum oxysulfide activated by europium, a luminescent material, is attached to the distal tip of an optical fiber. Ultraviolet (UV) radiation of irradiance I is sent through the fiber to excite luminescence. The emitted light of irradiance I_e is transmitted back through the same fiber. The irradiance I_e is related to the temperature. In this sensor, the input light irradiance should be continuously monitored to make sure that it remains constant. In order to overcome this problem, another type of luminescent material (magnesium fluorogermanate activated with manganese) is used, which is excited by pulses of UV light. It emits visible light, which decays with a characteristic time constant. The time constant varies linearly with temperature. By measuring the time constant, the temperature could be determined.

18.7.2.1.3 Blood Flow Measurement

A fiber-optic sensor for temperature measurement is also used for the measurement of blood flow. The method is based on thermodilution. A cold fluid (saline) is injected into the blood and the downstream temperature is monitored. It is assumed that there will be little exchange of heat between the cold fluid and the blood over a short time period. As the blood flows, the mixture of fluid and blood is colder than the blood and warmer than the fluid. One can calculate the relation of the temperature T_m of the mixture to the temperature T_i of the cold fluid and the temperature T_b of the blood. If all the parameters are known, the measurement of T_m is sufficient to determine the blood flow. According to theory, during steady-state hyperemia, the volumetric blood flow rate, Q_b, can be calculated from the equation

$$Q_b = (1.08)\, Q_i \left(\frac{T_b - T_i}{T_b - T_m} \right) \tag{18.59}$$

Here Q_i represents the volumetric infusion rate of the saline, T_b is the temperature of the blood before the start of saline infusion, T_i is the temperature of the infusate (saline) at the tip of the infusion catheter, and T_m is the temperature at the sensor during steady-state infusion (i.e., the temperature of the blood after complete mixing with the infused saline). The constant 1.08 is a correction factor to compensate for the difference in specific heat between the saline and the blood.

The thermodilution method can be performed conveniently using a catheter that includes a fiber-optic temperature sensor. The distal tip of the sensor, which includes the optode, is placed at the tip of the catheter. The catheter is inserted at the relevant point, such as the coronary sinus. Cold fluid (saline) is injected continuously through the catheter and the fiber-optic sensor measures the temperature T_m of the blood stream a few centimeters from the point where the fluid was injected. The thermodilution method is routinely used to assess cardiac output.

18.7.2.2 Indirect Chemical Sensors

Sensors to measure pH, blood gases, glucose content, and contents of other chemical substances fall under this category.

18.7.2.2.1 pH Measurement

Two methods are used for the measurement of pH with fiber-optic sensors: (1) measuring the absorption of an indicator and (2) measuring changes in the luminescence properties. A common method of determining pH is based on indicator substances, which change color when placed in either the acidic or the alkaline solutions. Such color change can be observed through an optical fiber. One of the first pH sensors was based on a dye, phenol red, as an indicator. It was found that the dye changes color in the physiological range of interest (pH range between 7.0 and 7.4).

The pH sensor uses two thin fibers, which are inserted into a hollow catheter and placed a few mm from the catheter tip. The optode consists of polymer gel microspheres of about 10 μm diameter to which the dye is attached. These spheres are mixed with smaller spheres of about 1 μm diameter packed near the tips of the fibers, at the end of the catheter, and encapsulated by a thin membrane. The tiny spheres assist in the scattering process.

Light sent through the input fiber is scattered by the tiny spheres. Part of the light is absorbed by the dye and the rest is transmitted through the output fiber. The amount of light transmitted back depends on the absorption in the dye.

The dye exhibits strong pH-dependent absorption in green (550 nm) and almost pH-independent low absorption in red (600 nm). Therefore, radiation from two

LEDs, one emitting a peak wavelength of 550 nm and another at 600 nm, is transmitted through the input fiber alternately. The returned light is measured by a photodetector. The green light is absorbed by the dye, whereas the red light absorption is not pH dependent and only serves the purpose of calibration. The ratio $I(550)/I(600)$ is a direct measure of the pH, where $I(550)$ and $I(600)$ are the irradiances of light at 550 nm and 600 nm received by the photodetector. The fiber-optic sensor is simple, compact, and battery operated. The sensor is found to provide pH resolution of 0.01 pH unit. The signal is temperature dependent, changing by about 0.02 pH unit/°C. The response time is approximately 45 seconds.

The other method of pH measurement is based on a luminescent dye whose luminescence is dependent on pH. One example is a water-soluble dye, hydroxypyrene trisulfide acid, which has two forms, acidic and basic. The excitation spectrum peaks at 410 nm for the acidic form and at 460 nm for the basic form. The luminescence irradiance at 520 nm is excited once at 410 nm and again at 460 nm. The ratio of the irradiances gives the relative concentrations of the acidic and the basic forms of the dye, which is then related to the pH of the solution. The optode for this sensor thus consists of a hydrophilic polymer to which the dye is bound. A coat of this is applied to the distal tip of the fiber. Light at two wavelengths, 410 nm and 460 nm, is sent alternately through the same single fiber to the optode. The luminescence at 520 nm is transmitted via the same fiber and is measured by a photodetector. The pH value is calculated electronically. The advantages of this dye are high quantum yield, both excitation and emission in the visible, and large Stokes shift. However, it is sensitive to changes in the ionic strength due to its high charge.

In addition to organic dyes, a number of dyes based on luminescent transition metal complex are potentially useful for pH determination. These dyes offer distinct advantages including absorption in the visible, long excited state lifetimes, high quantum yield, and photostability.

18.7.2.2.2 Measurement of Blood Gases

CO_2 measurement: Dye used for the measurement of pH can also be used for the determination of partial pressure of CO_2 (PCO_2) in the blood. The pH of a bicarbonate solution depends on the PCO_2, with which it is in equilibrium. Usually, an increase in PCO_2 will decrease the pH of the bicarbonate solution. Therefore, the measurement of pH can be related to PCO_2. The optode for PCO_2 measurement is different than that of pH measurement. In this optode, the ions are excluded from the probe and only gas molecules are allowed to enter. This is achieved by encapsulating the bicarbonate solution in a silicon matrix that provides the necessary isolation from the ions in the blood. There is an overcoat of thin opaque cellulose layer on the optode to protect it and its optical system from external light. By choosing the correct concentration of the bicarbonate and the dye, the PCO_2 can be determined in the range of 10 mm of Hg to 1000 mm of Hg with 1% accuracy.

Measurement of partial pressure of oxygen (PO_2): The PO_2 probe is similar in design to the pH probe. It is based on the principle of quenching of luminescence by oxygen. When fluorescent quenching dyes are irradiated by light at an appropriate wavelength, they fluoresce in a nonoxygen atmosphere for a given period of time. However, when the oxygen is present, the fluorescence is quenched, that is, the dye fluoresces for a shorter time. The period of dye fluorescence is inversely proportional to the partial pressure of oxygen in the environment. This follows the Stern–Volmer relationship, which is expressed as

$$\frac{I_f^o}{I_f} = 1 + k_q\ \tau_o [Q] \tag{18.60}$$

where I_f^o is the irradiance, or rate of fluorescence without the quencher; I_f is the irradiance, or rate of fluorescence with a quencher; k_q is the quencher rate coefficient; τ_o is the fluorescence lifetime of the dye in the absence of quencher; and $[Q]$ is the concentration of the quencher.

The optode contains inert beads for immobilizing the dye, which are enclosed in an oxygen-permeable hydrophobic sheet such as porous polypropylene.

In commercial instruments, two dyes are used. The luminescence of one of the dyes is insensitive to oxygen pressure and serves to calibrate the system, whereas the other dye serves to monitor oxygen.

Glucose monitoring: Accurate measurement of blood glucose is essential in the diagnosis and long-term management of diabetes. Glucose is the main circulating carbohydrate. In normal, fasting individuals, the concentration of glucose in blood is very tightly regulated, usually between 80 mg/100 mL and 90 mg/100 mL during the first hour or so following a meal. In those suffering from diabetes, insulin-regulated uptake is compromised, and blood glucose can reach concentrations ranging from 300 mg/mL to 700 mg/mL (hyperglycemia).

The fiber-optic glucose sensor is based on a method known as the competitive binding method. The optode consists of a hollow tube plugged at one end of the fiber bundle: the walls of the hollow tube are freely permeable to glucose. The tube is filled with a fluorescent dye, fluorescein, which is bound to a soluble glucose polymer. The inside of the hollow tube is coated with concanavalin A, a material that binds glucose. A bifurcated fiber bundle end is inserted into the open end of the hollow tube (optode), and the fiber bundle transmits light to the optode and back again.

In the absence of glucose, the binding compound binds the fluorescent polymer to the wall. When the optode is inserted into glucose solution, the glucose penetrates the optode and displaces the polymer from the walls. Thus, the dye concentration increases in the solution away from the walls. The luminescence takes place in the geometric cone (cone of light from the fiber bundle), which now has more concentration of dye. Under excitation, there is more luminescent light, which the bundle transmits to the detector. Therefore, the more glucose in the solution, the more of it displaces the polymer from the wall, and hence the detector receives higher amounts of luminescence irradiance.

Fiber optics provides the flexibility of designing very compact sensors for measurement, testing, and control. Most of the sensors are based on the principles utilized in conventional and bulk sensors. Fiber optics results in low cost and compact sensors.

QUESTIONS

18.1 Show that the coupling efficiency between two identical grade index fibers with parabolic refractive index variation due to lateral mismatch (shift by d) is given by

$$\eta = \frac{2}{\pi}\left\{\cos^{-1}\left(\frac{d}{2a}\right) - \frac{1}{3}\left(\frac{d}{2a}\right)\left[1-\left(\frac{d}{2a}\right)^2\right]^{1/2}\left[5-2\left(\frac{d}{2a}\right)^2\right]\right\}$$

where $2a$ is the core diameter.

18.2 Single-mode silica fiber is used to measure force, hydrostatic pressure, and temperature, respectively. Calculate the sensitivity of the interferometer when used with a wavelength of 1 μm. Given $E = 2 \times 10^{11}$ Pa, $\nu = 0.2$, $\mu_1 = 1.45$, $\frac{d\mu_1}{dT} = 7\times10^{-6}\ \text{K}^{-1}$, $\alpha = 5 \times 10^{-6}$/K, and external diameter of the fiber is 100 μm.

18.3 Calculate the fraction of light transferred from the emitting to receiving fiber in the two-fiber version of a Fotonic sensor where two fibers of radius a, and separation s are situated a distance z from the reflecting surface. {Hint: Consider the coupling between the image of the emitting fiber and the receiving fiber. There are three regions to be considered: (1) $R < a + s$, (2) $2s + 3a \geq R \geq s + a$, and (3) $R > 2s + 3a$, where $R = a + 2z\tan\left[\sin^{-1}(NA)\right]$. In the second region, where the radiation cone partially overlaps the receiving fiber core, assume that the circular edge of the cone may be replaced by a straight edge}. Sketch the coupling efficiency as a function of z for typical values of a and s.

18.4 It is proposed to use the amount of light transmitted between two fibers whose ends are offset longitudinally as an optical dipstick. Calculate how the amount of light transmitted between the fibers varies with the refractive index and the absorption coefficient of the intervening medium. If the sensor is lowered from air to water, by how much would you expect the coupled radiation to increase?

18.5 Derive the form of the output expected from an M-Z interferometer when beam irradiances in the two arms are unequal. What advantage is there in this situation using both possible outputs from the interferometer?

18.6 The light from a point object is focused with unit magnification onto the end of a fiber whose core diameter is $2a$ and numerical aperture NA. Show that the fiber may be moved backward and forward a distance $2fa$/R about the point of focus without altering the total amount of light entering the fiber core provided $\sin^{-1}(NA) \leq \tan^{-1}\left(\frac{R}{2f}\right)$. Here $2R$ is the diameter of the lens.

18.7 Show that the beat length L_B in a single-mode fiber in which the refractive indices for the two orthogonally polarized modes are μ_x and μ_y, respectively, is given by

$$L_B = \frac{\lambda_0}{|\mu_x - \mu_y|}$$

18.8 Quality of a PM fiber may be judged by the value of the polarization-holding parameter h. This is defined as the amount of polarization mode coupling per unit length. If $P_x(z)$ and $P_y(z)$ represent the mode powers at a point z in the fiber, then

$$\frac{dP_x(z)}{dz} = h\ P_y(z)$$

Assuming h to be small say 5×10^{-5}/m what percentage of P_y couples to P_x over a propagation distance of 1 km?

18.9 Assuming that the resolution in the measurement of angle of rotation is 0.05°, how many turns of the fiber should be used in the current transducer if a resolution of 10 A in current is desired? The Verdet constant of the fiber material is 4 rad/m T.

18.10 A silica fiber of length 10 cm is the sensing element of an accelerometer. A mass of 10 g is suspended at the end. Assuming the fiber diameter to be 125 μm, calculate the sensitivity of the interferometer. Take $\mu_1 = 1.50$, $E = 7.3 \times 10^{10}$ N/m^2 and operation at $\lambda = 633$ nm.

18.11 A 1.0m long fused silica fiber is to be used as a temperature-sensing element in an interferometer. The coefficient of thermal expansion of fused silica is $\alpha = 0.55 \times 10^{-6}$ / °C, its refractive index $\mu = 1.46$, and change of refractive index with temperature $\frac{d\mu}{dT} = 12.8 \times 10^{-6}$ / °C. What is the change in the optical path produced by a change of the fiber temperature by 1°C?

18.12 The smallest phase shift that can be measured in a fiber interferometer used for pressure sensing is 1.0 μrad. The normalized pressure sensitivity of a typical single-mode fused silica fiber coated with nylon is $\frac{\Delta\delta}{\delta\ \Delta P} = 3.2 \times 10^{-11}$ / Pa, where δ is the total phase shift produced by the fiber, $\Delta\delta$ is the change in the phase shift produced by a pressure change ΔP. What length of fiber should be used as a sensing element in a marine hydrophone, at a wavelength of 633 nm, to obtain adequate sensitivity to detect sea-state zero, i.e., 100 μPa at 1 kHz?

18.13 A fiber rotation sensor for a navigation application must be capable of detecting rotation rates equal to 0.1% of the earth's rotation rate (15°/hr). If phase measurement can be made with an accuracy of 0.1 μrad at an operating wavelength of 850 nm, how many turns of the fiber are required on a 200 mm diameter coil?

19 Length Measurement

19.1 INTRODUCTION

Length is one of the basic quantities whose measurement plays a vital role in science, engineering, trade, etc. Optical systems such as microscopes, telescopes, and interferometers are used for the length measurement; choice depends on the requirement. Microscopes and telescopes utilize mechanical scales for measurements. There are several optical techniques that are used for setting purposes thereby enhancing the accuracy of the measurement. Interferometers utilize a wavelength of light as the measuring stick. In this chapter, we discuss techniques and methods that use light wavelength as the measurement stick rather than calibrated mechanical scales. Earlier the interferometric measurement of length was generally confined to the Standards laboratory. However, the advent of the laser revolutionized the field and brought the technique to the shop floor and into everyday use. Prior to the laser, quasi-monochromatic sources were used in the interferometers. The techniques using these sources will be discussed first and then the techniques using a laser as a coherent source will be introduced. Further, the length measurement takes place over different ranges and hence techniques will be put into different groups. Measurement of displacement and its derivatives will also be considered under length measurement.

19.2 MEASUREMENT OF GAUGE BLOCKS AND SLIP GAUGES

Gauge blocks are metal or ceramic blocks that have been precision ground and lapped to a specific length. These are used as secondary standards of length, and hence their length is to be measured using interferometers. Their cross section is usually rectangular (35 mm × 9 mm). Any length can be realized by wringing together different gauges. For this reason, a set of gauge blocks is provided to a Standards laboratory. The length of these blocks ranges from 1 mm to 1 m. Both Fizeau and the Twyman–Green (T-G) interferometers in their various variants are used as gauge block interferometers. The T-G interferometers require lower temporal coherence than the Fizeau interferometers for the length measurement of gauge blocks.

19.2.1 METHOD OF EXACT FRACTIONS

To measure the length of a gauge block, the gauge block is wrung to an optically flat base plate, which is a part of the interferometer. The interference fringes are formed between waves reflected from the reference surface R and the upper surface of the gauge block, the reference surface, and the surface of the base plate as shown in Figure 19.1(a). Surface R' in Figure 19.1(b) is the image of the reference surface when the T-G interferometer is used for measurement. Figure 19.1(c) shows

DOI: 10.1201/9781003531845-19

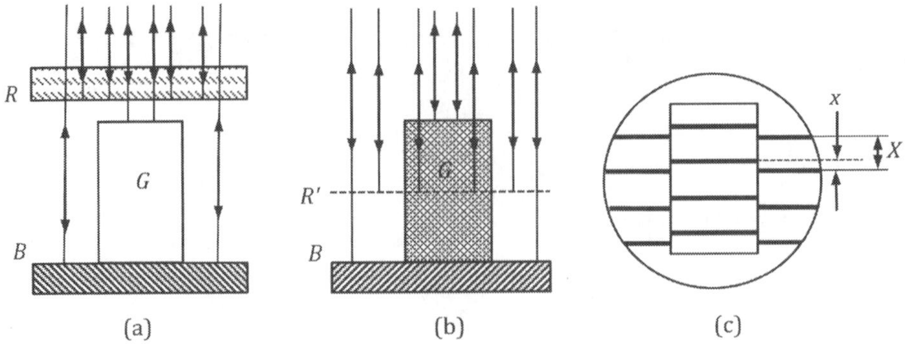

FIGURE 19.1 Measuring Gauge block with (a) Fizeau interferometer and (b) T-G interferometer. (c) Field of view.

the interference pattern in the field of view, which consists of two patterns arising when the reference surface is lightly tilted: the wedge thickness increases upward. These two fringe patterns are displaced with respect to each other, as the length of the gauge block is not an exact multiple of $\lambda/2$. The only information that is readily available by glancing at the fringe pattern is that the gauge length is not a multiple of half wavelength, $\lambda/2$, but is slightly bigger by a fraction. The ratio $\frac{x}{X} = \varepsilon$ is the fraction of $\lambda/2$ as shown in Figure 19.1(c) by which the gauge length is bigger from the half-integer value of the wavelength. The method of exact fractions is used to obtain the half-integer value. The method requires that the length must be approximately, within a few half wavelengths, known and the fraction should be measured at more than one wavelength. It is possible to know the approximate length by using accurate length comparators. Therefore, the length is known as $L \pm \Delta L$, where ΔL is the uncertainty of measurement by the length comparator.

Let us consider a gauge block interferometer that uses light source(s) to provide several wavelengths for measurement. Let the wavelengths be λ_1, λ_2, λ_3, and λ_4 and the corresponding measured fractions are ε_1, ε_2, ε_3, and ε_4. The length L of the gauge block can be expressed as

$$L = (N_1 + \varepsilon_1)\frac{\lambda_1}{2} = (N_2 + \varepsilon_2)\frac{\lambda_2}{2} = (N_3 + \varepsilon_3)\frac{\lambda_3}{2} = (N_4 + \varepsilon_4)\frac{\lambda_4}{2} \qquad (19.1)$$

where N_1, N_2, N_3, and N_4 are the integer values contained in the length corresponding to wavelengths λ_1, λ_2, λ_3, and λ_4, respectively. These integer values are not known. The method of coincidences is used to obtain these integer values and then length L is obtained by adding the fractions and multiplying the sum with the corresponding half wavelength.

A first approximation to N_1 is obtained by dividing the nominal value of L by the known value of $\lambda_1/2$ and taking the whole number of the quotient, which is N_1'. Addition to N_1' of the measured fraction gives the first approximation of N_1. This number is now multiplied by λ_1/λ_2, λ_1/λ_3, and λ_1/λ_4 so that we obtain the whole numbers and the fractions $N_2' + \varepsilon_2'$, $N_3' + \varepsilon_3'$, and $N_4' + \varepsilon_4'$ for each wavelength λ_2, λ_3, and λ_4, respectively. If these fractions ε_2', ε_3', and ε_4' do not agree, within the experimental error, with those observed experimentally, another value of N_1', differing from the first by 1 is tried. This process is continued by taking values of N_1' differing by 1 over the range $\pm 2\Delta L / \lambda_1$ until an agreement between the measured fractions and the calculated fractions is obtained. The agreement between the measured and the calculated fractions for all four wavelengths for that value N_1' plus the integer increments gives the value N_1. A few trials usually will lead to the correct result. The length is then obtained as $L = (N_1 + \varepsilon_1)\frac{\lambda_1}{2}$. Commonly used wavelengths in gauge block interferometry are 632.991162 nm (He-Ne laser), 644.02480 nm, 508.72379 nm, 480.12521 nm, and 467.94581 nm (all from a ^{114}Cd lamp). It may also be mentioned that the fraction can also be obtained by using the phase-shifting interferometry.

19.3 GAUGE BLOCK INTERFEROMETRY: COMPARISON WITH A STANDARD

By using either a Fizeau or a T-G interferometer equipped with a multi-wavelength source, we can find the length of any gauge block using the principle of exact fractions provided the length is much smaller than the coherence length. Both interferometers require a base plate to which the gauge is to be wrung. In many cases, an unknown gauge block is to be compared with a known one. Figure 19.2(a) shows the interferometric arrangement for the comparison of the gauge blocks that are not necessarily the same length. The interferometer uses multiple sources and the method of exact fractions.

It is seen that the waves reflected from faces G_3 and G_2, and from faces G_1 and G_4, interfere to produce the interference patterns. The fraction between the two interference patterns corresponds to the difference between the two gauge lengths. By measuring the fractions at least at three wavelengths, the difference between the two lengths is obtained using the method of exact fractions.

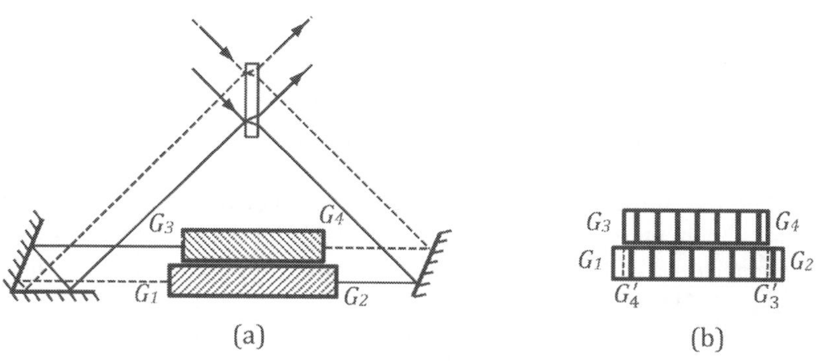

FIGURE 19.2 (a) Comparison of gauge blocks. (b) Field of view.

19.3.1 Single-Wavelength Interferometry for Gauge Blocks

The method of exact fractions was used to determine the integer fringe order uniquely. On the other hand, if the uncertainty in length as measured by an optical comparator is smaller than the wavelength of light, it is possible to measure length with a single wavelength.

The governing equation is

$$2(L \pm \Delta L) = m\lambda = (m_0 + \varepsilon)\lambda \tag{19.2}$$

where ΔL is the uncertainty of the comparator, m is the fringe order, m_0 is the integer fringe order, and ε is the fractional order. When the length is known within $\pm\Delta L$, it makes the right-hand side of the Equation (19.2) uncertain up to $u = \pm 2\Delta L / \lambda$. If $|u| < 1/2$, there will be no uncertainty in the integer m_0 due to uncertainty ΔL in the initial estimate of L. We need only to determine accurately ε consistent with the desired accuracy in the length measurement. It is, thus, obvious that we should have the wavelength $\lambda > 4|\Delta L|$ or $|\Delta L| < \lambda / 4$ for the technique to succeed. If we take an iodine-stabilized He-Ne laser operating at 3.39 μm in infrared as the source, then $\lambda/4 = 0.85$ μm. The present-day optical comparators have measurement uncertainties of less than 0.8 μm. Therefore, an infrared interferometer with a He-Ne laser operating at 3.39 μm either with electronic scanning of fringes or with phase shifting to measure ε with desired accuracy is an obvious choice for the measurement of the length of gauge blocks.

The range of length measurement can be further extended if the laser oscillates in two close wavelengths so that a synthetic wavelength λ_s could be used. Assuming oscillations at wavelength λ_1 and λ_2, the synthetic wavelength is given by

$$\lambda_s = \frac{\lambda_1 \lambda_2}{|\lambda_1 - \lambda_2|} \tag{19.3}$$

Therefore, the uncertainty in length ΔL should satisfy the following condition for the integer m_0 to be uniquely known

$$|\Delta L| < \lambda_s / 4 \tag{19.4}$$

The He-Xe laser provides two closely spaced wavelengths, namely 3.37 μm and 3.51μm, that oscillate simultaneously and can be stabilized. This then produces a synthetic wavelength of 84.49 μm; hence, the uncertainty in length is ≈21 μm. A length of several meters can be measured within this uncertainty. Hence, the length measurement can be carried out with desired accuracy if the fractional order is measured within the desired accuracy.

In one of the approaches, the gauge block of longer length is wrung to the base plate and an interferometer, say a T-G interferometer, is set up. The alignment of the interferometer is carried out with a visible laser. The irradiance at the detector plane is obtained as the sum of irradiances of two interference patterns belonging to two wavelengths. This is expressed as

$$I_T = 2I_{10}\left(1+\cos\frac{2\pi L}{\lambda_1}\right)+2I_{20}\left(1+\cos\frac{2\pi L}{\lambda_2}\right) \tag{19.5}$$

where I_{10} and I_{20} are the irradiances of each interfering beam belonging to wavelength λ_1 and λ_2, respectively. The signal i from the detector is high-pass filtered and squared. This results in

$$\begin{aligned} i^2 &\propto I_{01}^2\cos^2\frac{2\pi L}{\lambda_1}+I_{02}^2\cos^2\frac{2\pi L}{\lambda_2}+2I_{01}I_{02}\cos\frac{2\pi L}{\lambda_1}\cos\frac{2\pi L}{\lambda_2} \\ &= \frac{I_{01}^2}{2}\cos\frac{4\pi L}{\lambda_1}+\frac{I_{02}^2}{2}\cos\frac{4\pi L}{\lambda_2}+ \\ &\quad I_{01}I_{02}\left[\cos\left(2\pi L\frac{\lambda_1+\lambda_2}{\lambda_1\,\lambda_2}\right)+\cos\left(2\pi L\frac{|\lambda_1-\lambda_2|}{\lambda_1\,\lambda_2}\right)\right]+\frac{I_{01}^2+I_{02}^2}{2} \end{aligned} \tag{19.6}$$

By passing this signal through a low-pass filter, the first three terms in the Equation (19.6) can be eliminated. Therefore, the desired signal i' is

$$i' = I_{01}I_{02}\left[\cos\left(2\pi L\frac{|\lambda_1-\lambda_2|}{\lambda_1\,\lambda_2}\right)+C_0\right] = I_{01}I_{02}\left[\cos\left(\frac{2\pi L}{\lambda_s}\right)+C_0\right] \tag{19.7}$$

where $\lambda_s = \frac{|\lambda_1-\lambda_2|}{\lambda_1\,\lambda_2}$ and $C_0 = \frac{I_{01}^2+I_{02}^2}{2I_{01}I_{02}}$. The simultaneous interference of waves of different wavelengths λ_1 and λ_2 is made equivalent to the interference with a synthetic wavelength λ_s.

19.4 COMB GENERATION AND GAUGE BLOCK CALIBRATION

Before we discuss, comb generation let us dwell a little on a femtosecond laser. A passive cavity, in which the mirrors are separated by d, supports a large number of axial modes that satisfy the boundary condition, that is, the cavity length is an integral multiple of half wavelengths. This can be stated as $2d = m\lambda$. These modes are separated in frequency by $c/2d$, which is the reciprocal of the cavity round-trip time, where c is the velocity of light. In the presence of an active medium in the cavity, those modes that fall within the gain curve are supported and the output of the laser is multimodal or multi-wavelength. In a continuous multimodal laser, these modes oscillate independently and have random phases. The output is thus noisy with the irradiance fluctuating randomly. The phases of the modes can be manipulated by a technique called mode locking. A multimode laser in which the modes have a fixed relationship with each other is said to be mode locked. The output of a mode-locked laser would be a series of pulses. The mode locking can be done by some form of modulation and may involve passive, active, or self-phase modulation.

One of the convenient ways is to make use of the nonlinearity of the refractive index for self-phase modulation and hence the Kerr-lens effect is used for mode locking. Broadband active medium results in shorter pulses and the use of the Kerr-lens effect further broadens the spectrum resulting in still shorter pulses.

The heart of an optical comb is a Ti: sapphire laser that contains a titanium-doped aluminum oxide (Ti: Al_2O_3 or Ti: sapphire) crystal rod as an active medium. The Ti: sapphire laser has a very wide emission band spanning from 650 nm to 1100 nm but not broad enough to achieve one optical octave for the comb to be useful. Additional bandwidth is provided by self-phase modulation using the Kerr-lens effect or using a photonic fiber. The optical comb can be described by two independent parameters ν_r and ν_0, where ν_r is the mode spacing and ν_0 is the frequency offset of the optical comb from zero frequency as a whole, which is caused by the difference between the group and phase velocities of the pulses. Both these parameters fall in the microwave region of less than 1 GHz. These two parameters are stabilized to a well-established microwave standard such as the Cesium (Cs) clock. The frequency of the nth mode can be expressed as $\nu_n = n\nu_r + \nu_0$, where n is a very large number.

Figure 19.3 shows an optical comb in frequency domain, and irradiance pulses in time domain. The optical comb essentially consists of many laser modes equi-spaced in the frequency domain: it can be used like a ruler to determine large frequency difference and to measure length in dimensional metrology. Therefore, the individual laser modes should be available for measurement. This is achieved by an optical frequency generator, which is an ultra-stable tunable light source capable of producing or extracting the desired optical frequencies from the optical comb. The mode could be extracted directly by coarse filtering followed by fine filtering: fine filtering should have a bandwidth smaller than the repetition rate. The optical power of the extracted mode is too weak to be useful for interferometry, hence, it should be amplified. Alternately a working laser, an extra-cavity laser diode, with a wide wavelength tuning range of ~20 nm with narrow line width is used.

19.4.1 Measurement of Gauge Block with Optical Comb

The optical comb provides large number of wavelength sources (frequency sources) and hence the method of exact fractions can be used. The method is similar to one

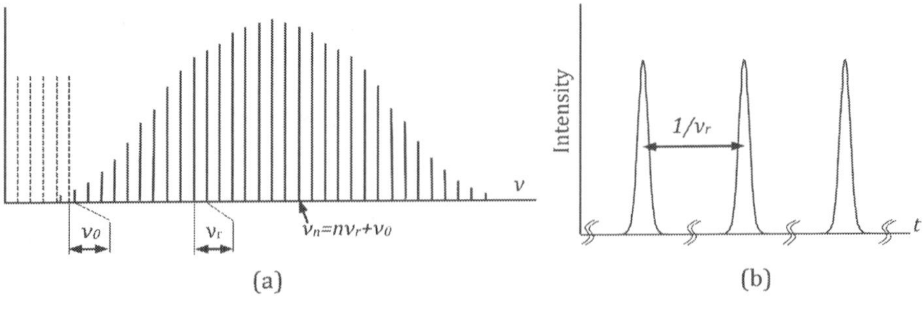

FIGURE 19.3 (a) Optical comb in frequency domain. (b) Laser pulses in time domain.

described earlier that uses several independent laser sources. With an optical comb there is no need for several lasers; only the working laser, which is tuned to the optical comb frequencies, is used.

The length is obtained using the following expression

$$L=(N_1+\varepsilon_1)\frac{\lambda_1}{2}=(N_2+\varepsilon_2)\frac{\lambda_2}{2}=(N_3+\varepsilon_3)\frac{\lambda_3}{2}=\cdots\ldots\ldots\ldots\ldots\ldots=(N_n+\varepsilon_n)\frac{\lambda_n}{2} \quad (19.8)$$

The length is measured by measuring fractions and finding the integer order as has been explained earlier. Length could also be obtained by using a synthetic wavelength. Because the inter-pulse frequency difference is $c/2d$, which could vary between 100 MHz and 1000 MHz depending on the cavity length, and the synthetic wavelength would be c/ν_r, assuming refractive index as unity. The group refractive index of the environment around the gauge block needs to be known. This procedure covers the length between 30 cm and 300 cm when adjacent modes are used for measurement.

19.4.2 Distance Measurement with Frequency Comb

The principle of the method can be explained with the help of Figure 19.4. The pulse train from a frequency comb laser is launched in the interferometer. The pulses have a constant phase difference so they can interfere with each other. Further, the frequency separation between the two consecutive pulses is $c/2d$, which lies between 100 MHz to 1000 MHz. Therefore, the inter-pulse distance, d_p, would be 30 cm to 300 cm. When the path difference between the two arms of the interferometer is an integral multiple of inter-pulse distance, two pulses will superpose forming an interference pattern, which can be detected. The spatial overlap of the two pulses, one each in each arm of the interferometer, is accomplished by translating the reflector in the short arm: the maximum translation of the reflector is limited to $d_p/2$ to determine the length without any ambiguity.

The interferometer can be used to measure large distances accurately. To perform this measurement, the long arm reflector is brought to the initial position, which is very close to the beam splitter, and the self-interference of the pulse is observed. Movement of the reflector will bring successive pulses to interference. The fractional part is obtained by moving the reflector in the short arm. If N such interference

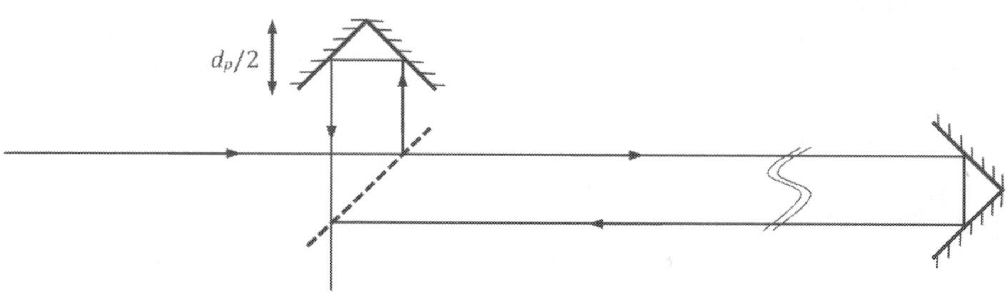

FIGURE 19.4 Schematic of an interferometer for length measurement with an optical comb.

maxima are observed and the reflector in short arm is translated by d_s to get the interference maximum, then the length L is $L = N\, d_p/2 + d_s$.

19.5 MODULATED FREQUENCY DISPLACEMENT SENSOR

A double heterostructure laser diode is a convenient and compact source for ranging as its output power can be modulated at a relatively high and constant frequency. In order to measure the distance, the DC current of the laser diode is modulated at a frequency f_m sinusoidally. The signal from the target is collected and received by a photodiode. The output current of the photodiode varies also sinusoidally but is phase-shifted due to the traverse of a distance by light, which is twice the distance of the target from the diode. Measurement of phase shift $\Delta\delta$ between the signal from an oscillator that modulates the current in the laser diode and the current from the photodiode are related to the distance L of the target through the relation

$$\Delta\delta = 2\pi f_m \frac{2L}{c} \tag{19.9}$$

where c is the velocity of light in free space. In fact, the refractive index of the air needs to be calculated from the measured values of temperature and pressure to obtain the velocity of light in air that should be used in place of c. The phase difference can be measured to an accuracy of ± 0.1 degree or better. Assuming laser diode modulation at 30 MHz, the unambiguous distance measurement up to 5 m with accuracy of ± 1.39 mm can be performed. In fact, longer distances can be unambiguously determined with lower frequency of modulation albeit lower accuracy. The accuracy of the method can be improved by measuring the phase shift at a lower frequency by heterodyning it with another signal of frequency different than f_m. It should be noted that the heterodyne technique preserves the phase shift versus distance relationship.

19.5.1 Frequency-Modulated Continuous Wave (FMCW) Laser Radar

Laser diodes have another interesting characteristic: their wavelengths can be tuned over a range by varying either the current or the temperature. This wavelength variation over a limited range can be used for the measurement of distance. Let us consider an experimental setup as shown in Figure 19.5(a). The light from a tunable laser is divided by a beam splitter: one part travels to a reference mirror and the other part to the target/object whose distance/range is to be measured. The returned beams are combined by the beam splitter and are incident on a photodetector. Mathematically, the amplitudes of the beams returned from the reference mirror and from the object at the plane of the detector can be expressed as

$$E_r = E_{r0} e^{i(2\pi\nu_r t + \phi_r)} \tag{19.10a}$$

$$E_t = E_{t0} e^{i(2\pi\nu_t t + \phi_t)} \tag{19.10b}$$

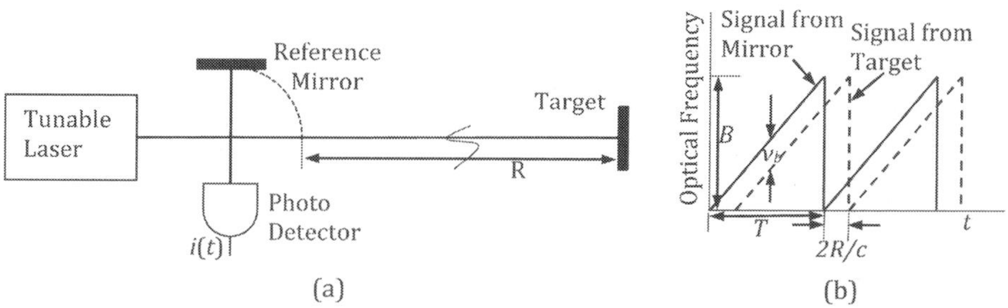

FIGURE 19.5 (a) Schematic of frequency-modulated continuous wave. (b) Linear optical frequency ramp.

where E_r, E_t are the amplitudes of the beams from the reference mirror and the target, and ν_r, ν_t are the frequencies of the beams at the detector plane. The output from the photodetector will be proportional to

$$i(t) \propto \left\{E_{r0}^2 + E_{t0}^2 + 2E_{r0}E_{t0}\cos\left[2\pi\left(\nu_t - \nu_r\right)t + \phi_t - \phi_r\right]\right\} \tag{19.11}$$

The detector outputs the signal at the beat frequency $\nu_b \left(= \nu_r - \nu_t\right)$. In practice the frequency of the source is linearly varied as shown in the Figure 19.5(b).

The frequency of the wave from the reference mirror at the detector plane can be written as

$$\nu_r(t) = \nu_0 + \frac{B}{T}t \tag{19.12}$$

Similarly, the frequency of the wave from the target mirror at the detector plane is written as

$$\nu_t(t) = \nu_0 + \frac{B}{T}\left(t - \frac{2R}{c}\right) = \nu_0 + \frac{B}{T}t - \frac{B}{T}\frac{2R}{c} \tag{19.13}$$

Therefore, the beat frequency ν_b is given by

$$\nu_b = \frac{B}{T}\frac{2R}{c} \Rightarrow R = \frac{T}{B}\frac{c\,\nu_b}{2} \tag{19.14}$$

The range R is measured by measuring the beat frequency, which is in the kilohertz range, and the ramp time, which is in the range 0.1 to 1 ms. Since the ramp period could be chosen arbitrarily, the range can be determined with good accuracy. It may be noted that there is a dead time equal to the round trip excursion time during which no beat frequency measurement can be made.

In a more sophisticated modulation scheme, the optical frequency is varied linearly as shown in Figure 19.6. The frequency-modulated continuous wave (FMCW) radar can be used to measure accurately the range and the axial velocity.

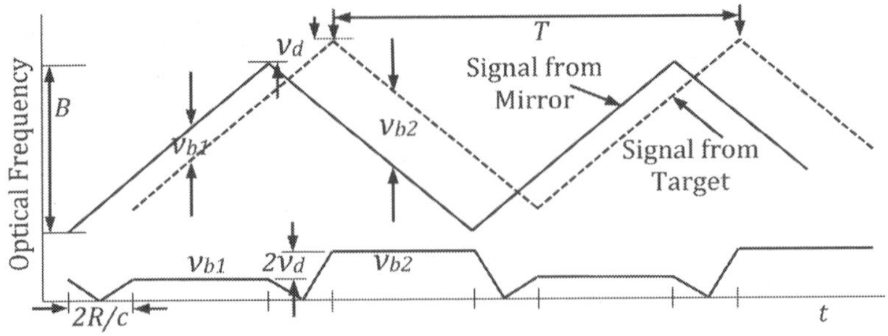

FIGURE 19.6 Modulation scheme to obtain both range and velocity of a target.

Signals from the reference mirror and the target are shown as solid and broken triangular lines at the detector plane. Due to the motion of the target, the returned signal is Doppler-shifted: the Doppler frequency is indicated by ν_d. In the presence of the motion of the target, the beat frequencies for the rising ramp and decreasing ramp are different. An average beat frequency ν_b is defined as $\nu_b = (\nu_{b1} + \nu_{b2})/2$ and the range is obtained from

$$R = \frac{T}{B}\frac{c\,\nu_b}{4} \tag{19.15}$$

The Doppler frequency ν_d is obtained as $\nu_d = (\nu_{b2} - \nu_{b1})/2$ and the radial velocity component v_r is obtained from

$$v_r = \frac{\lambda\,\nu_d}{2\cos\theta} \tag{19.16}$$

where λ is the wavelength of the laser, and θ is the angle between target velocity vector and the radar line of sight. This is valid when ν_d is less than the beat frequency.

19.6 DISPLACEMENT MEASUREMENT WITH INTERFEROMETRY

This is the direct method of length measurement in which the length is measured in terms of wavelength as a ruler. Use of long coherence sources such as stabilized lasers makes it possible to measure displacement. For this purpose, let us consider a T-G interferometer whose one of the mirrors is on a movable slide, and the displacement of the slide is to be measured. A fringe pattern of a suitable fringe width is obtained, and a cross wire is positioned at the center of a fringe in the field of view. A displacement of $\lambda/2$ given to the slide moves the fringe by one; the next fringe will now lie on the cross wire. If the mirror is mounted on the bed of a machine, the movement of the bed can be determined by counting the number of fringes moving past the cross wire and multiplying the number by $\lambda/2$. The direction of the movement of the bed governs the direction of the movement of the fringes. Because the number of fringes to be counted is very large (>31,000/cm for a He-Ne laser red wavelength), the counting is done electronically and displayed digitally.

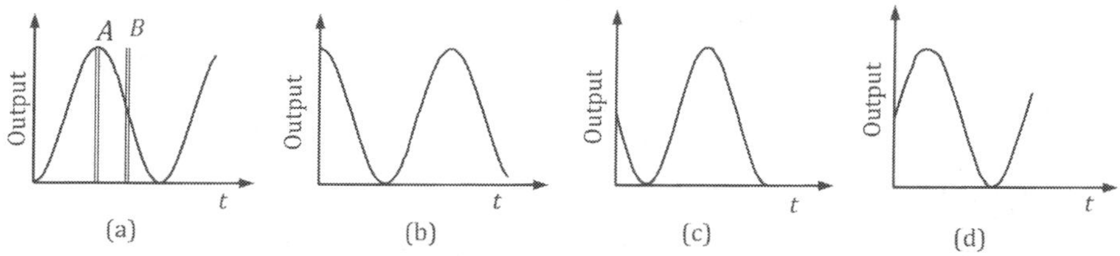

FIGURE 19.7 Signals for bidirectional counting. (a) Generation of quadrature signals with a pair of slits. (b) Reference signal. (c) Signal leads by $\pi/2$. (d) Signal lags by $\pi/2$.

A photodetector behind the slit receives the light flux passing through it and gives a sinusoidal signal as a function of displacement of the mirror; one cycle for every half wavelength displacement. A photodetector will show sinusoidal variation irrespective of the direction of displacement and count the cycles, which will always be added. A reversible counting technique is employed to overcome this difficulty. In order to achieve reversible counting, the interferometer should provide two outputs, one varying sinusoidally and the other co-sinusoidally with displacement. Two such signals are called the *quadrature signals*. One simple method to achieve this is to use two slits with separate detectors whose separation is adjusted to produce signals with 90° phase difference as shown in Figure 19.7. We can now study the phase relationship of the two signals in the two cases, namely when the fringes move to the left and when they move to the right. Taking the signal received from detector behind slit A as the reference signal as shown in Figure 19.7(b), and the signal from the slit B either leads or lags behind by 90° depending on the direction of motion as seen in Figure 19.7(c and d). An electronic processor can distinguish this difference and produce a command signal for the bidirectional counter either to add or subtract the incoming pulse in the register.

Figure 19.8 shows the processing of the quadrature signals. The sinusoidal signals are put into rectangular form by Schmitt triggers. The shaped signals are represented

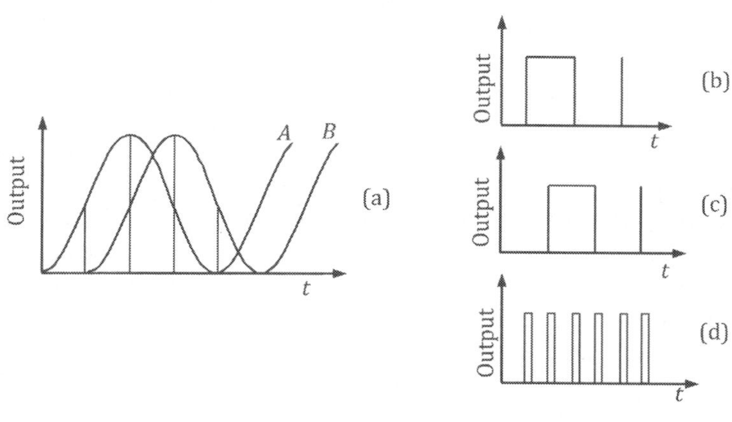

FIGURE 19.8 Processing of signals. (a) Signals from slits A and B. (b and c) Converted to digital signals. (d) Four pulses per cycle.

by the two digital states 1 and 0. The sequence of the two signals is representative of the direction of motion. One, two, or four pulses per cycle are produced depending on the least count requirement. When four pulses per cycle are produced, the least count is $\lambda/8$. The total count is converted into displacement by multiplying it with the least count.

The T-G interferometer with mirror reflectors requires extremely accurate guides for the machine bed or for the mount/slide, on which the mirror is placed, otherwise the fringe width would vary with displacement, or the fringes may disappear. This difficulty is overcome if the corner cube reflectors are used instead of the mirror reflectors. Because the corner cube reflector returns the beam in the same direction, it presents another difficulty. The interference between the retro-reflected beams by the corner cube reflectors results in an infinite width fringe and hence the two slits arrangement used earlier cannot be applied here. Instead, the polarization-based methods are used to obtain two signals in quadrature. There are several variants of these interferometers designed for different applications.

The fringe counting interferometers described earlier require that the irradiance variations in the interference pattern be centered around the triggering levels of the counter to generate rectangular pulses. These irradiance variations are about the DC level. If the irradiance of either of the beams or the source changes, the variations in irradiance may not cross the triggering levels, thus there will be malfunctioning of the instrument until the trigger levels are readjusted. This problem can also be handled by obtaining irradiance variations about zero irradiance. This can be achieved by obtaining four signals from the interferometer, which are then manipulated, to provide quadrature signals having zero mean values.

19.6.1 Two-Frequency Laser Interferometer for Displacement Measurement

A neater way to make the interferometer insensitive to irradiance variations is to use a two-frequency interferometer. This interferometer operates on the heterodyne principle. The schematic of the interferometer is shown in Figure 19.9. The laser emits two waves with frequencies ν_1 and ν_2, which are a few megahertz apart and are orthogonally polarized. A polarizing beam splitter, B_2, transmits ν_2 to the measuring corner cube C_2 and reflects ν_1 to the corner cube C_1. The reflected beams from both corner cube reflectors are combined by the beam splitter and directed to photodetector PD_2. A polarizer, P_2, brings these two beams to interfere. Because the frequencies of the two beams are different, resultant irradiance fluctuates at the rate of $(\nu_1 - \nu_2)$. The beam splitter, B_1, sends these two waves to another detector PD_1. A linear polarizer is placed before the detector and suitably oriented so that beat signal, $(\nu_1 - \nu_2)$, with high contrast is produced. The beat frequencies from detectors PD_1 and PD_2 are counted by two separate counters and their counts are subtracted to produce a net count. In the present situation, when both the corner cube reflectors are stationary, the net count will be zero.

When the corner cube C_2 is moved to make measurement, the frequency of the reflected beam is Doppler-shifted to $(\nu_2 \pm \Delta\nu)$ depending on its speed and the

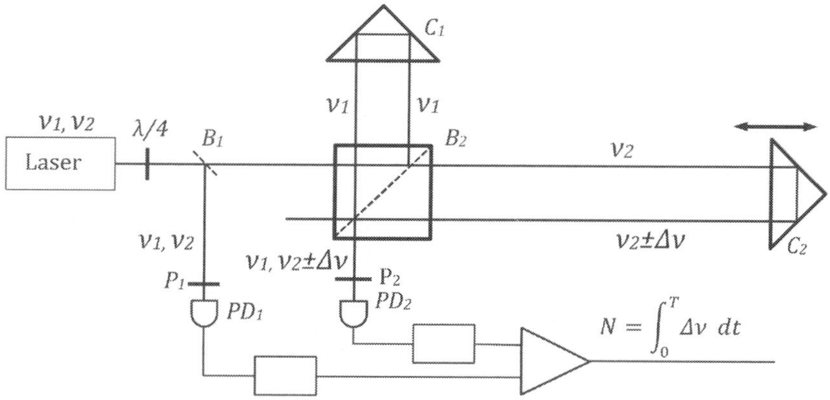

FIGURE 19.9 Two-frequency laser interferometer for measuring displacement.

direction of motion. If the corner cube C_2 is moving with a constant speed v away from the beam splitter, the frequency ν_2' of the reflected beam is given by

$$\nu_2' = \nu_2\left(1 - \frac{2v}{c}\right) \tag{19.17}$$

This will produce a beat frequency $\nu_1 - \nu_2' = \nu_1 - \nu_2 + \nu_2 \frac{2v}{c}$. These counts are accumulated over the period of measurement T and hence the net count N is $N = \nu_2 \frac{2v}{c} T$. It is now easy to show that this represents double the distance that is measured.

The Doppler shift is $\Delta\nu = \nu_2' - \nu_2 = -\nu_2 \frac{2v}{c} = -\frac{2v}{\lambda}$, where λ is the wavelength of light used in the interferometer. If the corner cube is moved for a period T, the total distance traveled is $L = v\,T$. The net accumulated counts over this period have been shown to be $N = \nu_2 \frac{2v}{c} T = \frac{2vT}{\lambda} = \frac{2L}{\lambda}$. Hence the length L is given by

$$L = \frac{N\lambda}{2} \tag{19.18}$$

Thus, the total displacement L is obtained by multiplying the accumulated net count N by $\lambda/2$; this is performed electronically, and the result is displayed. The displacement is measured with the least count of $\lambda/2$ ($\approx 0.3\ \mu$m). The resolution can be improved by electronic splitting of the sinusoidal signal. It may be noted that it is not essential that the corner cube C_2 moves with a constant velocity. In this interferometer, the detectors continuously generate an AC signal at the beat frequency even when the corner cube is stationary. This interferometer is, therefore, referred to as an AC interferometer. Because the counting pulses are not generated by triggering counter levels, loss of light due to any causes in the interferometer does not stop the functioning of the interferometer. Loss of light up to 90% of the rated value can be tolerated. A commercial instrument using Zeeman split He-Ne laser for measuring moderate to long distances with adequate accuracy is available.

19.7 ANGLE INTERFEROMETER

The interferometer is used to measure small angles and hence is easily adopted for the measurement of straightness. There are several variants of the angle interferometer. Figure 19.10 shows one of the variations, which utilizes a two-frequency laser. The reference beam and the measurement beam propagate side by side.

The polarization beam splitter, *PBS*, separates the two waves of frequency ν_1 and ν_2, which form the reference and the measurement beams. The heart of the interferometer is a pair of corner cube reflectors mounted on a single stand. This reflector slides on the surface. Any departure from the straightness results in an angular tilt of the reflector and consequently in a path length difference between the two beams. This path length change results in some counts from the detector.

The path length difference Δ is related to the angular tilt α of the reflector through the relation

$$\sin\alpha = \frac{\Delta}{p} \tag{19.19}$$

where p is the separation between the two corner cube reflectors. The beam splitter combines these two frequency-shifted beams, and the polarizer P aligns their polarization states resulting in a high contrast heterodyne signal from the detector. The number of counts N from the detector is given by

$$N = \frac{2\Delta}{\lambda} \tag{19.20}$$

Therefore, the angle α is given by

$$\sin\alpha = \frac{N\lambda}{2p} \tag{19.21}$$

It may be noted that one count here corresponds to a half wavelength path change. Further subdivision is done to increase the sensitivity. The resolution of commercial angle interferometers is typically in the order of 0.1 arcseconds.

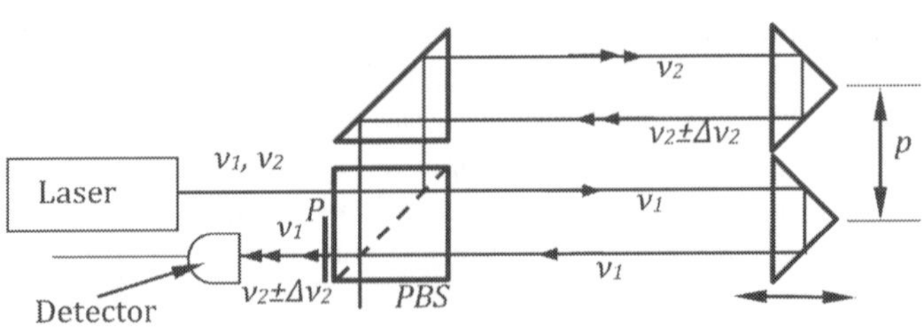

FIGURE 19.10 Schematic of an angle interferometer.

19.8 THE MOIRÉ TECHNIQUE FOR DISPLACEMENT MEASUREMENT

It has been shown that when two identical linear gratings of equal dark and clear spaces (pitch d) are superposed with their lines inclined at a small angle θ, the moiré fringes are formed, which have a spacing p ($=d/\theta$). The moiré fringes are straight and run almost perpendicular to the grating lines. The pitch of the moiré fringes can be controlled by the angle θ. It can be easily seen that if one of the gratings is displaced perpendicular to the grating lines, the moiré fringes also move normal to the direction of displacement. For a movement of one pitch of the grating, the moiré pattern displaces by one fringe. If we count the number of fringes passing across a reference mark, we can determine the displacement of the moving grating in multiples of the grating pitch. Thus, if one of the gratings is attached to a moving member, its displacement can be measured digitally as is done with an interferometer. At the same time, this technique is less sensitive to environmental conditions than are the interferometers. The moiré technique for displacement measurement is widely used in a variety of machine tools, numerically controlled machines, and coordinate measuring machines as well as in routine inspection instruments such as micrometers, comparators, etc.

An actual measurement system consists of a long measurement grating attached to the moving member and a small, fixed grating of the same pitch, called the index grating, as shown in Figure 19.11(a). The two gratings are mounted with a small gap of a fraction of a millimeter (~0.2 mm) determined by both geometrical and diffraction considerations. These two gratings enclose a small angle. A collimated beam illuminates the overlap area, and the transmitted light is received on a photodetector. As the main grating moves, the photodetector receives varying amounts of light flux, and a periodic signal is produced (one complete cycle for one pitch movement). Counting the periods and multiplying the counts with the pitch gives the displacement. A similar arrangement can be made with a reflective main grating.

A single detector cannot determine the direction in which the main grating is moving, which is a situation similar to that encountered in interferometers for displacement measurement. Signals in quadrature are essential for bidirectional counting. In the present case two signals are obtained by modifying the index grating as shown in Figure 19.11(b). The two parts have the same grating pitch as the main

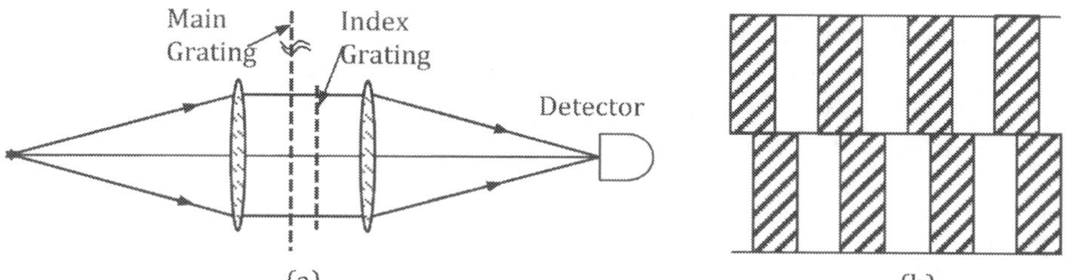

FIGURE 19.11 (a) Moiré technique for measuring displacement. (b) Index grating for bidirectional counting.

grating but are displaced by a quarter of a pitch with respect to each other. The lines of the main and index gratings are aligned parallel, and the light from the two areas of the index grating is collected and delivered to two separate detectors. The signals of these two detectors have the required 90° phase difference for bidirectional counting. These signals can be processed in the same way as is done in the case of fringe counting interferometers.

This scheme suffers from the same problem as encountered in fringe counting interferometers, namely the change in the average DC level of the signal due to fluctuations in the light irradiance. To overcome this problem, four signals are derived at 0°, 90°, 180°, and 270°. This is achieved by having four sets of lines instead of two on the index grating, with each set shifted by a quarter of the pitch. Four detectors receive light from the four areas, giving four signals as required. The alternate signal pairs are combined and amplified in a push-pull amplifier stage so that the modulations add and the DC levels cancel out, resulting in a pair of quadrature signals symmetrical about the mean level. Even when the variation in light irradiance is gradual because of overall reduction in the light transmission of the optical surfaces, this four-signal scheme operates satisfactorily. Because four signals are not derived from the same area of the grating surface, local nonuniform contamination on the surface can still affect the mean level. Proper sealing of the grating can overcome this problem over small to medium lengths. Alternately, the problem can be overcome by generating four signals from a common area of the grating. Two quadrature signals may be treated electronically to obtain four counting pulses per period, as is done in the case of fringe counting interferometers. Thus, with a grating of 25 lines per millimeter, a resolution of 0.01 mm can be obtained. Further subdivision is required to improve resolution.

19.9 DISPLACEMENT DISTRIBUTION MEASUREMENT

It has been shown that large and medium displacements can be measured with an interferometer and its various variants. It requires an optical element on the carriage or an optical surface standing on its own. Essentially, it is a point-wise measurement requiring good quality optical elements. Measurement of displacement of a real object requires attaching an optical element to it. The more interesting problem from the point of view of stress analysis to obtain displacement distribution on an object by an application of load and/or study of vibration modes of real objects cannot be tackled with the classical interferometry. Hologram interferometry (HI) and speckle interferometry are used to study deformations of objects under load, and also for vibration studies. Rotating objects with their motion frozen can also be studied by these techniques.

19.9.1 Hologram Interferometry (HI)

The governing equation of HI is

$$\delta = \left(\bar{k}_2 - \bar{k}_1\right) \cdot \Delta\bar{L} \tag{19.22}$$

where δ is the phase difference between the interfering waves; $\bar{k}_1, \bar{k}_2$ are the propagation vectors of the illumination and the observation waves; and $\Delta\bar{L}\left(\Delta\bar{L} = \hat{i}\ u + \hat{j}\ v + \hat{k}\ w\right)$ is the deformation vector at a point on the object; u, v, and w are the components of the deformation vector. If the object is illuminated by a plane wave, which usually is the case when small objects or small models are studied, $\bar{k}_1$ is a constant. For studying larger objects, a spherical wave is used for illumination.

Assuming double-exposure HI, the bright fringes are formed wherever

$$\delta = 2m\pi:\ m = 0, \pm1, \pm2, \ldots \Rightarrow \left(\bar{k}_2 - \bar{k}_1\right)\cdot\Delta\bar{L} = 2m\pi \tag{19.23}$$

The objective is to obtain the components of the deformation vector. In order to obtain these components, we need to set up at least three equations. These can be set up by changing the direction of observation and counting the number of fringes passing through a point when the direction of observation is varied. The solution of these equations gives the components of the deformation vector at that point.

Using phase shifting we can obtain phase difference at different points on the object. Therefore, the experimental setup is arranged to be sensitive to one component of the deformation vector. When the propagation vectors of the observation and the illumination waves are anti-parallel $\left(\bar{k}_2 = -\bar{k}_1 = \bar{k}\right)$ and the observation propagation vector is in the direction of the local surface normal, the fringe formation is governed by

$$2\bar{k}\cdot\Delta\bar{L} = 2m\pi \Rightarrow w = \frac{m\lambda}{2} \tag{19.24}$$

where w is the out-of-plane component of the deformation vector. The experimental arrangement is sensitive to the out-of-plane component only. In a general case, the object is illuminated with an oblique collimated beam at an angle of incidence θ and the observation is along the local normal, then

$$w = \frac{m\lambda}{(1+\cos\theta)} \tag{19.25}$$

It is assumed that the object undergoes only out-of-plane deformation.

The arrangement can be made sensitive to the in-plane component of deformation only by illuminating the object symmetrically about the local normal and making observations along the local normal. Mathematically,

$$\delta = \left(\bar{k}_2 - \bar{k}_1\right)\cdot\Delta\bar{L} - \left(\bar{k}_2 - \bar{k}_1'\right)\cdot\Delta\bar{L} = \left(\bar{k}_1' - \bar{k}_1\right)\cdot\Delta\bar{L} \tag{19.26}$$

When the illumination beams lie in the y–z plane, and make an angle $\pm\theta$ with the z-axis (local normal), the phase difference given by Equation (19.26) can be expressed as

$$\frac{2\pi}{\lambda}2v\sin\theta = 2m\pi \Rightarrow v = \frac{m\lambda}{2\sin\theta};\ m = 0, \pm1, \pm2, \ldots. \tag{19.27}$$

The fringes represent the y-component of displacement. The sensitivity is governed by the angle enclosed by the illumination beams.

19.9.2 Measurement of Amplitude of Vibration

Assume that the object is vibrating with an amplitude $w(x, y)$ at a frequency ω. It is illuminated at an angle θ with the local normal and a time-averaged hologram is recorded on a plate placed normal to the local normal. On reconstruction of the hologram, the irradiance distribution on the object is given by

$$I_{obj}(x,y) = I_0(x,y)\ J_0^2\left[\frac{2\pi}{\lambda}\ w(x,y)(1+\cos\theta)\right] \tag{19.28}$$

where $I_0(x, y)$ is the irradiance distribution on the object when it is stationary. Thus the object irradiance distribution is modulated by J_0^2 function; the irradiance is brightest where the amplitude of vibration $w(x, y)$ is zero, i.e., $w(x, y) = 0$. The minima of irradiance (dark fringes) occur at places where the J_0^2 function is zero. This occurs when

$$\frac{2\pi}{\lambda}\ w(x,y)(1+\cos\theta) = 2.4048,\ 5.5200,\ 8.6537,\ 11.7915\ldots \tag{19.29}$$

Therefore, the amplitude of vibration can be obtained by locating the minima of the irradiance pattern. Because the J_0^2 function decreases for higher values of its argument, consequently the irradiance modulation tends to become small for higher values of amplitude of vibration. For example, if we consider time-averaged recording of a circular diaphragm rigidly clamped at the boundary and subjected to vibration in the fundamental mode, the reconstructed image will consist of circular fringes; the brightest fringe will be at the periphery and the dark circular fringes will correspond to the values of the amplitude of vibration as given by Equation (19.29). The contrast of the fringes will keep on decreasing with increasing amplitude of vibration. Hence, it is not possible to measure large amplitudes.

On the other hand, if the frequency of the reference beam is shifted by $n\omega$, (where n is an integer) and a time-averaged hologram is recorded of an object vibrating with an amplitude $w(x, y)$ at frequency ω, the reconstructed object irradiance will be modulated by the J_n^2 function. In other words, the irradiance distribution of the reconstructed object will be given by

$$I_{obj}(x,y) = I_0(x,y)\ J_n^2\left[\frac{2\pi}{\lambda}\ w(x,y)(1+\cos\theta)\right] \tag{19.30}$$

The fringe contrast does not appreciably decrease with the increase in the amplitude of vibration and hence large values of the amplitude can now be measured. The phase of vibration is not known.

Let us now consider another possibility in which the phase of the reference wave also oscillates with an amplitude A_R at the frequency of excitation ω of the object. This can be achieved by mounting the folding mirror in the reference arm on a PZT and exciting it with the same voltage signal that is used to drive the object. However, if there is a phase difference Δ between the excitation of the object and that of the mirror in the reference arm, the reconstructed object from a time-averaged hologram will have an irradiance distribution as

$$I_{\text{obj}}(x,y) = I_0(x,y)\ J_0^2\left\{\frac{2\pi}{\lambda}(1+\cos\theta)\left[\ w^2 + A_R^2 - 2\ wA_R\cos\Delta\right]^{1/2}\right\} \quad (19.31)$$

When there is no excitation of the mirror in the reference arm, the object is covered with J_0^2 fringes as discussed earlier. However, if the mirror is excited in phase with the object, the irradiance distribution on the object will be

$$I_{\text{obj}}(x,y) = I_0(x,y)\ J_0^2\left\{\frac{2\pi}{\lambda}(1+\cos\theta)(w - A_R)\right\} \quad (19.32)$$

The object is again covered with J_0^2 fringes but the amplitude in the argument is $(w - A_R)$ instead of w. Accordingly, the brightest fringe will be formed where $w = A_R$, i.e., where the amplitude of the reference mirror matches with that of the object and is in phase with that of the mirror. This way the amplitude of vibration up to 10 μm can be measured. As Δ can be varied, it is possible to trace out areas of the object vibrating in the same phase as that of the mirror, thereby mapping the phase distribution of the object vibration.

19.9.3 Electronic Detection: ESPI/DSPI and Speckle Photography

The methods described in this section to measure the displacement and the vibration amplitude can be easily performed using electronic detection, i.e., the holographic plate/film is replaced by an array detector (charge-coupled device [CCD] array) and an imaging system is used to image the object on the array. However, due to the poor resolution of CCD compared with the holographic plate, the reference beam is added axially, i.e., both the object and the reference beams travel along the axis of the imaging system. Unlike in HI with recording on the holographic plate, the two exposures in electronic speckle pattern interferometry (ESPI) are independently handled and hence the quasi-constant DC term can be removed through electronic subtraction. Further, the object and the reference beams are of nearly the same irradiance to produce speckles of high contrast. This falls under the generic term electronic speckle pattern interferometry. Phase shifting is performed by mounting the mirror on a PZT in the reference arm and applying appropriate voltage to introduce the desired phase shift.

ESPI/digital speckle pattern interferometry (DSPI) has been discussed along with its applications in an earlier chapter. Speckle photography (SP), which is sensitive to in-plane deformation components, has also been described earlier.

19.10 MOIRÉ TECHNIQUES

The moiré phenomenon has high sensitivity for the in-plane displacement measurement and poor sensitivity for the out-of-plane displacement measurement. When the moiré phenomenon is used for metrology, one of the gratings is mounted on the object, which is subjected to deformation. Moiré fringes are then observed between the deformed grating and the reference grating (undeformed grating). One can also obtain a moiré pattern between two deformed gratings (i.e., when two deformed states of the object are compared).

19.10.1 Measurement of In-Plane Displacement/Deformation

A measurement grating of pitch a is bonded to the object and a reference grating of the same pitch is aligned parallel to it. In-plane deformation in the direction of the grating vector (direction normal to the grating elements) causes the period of the bonded grating to change from a to b. The moiré fringes formed will have a period $d = a\ b / |a-b|$. The normal strain ε as measured by the moiré method is $|a-b| / a$. Therefore, $\varepsilon = a / d \approx b / d$. Thus, the normal strain ε is obtained as the ratio of the grating period to that of the moiré pattern. At this juncture, it may be noted that the moiré method measures the Lagrangian (engineering) strain. However, if the deformation is small, the Lagrangian and the Eulerian strains are practically equal.

The shear strain is obtained likewise. The measurement grating (grating bonded on the object), as a result of shear, gets inclined at an angle θ with the reference grating resulting in the formation of a moiré pattern. Moiré fringes are formed wherever

$$x(1-\cos\theta)+y\sin\theta = pb \tag{19.33}$$

The period d of the moiré fringes for very small rotation is b/θ. In fact, the shear strain, when the rotation is small and also when strains are small, is equal to θ. Thus, the shear strain γ is given by

$$\gamma = b / d \tag{19.34}$$

The shear strain is also obtained as the ratio of the pitch of the grating to that of the moiré fringe pattern. This is valid for the homogeneous normal strain and simple shear strain.

19.10.2 Two-Dimensional In-Plane Displacement Measurement

The analysis of the moiré pattern becomes quite easy when it is recognized that the moiré fringes are the contours of the constant displacement component, the so-called isothetics. The zeroth-order moiré fringe runs through regions where the periods of the reference grating and the measurement grating are equal (i.e., the displacement component is zero). Similarly, the Nth-order moiré fringe runs through the regions where the displacement component is N times the period, a, of the grating. If the reference grating has its grating vector along the x-direction, the moiré fringes represent loci of constant u displacement (i.e., $u = N\ a$). If the v component of the displacement is to be measured, both the reference grating and the measurement grating are aligned to have their grating vectors along the y-direction. The moiré fringes are now the loci of the constant v component, hence, $v = N'\ a$, where N' is the moiré fringe order. To obtain both u and v components simultaneously, a cross-grating may be used. The u and v components are isolated by optical filtering. For generating the displacement curve accurately, one needs a large number of data points, hence methods have been found to increase the number of moiré fringes, and hence the data points, for the same loading conditions. This can be accomplished either by pitch mismatch, or angular mismatch, or their combination. The strains can also be obtained by shearing. The moiré patterns representing the displacement fringes are sheared to obtain the fringes corresponding to the strain.

19.10.3 High Sensitivity In-Plane Displacement Measurement

The sensitivity of in-plane displacement measurement depends on the period of the grating. In moiré work, low-frequency gratings are usually employed, and the analysis is based on the geometrical optics. However, with fine-period gratings, though the sensitivity is increased, the diffraction effects are prominent, and consequently quasi-monochromatic, spatially coherent light needs to be used. Indeed, with laser radiation, in-plane displacement measurement with a very high degree of accuracy can be made. A cross-grating, say 1200 lines/mm in either direction, recorded holographically and aluminum-coated for reflection, is bonded onto the surface under test. The use of high-density gratings puts a limitation on the size of the object that could be examined. The grating is illuminated by four collimated beams as shown in Figure 19.12. Two beams lie in the y–z plane and the other two beams lie in the x–z plane and make angles such that first-order diffracted beams propagate along the z-axis. These beams, on interference, generate fringe patterns characteristic of u and v families of the displacement.

To understand the working of the technique, let us consider a 1D grating with its grating vector along the x-direction. The grating is recorded holographically by interference between two plane waves, one propagating along the z-axis and the

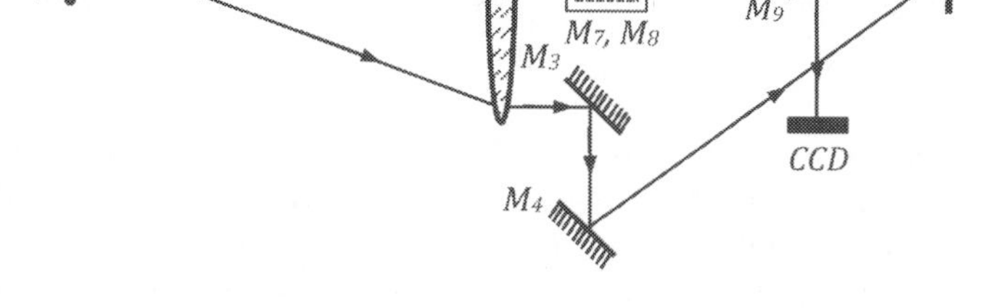

FIGURE 19.12 High-sensitivity in-plane components measurement setup: mirrors M_5, M_6, M_7, and M_8 generate two beams in the x–z plane.

other making an angle θ with this but lying in the x–z plane. The period b of the grating is given by $b \sin\theta = \lambda$. The spatial frequency μ of the grating is given by $\mu = \sin\theta/\lambda$. This grating is bonded onto the surface of the object. When this grating is illuminated normally by a collimated beam, the first-order diffracted beams make angles of θ and $-\theta$ with the normal to the grating. Alternately, when the grating is illuminated at an angle θ, one of the diffracted beams will propagate along the normal to the grating. Let us consider that the object is loaded resulting in distortions in the grating period. Let the modified spatial frequency of the distorted grating be $\mu(x)$. The grating function can be expressed by $t(x)=\frac{1}{2}\left[1+\cos 2\pi\mu(x)x\right]$. The grating is illuminated symmetrically at angles of θ and $-\theta$ by two collimated beams. The collimated beams can be represented by $R \exp(2\pi i\mu x)$ and $R \exp(-2\pi i\mu x)$, where R is the amplitude of the beam. These collimated beams will be diffracted by the grating and the diffracted field can be expressed as

$$R\left(e^{2\pi i\mu x}+e^{-2\pi i\mu x}\right)\frac{1}{2}\left[1+\cos 2\pi\mu(x)x\right] \tag{19.35}$$

Collecting terms of interest (i.e., those terms representing beams that propagate along the z-axis), we obtain the resultant amplitude of the beams as

$$A=\frac{1}{4}R\left(e^{2\pi i[\mu-\mu(x)]x}+e^{-2\pi i[\mu-\mu(x)]x}\right) \tag{19.36}$$

The irradiance distribution in the image as recorded by the CCD camera is given by

$$I(x)=\frac{1}{4}R^2 \cos^2 2\pi\left[\mu-\mu(x)\right]x \tag{19.37}$$

This represents a moiré pattern. The moiré fringes are formed wherever

$$2\pi\left[\mu-\mu(x)\right]x = N\pi \Rightarrow x = N\,\frac{1}{2}\frac{b\,b(x)}{|b-b(x)|} \tag{19.38}$$

where $b(x)$ is the period of the deformed grating and N is the order of the moiré fringe.

The moiré fringe width d is given by

$$d=\frac{1}{2}\frac{b\,b(x)}{|b-b(x)|} \tag{19.39}$$

It is to be noted that the sensitivity of the method is twice as large as would be obtained with the same grating using the conventional method. This is due to the fact that the deformed grating in +1 and –1 diffraction orders is being compared. This increase in sensitivity by a factor of two has been explained by Post as due to moiré formation between a grating on the test surface and a virtual grating of twice the frequency formed by interference between the two beams, thus providing a multiplicative factor of two. Use of higher diffraction orders results in increased fringe multiplication.

An arrangement to measure u and v components of the displacement simultaneously uses a high-frequency cross-grating bonded onto the surface of the object. The grating bonded on the object is illuminated simultaneously and symmetrically in the x–z and the y–z planes. Four beams generated in ±1 orders propagate axially along the z-direction. The moiré fringes representing u and v displacement components are then obtained by interference of these four beams.

19.10.4 Measurement of Out-of-Plane Component

The moiré technique is well suited for the measurement of in-plane displacement components: the sensitivity is controlled by the period of the grating. Further, the techniques used for moiré formation are based either on pitch mismatch or on angular mismatch. Therefore, moiré formation while measuring out-of-plane displacement will also be governed by these techniques. Consequently, the moiré method for out-of-plane measurement is not as sensitive as it is for the in-plane measurement. The out-of-plane displacement and the surface topography can be measured by one of the following, the shadow moiré method and projection moiré method.

19.10.4.1 The Shadow Moiré Method

As the name suggests, the moiré pattern is formed between the grating and its shadow on the object. The shadow grating will be distorted by the object topography and, hence, moiré fringes between the distorted and the reference gratings are observed.

19.10.4.1.1 Parallel Illumination and Parallel Observation

Figure 19.13(a) shows the arrangement for the shadow moiré method that relies on the parallel illumination and the observation at infinity. A reference grating of period a is placed on the object. Without loss of generality, we may assume that point A

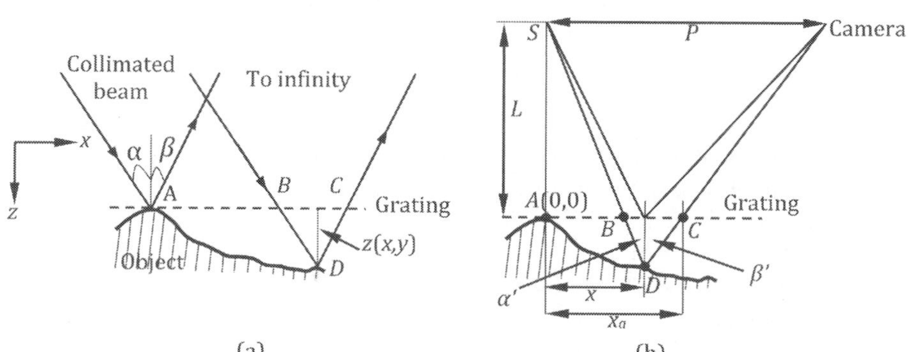

FIGURE 19.13 Schematic for shadow moiré method (a) parallel illumination and (b) divergent illumination.

on the object surface is in contact with the grating. The grating is illuminated by a collimated beam incident at an angle α with the normal to the grating surface (i.e., the z-axis). It is viewed from infinity at an angle β. It is obvious that the grating elements contained in a distance AB occupy a distance AD on the object surface. The elements on AD will form a moiré pattern with the grating elements contained in distance AC. Let us assume that AB and AC have p and q grating elements, respectively. So, $AB = p\,a$, and $AC = q\,a = p\,b$ where b is the period of the grating formed due to shadow. From geometry,

$$BC = AC - AB = (q - p)\; a \tag{19.40}$$

Nth-order moiré fringe will be formed when pitch mismatch over the p elements accumulates to N times the pitch of the reference grating, i.e.,

$$(b-a)p = (q-p)a = N\;a\text{: } N = 0, \pm 1, \pm 2, \pm 3, \ldots\;\ldots\;\ldots \tag{19.41}$$

Equation (19.41) can be rewritten in terms of depth $z(x, y)$ as measured from the grating plane as

$$z(x,y)(\tan\alpha + \tan\beta) = N\;a \tag{19.42}$$

or

$$z(x,y) = \frac{N\;a}{\tan\alpha + \tan\beta} \tag{19.43}$$

This is the governing equation of this method. It is seen that the moiré fringes are contours of equal depth measured from the grating. When the viewing is along the normal to the grating, i.e., $\beta = 0$, then $z(x, y) = N\,a/\tan\alpha$. Alternately, the grating may be illuminated normally and viewed obliquely. Then $z(x, y) = N\,a/\tan\beta$. The assumption that the viewing is done by a collimated beam is not always valid. However, when the object under study is small and the camera is placed sufficiently far away, this requirement is nearly met. This method is not suited for large objects due to the limitations on the size of the collimated beam.

19.10.4.1.2 Spherical Wave Illumination and Camera at Finite Distance

The assumption made earlier that both the source and the camera are at infinity limits the application of the method to the study of small objects. However, when divergent illumination is used, larger objects can be studied. In general, the source and the camera may be located at different distances from the reference grating. However, a special case where the source and the camera are at equal distances from the grating is of considerable practical importance and hence is discussed here in detail.

Let the point source S and the camera be at a distance L from the grating surface and their inter-separation be P, as shown in Figure 19.13(b). The object is illuminated by a divergent wave from a point source. As before, the number of grating elements p contained in AB on the grating are projected onto AD on the object surface. These elements interact with the elements q in AC, thus producing a moiré pattern.

If we assume that *Nth*-order moiré fringe is observed at the point *D*, then

$$BC = AC - AB = (q - p)a = Na$$

But $BC = z(x, y)(\tan\alpha' + \beta')$, where $z(x, y)$ is the depth of the point *D* from the grating. Therefore, we obtain

$$z(x, y) = \frac{N\,a}{\tan\alpha' + \tan\beta'} \quad (19.44)$$

Here, α' and β' vary over the surface of the grating or over the surface of the object. From Figure 19.13(b), we have $\tan\alpha' = x / [L + z(x, y)]$ and $\tan\beta' = (P - x)/[L + z(x, y)]$. Substituting these in Equation (19.44), we obtain

$$z(x, y) = \frac{N\,a}{\dfrac{x}{L + z(x, y)} + \dfrac{P - x}{L + z(x, y)}} = Na\,\frac{L + z(x, y)}{P} \quad (19.45)$$

Rearranging Equation (19.45), we obtain

$$z(x, y) = \frac{Na\,L}{P - Na} = \frac{Na}{\dfrac{P}{L} - \dfrac{Na}{L}} \quad (19.46)$$

The ratio *P*/*L* is called the base-to-height ratio. This is an extremely simple formula. In fact, it is this simplicity that makes the technique attractive over the one in which the source and the camera are placed at different distances from the grating. The distance $\Delta z(x, y)$ between adjacent moiré fringes (i.e., $\Delta N = 1$) is

$$\Delta z(x, y) = \frac{a\,L}{P}\left(1 + \frac{z}{L}\right)^2 \quad (19.47)$$

It is seen that the fringe spacing is not constant but increases with the depth. If $\frac{z}{L} \ll 1$, then the fringe spacing is constant and is given by $\Delta z(x, y) = aL/P$. The moiré fringes then represent the true depth contours.

Though the method is applicable to the study of large objects, it is essential to correct for the perspective error. This error arises as the actual coordinates (x, y) of a point appear as (x_a, y_a). From Figure 19.13(b), the actual coordinates (x, y) and apparent coordinates (x_a, y_a) are related as

$$\frac{x_a - x}{z(x, y)} = \frac{P - x_a}{L}$$

or

$$x = x_a - \frac{z}{L}(P - x_a) = x_a\left(1 + \frac{z}{L}\right) - P\frac{z}{L} \quad (19.48a)$$

Similarly

$$y = y_a\left(1+\frac{z}{L}\right) - P\frac{z}{L} \tag{19.48b}$$

Using this arrangement, the shadow moiré method can be applied to the study of very large structures, obviously limited by the size of the grating.

Because the fringe spacing is not constant over the object depth, the sensitivity of the method may not be adequate to map out surfaces that are strongly curved or are very steep. Therefore, the use of a composite grating has been suggested. The composite grating consists of two parallel-superposed gratings with two discretely different periods. This, therefore, provides two sensitivities.

One of the problems of the shadow moiré method is that the moiré fringes do not localize on the surface of the object. Also, the contrast of the fringes is not constant. Fringes of maximum contrast are obtained when the surface under test is nearly flat. The moiré fringes become fuzzy when objects of steep curvatures are examined. As a caution, the objects with steep curvature should be placed as close to the grating as possible, and the depth of field of the photographic objective should be large enough to focus the grating and its shadow simultaneously.

19.10.4.2 Projection Moiré

There are several differences between the shadow moiré and the projection moiré techniques. In projection moiré, a higher frequency grating is used, and the grating is projected on the object surface and reimaged on a reference grating that is identical to the projection grating and aligned parallel to it. The moiré fringes are formed at the plane of the reference grating. The projection moiré method also provides a certain flexibility not possible in the shadow moiré method. We will examine the projection moiré method under the following two cases:

a. The optical axes of the projection and imaging (observation) systems are parallel.
b. The optical axes are inclined with each other.

19.10.4.2.1 Optical Axes Parallel

Following Figure 19.14(a), grating G_1 of period b is imaged on a reference plane R where its period is Mb, with M being the magnification of the projection system. If the object surface was plane and located on this reference plane, the projected grating would have a constant period. This grating is reimaged on the reference grating G_2. Assuming the projection and the imaging systems to be identical, the pitch of the imaged grating will be equal to that of G_2 and the grating elements will lie parallel to each other due to initial alignment. Hence, no moiré pattern will be formed. However, if the surface is curved, the projection grating on the surface will have a varying period and hence a moiré pattern is formed. We can examine this exactly in the same manner as applied to the shadow moiré, i.e., the grating of period Mb at the reference plane is illuminated by a spherical

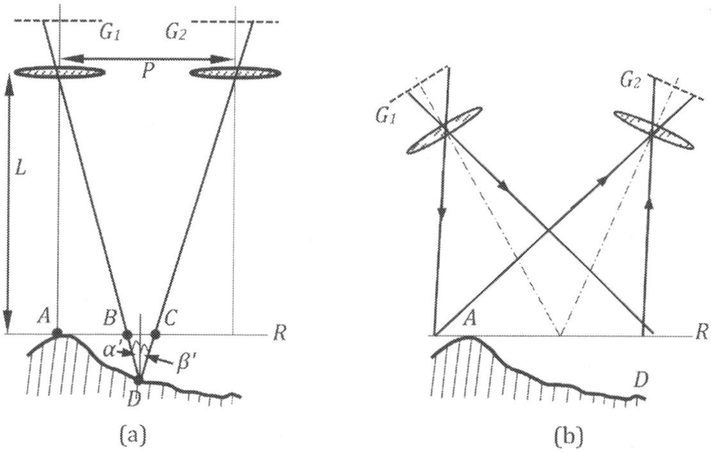

FIGURE 19.14 Projection moiré method: (a) axes parallel and (b) axes inclined.

wave from the exit pupil of the projection lens. As described in the shadow moiré method, we can write

$$z(x,y)=\frac{N\ Mb}{\tan\alpha'+\tan\beta'}=\frac{N\ Mb}{\dfrac{x}{L+z(x,y)}+\dfrac{P-x}{L+z(x,y)}}=NMb\,\frac{L+z(x,y)}{P} \tag{19.49}$$

On rearranging Equation (19.49), we obtain

$$z(x,y)=\frac{N\ MbL}{P-N\ Mb} \tag{19.50}$$

Writing for the magnification of the imaging system as $M = (L-f)/f$, where f is the focal length, Equation (19.50) is rewritten as

$$z(x,y)=\frac{N\ bL(L-f)}{fP-N\ b(L-f)} \tag{19.51}$$

Here, $z(x, y)$ corresponds to the depth of the object at the point where the Nth-order moiré fringe is formed. It is seen that the moiré fringe spacing is not constant. However, when $Mb \ll P$, the moiré fringes are equally spaced.

19.10.4.2.2 Optical Axes Inclined

The disadvantage of the method with optical axes being parallel is that only very small objects can be studied. However, large objects also can be studied when the optical axes of the projection system and the imaging system are inclined as shown in Figure 19.14(b). The grating can be imaged on the reference plane R where the Scheimpflug condition is satisfied. However, the magnification is not constant and, consequently, the period of the grating on the reference plane varies. This problem can be solved by using a special grating whose period remains constant on projection at the reference plane.

The technique can be used for the comparison of two objects if, instead of the reference grating, a photographic record of the projected grating is made first with one object, and then with the object replaced by another object. The records could be made on the same film (additive moiré) or on two separate films (multiplicative moiré). Likewise, one can study the out-of-plane deformation of the object by recording gratings corresponding to two states of the object.

As may be obvious, the grating plays a very important role: the sensitivity depends on the grating period. Therefore, several gratings may be required to maintain accuracy over a steeply curved surface. Further, for phase shifting, the grating needs to be shifted. Use of programmed LCD panels for generating gratings of variable periods, sinusoidal or binary transmission function, and being arbitrarily inclined makes the measurement and the automatic shape determination a lot more convenient.

19.10.5 Measurement of Amplitudes of Vibration

Both the shadow moiré and the projection moiré methods can be used to measure the amplitude of vibration. However, we describe a method in which a grating is produced on the surface of the object by the superposition of two collimated beams of wavelength λ. The collimated beams lie in the x–z plane and enclose a small angle $\Delta\theta$. One of the beams makes an angle θ with the normal to the reference plane. The irradiance distribution on the object surface can be expressed as

$$I(x,y) = I_0\left[1+\cos\frac{2\pi}{a}(x\cos\theta + z\sin\theta)\right] \tag{19.52}$$

where $a = \lambda/2\sin\Delta\theta$.

Let us now assume that the object surface executes a simple harmonic motion of amplitude $w(x, y)$ at frequency ω. We can express this as

$$z(x,y) = z_0 + w(x,y)\sin\omega t \tag{19.53}$$

where z_0 represents the static position of the surface. Due to surface motion, the phases of the two interfering beams at the surface will change with time. Consequently, the instantaneous irradiance distribution in the record will be

$$I(x,y\!:t) = I_0\left\{1+\cos\frac{2\pi}{a}\left[x\cos\theta + \left(z_0 + w(x,y)\sin\omega t\right)\sin\theta\right]\right\} \tag{19.54}$$

This irradiance distribution is integrated over a period T much larger than the period of vibration $2\pi/\omega$, that is,

$$I(x,y) = \frac{1}{T}\int_0^T I(x,y:t)\,dt = I_0\left\{1+J_0\left[\frac{2\pi}{a}w(x,y)\sin\theta\right]\cos\frac{2\pi}{a}\left[x\cos\theta + z_0\sin\theta\right]\right\} \tag{19.55}$$

where $J_0(x)$ is the Bessel function of zeroth order. This again represents a grating with a modulation given by the Bessel function. Information about the vibration amplitude $w(x, y)$ can be extracted from this record by optical filtering. We assume that the record has an amplitude transmittance proportional to the irradiance distribution. The record is placed at the input plane of a Fourier transform (FT) processor and one of the first orders is allowed for image formation. The irradiance distribution at the output plane is proportional to

$$J_0^2\left[\frac{2\pi}{a}w(x,y)\sin\theta\right] \tag{19.56}$$

The irradiance distribution exhibits maxima and minima. The amplitude of vibration is obtained as explained for time-averaged HI in Chapter 5.

19.10.6 Reflection Moiré Method

Unlike the shadow moiré and the projection moiré methods, the reflection moiré method yields information about the slope and the curvature. In the problem of flexure of thin plates, the second derivatives of the deflection are related to the bending moments and twist. The deflection data obtained from the shadow moiré, or the projection moiré need to be differentiated twice in order to obtain curvature, resulting in inaccuracies. Therefore, there is a need to obtain slope or curvature data so that either a single differentiation is performed, or it is eliminated altogether. The reflection moiré method serves this purpose. The only disadvantage is that the specimen has to be specular.

Figure 19.15 is a schematic of the experimental setup for the reflection moiré, also known as the Ligtenberg's arrangement. The object is taken as an edge clamped plate. The surface of the plate is mirror-like so that a virtual image of the reference grating is formed. Consider a ray from point D on the grating, which after reflection from a point P on the object, meets at point I at the image plane as shown in Figure 19.15. In other words, point I is the image of point D as formed by reflection on the plate surface. When the plate is deformed, point P moves to point P'. If the

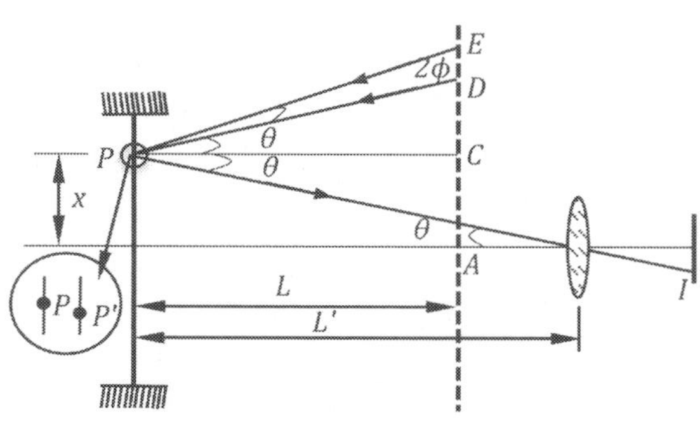

FIGURE 19.15 Reflection moiré: the Ligtenberg arrangement.

local slope is ϕ, point I now receives the ray from point E; the image of point E is formed on I again. In reality, an image of the grating as seen on reflection from the plate is recorded. This image itself may contain distortions due to nonflatness of the plate surface. It can be shown that, due to self-healing, these distortions do not influence the moiré formation. After the plate is deformed (loaded), the image is again recorded on the same photographic film/plate. Due to loading, the grating image will be further distorted, thus forming a moiré pattern.

Following the arguments presented earlier, a grating AE, after loading, is superposed on another grating AD leading to the moiré formation due to pitch mismatch. Let there be p grating elements in AE, and q elements in AD. If the Nth-order moiré fringe appears at the point I, then

$$ED = AE - AD = p\ b - q\ b = (p - q)b = N\ b \tag{19.57}$$

But the distance ED can be expressed as

$$ED = EC - DC = L\ \tan(\theta + 2\phi) - L\ \tan\ \theta = L\frac{\tan\theta + \tan 2\phi}{1 - \tan\theta \tan 2\phi} - L\tan\theta$$

Assuming slope ϕ to be very small, which usually is the case, we write $\tan 2\phi \approx 2\phi$. Therefore

$$ED = L\frac{2\phi\ (1 + \tan^2\theta)}{1 - 2\phi\ \tan\theta} = N\ b \tag{19.58}$$

Further, $2\phi\ \tan\theta \ll 1$, and slope ϕ is the partial x-slope, i.e., $\phi = \frac{\partial w}{\partial x}$, we have

$$\frac{\partial w}{\partial x} = \frac{N\ b}{2L(1 + \tan^2\theta)} = \frac{N\ b}{2L\left(1 + \dfrac{x^2}{L^2}\right)} \tag{19.59}$$

The slope ϕ depends on the value of x. In order to eliminate the dependence of $\frac{\partial w}{\partial x}$ on x, a curved grating is used. However, the Ligtenberg method utilizing curved gratings also suffers from several disadvantages: the large dimensions of the curved surface gratings, the necessity of using gratings of low spatial frequency to obtain a moiré pattern of good contrast, limitation to static loading problems, and use of models of relatively large flexures. The disadvantages of the Ligtenberg method are largely removed by the Rieder–Ritter arrangement. Both x-partial slope $\frac{\partial w}{\partial x}$ and y-partial slope $\frac{\partial w}{\partial y}$ are simultaneously recorded when a cross-grating is used. Optical filtering is used to separate out fringe patterns depicting x-partial slope and y-partial slope.

Further, as with other moiré methods, the sensitivity is dependent on the grating period. It is therefore desirable to have an arrangement in which the grating period can be varied easily. Figure 19.16 shows an arrangement where a grating G_V is projected onto a ground-glass screen D by the projection lens L_P. The projected grating G_P is imaged using the Rieder-Ritter arrangement. The pitch of the grating can be varied by changing the magnification during the projection of the grating.

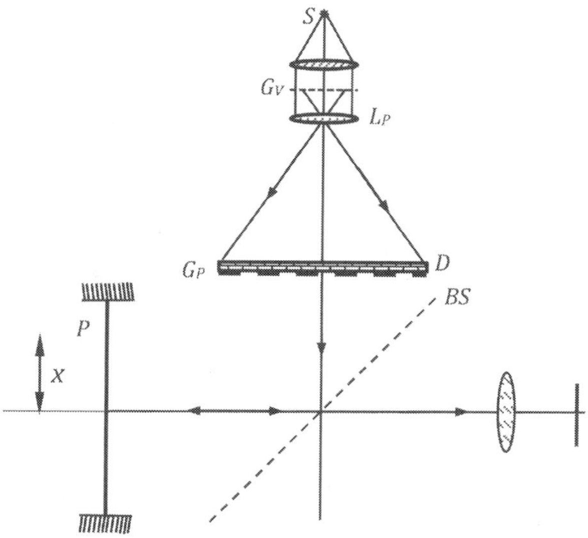

FIGURE 19.16 Reflection moiré: grating with variable pitch.

Several other arrangements have been suggested to obtain slope and curvature. A diffraction grating is placed at or near the focal plane of an imaging lens. The diffraction grating produces sheared fields. Double exposure, before and after the plate loading, is performed. In another interesting arrangement, an inclined plane parallel plate is placed near the focal plane of an imaging lens. The image is recorded in reflection: two reflected beams, which are sheared, produce a grating in the image plane. Loading of the plate deforms this grating. Therefore, the double-exposure recording, before and after plate loading, results in the formation of a moiré pattern, which is due to the slope variation. When two plates are used in tandem and properly aligned such that three sheared beams participate in interference, the curvature fringes are obtained.

These methods are coherent methods requiring laser radiation. The permissible size of the object depends on the collimating optics and hence these methods are limited to small to moderate size objects. As mentioned in the beginning of this section, the surface of the plate (object) need not be optically flat. The departure from flatness causes some distortions in the grating, but the moiré phenomenon has a self-healing property thereby these distortions cancel out, and the moiré fringes are due alone to the deformation of the plate caused by loading.

19.10.7 Slope Determination for Dynamic Events

So far, we have described methods that are applicable only to static problems. Pulse illumination is frequently used for the examination of dynamic processes. The grating images corresponding to successive states of the object are recorded. However, it would be necessary to record the reference grating each time as well in order to

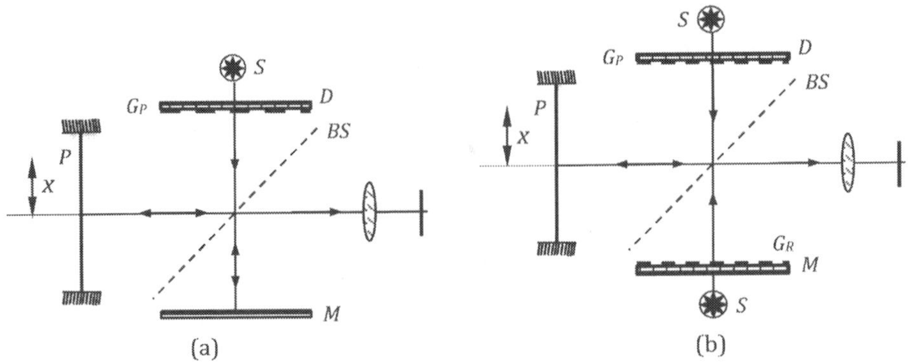

FIGURE 19.17 Arrangement for slope measurement. (a) Reference grating from mirror reflection. (b) Separate grating as reference.

generate the moiré pattern. This makes these methods rather troublesome to apply to the study of dynamic events.

Several methods have been developed whereby the images of the reference grating G_R and that of the deformed grating from the surface under investigation are recorded simultaneously. In one of the arrangements as seen in Figure 19.17(a), the beam splitter (*BS*) provides two paths for the imaging. One image of grating G_P is formed by reflection from the object surface *P*, while the other image is formed by reflection from a mirror surface *M*. (This image may be considered as reference grating G_R.) Therefore, both images of the grating can be recorded in a single exposure (pulse illumination) and hence an additive moiré pattern can be observed for the various states of the object. This arrangement, however, does not exploit the healing property of the moiré phenomenon. Because the object surface in its initial state and the mirror surface are not identical in general, the grating formed by reflection from the object surface will carry some distortions, which would show up as moiré fringes even for the unloaded state of the object. In other words, the initial fringe-free field is not obtained. This problem is solved in the arrangement shown in Figure 19.17(b). The grating G_R is a replica of the image formed by reflection from the object surface. Therefore, the images of gratings G_P and G_R will be identical, and an initial fringe-free field is obtained. However, for each object surface, a new grating G_R is to be produced and aligned before the start of an experiment.

19.11 DIGITAL IMAGE CORRELATION

With the advent of the laser, two whole-field measurement techniques, namely, HI and speckle metrology, evolved about the same time to make deformation measurements on real objects with very high accuracy. Speckle metrology included two different techniques: SP and ESPI. ESPI, like HI, was very sensitive to the out-of-plane component of the deformation vector, whereas SP was sensitive to in-plane components. The information from speckle photographs was extracted by optical filtering. With advancements in CCD (array detector) technology and the availability of adequate computing power in desktops, researchers developed techniques with which displacement components could be obtained digitally rather than optically. These came to be known as the digital image correlation (DIC) technique, digital image method (DIM), digital speckle correlation method (DSCM), and texture correlation (TC) and as well as a whole host of other names. DIC can be performed with laser speckles and is known as digital speckle photography (DSP). It can also be performed with white light speckles and artificially created speckles. In order to create artificial speckles, the specimen is given a thin coat of black paint, which when dry is sprinkled with a white paint. Two-dimensional DIC measures in-plane components. It involves three steps: (1) the specimen and experimental preparation, (2) recording images of a planar object before and after loading, and (3) processing the images to obtain displacement components. In order to measure both in-plane and out-of-plane deformation components, two cameras are used to capture images. It is then termed as 3D DIC or stereo DIC.

The specimen is generally planar and is placed parallel to the CCD plane. It should remain parallel during loading when the image is captured. The basic principle of 2D DIC is the tracking of the same points (or pixels) between the images recorded before and after loading. The image taken before loading is the reference image, whereas the image taken after loading is the deformed image. Let us consider a point $P(x_0, y_0)$ whose location in the deformed image is to be tracked. A reference subset of $(2N + 1) \times (2N + 1)$ pixels centered on $P(x_0, y_0)$ is chosen. On loading, the point $P(x_0, y_0)$ moves to its new position $P'(x_0', y_0')$ and a neighboring point $Q(x_i, y_j)$ moves to point $Q'(x_i', y_j')$ as shown in Figure 19.18.

The coordinates of a point $Q(x_i, y_j)$ around the subset center $P(x_0, y_0)$ can be mapped to a point $Q'(x_i', y_j')$ in the deformed subset according to the displacement mapping function:

$$x_i' = x_i + \xi(x_i, y_j) \tag{19.60a}$$

$$y_j' = y_j + \eta(x_i, y_j) \tag{19.60b}$$

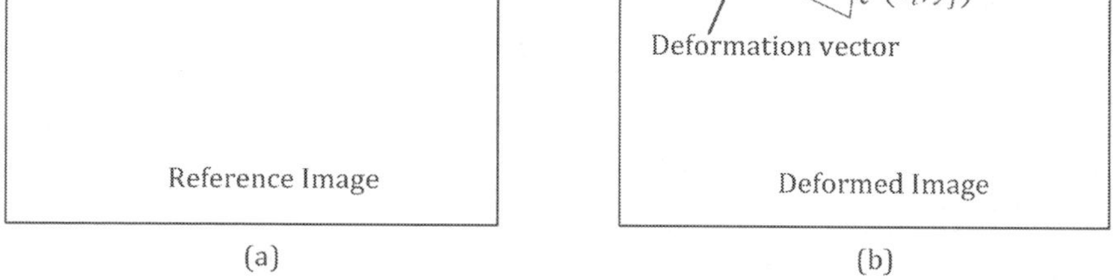

FIGURE 19.18 (a) Reference image with reference subset. (b) Deformed image with deformed subset.

where the indices i and j take values from $-N$ to N, and the mapping functions $\xi\left(x_i, y_j\right)$ and $\eta\left(x_i, y_j\right)$ are expressed as

$$\xi\left(x_i, y_j\right) = u\left(x_0, y_0\right) + \left.\frac{\partial u}{\partial x}\right|_{x_0, y_0} \left(x_i - x_0\right) + \left.\frac{\partial u}{\partial y}\right|_{x_0, y_0} \left(y_j - y_0\right) = u + u_x\ \Delta x + u_y\ \Delta y \tag{19.61a}$$

$$\eta\left(x_i, y_j\right) = v\left(x_0, y_0\right) + \left.\frac{\partial v}{\partial x}\right|_{x_0, y_0} \left(x_i - x_0\right) + \left.\frac{\partial v}{\partial y}\right|_{x_0, y_0} \left(y_j - y_0\right) = v + v_x\ \Delta x + v_y\ \Delta y \tag{19.61b}$$

where u and v are the in-plane components of deformation along x- and y-axes. In order to evaluate the degree of similarity between the reference subset and the deformed subset, various kinds of correlation criteria have been used. We discuss normalized discreet cross-correlation between the reference subset $S_1(i,j)$ and the deformed subset $S_2(i, j)$, which is expressed as

$$C_{S_1 S_2}\left(u, v\right) = \sum_{i=-M}^{M} \sum_{j=-M}^{M} \frac{S_1\left(i, j\right)\ S_2\left(i+u, j+v\right)}{\overline{S_1}\ \overline{S_2}} \tag{19.62}$$

where $\overline{S_1}$ and $\overline{S_2}$ are the mean values of the reference subset and the deformed subset, respectively, and defined as

$$\overline{S_1} = \sqrt{\sum_{i=-M}^{M} \sum_{j=-M}^{M} \left[S_1\left(i, j\right)\right]^2} \tag{19.63a}$$

$$\overline{S_2} = \sqrt{\sum_{i=-M}^{M} \sum_{j=-M}^{M} \left[S_2\left(i+u, j+v\right)\right]^2} \tag{19.63b}$$

This operation means that the reference subset $S_1(i, j)$ is swept over the deformed image and the statistical agreement between the region of the deformed image and the reference subset will determine whether there is any movement or not. Having thus determined the deformed subset, the location of the center, $P'\left(x_0', y_0'\right)$, of the deformed subset is found. The displacement between points $P\left(x_0, y_0\right)$ and $P'\left(x_0', y_0'\right)$ is the deformation vector with components u and v. The process is continued with another reference subset and corresponding deformation vector found. This is carried over for different reference subsets until the displacement over the entire image is extracted. The reference subsets are usually chosen to have either 3×3 pixels or 5×5 pixels.

Because DIC uses imaging, the lens parameters are to be evaluated. The parameters are the focal lengths, tangential and sagittal distortion, and location of principal planes. This is done by the process of calibration wherein a known test pattern is imaged and all the intrinsic parameters calculated.

QUESTIONS

19.1 The length L is measured using the formula

$$L = (m + \varepsilon)\frac{\lambda}{2}$$

Show that the uncertainty in length measurement is given by

$$s_L = \sqrt{\left(\frac{L}{\lambda} s_\lambda\right)^2 + \left(\frac{\lambda}{2} s_\varepsilon\right)^2}$$

where S_L, S_λ, and S_ε are the standard deviations for length, wavelength, and fraction measurement sets, respectively

19.2 Suggest an experimental configuration to obtain moiré fringes representing curvature for a tightly clamped thin circular plate using reflection moiré method.

19.3 In an experiment with a T-G interferometer equipped with a Cd spectral lamp, the fractions observed with the wavelengths 644.02480 nm, 508.72379 nm, and 467.94581 nm were 0.22, 0.73, and 0.19, respectively. The nominal length of the gauge block as measured with a comparator was found to be 5.062 ± 0.002 mm. What is its correct length?

19.4 A double-exposure hologram of a flat circular diaphragm clamped at the edges is recorded by illuminating it at an angle of 45° with the local normal with a collimated beam of He-Ne laser (λ = 633 nm) and changing the pressure between exposures. The hologram is reconstructed by illuminating it normally and 10 circular fringes are observed. What is the maximum deflection of the diaphragm in the center? What is the value of the applied pressure? Obtain the deflection profile, assuming it to be made of steel with Poisson's ratio $\nu = 0.3$ and Young's modulus $E = 200$ GPa (200×10^9 N/m^2), the thickness and diameter being 1 mm and 20 mm, respectively.

19.5 Figure 19.19 shows the vibration mode of an Indian musical instrument excited at 2983 Hz using time-averaged HI with a He-Ne laser. Obtain the deflection profile along the line, assuming that the right lobe shows the outward deflection.

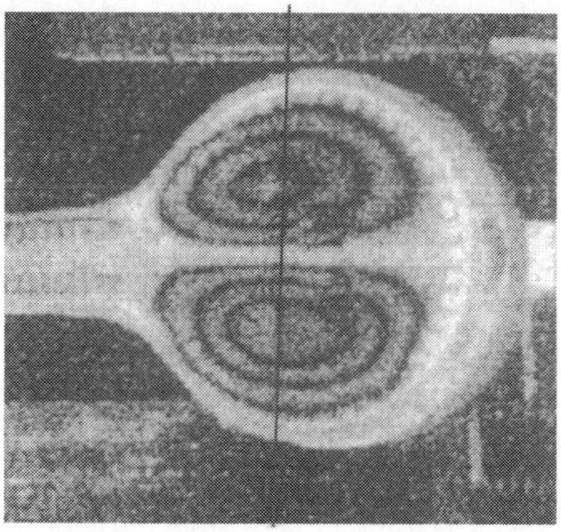

FIGURE 19.19

19.6 A diaphragm of thickness t and diameter $2R$ is centrally loaded. The deflection profile is given by an equation

$$w(x,y)=\frac{PR^4}{16\pi\ D}\left(1-\frac{r^2}{R^2}+2\frac{r^2}{R^2}\ln\frac{r}{R}\right)$$

where D is the plate stiffness defined as $D=\frac{E\,t^3}{12\left(1-\nu^2\right)}$ with E and ν being the Young's modulus and Poisson's ratio of the material of the diaphragm. P is the concentrated load applied to the diaphragm. A speckle shear interferogram is obtained for this diaphragm, which is shown in Figure 19.20. The central deflection is 10.5 μm and the shear is 10% of the radius of the diaphragm. Obtain the slope profile along the line and compare it with a theoretically calculated slope profile. Take 633 nm as the wavelength.

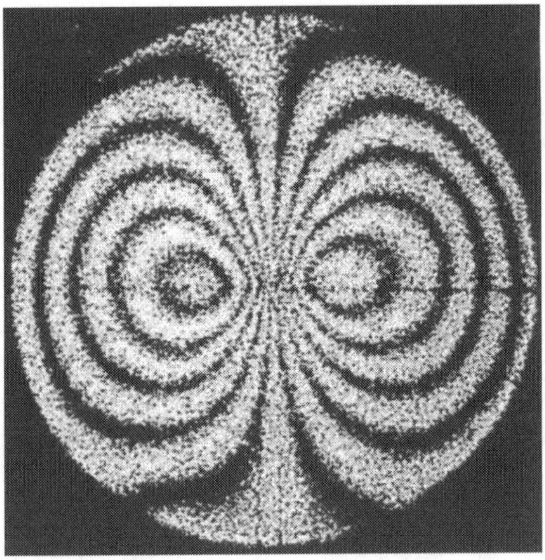

FIGURE 19.20

19.7 A circular diaphragm of diameter $2R$ and thickness t is loaded by applying pressure P (760 mm of Hg) and the deflection profile is measured using ESPI at 633 nm. Figure 19.21 shows the interference pattern obtained with ESPI with a 4.5-μm central deflection. From the Figure 19.21 plot the deflection profile for a diaphragm of 30 mm diameter. Given that the thickness of the diaphragm is 1 mm and the Poisson's ratio of the material of the diaphragm is 0.35, calculate the Young's modulus of the material of the diaphragm.

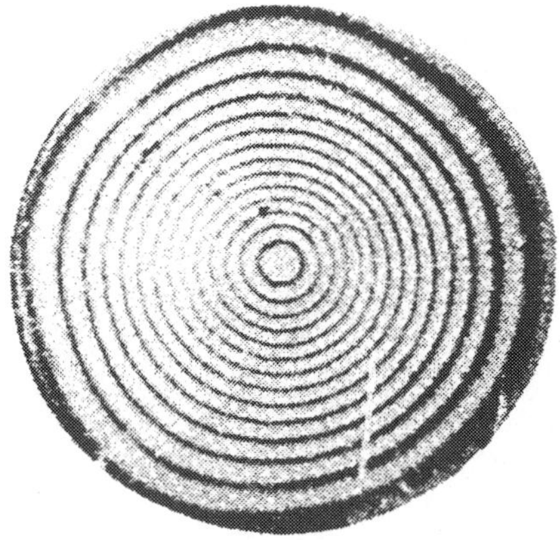

FIGURE 19.21

19.8 An angle interferometer in which the corner cube reflectors are separated by 15.0 mm is used with a He-Ne laser ($\lambda = 632.8$ nm). Assuming that the path difference of $\lambda/10$ can be measured, what is the minimum angle the interferometer could measure?

Bibliography

BOOKS

1. M. Frocht, *Photoelasticity*, John Wiley and Sons, New York (1941).
2. H. T. Jessop and F. C. Harris, *Photoelasticity: Principles and Methods*, Dover, New York (1949).
3. C. Candler, *Modern Interferometers*, Hilger and Watts, London (1951).
4. S. Tolansky, *An Introduction to Interferometry*, Longmans, London (1955).
5. E. G. Coker and L. N. G. Filon, *A Treatise on Photoelasticity*, Cambridge University Press, New York (1957).
6. S. Timoshenko, *Theory of Plates and Shells*, McGraw-Hill, New York (1959).
7. W. T. Welford, *Geometrical Optics*, North-Holland, Amsterdam (1962).
8. A. Sommerfeld, *Optics*, Academic Press, New York (1964).
9. A. J. Durelli and W. F. Riley, *Introduction to Photomechanics*, Prentice-Hall, Englewood Cliffs, New Jersey (1965).
10. W. J. Smith, *Modern Optical Engineering*, McGraw-Hill, New York (1966).
11. R. S. Longhurst, *Geometrical and Physical Optics*, Longmans, London (1967).
12. A. S. Holister, *Experimental Stress Analysis*, Cambridge University Press, New York (1967).
13. J. W. Goodman, *Introduction to Fourier Optics*, McGraw-Hill Book Company, New York (1968), (1996).
14. S. Tolansky, *Multiple-beam Interferometry of Surfaces and Films*, Dover, New York (1970).
15. M. Françon and S. Mallick, *Polarization Interferometers: Applications in Microscopy and Macroscopy*, Wiley-Interscience, London (1971).
16. W. T. Cathey, *Optical Information Processing and Holography*, John Wiley, New York (1974).
17. R. K. Erf (Ed.), *Holographic Non-Destructive Testing*, Academic Press, New York (1974).
18. A. Kuske and G. Robertson, *Photoelastic Stress Analysis*, John Wiley and Sons, New York (1974).
19. J. C. Dainty (Ed.), *Laser Speckle and Related Phenomena*, Springer-Verlag, Berlin (1975).
20. F. A. Jenkins and H. E. White, *Fundamentals of Optics*, McGraw-Hill, New York (1976).
21. D. F. Melling and J. H. Whitelaw, *Principles and Practice of Laser-Doppler Anemometry*, Academic Press, London (1976).
22. H. M. Smith, *Holographic Recording Materials*, Springer-Verlag, Berlin (1977).
23. Ju. I. Ostrovskij, M. M. Butusov and G. V. Ostrovskaja, *GolografiCeskaja interferometrija* (Holographic Interferometry), Nauka Publishing House, Moskva (1977).
24. D. Casasent (Ed.), *Optical Data Processing*, Springer-Verlag, Berlin (1978).
25. J. W. Dally and W. F. Riley, *Experimental Stress Analysis*, McGraw-Hill, New York (1978).
26. R. K. Erf (Ed.), *Speckle Metrology*, Academic Press, New York (1978).
27. J. D. Gaskill, *Linear Systems, Fourier Transforms and Optics*, John Wiley, New York (1978).

28. H. Aben, *Integrated Photoelasticity*, McGraw-Hill, New York (1979).
29. M. Françon (Ed.), *Laser Speckle and Applications in Optics*, Academic Press, New York (1979).
30. W. Schumann and M. Dubas, *Holographic Interferometry*, Springer Series in Optical Sciences, Vol. 16, Springer Verlag, Berlin Heidelberg (1979).
31. C. M. Vest, *Holographic Interferometry*, Interscience, New York (1979).
32. Yu. I. Ostrovsky, M. M. Butusov and G. V. Ostrovskaya, *Interferometry by Holography*, Springer Series in Optical Sciences, Vol. 20, Springer Verlag, Berlin Heidelberg (1980).
33. N. Abramson, *The Making and Evaluation of Holograms*, Academic Press Inc., London (1981).
34. S. H. Lee (Ed.), *Optical Data Processing - Fundamentals*, Springer-Verlag, Berlin (1981).
35. K. Leonhardt, *Optische Interferenzen*, Wissenschaftliche Vergesellschaft, Stuttgart (1981).
36. G. Wernicke and W. Osten, *Holografische Interferometrie*, Physik-Verlag, Weinheim (1982).
37. R. J. Jones and C. Wykes, *Holographic and Speckle Interferometry*, Cambridge University Press, Cambridge (1983), (1989).
38. L. S. Srinath, *Scattered Light Photoelasticity*, Tata McGraw-Hill, New Delhi (1983).
39. W. H. Steel, *Interferometry*, Cambridge University Press, Cambridge, UK (1983).
40. B. Culshaw, *Optical Fibre Sensing and Signal Processing*, Peregrinus, London (1984).
41. E. L. Dereniak and D. G. Crowe, *Optical Radiation Detectors*, John Wiley & Sons, New York (1984).
42. J. R. Meyer-Arendt, *Introduction to Classical and Modern Optics* (2nd Edition), Prentice Hall, Inc., Englewood Cliffs, New Jersey (1984).
43. W. Schumann, J. P. Zürcher and D. Cuche, *Holography and Deformation Analysis*, Springer Series in Optical Sciences, Vol. 46, Springer Verlag, Berlin Heidelberg (1985).
44. R. S. Sirohi, *A Course of Experiments with He-Ne Laser*, Wiley Eastern Publishers, Ltd., New Delhi (1985).
45. K. J. Gasvik, *Optical Metrology*, John Wiley and Sons, Chichester (1987).
46. O. D. D. Soares (Ed.), *Optical Metrology*, Martinus Nijhoff Publishers, Dordrecht (1987).
47. A. Lagarde (Ed.), *Optical Methods in Mechanics of Solids*, Society for Experimental Mechanics, Inc., Bethel, Connecticut (1989).
48. J. F. Doyle and J. W. Phillips (Eds.), *Manual of Experimental Stress Analysis*, Society for Experimental Stress Analysis, Westport, Connecticut (1989).
49. O. Svelto, *Principles of Lasers*, Plenum Press, New York (1989).
50. J. Wilson and J. F. B. Hawkes, *Optoelectronics: An Introduction* (2nd Edition), Prentice Hall, New York (1989).
51. J. M. Vaughan, *The Fabry-Perot Interferometer*, Adam Hilger, Bristol (1989).
52. O. Kafri and I. Glatt, *The Physics of Moiré Metrology*, John Wiley and Sons, New York (1990).
53. M. P. Petrov, S. I. Stepanov and A. V. Khomenko, *Photorefractive Crystals in Coherent Optical Systems*, Springer-Verlag, Berlin (1991).
54. W. Osten, *Digital Processing and Evaluation of Interference Images*, Akademie Verlag, Berlin (1991).
55. Yu. I. Ostrovsky, V. P. Schepinov and V. V. Yakovlev, *Holographic Interferometry in Experimental Mechanics*, Springer Series in Optical Sciences, Vol. 60, Springer Verlag, Berlin Heidelberg (1991).
56. R. S. Sirohi and Mahendra P. Kothiyal, *Optical Components, Systems, and Measurement Techniques*, Marcel Dekker, Inc., New York (1991).

57. R. S. Sirohi (Ed.), Selected Papers on *Speckle Metrology*, Milestone Series MS 35, SPIE, Bellingham, Washington (1991).
58. P. Das, *Lasers and Optical Engineering*, Narosa Publishing House, New Delhi, India (1992).
59. G. Indebetouw and R. Czarnek (Eds.), Selected Papers on Optical Moiré and Applications, MS64, SPIE Publishing Press, Bellingham, Washington (1992).
60. J. T. Luxon and David E. Parker, *Industrial Lasers and Their Applications*, Prentice Hall, Englewood Cliffs, New Jersey (1992).
61. H. I. Bjelkhagen, *Silver-Halide Recording Materials: For Holography and Their Processing*, Springer-Verlag, Berlin (1993).
62. A. S. Kobayashi (Ed.), *Handbook on Experimental Mechanics*, Society for Experimental Mechanics, Bethel, Connecticut (1993).
63. K. Patorski and M. Kujawińska, *Handbook of the Moiré Fringe Technique*, Elsevier, Amsterdam (1993).
64. D. W. Robinson and G. T. Reid (Eds.), *Interferogram Analysis-Digital Fringe Measurement Techniques*, Institute of Physics Publishing, Bristol, UK (1993).
65. R. S. Sirohi (Ed.), *Speckle Metrology*, Marcel Dekker, New York (1993).
66. R. S. Sirohi, *Wave Optics and Applications*, Orient Longmans, Hyderabad, India (1993).
67. D. C. Williams (Ed.), *Optical Methods in Engineering Metrology*, Chapman & Hall, London (1993).
68. D. Post, B. Han and P. Ifju, *High Sensitivity Moiré*, Springer-Verlag, New York (1994).
69. P. K. Rastogi (Ed.), *Holographic Interferometry, Principles and Methods, Springer* Series in Optical Sciences Vol. 68, Springer Verlag, Berlin Heidelberg (1994).
70. G. L. Cloud, *Optical Methods of Engineering Analysis*, Cambridge University Press, Cambridge, UK (1995).
71. S. S. Jha (Ed.), *Perspectives in Optoelectronics*, World Scientific Company, Singapore (1995).
72. E. Uiga, *Optoelectronics*, Prentice Hall, Englewood Cliffs, New Jersey (1995).
73. H. I. Bjelkhagen (Ed.), *Holographic Recording Materials*, SPIE Milestone MS 130, SPIE Optical Engineering Press, Bellingham, Washington (1996).
74. P. Hariharan, *Optical Holography*, Cambridge University Press, Cambridge, UK (1996).
75. G. C. Holst, *CCD Arrays, Cameras, and Displays*, JCD Publishing and SPIE Optical Engineering Press, Bellingham, Washington (1996).
76. T. Kreis, *Holographic Interferometry, Principles and Methods*, Akademie Verlag Series in Optical Metrology, Vol. 1, Akademie Verlag, Berlin (1996).
77. P. Meinlschmidt, K. D. Hinsch and R. S. Sirohi (Eds.), *Selected Papers on Electronic Speckle Pattern Interferometry*, Milestone Series MS 132, SPIE, Bellingham, Washington (1996).
78. V. P. Shchepinova and V. S. Pisarev, *Strain and Stress Analysis by Holographic and Speckle Interferometry*, John Wiley & Sons, England (1996).
79. J. E. Stewart, *Optical Principles and Technology for Engineers*, Marcel Dekker, Inc., New York (1996).
80. N. H. Abramson, *Light in Flight or The Holodiagram; The Columbi Egg of Optics*, Vol. PM 27, SPIE Press, Bellingham, Washington (1997).
81. P. K. Rastogi (Ed.), *Optical Measurement Techniques and Applications*, Artech House, Inc., Boston, Massachusetts (1997).
82. F. T. S. Yu and X. Yang, *Introduction to Optical Engineering*, Cambridge University Press, Cambridge, UK (1997).

83. C. Ghiglia and M. D. Pritt, *Two Dimensional Phase Unwrapping: Theory, Algorithms and Software*, John Wiley and Sons, Inc., New York (1998).
84. D. Malacara, M. Servin and Z. Malacara, *Optical Testing: Analysis of Interferograms*, Marcel Dekker, Inc., New York (1998).
85. P. L. Polavarapu (Ed.), *Principles and Applications of Polarization-Division Interferometry*, John Wiley and Sons, Inc., New York (1998).
86. R. S. Sirohi and K. D. Hinsch, Selected Papers on *Holographic Interferometry*, SPIE Milestone Series MS144, SPIE Publishing, Bellingham, Washington (1998).
87. M. Born and E. Wolf, *Principles of Optics*, Cambridge University Press, Cambridge, UK (1999).
88. K. Ramesh, *Digital Photoelasticity*, Springer-Verlag, Berlin (2000).
89. P. K. Rastogi (Ed.), *Digital Speckle Pattern Interferometry and Related Techniques*, Wiley-VCH, Weinheim (2000).
90. O. Feldmann and F. Mayinger (Editors), *Optical Measurements: Techniques and Applications*, Springer-Verlag, Berlin (2001).
91. P. R. N. Childs, *Practical Temperature Measurement*, Butterworth-Heinemann, Oxford, UK (2001).
92. P. Hariharan, *Optical Interferometry*, Academic Press, New York USA (2003).
93. W. Steinchen and L. Yang, *Shearography: Theory and Application of Digital Speckle Pattern Shearing Interferometry*, PM 100m Optical Engineering Press, SPIE, Bellingham Washington DC (2003).
94. C. A. Walker (Ed.), *Handbook of Moiré Measurement*, Taylor and Francis, New York (2003).
95. D. Malacara, M. Servín and Z. Malacara, *Interferogram Analysis for Optical Testing* (2nd Edition), CRC Press, Boca Raton, Florida (2005).
96. T. Kreis, *Handbook of Holographic Interferometry: Optical and Digital Methods*, Wiley-VCH Verlag GmbH, Weinheim (2005).
97. W. Osten (Ed.), *Optical Inspection of Microsystems*, Taylor and Francis, New York (2006).
98. G. C. Holst and T. S. Lomheim, *CMOS/CCD Sensors and Camera Systems*, JCD Publishing and SPIE Optical Engineering Press, Bellingham, Washington (2007).
99. E. Hecht and A. R. Ganesan, *Optics* (4th Edition), Dorling Kindersley (India) Pvt. Ltd., Delhi (2008).
100. R. S. Sirohi, *Optical Methods of Measurement: Wholefield Techniques* (2nd Edition), CRC Press, Boca Raton, Florida (2009).
101. T. Yoshizawa (Ed.), *Handbook of Optical Metrology: Principles and Applications*, CRC Press, Taylor and Francis, Boca Raton, Florida (2009).
102. A. Ghatak, *Optics*, McGraw-Hill, New York (2010).
103. W. Osten and N. Reingand (Eds.), *Optical Imaging and Metrology: Advanced Technologies*, Wiley-VCH Verlag & Co. KGaA, Weinheim (2012).
104. K. Harding (Ed.), *Handbook of Optical Dimensional Metrology*, CRC Press, Taylor & Francis Group, LLC, Boca Raton, Florida (2013).
105. K. Ramesh, *Developments in Photoelasticity-A renaissance*, IOP Publishing Ltd, Bristol (2021).
106. W. Gao and Y. Shimizu, *Optical Metrology for Precision Engineering*, Walter de Gruyter GmbH, Berlin/Boston (2022).
107. W. Osten and Y. Bai (Eds.), *Holography and Speckle-A Review of 60 Successful Years*, Springer Nature, Switzerland (2023).

JOURNAL PAPERS

1. E. Reissner, On bending of elastic plates, Q. Appl. Math., 5, 55–68, (1947).
2. J. M. Burch, Scatter fringes of equal thickness, Nature, 4359, 889–890, (1953).
3. E. Archbold, J. M. Burch, A. E. Ennos and P. A. Taylor, Visual observations of surface vibration nodal patterns, Nature (London), 222, 263–265, (1969).
4. L. A. Selke, Theoretical elastic deflections of a thick horizontal circular mirror on a ring support, Appl. Opt., 9, 149–153, (1970).
5. J. A. Leendertz, Interferometric displacement measurement on scattering surface utilizing speckle effect, J. Phys. E: Sci. Instrum., 3, 214–218, (1970).
6. J. D. Barnett, S. Block and G. J. Piermarini, Optical fluorescence system for quantitative pressure measurement in the diamond-anvil cell, Rev. Sci. Instrum., 44(1), 1–9, (1973).
7. N. Abramson, Sandwich hologram interferometry: A new dimension in holographic comparison, Appl. Opt., 13(9), 2019–2025, (1974).
8. N. Abramson, Sandwich hologram interferometry. 2: Some practical calculations, Appl. Opt.,14(4), 981–984, (1975).
9. K. A. Stetson, The use of an image derotator in hologram interferometry and speckle photography of rotating objects, Exp. Mech., 18, 67–73, (1978).
10. M. C. Gerchman and G. C. Hunter, Differential technique for accurately measuring the radius of curvature of long radius concave optical surfaces, Proc. SPIE, 192, 75–84, (1979).
11. K. Itoh, Analysis of the phase unwrapping algorithm, Appl. Opt., 21, 2470–2470, (1982).
12. R. Swanepoel, Determination of the thickness and optical constants of amorphous silicon, J. Phys. E: Sci. Instrum., 16, 1214–1222, (1983).
13. 17th Conférence Générale des Poids et Mesures (CGPM) - Resolution 1 of the CGPM (1983): Definition of the metre". Bureau international des poids et mesures (BIMP).
14. E. Moreels, C. de Greef and R. Finsy, Laser light refractometer, Appl. Opt., 23, 3010–3013, (1984).
15. B. S. Fritz, Absolute calibration of an optical flat, Opt. Eng., 23(4), 379–383, (1984).
16. Y. Nakano and K. Murata, Talbot interferometry for measuring the focal length of a lens, Appl. Opt., 24(19), 3162–3166, (1985).
17. C. Joenathan and R. S. Sirohi, Elimination of errors in speckle photography, Appl. Opt., 25, 1791–1794, (1986).
18. I. T. Young, R. Zagers, L. J. van Vliet, J. Mullikin, F. Boddeke and H. Netten, Depth-of-focus in microscopy, in: SCIA'93. Proc. of 8th Scandinavian Conference on Image Analysis, Tromso, Norway, pp. 493–498, (1993).
19. J. A. Quiroga and A. González-Cona, Phase measuring algorithms for extraction of isochromatics of photoelastic fringe patterns, Appl., Opt., 36, 8397–8402, (1997).
20. I. Powell and E. Goulet, Absolute figure measurements with a liquid-flat reference, Appl. Opt., 37(13), 2579–2588, (1998).
21. C. Shakher and A. K. Nirala, A review on refractive index and temperature profile measurements using laser-based interferometric techniques, Opt. Lasers Eng., 31, 455–491, (1999).
22. J. M. Bioucas-Dias and G. Valadao, Phase unwrapping via graph cuts, IEEE Trans. Image Process., 16, 698–709, (2007).
23. L.-Y. Liao, F. C. Bráulio de Albuquerque, Robert E. Parks and José M. Sasian, Precision focal-length measurement using imaging conjugates, Opt. Eng., 51(11), 113604, (2012).

24. C. Pruss, G. B. Baer, Johannes Schindler and Wolfgang Osten, Measuring aspheres quickly: Tilted wave interferometry, Opt. Eng., 56(11), 111713, (2017), doi: 10.1117/1.OE.56.11.111713.
25. R. Sirohi, Shearography and its applications - A chronological review, Light: Advanced Manufacturing, 3(1), 35–64, (2022).
26. C. Schober, R. Beisswanger, A. Gronle, C. Pruss and W. Osten, Tilted wave Fizeau interferometer for flexible and robust asphere and freeform testing, Light: Advanced Manufacturing, 3(4), 687–698, (2022).
27. V. Quaresima, M. Ferrari and F. Scholkmann, Ninety years of pulse oximetry: History, current status, and outlook, JBO, 29(S3), S33307-1–S33307-34, (2024).

Index

F

G

H

I

P

T

U

V

W

Y

Z

FSC
www.fsc.org
MIX
Papier | Fördert gute Waldnutzung
FSC® C083411

Kolibri 360 GmbH
Ferdinand-Jühlke-Straße 7
99095 Erfurt, Deutschland
produktsicherheit@kolibri360.de